The seminal formula of Gross and Zagier relating heights of Heegner points to derivatives of the associated Rankin L-series has led to many generalizations and extensions in a variety of different directions, spawning a fertile area of study that remains active to this day.

This volume, based on a workshop on Special Values of Rankin L-Series held at the MSRI in December 2001, is a collection of articles written by many of the leading contributors in the field, having the Gross–Zagier formula and its avatars as a common unifying theme. It serves as a valuable reference for mathematicians wishing to become better acquainted with the theory of complex multiplication, automorphic forms, the Rankin–Selberg method, arithmetic intersection theory, Iwasawa theory, and other topics related to the Gross–Zagier formula.

Mathematical Sciences Research Institute
Publications

49

Heegner Points and Rankin L-Series

Mathematical Sciences Research Institute Publications

1 Freed/Uhlenbeck: *Instantons and Four-Manifolds*, second edition
2 Chern (ed.): *Seminar on Nonlinear Partial Differential Equations*
3 Lepowsky/Mandelstam/Singer (eds.): *Vertex Operators in Mathematics and Physics*
4 Kac (ed.): *Infinite Dimensional Groups with Applications*
5 Blackadar: *K-Theory for Operator Algebras*, second edition
6 Moore (ed.): *Group Representations, Ergodic Theory, Operator Algebras, and Mathematical Physics*
7 Chorin/Majda (eds.): *Wave Motion: Theory, Modelling, and Computation*
8 Gersten (ed.): *Essays in Group Theory*
9 Moore/Schochet: *Global Analysis on Foliated Spaces*
10–11 Drasin/Earle/Gehring/Kra/Marden (eds.): *Holomorphic Functions and Moduli*
12–13 Ni/Peletier/Serrin (eds.): *Nonlinear Diffusion Equations and Their Equilibrium States*
14 Goodman/de la Harpe/Jones: *Coxeter Graphs and Towers of Algebras*
15 Hochster/Huneke/Sally (eds.): *Commutative Algebra*
16 Ihara/Ribet/Serre (eds.): *Galois Groups over* $\mathbb{Q}$
17 Concus/Finn/Hoffman (eds.): *Geometric Analysis and Computer Graphics*
18 Bryant/Chern/Gardner/Goldschmidt/Griffiths: *Exterior Differential Systems*
19 Alperin (ed.): *Arboreal Group Theory*
20 Dazord/Weinstein (eds.): *Symplectic Geometry, Groupoids, and Integrable Systems*
21 Moschovakis (ed.): *Logic from Computer Science*
22 Ratiu (ed.): *The Geometry of Hamiltonian Systems*
23 Baumslag/Miller (eds.): *Algorithms and Classification in Combinatorial Group Theory*
24 Montgomery/Small (eds.): *Noncommutative Rings*
25 Akbulut/King: *Topology of Real Algebraic Sets*
26 Judah/Just/Woodin (eds.): *Set Theory of the Continuum*
27 Carlsson/Cohen/Hsiang/Jones (eds.): *Algebraic Topology and Its Applications*
28 Clemens/Kollár (eds.): *Current Topics in Complex Algebraic Geometry*
29 Nowakowski (ed.): *Games of No Chance*
30 Grove/Petersen (eds.): *Comparison Geometry*
31 Levy (ed.): *Flavors of Geometry*
32 Cecil/Chern (eds.): *Tight and Taut Submanifolds*
33 Axler/McCarthy/Sarason (eds.): *Holomorphic Spaces*
34 Ball/Milman (eds.): *Convex Geometric Analysis*
35 Levy (ed.): *The Eightfold Way*
36 Gavosto/Krantz/McCallum (eds.): *Contemporary Issues in Mathematics Education*
37 Schneider/Siu (eds.): *Several Complex Variables*
38 Billera/Björner/Green/Simion/Stanley (eds.): *New Perspectives in Geometric Combinatorics*
39 Haskell/Pillay/Steinhorn (eds.): *Model Theory, Algebra, and Geometry*
40 Bleher/Its (eds.): *Random Matrix Models and Their Applications*
41 Schneps (ed.): *Galois Groups and Fundamental Groups*
42 Nowakowski (ed.): *More Games of No Chance*
43 Montgomery/Schneider (eds.): *New Directions in Hopf Algebras*
44 Buhler/Stevenhagen (eds.): *Algorithmic Number Theory*
45 Jensen/Ledet/Yui: *Generic Polynomials: Constructive Aspects of the Inverse Galois Problem*
46 Rockmore/Healy (eds.): *Modern Signal Processing*
47 Uhlmann (ed.): *Inside Out: Inverse Problems and Applications*
48 Gross/Kotiuga: *Electromagnetic Theory and Computation: A Topological Approach*
49 Darmon/Zhang (ed.): *Heegner Points and Rankin L-Series*

Volumes 1–4 and 6–27 are published by Springer-Verlag

Heegner Points and Rankin L-Series

Edited by

Henri Darmon
McGill University

Shou-Wu Zhang
Columbia University

CAMBRIDGE
UNIVERSITY PRESS

Henri Darmon
McGill University
Mathematics Department
805 Sherbrooke St. West
Montreal, QUE H3A 2K5
Canada
darmon@math.mcgill.ca

Shou-Wu Zhang
Department of Mathematics
Columbia University
New York, NY 10027
United States
szhang@math.columbia.edu

The Mathematical Sciences Research Institute wishes to acknowledge support by the National Science Foundation. This material is based upon work supported by NSF Grant 9810361.

CAMBRIDGE UNIVERSITY PRESS
Cambridge, New York, Melbourne, Madrid, Cape Town, Singapore,
São Paulo, Delhi, Dubai, Tokyo, Mexico City

Cambridge University Press
The Edinburgh Building, Cambridge CB2 8RU, UK

Published in the United States of America by Cambridge University Press, New York

www.cambridge.org
Information on this title: www.cambridge.org/9780521158206

Paperback edition 2010

A catalogue record for this publication is available from the British Library

ISBN 978-0-521-83659-3 Hardback
ISBN 978-0-521-15820-6 Paperback

Heegner Points and Rankin L-Series
MSRI Publications
Volume **49**, 2004

Contents

Preface ix
HENRI DARMON AND SHOU-WU ZHANG

Heegner Points: The Beginnings 1
BRYAN BIRCH

Correspondence 11
BRYAN BIRCH AND BENEDICT GROSS

The Gauss Class Number Problem for Imaginary Quadratic Fields 25
DORIAN GOLDFELD

Heegner Points and Representation Theory 37
BENEDICT GROSS

Gross–Zagier Revisited 67
BRIAN CONRAD (with an appendix by W. R. MANN)

Special Value Formulae for Rankin L-Functions 165
VINAYAK VATSAL

Gross-Zagier Formula for GL(2), II 191
SHOU-WU ZHANG

Special Cycles and Derivatives of Eisenstein Series 243
STEPHEN KUDLA

Faltings' Height and the Derivatives of Eisenstein Series 271
TONGHAI YANG

Elliptic Curves and Analogies Between Number Fields and Function Fields 285
DOUG ULMER

Heegner Points and Elliptic Curves of Large Rank over Function Fields 317
HENRI DARMON

Periods and Points Attached to Quadratic Algebras 323
MASSIMO BERTOLINI, HENRI DARMON AND PETER GREEN

Contents

Heegner Points and Rankin L-Series
MSRI Publications
Volume **49**, 2004

Preface

Modular curves and their close relatives, Shimura curves attached to multiplicative subgroups of quaternion algebras, are equipped with a distinguished collection of points defined over class fields of imaginary quadratic fields and arising from the theory of complex multiplication: the so-called *Heegner points*[1]. It is customary to use the same term to describe the images of degree zero divisors supported on these points in quotients of the jacobian of the underlying curve (a class of abelian varieties which we now know is rich enough to encompass all elliptic curves over the rationals). It was Birch who first undertook, in the late 70's and early 80's, a systematic study of Heegner points on elliptic curve quotients of jacobians of modular curves. Based on the numerical evidence that he gathered, he observed that the heights of these points seemed to be related to first derivatives at the central critical point of the Hasse–Weil L-series of the elliptic curve (twisted eventually by an appropriate Dirichlet character). The study initiated by Birch was to play an important role in the number theory of the next two decades, shedding light on such fundamental questions as the Gauss class number problem and the Birch and Swinnerton-Dyer conjecture.

The study of Heegner points took off in the mid-1980's thanks to two breakthroughs. The first of these was the Gross–Zagier formula, which, confirming Birch's observations, expressed the heights of Heegner points in the jacobian of the modular curve $X_0(N)$ in terms of the first derivative at the central point of an associated Rankin L-series. The second came a few years later, when Kolyvagin showed how the system of Heegner points on an elliptic curve controls the size and structure of its Selmer group. Taken together, these two insights led to a complete proof of the Birch and Swinnerton-Dyer conjecture (in its somewhat weaker form stipulating an equality between the rank of the elliptic curve, and the order of vanishing of its L-series at $s = 1$) for all modular elliptic curves over $\mathbb{Q}$ whose L-series has at most a simple zero at $s = 1$. The argument yielded a proof of the Shafarevich–Tate conjecture for these curves as well. The subsequent proof of the Shimura–Taniyama conjecture in 1994 showed that the result of Gross–Zagier and Kolyvagin applies unconditionally to all elliptic curves over the rationals.

[1]In the literature, Heegner points are often required to satisfy additional restrictions; in this preface the term is used more loosely, and synonymously with CM point.

Both the Gross–Zagier formula and the techniques introduced by Kolyvagin have proved fertile, and generalizations in a variety of different directions were actively explored throughout the 1990's. The organizers of the MSRI workshop on Special Values of Rankin L-series held in December 2001 felt the time was ripe to take stock of the new questions and insights that have emerged in the last decade of this study. The workshop focussed on several topics having Heegner points and Rankin L-series as a common unifying theme, grouped roughly under the following overlapping rubrics.

The original Gross–Zagier formula: Birch's partly historical article in this volume describes the study of Heegner points (both theoretical, and numerical) that preceded the work of Gross and Zagier. The editors are pleased to reproduce in this volume the correspondence between Birch and Gross in the months leading up to this work, which gives an enlightening snapshot of the subject at a time when it was about to undergo a major upheaval. Revisiting the height calculations of Gross and Zagier, the article by Conrad and Mann supplies further background on some of its more delicate aspects. Gross's contribution to this volume places the Gross–Zagier formula (together with its natural generalization for Heegner points on Shimura curves) in a broader conceptual setting in which the language of automorphic representations plays a key role. This larger perspective is useful in understanding Zhang's article which treats extensions of the Gross–Zagier formula for totally real fields.

Analytic applications: An immediate consequence of the Gross–Zagier formula is the first (and still, at present, the only) class of examples of L-series whose order of vanishing at $s = 1$ can be proved to be ≥ 3. These examples are produced by finding elliptic curves E over $\mathbb{Q}$ whose L-function $L(E/\mathbb{Q}, s)$ vanishes to odd order because of the sign in its functional equation, and whose associated Heegner point on $E(\mathbb{Q})$ is of finite order so that $L'(E/\mathbb{Q}, 1) = 0$. In his article, Goldfeld explains how the existence of such an L-function yields effective lower bounds for the growth of class numbers of imaginary quadratic fields. Before the work of Goldfeld, such bounds were only known ineffectively due to the possible existence of Siegel zeroes. Goldfeld's effective solution of the Gauss class number problem was one of the striking early applications of the Gross–Zagier formula.

Extensions to totally real fields: Experts were always aware of the potential generalizations of the Gross–Zagier formula to the context where the rational numbers are replaced by a totally real field F of degree d over $\mathbb{Q}$, although generalizations of this type gave rise to substantial technical difficulties. In this setting one is led to consider cuspidal automorphic forms on $\mathrm{GL}_2(F)$. A generalization of the Shimura–Taniyama conjecture predicts that any elliptic curve E over F (or, more generally, any two-dimensional ℓ-adic representation of $G_F := \mathrm{Gal}(\overline{F}/F)$ arising from the ℓ-adic Tate module of an abelian variety over F) corresponds to a particular type of automorphic form—a holomorphic Hilbert modular form of parallel weight 2. These forms can be described as differential d-forms on a d-dimensional Hilbert modular variety X. Such a variety admits an analytic description as a union of quotients of $\mathcal{H}^d$ by the action of appropriate subgroups of the Hilbert modular group $\mathrm{SL}_2(\mathcal{O}_F)$. Although these varieties are equipped with a supply of CM points, there is no direct generalization of the Eichler–Shimura construction which, when $d = 1$, realizes E as a quotient of $\mathrm{Jac}(X)$. Hence, CM points on X do not give rise to points on E.

There is nonetheless an extension of the Heegner point construction for totally real fields which is slightly less general and relies crucially on the notion of Shimura curves. These notions are discussed at length in the articles by Gross and Zhang in this volume. Let us briefly introduce the relevant definitions.

Given a set T of places of F of even cardinality, denote by B_T the quaternion algebra ramified exactly at the places $v \in T$. Let S be a finite set of places of F of odd cardinality containing all the archimedean places of F. One may attach to the data of S and an auxiliary choice of level structure a Shimura curve $X_{/F}$ which admits a v-adic analytic description, for each place $v \in S$, as a quotient of the v-adic upper half plane $\mathcal{H}_v$ by the action of an appropriate arithmetic subgroup of $B^{\times}_{S-\{v\}}$. It can also be described, following work of Shimura, in terms of the solution to a moduli problem, leading to a canonical model for X over F. Using the moduli description of the Shimura curve X, one can see that it admits a collection of CM points analogous in many respects to the Heegner points on $X_0(N)$ which provide the setting for the original Gross–Zagier formula.

If E is any elliptic over F having a suitable type of (bad) reduction at the non-archimedean places of S, a generalization of the Shimura–Taniyama conjecture predicts that E is isogenous to a factor of the jacobian of such a Shimura curve X. One thus has a direct analogue of the modular parametrization over any totally real field F, which should apply to many (albeit not all) elliptic curves over F.

The Heegner points arising from such Shimura curve parametrizations provide the setting for Zhang's extension of the Gross–Zagier formula to elliptic curves (and abelian varieties of "GL_2 type") over totally real fields that is the theme of his contribution in these proceedings.

Iwasawa theory: In late 70's Mazur formulated a program for studying the arithmetic of elliptic curves over a $\mathbb{Z}_p$-extension of the ground field along the lines pioneered by Iwasawa. The most natural setting is the one where E is defined over $\mathbb{Q}$ and one examines its Mordell–Weil group (and its p-power Selmer groups) over the cyclotomic $\mathbb{Z}_p$-extension of $\mathbb{Q}$. But Mazur also singled out the study of the Mordell–Weil groups of E over the anti-cyclotomic $\mathbb{Z}_p$-extension of an imaginary quadratic field because of the new phenomena manifesting themselves in this case through the possible presence of Heegner points. To give the flavor of the kind of results that Cornut and Vatsal obtain in a special case, let K be an imaginary quadratic field, let p be a prime which (for simplicity) we assume does not divide the conductor of E, and let K_∞ be the anticyclotomic $\mathbb{Z}_p$-extension of K, which is contained in the somewhat larger extension $\tilde{K}_\infty$, the union of all ring class fields of K of p-power conductor. An elementary argument shows that the collection of Heegner points of p-power conductor contained in $E(\tilde{K}_\infty)$ is either empty, or is infinite and generates a group of infinite rank, depending on the sign in the functional equation for $L(E/K, s)$. Mazur conjectures that the same is true for the traces of these points to $E(K_\infty)$. Cornut and Vatsal were able to prove Mazur's conjecture, as well as a non-triviality result for a p-adic L-function attached to E and K_∞ when the collection of Heegner points on $E(\tilde{K}_\infty)$ is empty.

The results of Cornut and Vatsal, when combined with those of Bertolini and Darmon on the anticyclotomic main conjecture of Iwasawa theory building on the methods of Kolyvagin, show that the Mordell–Weil group $E(K_\infty)$ is finitely generated when the sign in the functional equation for $L(E/K, s)$ is 1, and that, when this sign is -1, the rank of $E(L)$ grows like $[L : K]$ up to a bounded error term as L ranges over the finite subextensions of K_∞, the main contribution to the growth of the rank being accounted for by Heegner points. Thus the study of Heegner points over the anticyclotomic tower has yielded an almost complete understanding of Mazur's program in that setting.

Vatsal's article in this volume is motivated by the desire to extend his non-vanishing results for Heegner points and p-adic L-functions to the more general setting of totally real fields.

Higher dimensional analogues: Heegner points admit a number of higher-dimensional analogues, such as the arithmetic cycles on Shimura varieties "of orthogonal type" which are considered in Kudla's article. In his contribution Kudla surveys his far-reaching program of relating heights (in the Arakelov sense) of these cycles to the derivatives of associated Eisenstein series; while a tremendous amount of mathematics remains to be developed in this direction, substantial inroads have already been made, as is explained in Tonghai Yang's article in which one of the simplest cases of Kudla's program is worked out.

Function field analogues: Modular curves admit analogues in the function field setting: the so-called Drinfeld modular curves. Ulmer's article surveys recent work aimed at extending the ideas of Gross–Zagier and Kolyvagin to the function field setting, with the ultimate goal of proving the Birch and Swinnerton-Dyer conjecture for elliptic curves whose L-function has at most a simple zero. An interesting issue that is discussed in Ulmer's contribution is the possibility of constructing elliptic curves of arbitrarily large rank over the "ground field" $\mathbb{F}_p(T)$ with non-constant j-invariant. As is remarked in the note following Ulmer's contribution, Ulmer's Mordell–Weil groups of arbitrarily large rank might be accounted for, and generalized, via a suitable Heegner point construction. This stands in marked contract to the case of elliptic curves over $\mathbb{Q}$, where Heegner points are expected (and in many cases known) to yield only torsion points when the rank is strictly greater than one, and where the rank of elliptic curves is not even known to be unbounded.

Conjectural variants: Heegner points can be viewed as the elliptic curve analogue of special units such as circular or elliptic units, whose logarithms are related to first derivatives of Artin L-series at $s = 0$, just as the heights of Heegner points encode first derivatives of Rankin L-series via the Gross–Zagier formula. The article by Bertolini, Darmon and Green describes several largely *conjectural* analytic constructions of "Heegner-type" points which might be viewed as the elliptic curve analogue of Stark units. This is why the term "Stark–Heegner points" has been coined to describe them. Two cases of "Stark–Heegner" type constructions are considered in detail: one, p-adic analytic, of points over ring class fields of real quadratic fields, and the second, complex analytic, of points over ring class fields of so-called "ATR" extensions of a totally real field in terms of periods of Hilbert modular forms. Providing evidence (both experimental, and theoretical) for the conjectures on Stark–Heegner points, and understanding the relation between these points and special values of Rankin L-series, represent interesting challenges for the future.

The organizers thank the NSF for funding their workshop on Rankin L-series, the MSRI for hosting it, and all the participants who made it such a stimulating and pleasant event. Special thanks go to Samit Dasgupta for taking the notes of the workshop lectures which are now posted on the MSRI web site, and to Silvio Levy for his editorial assistance in putting together these proceedings.

Henri Darmon
Shou-Wu Zhang

Heegner Points and Rankin L-Series
MSRI Publications
Volume **49**, 2004

Heegner Points: The Beginnings

BRYAN BIRCH

1. Prologue: The Opportune Arrival of Heegner Points

Dick Gross and I were invited to talk about Heegner points from a historical point of view, and we agreed that I should talk first, dealing with the period before they became well known. I felt encouraged to indulge in some personal reminiscence of that period, particularly where I can support it by documentary evidence. I was fortunate enough to be working on the arithmetic of elliptic curves when comparatively little was known, but when new tools were just becoming available, and when forgotten theories such as the theory of automorphic function were being rediscovered. At that time, one could still obtain exciting new results without too much sophisticated apparatus: one was learning exciting new mathematics all the time, but it seemed to be less difficult!

To set the stage for Heegner points, one may compare the state of the theory of elliptic curves over the rationals, E/Q for short, in the 1960's and in the 1970's; Serre [15] has already done this, but never mind! Lest I forget, I should stress that when I say "elliptic curve" I will always mean "elliptic curve defined over the rationals".

In the 1960's, we were primarily interested in the problem of determining the Mordell-Weil group $E(Q)$, though there was much other interesting apparatus waiting to be investigated (cf Cassels' report [7]). There was a good theory of descent, Selmer and Tate-Shafarevich groups, and so forth: plenty of algebra. But there was hardly any useful analytic theory, unless the elliptic curve had complex multiplication; $E(C)$ was a complex torus, beautiful maybe, but smooth and featureless, with nothing to get hold of. One could define the L-function

$$L_S(E,s) \sim \prod_{p \notin S} (1 - a_p p^{-s} + p^{1-2s})^{-1}$$

(where S is a set of "bad" primes), and Hasse had conjectured that this is an analytic function with a good functional equation; but most of us could only prove this when the curve had complex multiplication.

By the 1970's, everything had changed. Shimura showed that for *modular* elliptic curves (that is, elliptic curves parametrised by functions on $X_0(N)$) the L-function automatically has a good functional equation. For Peter Swinnerton-Dyer and myself, the turning point came when Weil wrote to Peter [A], stressing the importance of the functional equation, which (Weil said) *is of the form*

$$\Lambda(s) = (2\pi)^{-s}\Gamma(s)L(s) = C.N^{1-s}\Lambda(2-s)$$

in all known cases (and, conjecturally, also in all unknown cases). Here, N is the analytic conductor of the curve, conjecturally the same as the algebraic conductor defined by Serre and Tate. Weil went on to point out that this conjecture of Hasse's was known for modular elliptic curves (Weil actually called them Eichler-Shimura curves) as well as for curves with complex multiplication. Though Weil didn't actually say it explicitly, we knew that he was advising us to concentrate on *modular* elliptic curves. The next year, Weil [20] proved that the functional equations conjectured by Hasse for an elliptic curve over Q were valid *only* if the curve was modular. From then on, it was clear that in all work that needed $L_E(s)$ to be well behaved, one might as well assume that the elliptic curve E/Q under consideration was modular. We referred (for instance, in letters [B] between Peter and John Tate) to the hypothesis, implicit in Weil's letter and paper, that every elliptic curve over Q really was modular, as the "Weil conjecture"; years later we learnt that this had been suggested much earlier, by Taniyama [18]. (I hope these remarks, and the slightly earlier references, are a helpful amplification of the very accurate account of the history of this conjecture given by Serre [15].)

Almost on cue, Heegner points came along, specifically on modular curves! Suddenly, instead of being a featureless homogeneous space, $E(C)$ was a highly structured object, studded all over with canonically defined families of points, with coordinates in known number fields. In studying $E(Q)$, instead of searching for structure, one had the much more hopeful task of analysing a situation where there was almost too much of it.

And sure enough, the theorems rolled in, though not immediately. There was about a ten year gap between the repopularisation of Heegner points and anyone making proper use of them! That will be what Dick talks about. My job is to tell you where these points came from.

2. Prehistory

In this context, "prehistory" means the latter half of the nineteenth century; and it is summarised in Weber's *Algebra* [19], one of the great books of mathematics. I believe that Weber remained the most up-to-date book on the arithmetic of modular functions until Shimura's book [16] was published in 1971; certainly it was the best I could find in 1966.

The story starts with the modular function, $j(z)$, characterised by its values at i and ρ, its pole at ∞ and its functional equation $j(M(z)) = j(z)$ for any unimodular integral transformation $z \to M(z) := (az+b)/(cz+d)$; that is, j is invariant by the modular group $\Gamma(1)$. So j may be regarded as a function of similarity classes of lattices: if $\Lambda = \Lambda(\omega_1, \omega_2)$ is a lattice with basis (ω_1, ω_2) then $j(\Lambda) := j(\omega_1/\omega_2)$ does not depend on the choice of basis. If $f(z)$ is another function invariant by the modular group then f is a rational function of j, and if f is invariant by a group commensurable with $\Gamma(1)$ then $f(z), j(z)$ are algebraically dependent. In particular, if N is any natural number, the function $j_N(z) := j(Nz)$ is invariant by the conjugate $\left(\begin{smallmatrix} N & 0 \\ 0 & 1 \end{smallmatrix}\right)\Gamma(1)\left(\begin{smallmatrix} N & 0 \\ 0 & 1 \end{smallmatrix}\right)^{-1}$ of $\Gamma(1)$, the intersection $\Gamma_0(N)$ of these two conjugate subgroups has finite index in both, and so j, j_N are related by a polynomial equation $F_N(j(z), j_N(z)) = 0$ with $F_N(X,Y) \in \mathbb{Z}[X,Y]$. We recognise $F_N(X,Y) = 0$ as the modular curve $Y_0(N)$, the quotient of the upper half plane by $\Gamma_0(N)$. Functions invariant by $\Gamma_0(N)$ are rational functions of j and j_N. If p is a prime, we have Kronecker's congruence

$$F_p(X,Y) \equiv (X^p - Y)(X - Y^p) \pmod{p}.$$

A beautiful discovery was the theory of "complex multiplication". Suppose that ω is a complex quadratic surd satisfying a primitive equation $A\omega^2 + B\omega + C = 0$ with A, B, C integers; the discriminant of ω is $D(\omega) := B^2 - 4AC < 0$. Then $j(\omega)$ is an algebraic integer: the reason is that we can find ω' in the lattice $\Lambda(1,\omega)$ so that the lattices $\Lambda(1,\omega')$ and $\Lambda(1, N\omega')$ are the same for some N. Further, the field $K(D) := Q(\omega, j(\omega))$ in which $j(\omega)$ lives depends only on $D(\omega)$, not on ω, and the degree $[K(D) : Q(\omega)]$ is equal to the class number of the ring $R(D) := Z\left[\frac{D + D^{1/2}}{2}\right]$. In fact, one may regard the ideals A of this ring as lattices, so it makes sense to evaluate j at an ideal class A, and then when A runs through the classes of $R(D)$ the values of $j(A)$ are all conjugate. The field $K(D)$ is called the *ring class field* corresponding to the ring $R(D)$; in particular, if Δ is a field discriminant (discriminant of the ring of integers of $Q(\Delta^{1/2})$) then $K(\Delta)$ is simply the class field of $Q(\Delta^{1/2})$, and the fields $K(s^2\Delta)$ are extensions of $K(\Delta)$ of predictable degrees.

Complex multiplication was the beginning of class field theory, and nowadays it is often treated as a particular case of the general theory. But that is really the wrong way round: the theory presents us with the explicit field $K(D)$ constructively, at the very start, and the beautiful "class field" properties of $K(D)$ are more easily obtained directly; it is the subject of Weber's book. (The language in Weber is now unfamiliar, so that the arguments seem more complicated than they actually are. I should perhaps add that until the Brighton conference in 1965, published as [8], the apparatus of class field theory was much more forbidding than was Weber's *Algebra*.)

The theory of complex multiplication as developed by Weber tells us about the field $j(A)$ lives in when A is an ideal of a given complex quadratic ring. When one reads Weber, one sees that he aims to work in rather greater generality. He considers other modular functions, invariant by various subgroups of $\Gamma(1)$;

for instance, he defines particular functions $\gamma_2, \gamma_3, \sigma(x)$ which satisfy $\gamma_2^3 = j$, $\gamma_3^2 = j - 1728$, $\sigma^{24} - 16 = \sigma^8\gamma_2$ and $(\sigma^{24}-64)(\sigma^{24}+8)^2 = (\gamma_3\sigma^{12})^2$ (I use Heegner's later notation for these functions); and he proves by various contortions that if $D(\omega)$ satisfies various congruence conditions then evaluating these functions at ω gives values in the ring class field $K(D)$, not as one might expect in some proper extension. For instance, if $(3, D) = 1$ then $\gamma_2(\omega) \in K(D)$, so $j(\omega)$ is a cube—e.g, $j(\sqrt{-2}) = 8000$, and $j(\frac{1+\sqrt{-163}}{2}) = -640320^3$; if D is odd then $\gamma_3(\omega) \in K(D)$ and if $D \equiv 5(\text{mod}8)$ then $\sigma^6(\omega) \in K(D)$. Unhappily, though Weber was aiming for a more general theory, he seemed only to succeed in constructing a plethora of other special cases. Many beautiful numbers were calculated, but everything was far too particular, and the theory too complicated: it was too far ahead of the rest of mathematics. New mathematical concepts were needed before a civilised theory of automorphic functions could be developed.

Quite abruptly, the theory of modular functions dropped completely out of fashion; Hecke did important work and so did Rankin (hence the title of this workshop), but it is hardly an exaggeration to say that for half a century most mathematicians hardly knew that the theory of modular functions had ever existed.

3. Heegner

So we may jump directly to Heegner's paper [11] of 1952. Heegner was a fine mathematician, with a rather low-grade post in a gymnasium in East Berlin; he clearly knew Weber's book well. He was interested in the congruence number problem: recollect that m is a congruence number if it is the area of a rightangled triangle with rational sides (most people call this a Pythagorean triangle; Heegner called it a Harpedonapten triangle). In his famous, very eccentrically written, paper he begins with a historical introduction concerning the congruence number problem, then he quotes various things from Weber and proves some highly surprising theorems showing that the congruence number problem is soluble for certain families of m; and then he suddenly (correctly but over succinctly) solves the classical class number one problem (see also [1] and [17]). Unhappily, in 1952 there was noone left who was sufficiently expert in Weber's *Algebra* to appreciate Heegner's achievement.

Heegner proved that if p is a prime congruent to 5 or 7 modulo 8 then p is a congruence number, and if p is congruent to 3 or 7 modulo 8 then $2p$ is a congruence number. The proofs are similar, I will sketch his proof that $2p$ is congruence when $p \equiv 3(\text{mod } 8)$, since it is the simplest. A typical Pythagorean triangle has sides $2rst$, $r(s^2-t^2)$ with rational r, s, t, so $2p$ is a congruence number if there are rational r, s, t with $2p = r^2st(s^2 - t^2)$. For this it is clearly enough that the elliptic curve

$$E \; : \; -py^2 = x(x^2 - 64) \tag{1}$$

should have a nontrivial rational point, and for this it is enough that the Diophantine equation

$$-pu^2 = v^4 - 64 \tag{2}$$

is soluble in rational u, v. Referring to Weber, we see that if $p \equiv 3 (\mathrm{mod}\ 8)$ there is a solution of (2) with $(u, v) \in K(-p)$ given by $\sqrt{-p}u = \gamma_3\sigma^{12}/(\sigma^{24} + 8)$, $v = \sigma^6$, with the functions evaluated at $\omega = \frac{-3+\sqrt{-p}}{2}$; and it is easy to check that u, v are real. If now p is prime, the class number is odd, so the classfield $K(-p) \cap \mathbb{R}$ has odd degree over the rationals. So to prove Heegner's theorem that twice every prime congruent to 3 modulo 8 is a congruence number, it is enough to show that if (2) has a point in an extension of odd degree then it has a rational point. Nowadays, this would be done by saying that a solution of (2) gives a point of E in the nontrivial coset C of $E(\mathbb{R})/2E(\mathbb{R})$, and adding up an odd number of points of C gives a point in C, which has to be nontrivial. Heegner uses a characteristically offbeat method; it is hardly known and has the advantage of being good for explicit computation over Q, so I quote it:

HEEGNER'S LEMMA. *Suppose that $f(X)$ is a quartic over a field L whose leading coefficient is not a square in L, and that $Y^2 = f(X)$ has a solution in a field M with M/L an extension of odd degree d. Then $Y^2 = f(X)$ has a solution in L.*

If not, we may suppose that M is the extension of least odd degree in which there exist x, y satisfying $y^2 = f(x)$. We may suppose that $y \in L(x)$, else it would need an extension of even degree, so $L(x) = M$, so x is a root of $g(X) = 0$ where g is a polynomial over L of degree $d \geq 3$; and $y = h(x)$ where h is a polynomial over L of degree $s \leq d - 1$. We see that $h^2 - f$ is of degree $\max(4, 2d - 2)$ since the leading coefficients cannot cancel; and $h^2 - f$ is divisible by g, so $h^2 - f = gk$ where k is a polynomial over L of degree $\max(4 - d, d - 2)$ which is certainly odd and less than d. But now if θ is a root of $k(X) = 0$, we see that $x = \theta, y = h(\theta)$ gives a point of $y^2 = f(x)$ in a smaller extension than M.

Heegner's paper was written in an amateurish and rather mystical style, so perhaps it was not surprising that at the time noone tried very hard to understand it. It was thought that his solution of the class number problem contained a gap, and though his work on the congruence number problem was clearly correct, noone realised that it contined the germs of a valuable new method. Sadly, he died in obscurity.

4. Simplification and Generalisation

Looking back at old diaries and suchlike, I find that I first saw Heegner's paper in 1966 (a little later than Stark, he tells me); I had been told it was wrong, but so far as I could see, it followed from results in Weber's *Algebra*; and his results on points on elliptic curves were exciting. It took a while to decide he was right (one had to read Weber first, and I hadn't even got good German) but this was achieved by the end of 1967 (see [2] and [3]). It took very much longer to understand it properly, maybe until 1973; it was necessary to both simplify and generalise. One needed to replace Heegner's rather miraculous construction of rational points on certain elliptic curves by a theorem that modular elliptic curves, indeed modular curves, are born with natural points on them, defined over certain classfields.

One also wanted to relate these points to something else – maybe to L_E. I persuaded Nelson Stephens (while he held an Atlas Fellowship) to compute the functions γ_2, γ_3 for discriminants D up to 1580, prime to 6. (He computed for even D too, but for the sake of exposition let us restrict to odd D.) We know that for such discriminants $D^{1/2}\gamma_3$ and γ_2 are in the class field $K(D)$, so we get points $P(\omega)$ on the curve

$$y^2 = x^3 - 1728.$$

(It was exceptionally easy to compute the points $P(\omega)$ as complex points, as one simply integrated the differential $\eta^4(6z)$.) Summing over the ideal class group of $R(D)$, we get a rational point $u(D, 1)$ of the curve

$$E_D : \; Dy^2 = x^3 - 1728.$$

More generally, if we take χ as a genus character of the classgroup, then the sum $\sum \chi(\omega)P(\omega)$ gives a point $u(e, f)$ of the curve

$$E_e : \; ey^2 = x^3 - 1728,$$

where the factorisation $D = ef$ depends on χ. The computations were consistent with a formula

$$\hat{h}(u(e, f)) = 2^A 3^B L(E_f, 1) L'(E_e, 1) / \sqrt{-3/ef}\, \Omega^2$$

where the exponents A, B were explicit and not very interesting (but we did not understand them at the time); note that $u(e, f)$ was trivial when $E_e(Q(\sqrt{D}))$ had rank more than 1. We told people, the above formula is quoted from a 1973 Harvard seminar [C] (unfortunately Dick Gross was away in Oxford that term), but as we did not understand what we were doing, we did not publish these computations till years later [5]. Nowadays, we know that $ey^2 = x^3 - 1728$ is the "wrong" model, which explains the unwanted factors $2^A 3^B$.

Meanwhile, we realised what one should be doing in a general case. One wants points on a modular curve, with coordinates in smaller fields than one would expect, and in the first instance one finds points on $X_0(N)$ itself rather than on the elliptic curves it covers. Once one realises this, the problem becomes fairly simple. $X_0(N)$ is the completion of the upper half plane factored by $\Gamma_0(N)$, it is parametrised by $j(z)$ and $j_N(z) = j(Nz)$, so we may take a typical point of $X_0(N)$ as $P(z) := (j(z), j(Nz))$. If we take ω as a quadratic surd with discriminant D, then $j(\omega) \in K(D)$ and usually $N\omega$ will have discriminant N^2D and $j(N\omega) \in K(N^2D)$ so that $P(\omega)$ is defined over the field $K(N^2D)$ which is big and useless; but it is actually easy to persuade ω and $N\omega$ to have the same discriminant. Simply take ω as a root of an equation of shape

$$NA\omega^2 + B\omega + C = 0,$$

then ω and $N\omega$ both have the discriminant $D = B^2 - 4NAC$, and $P(\omega) \in X_0(N)(K(D))$. Note that there is enormous freedom in choosing D.

It is inelegant to evaluate functions at complex numbers, when really they depend only on ideal classes. Fix N as a conductor, choose D as a negative discriminant so that $D = B^2 - 4NAC$ is soluble, and write $R(D)$ for the corresponding quadratic ring $Z\left[\frac{D+\sqrt{D}}{2}\right]$; then there is a primitive ideal n of $R(D)$ with norm N, fix such an ideal. Then for every ideal a of R, $P(n,a) := (j(a), j(\overline{n}a))$ is a *Heegner point* of $X_0(N)(K(D))$; we can go to and fro between the notations $P(n,a)$ and $P(\omega)$—to every pair of ideal classes $(a, \overline{n}a)$ there corresponds a coset (ω) modulo $\Gamma_0(N)$.

Suppose now that E is an elliptic curve over Q, covered by $X_0(N)$; write $\phi : X_0(N) \to E$ for the covering map. Then $\phi(P(n,a))$ is a point of $E(K(D))$, and taking $u(D,1) := \sum_{(a)} P(n,a)$ as the $K(D)/Q(\sqrt(D)$ trace, we obtain a point of $E(Q(\sqrt{D}))$. More generally, if χ is a genus character, $u(e,f) := \sum \chi(a)\phi(P(n,a))$ is a point of $E(Q(\sqrt{e}))$, where e, f are determined by χ with $D = ef$; we may call the $u(e,f)$ Heegner points too. The elliptic curve E will correspond to a differential $f(z)\,dz$ on $X_0(N)$, and then the period lattice $\Lambda(E)$ of E is easily calculable as $\int_{H_1(X_0(N))} f(z)\,dz$, and $\phi(P(\omega))$ can be calculated as $\int_\omega^{i\infty} f(z)\,dz \in E(C) = C/\Lambda(E)$. So we are in good shape for actually computing the Heegner points $u(e,f)$, at least their elliptic parameters.

This is essentially the point that had been reached in 1973–75. I lectured in Rome, Paris, Kyoto, Moscow and Harvard; and the Rome talk was summarised as a short note [4]; but there was very little immediate feedback (I missed Kurčanov's paper [13]). With hindsight, I should have realised that the theory of Heegner points was a natural extension of Weber's theory of complex multiplication, worth developing for its own sake (and indeed the functorial properties of Heegner points have turned out to be immensely valuable, in particular for Kolyvagin's Euler systems [12]); but I didn't, and indeed was discouraged. There was undue concentration on the original application of Heegner points, the construction of rational points on elliptic curves and (harder) proving that the points one had constructed were non-trivial. I gave another method on these lines, Barry Mazur (in [14]) gave one which worked beautifully for quadratic twists of $X_0(11)$, and Dick Gross (in II of [9]; [9] was not published until after the discovery of the Gross-Zagier theorem, but I think the ideas of II came a year or so earlier) gave a third. Nowadays, people tend to say that there is an adequate criterion for the nontriviality of the rational Heegner point using Gross-Zagier, "one just has to check that $L'(E,1) \neq 0$"; but I've never understood why computing $L'(E,1)$ should be considered easier than the direct computation of the Heegner point as a point of $E(C) = C/\Lambda$ (of course, Gross-Zagier shows that if one Heegner point of a given curve is non-trivial then they almost all are).

The involvement of Dick Gross marked the turning point: at last someone young enough and bright enough was thinking seriously about Heegner points! It is the logical point at which to end this lecture, and hand over to him. But it is happier to end with a bang rather than a whimper, so I need a final paragraph.

5. 1982

In 1981 and 1982, Nelson Stephens arranged a sabbatical year, and we planned a massive computation of the Heegner points of modular curves, to see what they would tell us. We actually did those computations [6], and very illuminating the results would have been — but they were anticipated by far more exciting developments.

In 1982, I got several letters from Dick (to which I replied with increasing delight).

March 1st. Dick's first letter [D] begins "I recently found an amusing method to study Heegner points on $J_0(N)$."

This included the method in II of [9]; it was exciting, because for the first time it related the index of the Heegner point in $E(Q)$ to the order of the Tate-Shafarevich group. He conjectures a not-quite-correct form of Gross Zagier and proves a tiny bit of it. This letter was only a foretaste of what was to come.

My reply [E] included "you seem to be opening so many doors that I'm almost afraid to push", which Dick correctly translated from British to American as "Get shoving, you lucky so-and-so".

May 14th. Dick's second letter [F]; it was wonderful. "I noticed some really amazing things, like the following:

1) The product $L'(E^{(\chi)},1)L(E^{(\chi')},1)$ is just the derivative at $s = 1$ of the L-series $L(E \otimes \operatorname{Ind}_F^Q \chi, s) \ldots$

2) The L-series $L(E\otimes\operatorname{Ind}\chi, s)$ has a beautiful integral expression by Rankin's method..."

and so on for four beautiful pages, culminating with "So all one has to do is prove the formula

$$L'(f \otimes \operatorname{Ind}\chi, s) = \langle y_\chi, y_{\chi^{-1}}\rangle_f \cdot \int_X \omega_f \wedge \overline{\omega}_f/\sqrt{D_F}.\text{"}^1$$

After that there was no going back! I replied, and got a third letter dated September 17 asking for more data; could Nelson Stephens and I supply concrete

[1]Readers of the September 6 letter printed on page 17 of this volume will see that it does not comment directly on the ideas of May 14th; instead it describes the results of relevant computations (cf. [4]), and also makes detailed comments on a preliminary version of [9].

evidence supporting the Gross-Zagier theorem-to-be? By that time, we had plenty, so I supplied it.

On December 9th, I got the news "Dear Bryan, Working with Don Zagier, I think I've assembled a proof ...." And the rest is in print.

References

[1] Alan Baker, "Linear forms in the logarithms of algebraic numbers", *Mathematika* **13** (1966), 204–216.

[2] B. J. Birch, *Diophantine analysis and modular functions*, Conference on Algebraic Geometry, Tata Institute, Bombay, 1968.

[3] B. J. Birch, "Elliptic curves and modular functions", pp. 27–32 in *Symposia Mathematica* **4**, Istituto Nazionale di Alta Matematica, New York, Academic Press, 1970.

[4] B. J. Birch, "Heegner points of elliptic curves", pp. 441–445 in *Symposia Mathematica* **15**, Istituto Nazionale di Alta Matematica, New York, Academic Press, 1975.

[5] B. J. Birch and N. M. Stephens, "Heegner's construction of points on the curve $y^2 = x^3 - 1728e^3$", in *Séminaire de Théorie des Nombres*, Paris 1981–2, edited by Marie-José Bertin, *Progress in mathematics* **38**, Boston, Birkhäuser, 1983.

[6] B. J. Birch and N. M. Stephens, "Computation of Heegner points", pp. 13–41 in *Modular forms*, edited by R. A. Rankin, Chichester, Ellis Horwood, 1984.

[7] J. W. S. Cassels, "Diophantine equations with special reference to elliptic curves", *J. London Math. Soc.* **41** (1966), 193–291.

[8] J. W. S. Cassels and A. Fröhlich (editors), *Algebraic number theory* (Brighton, 1965), London, Academic Press, 1967.

[9] B. H. Gross, "Heegner points on $X_0(N)$", pp. 87–105 in *Modular forms*, edited by R. A. Rankin, Chichester, Ellis Horwood, 1984.

[10] B. H. Gross and D. Zagier, "Heegner points and derivatives of L-series", *Invent. Math.* **84** (1986), 225–320.

[11] Kurt Heegner, "Diophantische Analysis und Modulfunktionen", *Math. Zeitschrift* **56** (1952), 227–253.

[12] V. A. Kolyvagin, "Finiteness of $E(Q)$ and $Ш(E/Q)$ for a class of Weil curves", *Izv. Akad. Nauk SSSR* **52** (1988).

[13] P. K. Kurčanov, "The zeta-function of elliptic curves over certain abelian extensions of imaginary quadratic fields", *Mat. Sbornik* (N.S.) **102 (144)** (1977), 56–70.

[14] Barry Mazur, "On the arithmetic of special values of L-functions", *Invent. Math.* **55** (1979), 207–240.

[15] J.-P. Serre, "Lettre à David Goss, 30 mars 2000", pp. 537–9 in *Wolf Prize in Mathematics*, vol. 2, edited by S.-S. Chern and F. Hirzebruch, River Edge, NJ, 2001.

[16] Goro Shimura, *Introduction to the arithmetic theory of automorphic functions*, Tokyo, Iwanami Shoten and Princeton, University Press, 1971.

[17] H. M. Stark, "A complete determination of the complex quadratic fields with class-number one", *Michigan Math. J.* **14** (1967), 1–27.

[18] Yutaka Taniyama, Problem 12 in the Japanese version of the *Proceedings of the International Symposium on Algebraic Number Theory*, Tokyo and Nikko, 1955; see also p. 399 in J.-P. Serre, *Collected papers*, v. 3, New York, Springer, 1983.

[19] H. Weber, *Lehrbuch der Algebra*, Braunschweig, Vieweg, 1908 (especially volume III).

[20] André Weil, "Über die Bestimmung Dirichletscher Reihen durch Funktionalgleichungen", *Math. Annalen* **168** (1967), 149–156.

Letters and Manuscripts

[A] Letter from André Weil to Peter Swinnerton-Dyer, dated July 24, 1965.

[B] Letter from John Tate to Peter Swinnerton-Dyer, dated November 5, 1965, and reply from Peter Swinnerton-Dyer to John Tate in December.

[C] Harvard Seminar ("Mazur-Birch seminar"), Fall 1973.

[D] Letter from Dick Gross to Bryan Birch, dated March 1, 1982 (page 11 of this volume).

[E] Letter from Birch to Gross, dated May 6, 1982 (page 13 of this volume).

[F] Letter from Gross to Birch, dated May 14, 1982 (page 14 of this volume).

[G] Letter from Gross to Birch, dated Sept 17, 1982 (page 21 of this volume).

[H] Letter from Gross to Birch, dated December 1, 1982 (page 22 of this volume).

Bryan Birch
Mathematical Institute
24-29 St Giles'
Oxford OX1 3LB
United Kingdom
birch@maths.ox.ac.uk

Heegner Points and Rankin L-Series
MSRI Publications
Volume **49**, 2004

Correspondence

BRYAN BIRCH AND BENEDICT GROSS

Gross to Birch: March 1, 1982

Dear Birch,

I recently found an amusing method to study Heegner points on $J_0(N)$. Let E be an elliptic curve over $\mathbb{Q}$ of level N, together with a parametrization $J_0(N) \to_\pi E$. Let K be a quadratic field of discriminant d_K prime to N; let χ be the associated quadratic Dirichlet character and E^χ the twisted curve.

Let F be an imaginary quadratic field in which all prime factors of N split and choose an integral ideal $\mathfrak{n}$ with $\mathfrak{n}\bar{\mathfrak{n}} = N$ and $(\mathfrak{n}, \bar{\mathfrak{n}}) = 1$. Assume further that d_χ divides d_F, so K is contained in H, the Hilbert class field of F. The modular data $x = (\mathbb{C}/\mathcal{O}_F, \ker \mathfrak{n})$ defines a point of $X_0(N)$ rational over H and the divisor $e_f = \pi\big(\sum_{\mathrm{Aut}(H)} \chi(\sigma)\sigma x\big)$ gives a point of $E(K)^-$, or equivalently, a rational point on E^χ. One can check that e_F is killed by 2 whenever the sign in the functional equation for E^χ is $+1$. Do your computations support the following?

CONJECTURE. *e_F has infinite order iff* rank $E^\chi(\mathbb{Q}) = 1$. *If this is the case and π is a strong Weil parametrization, let M denote the subgroup generated by the points e_F. Then $(E^\chi(\mathbb{Q}) : M)^2 = \mathrm{Card}(\text{Ш}(E^\chi/\mathbb{Q}))$.*

I think I can prove that the point e_F has infinite order whenever the image of the cuspidal group on E has order divisible by $p \geq 3$ *and* certain p-class groups are trivial. In all these cases, the rank is 1.

Here is a simple case which illustrates the method. Let $E = J_0(11)$ and let K be a *real* quadratic field in which the prime 11 is inert. Choose F as above, and let K' denote the other imaginary quadratic field contained in FK.

PROPOSITION. *If $h_K h_{K'} \not\equiv 0 \pmod 5$ then $E^\chi(\mathbb{Q}) \simeq \mathbb{Z}$, $e_F \neq 0$ in $E^\chi(\mathbb{Q})/5E^\chi(\mathbb{Q})$, and $\text{Ш}(E^\chi)_5 = (0)$.*

PROOF. A 5-descent, combined with the fact that $h_\chi \not\equiv 0 \pmod 5$, gives an exact sequence

$$0 \to E^\chi(\mathbb{Q})/5E^\chi(\mathbb{Q}) \to^\delta \mathcal{O}_K^*/(\mathcal{O}_K^*)^5 \to \text{Ш}(E^\chi/\mathbb{Q})_5 \to 0$$

with $\mathcal{O}_K^*/(\mathcal{O}_K^*)^5 \simeq \mathbb{Z}/5$. It will suffice to show $\delta(e_F) \neq 0$ in $\mathcal{O}_K^*/(\mathcal{O}_K^*)^5$, as $E^\chi(\mathbb{Q})_5 = (1)$.

But the map δ arises from the cohomology of the covering $\tilde{E} \to E$ with fibre μ_5 which is obtained by taking a 5th root of the modular unit $f(z) = \Delta(z)/\Delta(11z)$. Some calculation then gives $\delta(e_F) = \delta(\sum_{\mathrm{Aut}(H)} \chi(\sigma)\sigma x) \equiv (\prod_{\mathfrak{a}} \Delta(\mathfrak{a})\Delta(\mathfrak{a}^{-1})^{\chi(\mathfrak{a})})^2$ mod $(\mathcal{O}_K^*)^5$, where the product is taken over ideals representing the classes of $\mathcal{O}_F$. Kronecker's limit formula can be used to show that the "elliptic unit" $\prod \Delta(\mathfrak{a})\Delta(\mathfrak{a}^{-1})^{\chi(\mathfrak{a})}$ is equal to the $24h_K h_{K'}$ power of a fundamental unit for K. By our assumptions on h_K and $h_{K'}$, we see $\delta(e_F) \neq 0$.

Could you send me a preprint of your height computations, and anything else you might have on the subject?

Best wishes,

Dick Gross

Birch to Gross: May 6, 1982

Dear Dick,

I received your letter of March 1. I thought it was beautiful and decided to reply when I'd sorted it all out and had time to spare — which never happens, so I must apologize for not replying sooner.

First, the conjecture on your first page is insufficiently elaborate — in your notation there are three fields involved K, F and K', with corresponding characters χ, $\chi\chi'$ and χ'. For each of them, there corresponds a twisted elliptic curve defined over $\mathbb{Q}$. The correct conjecture appears to be

$$\begin{matrix}\text{canonical}\\ \text{height}\end{matrix}\begin{pmatrix}\text{point of } E^{(\chi)}(\mathbb{Q})\\ \text{given by } e_F\end{pmatrix} \doteq \frac{L'(E^{(\chi)},1)L'(E^{(\chi')},1)}{\begin{matrix}\text{real}\\ \text{period}\end{matrix}(E^{(\chi)})\ \begin{matrix}\text{real}\\ \text{period}\end{matrix}(E^{(\chi')})}$$

where $\doteq$ means "equal except for one or two stray factors like 2 or 3 that come in because I presumably haven't got quite the correct model."

Accordingly, your index

$$(E^{(\chi)}(\mathbb{Q}):M)^2 \doteq \mathrm{Card}(\text{Ш}(E^{(\chi)}/\mathbb{Q}))\cdot\mathrm{Card}(\text{Ш}(E^{(\chi')}/\mathbb{Q}))$$
$$\cdot\begin{pmatrix}\text{factors coming from the}\\ \text{bad primes of } E^{(\chi)}\end{pmatrix}\cdot\begin{pmatrix}\text{factors coming from the}\\ \text{bad primes of } E^{(\chi')}\end{pmatrix};$$

and I guess the torsion comes into it too. The significant point is that one needs the contribution from the "other" twist $E^{(\chi')}$; the less significant point is, of course, that there are more junk factors than you acknowledged. The conspicuous "corollary of the conjecture" is that e_F should have infinite order precisely when $E^{(\chi)}(\mathbb{Q})$ has rank 1 and $E^{(\chi')}(\mathbb{Q})$ has rank 0; if rank $E^{(\chi)}(\mathbb{Q}) \geq 3$ or rank $E^{(\chi')}(\mathbb{Q}) \geq 2$ the thing, as a matter of experience, is trivial.

I found your second page very nice indeed, and the more I think about it the more I like it; there seem to be an awful lot of doors it may open, and I'm almost reluctant to push them in case there is a block that I can't see! It's so much more down to earth than Barry's ideas, and seems to give a lot more information. I feel foolish not to have thought of something like it years ago — but then if one is set on points of order 2, the fact that the image of δ in the sequence

$$0 \to E^{(\chi)}(\mathbb{Q})/2E^{(\chi)}(\mathbb{Q}) \xrightarrow{\delta} \mathcal{O}_K^\times/\mathcal{O}_K^{\times 2}$$

is automatically trivial is a very thick hedge!

Yours,

Bryan Birch

Gross to Birch: May 14, 1982

Dear Bryan,

Thanks for your note. In the interim between my letter and yours, I had noticed the point that I was neglecting the field K', as you kindly pointed out. Then I noticed some really *amazing* things — like the following:

1) The product $L'(E^{(\chi)},1)L(E^{(\chi')},1)$ is just the derivative at $s=1$ of the L-series $L(E \otimes \mathrm{Ind}_F^Q \chi, s)$, where χ is viewed as a character of $\mathrm{Gal}(\overline{F}/F)$ and $\mathrm{Ind}\,\chi$ is the 2-dimensional induced representation to $\mathrm{Gal}(\overline{\mathbb{Q}}/\mathbb{Q})$.
2) The L-series $L(E\otimes\mathrm{Ind}\,\chi, s)$ has a beautiful integral expression — by Rankin's method — as $L(\mathrm{Ind}\,\chi, s)$ is the L-series of a modular form of wt 1 !
3) The product of periods: *real period*$(E^{(\chi)})$ · *real period*$(E^{(\chi')})$ is *equal* to the integral $\int_{E(\mathbb{C})} \omega \wedge \overline{\omega}/\sqrt{d_F}$, as the characters χ and χ' have opposite parity. This integral can also be expressed as a Petersson inner product when E is a Weil curve (ω is a Néron differential on E).

Anyhow, this led me to drop the restriction of considering only *elliptic curves over* $\mathbb{Q}$, and I've arrived at the following crazy business.

Let $f = \sum_{n=1}^{\infty} a_n q^n$ be a newform of wt 2 on $\Gamma_0(N)$ with coefficients in a subfield of $\mathbb{R}$; let $\omega_f = 2\pi i f(z)\,dz$ be the corresponding holomorphic 1-form on $X = X_0(N)$.

Let $J = J_0(N)$; there is a canonical symmetric pairing $\langle\ ,\ \rangle$ on $J(\overline{\mathbb{Q}}) \times J(\overline{\mathbb{Q}})$ with values in $\mathbb{R}$ which is obtained by composing the Poincaré height on $J \times J^\vee$ with the standard isomorphism $J \simeq J^\vee$. We can use the action of the Hecke algebra on J to *refine* this to a pairing $\langle\ ,\ \rangle_f : J(\overline{\mathbb{Q}}) \times J(\overline{\mathbb{Q}}) \to \mathbb{R}$ which satisfies $\langle T_l x, y\rangle_f = \langle x, T_l y\rangle_f = a_l \langle x, y\rangle_f$ for all $l \nmid N$. If f has rational coefficients, so corresponds to the Weil curve $X \to_\pi E$, then $\langle x, y\rangle_f = \frac{1}{\deg \pi}\langle \pi x, \pi y\rangle_E$.

Now let F be an imaginary quadratic field in which all primes dividing N *split*, and fix a factorisation $N = \mathfrak{n}\cdot\overline{\mathfrak{n}}$ with $(\mathfrak{n},\overline{\mathfrak{n}}) = 1$. The modular data $x = (\mathbb{C}/\mathcal{O}_F, \ker \mathfrak{n})$ defines a point of X which is rational over H, the Hilbert class field of F. Let y be the class of the divisor $\frac{1}{\delta}\{(x) - (\infty)\}$ in $J(H)\otimes\mathbb{Q}$, where $\delta = 1$ unless $F = \mathbb{Q}(i), \mathbb{Q}(\rho)$ in which case $\delta = 2, 3$. If χ is any character of $\mathrm{Gal}(H/K)$ we let $y_\chi = \sum \chi^{-1}(\sigma) y^\sigma$ in $(J(H)\otimes\mathbb{C})^\chi$.

By studying the functional equation of the L-series $L(f\otimes\mathrm{Ind}\,\chi, s)$ one can show that $L(f\otimes\mathrm{Ind}\,\chi, 1) = 0$. The "crude" conjecture then becomes

The following are equivalent: a) $\pi(y_\chi) \neq 0$ in $(A_f(H) \otimes \mathbb{C})^\chi$

b) $\dim_{\mathbb{C}}(A_f(H) \otimes \mathbb{C})^\chi = \dim A_f$

c) $L'(f \otimes \operatorname{Ind}\chi, 1) \neq 0$

Here A_f is the quotient of J determined by the newform f: $\pi : J \to A_f$.

Right now, all my descent evidence only helps to show b) $\Longrightarrow$ a). But I think that the equivalence a) $\Longleftrightarrow$ c) may be tractable. "All one has to do" is to prove the formula:

$$L'(f \otimes \operatorname{Ind}\chi, 1) = \frac{1}{\sqrt{d_F}} \int_X \omega_f \wedge \overline{\omega_f} \cdot \langle y_\chi, y_{\chi^{-1}} \rangle_f. \qquad (*)$$

I hope I've got the fudge factors right — it seems to check against some hand computation, but I'd like to put it up against your tables. I should remark that there is a similar conjecture for $L'(f \otimes \operatorname{Ind}\chi, 1)$ when χ is a ring class character of F of conductor prime to N. Here one uses the Heegner point x constructed from the corresponding order.

Formula $(*)$ for all characters χ is equivalent to

$$L'(f, \sigma, 1) = \frac{1}{\sqrt{d_F}} \int_X \omega_f \wedge \overline{\omega_f} \cdot h_F \langle y, y^\sigma \rangle_f \qquad (**)$$

for all elements $\sigma \in \operatorname{Gal}(H/F)$. Here $L(f, \sigma, s) = \frac{1}{h_F} \sum_\chi L(f \otimes \operatorname{Ind}\chi, s)\chi^{-1}(\sigma)$ is a "partial L-function", which has an even *nicer* analytic expression via Rankin's method. Namely, let

$$g_\sigma = \frac{1}{2\delta}\left(1 + \sum_{m,n}{}' q^{B(m,n)}\right),$$

where $B(x, y) = ax^2 + bxy + cy^2$ is a binary quadratic form in the class of σ. Then g_σ is a modular form of wt 1 and character ε on $\Gamma_0(d_F)$, where ε is the quadratic character corresponding to the extension $F/\mathbb{Q}$. Put $M = Nd_F$ and define the Eisenstein series

$$E(z, s) = \sum_{c,d}{}' \frac{\varepsilon(d) y^s}{|cMz + d|^{2s}(cMz + d)};$$

this has wt 1 and character ε on $\Gamma_0(M)$ (but isn't holomorphic when $s \neq 0$.) Modulo powers of π and simple Γ-factors, one has the equality

$$L(f, \sigma, s) = \int_{\mathcal{H}/\Gamma_0(M)} \overline{f(z)} g_\sigma(z) E(z, s-1)\, dx\, dy.$$

At this point things become even more interesting, as I believe the integral breaks into h pieces (after dealing with some imprimitivity factors) which one can hope correspond to the *local* heights of the points y and y^σ at the archimedean places of H. Anyhow, it begins to look more like the sort of statement an analytic number theorist can deal with, I may even dare a crack at it myself!

Best wishes,

Dick

P.S. the relation b) $\Longleftrightarrow$ c) is pretty much your conjecture with Swinnerton-Dyer, as $L(A_f/H,s) = \prod_\alpha\left(\prod_\chi L(f^\alpha \otimes \operatorname{Ind}\chi, s)\right)$, where the f^α are the conjugates of the modular form f. Note that the crude conjecture implies

$$L'(f \otimes \operatorname{Ind}\chi, 1) \neq 0 \Longleftrightarrow L'(f^\alpha \otimes \operatorname{Ind}\chi, 1) \neq 0.$$

P.P.S. Don't despair at 2, just don't project to the ε-component of $E(F)$! For example, let $X = X_0(17)$ and F where $(17) = \wp.\overline{\wp}$ is split. The modular unit $h(z) = \{\Delta(z)/\Delta(17z)\}^{1/4}$ has divisor $4\{(0) - (\infty)\}$ on X and the map $J(H) \to^{\delta} H^*/(H^*)^4$ defined on divisors prime to $0, \infty$ by $\delta(\mathfrak{a}) = h(\mathfrak{a})$ mod $(H^*)^4$ is a group homomorphism. The image of the Heegner point y is equal to $\{\Delta(\mathscr{O})/\Delta(\wp)\}^{1/4}$, provided $F \neq \mathbb{Q}(i)$, this modular unit and all its conjugates generate the ideal $\wp^3$. Hence $\delta(y_1) = \prod_\sigma \delta(y^\sigma)$ is an element of $F^*/(F^*)^4$ which generates the ideal $\wp^{3h}$. If $4 \nmid h$ this is *not* a fourth power. And — again assuming $F \neq \mathbb{Q}(i)$ — it is not the image of a torsion point on $J(F)$. Therefore, y_1 has infinite order.

Birch to Gross: around September 6, 1982

Dear Dick,

Thank you for your stuff — comments at end of letter, at least preliminary ones.* But no doubt you would prefer to know the present state of play as regards computations.

Nelson (Stephens) has been calculating Heegner points wholesale, dealing with all discriminants $-D$ with $D < 1000$ for various curves. So far, I have good data for the curves 11B, 17C, 19B, 26D, 37A, 43A, 57E, 67A, 76A; plus a few others I haven't had time to sort out yet (the data are fairly bulky) 79A, 109A.

Let us set up notation: suppose our curve is E with conductor N, so that E is associated with a normalized (i.e. $a_1 = 1$) differential ω on $X_0(N)$; $-D$ is the discriminant of your ring $\mathscr{O}$ so that $-D = efm^2$ where e, f are 1 or discriminants of quadratic fields, $(e, f)m$ is your e, and the game gives rise to a point of $E(\mathbb{Q}(\sqrt{e}))$ which you have been denoting by e_χ (at any rate for $\chi \neq \chi_0$). In case χ is not principal, we are led to a rational point of the twist $E^{(e)}$, which Nelson denotes by $P(e, f, m, \omega, N)$; in the principal case we are liable to have to take $e_\chi - e_\chi^\tau$ like you do, and are led to a point $P^*(e, f, m, \omega, N)$ "which is twice as big as it ought to be."

For the purpose of this letter, let us restrict ourselves to the cases

$$(em, fm, N) = 1$$

in order to avoid nonsense and complication (but in case N is not square free, this seems to be an undesirable restriction — e.g., for $N = 76$ we certainly need to allow $m = 2$). At this stage, there is no need to take the "imprimitivity index" m as 1. We throw away the "rubbish" cases that are automatically trivial by your Lemma 11.1 — and the conjectures that follow do not apply to such rubbish.

What I've actually tabulated is a near integer M. In case the point

$$P(e, f, m, \omega, N)$$

is torsion, $M = 0$; otherwise, it has always happened that rank $E^{(e)}(\mathbb{Q}) = 1$ (as predicted by conjectures made on the basis of this evidence!), we *fix* a generator $Q(e, E)$ of $E^{(e)}(\mathbb{Q})$, and we define M by

$$P(e, f, m, \omega, N) = \frac{w}{2} M(e, f, m, \omega, N) Q(e, E), \text{ in case } \chi \text{ is not principal}$$

$$P^*(e, f, m, \omega, N) = 2\frac{w}{2} M(e, f, m, \omega, N) Q(e, E), \text{ in case } \chi \text{ is principal}$$

*Gross had sent Birch an early version of his manuscript "Heegner points on $X_0(N)$".

where of course $\frac{w}{2}$ is the usual factor

$$\frac{w}{2} = \begin{cases} 2 & \text{if } efm^2 = -4, \\ 3 & \text{if } efm^2 = -3, \\ 1 & \text{otherwise.} \end{cases}$$

So apart from cases with e or f equal to 1, $M \in \mathbb{Z}$; and though the sign of M is meaningless, the sign of a ratio $M(e, f_1, m_1, \omega, N)/M(e, f_2, m_2, \omega, N)$ is well-determined. Unfortunately, Nelson's present tables only seem to give the sign reliably when $e > 0$.

Our main conjecture is of course that

$$\text{canonical height}\,(P(e, f, 1, \omega, N)) = L^{*\prime}(E^{(e)}, 1)L^*(E^{(f)}, 1)$$

where the L-functions on the right are suitably normalized — one needs to divide by the real period, allow for the torsion, and do a little fudging (e.g. by $\frac{w}{2}$). So far, we have very few computations of $L^{*\prime}$, so we can't be absolutely specific — the [burden] of the conjecture at present is that the amount of fudging will be slight and predictable! In particular, Ш should *not* come into the formula. Particular cases, more easily verified experimentally, are

(1) $P(e, f, 1, \omega, N)$ is torsion if $L^{*\prime}(E^{(e)}, 1) = 0$ (i.e. rank $E^{(e)}(\mathbb{Q}) \geq 3$) or $L^*(E^{(f)}, 1) = 0$ (i.e. rank $E^{(f)}(\mathbb{Q}) \geq 2$).
(2) For fixed e and variable f,

$$\frac{h(P(e, f_1, 1, \omega, N))}{h(P(e, f_2, 1, \omega, N))} = \frac{L^*(E^{(f_1)}, 1)}{L^*(E^{(f_2)}, 1)}$$

For the curve $y^2 = x^3 - 1728$, the evidence is about 10 years old by now!

Experimental facts gleaned from our present computations are as follows:

(1) If $m > 1$, $P(e, f, m, \omega, N)$ is a predictable multiple of $P(e, f, 1, \omega, N)$; and this can be proved, I think. End of story! (Use Hecke operators, cf. your § 6).
So it is nearly enough to consider the case $m = 1$ — at any rate for square-free N.
(2) For fixed ω, N and any e_1, e_2, f_1, f_2 for which the relevant Heegner points all exist

$$M(e_1, f_1, 1, \omega, N)M(e_2, f_2, 1, \omega, N) = \pm M(e_1, f_2, \cdots)M(e_2, f_1, \cdots)$$

and the sign is + (whenever e_1, e_2 are positive).
(3) $M^2(e_1, f_1, 1, \omega, N)L^*(E^{(f_2)}, 1) = M^2(e_2, f_2, 1, \omega, N)L^*(E^{(f_1)}, 1)$.
(4) For certain curves (maybe all curves without 2-torsion?) the parity of $M(e, f, m, \omega, N)$ depends *only* on $-D = efm^2$, ω, N, and *not* on e, f, m separately. Nelson can prove this in some cases, and I can in others. Combined with (1) above, it is quite a strong criterion for proving Heegner points non-trivial, *different* from those already known.

(Corollary: Subject to various conditions, for square free e

$$L^*(E^{(e)}, 1) \equiv a_e \bmod 2,$$

where of course a_n is eigenvalue of Hecke T_n. Is this well known?)

My guess is that (1), (4), and probably (2) are accessible with present techniques — but not (3).

Our evidence for (1) and (2) is all the curves I've mentioned, for all $-D \geq -1000$; the evidence for the "+" in (2) is only for the curves 17, 19, 37, 67, only in the cases $e > 0$. We've verified (3) in all cases, except that $L^{(f)}(1)$ has usually only been calculated for f up to 200 or so.

Comments on your M/S* — nothing very useful to say, but you asked!

p. 1. It still isn't clear to me that your method yields all the known results when E has (cuspidal) 2-torsion.

Chapter I. Doesn't claim to contain anything new, but it's a very nice exposition. I like the statement of Lemma 11.1 — it is straightforward junk-dunking, but more elegant than my own explicit statement in terms of quadratic residues and the like. § § 4-6 are nice and clear too.

A very small point — don't call c the *level*, it would be waste of a good word if it were spare, and it isn't — the level is N! After all, c has a perfectly good name (the conductor of $\mathcal{O}$) if it needs one.

More material points:

(i) $(c, N) = 1$ is just a bit too restrictive in general, e.g. one needs to allow $c = 2$ when $N = 76$.

(ii) I guess 11.2 is correct, but it is hard to produce any evidence for cases with $\chi^2 \neq 1$. I like my conjectures experimentally verified as well as theoretically hyperplausible!

Chapter II. This seems rather more technical than I would have expected, but I have to admit that I've only played with the cases where χ is quadratic, when the L-series are very easy to write down.

p. 16 Conjecture 17.1. Do you *really* think this is accessible? After all, one doesn't even know how to prove $L'(1) = 0$.

"all evidence". Is there any evidence other than Stephens–Birch?

Chapter III (Very nice too.)

§ 20. I note that this is Barry's proof with the mappings made explicit!

§ 22. The case N prime is pegged rather closely to the Eisenstein component theory, and the case N composite is rather vague. Being a very down-to-earth sort of person (who isn't?) I prefer to start at the other end. Just at the moment, the most general formulation seems to be that *you have a theorem whenever you*

---

*See note on page 17.

have a nice function supported by the cusps—for instance, for $N = 26$ one writes down

$$\left(\frac{\eta(2z)\,\eta(26z)}{\eta(z)\,\eta(13z)}\right)^{12}$$

and the rest is automatic, whether N is prime or composite matters little. There is a prehistoric paper of Morris Newman that deals with explicit functions of shape $\prod \eta(dz)^{r_d}$; but I guess that the torsion doesn't all come from functions of this particular shape, and of course (e.g. in § 23) you don't need quite such special functions to apply the Kronecker formula.

I'm not clear that your argument includes the $n = 2$ case.

Incidentally, you will be amused to hear that in the cases e.g. for $N = 11$ where you predict that e_χ is trivial in $E^{(\chi)}/5E^{(\chi)}$ the computations, sure enough, give M divisible by 5—but *not* (usually) trivial.

Have sent a copy of your draft to Nelson Stephens—hope this is O.K. Will try to keep you posted, despite my well known reluctance to set pen to page.

Bryan

Gross to Birch: September 17, 1982

Dear Bryan,

Thanks for your letter. Your comments were helpful, and I'll incorporate them into the next version. Some questions.

1) I can't see how to handle the case where $gcd(c, N) \neq 1$, as I can't find a nice formula for the sign in the functional equation of $L(f \otimes g_\chi, s)$ when χ is primitive. Can you?

2) When I proposed conjecture 17.1, I didn't know about your wholesale data. It was motivated by the theoretical evidence in chapter III and some retail computational checks (about 1,000 in number) I made with Joe Buhler. We seem to get:

$$2^r \cdot \delta^2 \cdot D^{1/2} L'(f \otimes g_\chi, 1) = (\omega_f, \omega_f)\langle \nu_{\chi,f}, \nu_{\chi,f}\rangle$$

where $r = \text{Card}\{p \mid \gcd(N, D)\}$ and $\delta = w/2$. I hope this is in agreement with your tables.

Of course, I never would have considered the possibility of such an identity if I hadn't once seen an old paper of yours on $y^2 = x^3 - 1728$. I seem to remember some spurious powers of 2 and 3 in your formulae there; perhaps that's because you're on a curve isogenous to $X_0(36)$. Could you send me a copy of this manuscript if it still exists? Also, I'd appreciate a summary of Stephens' data on $X_0(N)$ for $N = 11, 17, 19$. (if the tables can be brought down to size).

3) You're right about chapter III — once you find the right modular unit you're in like Flynn. The functions $\prod_{dN} \eta(dz)^{m(d)}$ give the *rational* cuspidal group; the entire group of modular units can be quite a mess to determine. To preserve my sanity, I restricted to those cases where the Galois eigenfunctions of the cuspidal group were obviously cyclic.

4) My method works at $p = 2$ only when $\chi = 1$. But yours is nicer anyhow. Could you send me a write-up of your recent parity result — that looks neat.

Best,

Dick

Gross to Birch: December 1, 1982

Dear Bryan,

Working with Don Zagier, I think I've assembled a proof of the identity following conjecture 17.1 in my paper on Heegner points. Up to now we've been assuming that both N and D are prime, but I'd be surprised if the techniques didn't work in the general case. The method is more or less as I suggested in my letter of May 14 ; one uses Rankin's method to obtain explicit formulae for the derivatives of the L-series and stare at these long enough until one begins to see the local heights of Heegner points emerging. *Something* should actually be written down by the late Spring, and you'll get the first copy.

Two requests: would you mind if we referred to the identity and the resulting 17.1 in the next write-up as the conjecture of Birch (or of Birch/Stephens). I know you only make conjectures with *lots* of evidence, and only really *believed* it when $\chi^2 = 1$ and f came from an elliptic curve, but you were the one who discovered this amazing phenomenon, and without the security blanket of your evidence, I would never have dared a proof.

Second: could you send us some of your computations on $X_0(11)$, $X_0(17)$, and $X_0(19)$? The fun of the subject seems to me to be in the *examples.*

Best wishes,

Dick

Birch to Gross: December 27, 1982

Dear Dick,

Wonderful news. Does this mean that in particular you can show $L' = 0$ when it ought to (thus fulfilling Dorian Goldfeld's requirements?).

Will send O/P when Xmas recedes — at the moment all offices, not to mention the mail system, are inert.

Yours,

Bryan

BRYAN BIRCH
MATHEMATICAL INSTITUTE
24-29 ST GILES'
OXFORD OX1 3LB
UNITED KINGDOM
birch@maths.ox.ac.uk

BENEDICT GROSS
DEPARTMENT OF MATHEMATICS
HARVARD UNIVERSITY
ONE OXFORD STREET
CAMBRIDGE, MA 02138
UNITED STATES
gross@math.harvard.edu

Heegner Points and Rankin *L*-Series
MSRI Publications
Volume **49**, 2004

The Gauss Class Number Problem for Imaginary Quadratic Fields

DORIAN GOLDFELD

1. Introduction

Let $D < 0$ be a fundamental discriminant for an imaginary quadratic field $K = \mathbb{Q}(\sqrt{D})$. Such fundamental discriminants D consist of all negative integers that are either $\equiv 1 \pmod 4$ and square-free, or of the form $D = 4m$ with $m \equiv 2$ or $3 \pmod 4$ and square-free. We define

$$h(D) = \#\left\{\frac{\text{group of nonzero fractional ideals } \frac{\mathfrak{a}}{\mathfrak{b}}}{\text{group of principal ideals } (\alpha),\ \alpha \in K^\times}\right\},$$

to be the cardinality of the ideal class group of K. In the *Disquisitiones Arithmeticae*, (1801) [G], Gauss showed (using the language of binary quadratic forms) that $h(D)$ is finite. He conjectured that

$$h(D) \longrightarrow \infty \quad \text{as} \quad D \longrightarrow -\infty,$$

a result first proved by Heilbronn [H] in 1934. The *Disquisitiones* also contains tables of binary quadratic forms with small class numbers (actually tables of imaginary quadratic fields of small class number with even discriminant which is a much easier problem to deal with) and Gauss conjectured that his tables were complete. In modern parlance, we can rewrite Gauss' tables (we are including both even and odd discriminants) in the following form.

$h(D)$	1	2	3	4	5
# of fields	9	18	16	54	25
largest $\|D\|$	163	427	907	1555	2683

The problem of finding an effective algorithm to determine all imaginary quadratic fields with a given class number h is known as the Gauss class number

This research is partially supported by the National Science Foundation.

h problem. The Gauss class number problem is especially intriguing, because if such an effective algorithm did not exist, then the associated Dirichlet L-function would have to have a real zero, and the generalized Riemann hypothesis would necessarily be false. This problem has a long history (see [Go2]) which we do not replicate here, but the first important milestones were obtained by Heegner [Heg], Stark [St1; St2], and Baker [B], whose work led to the solution of the class number one and two problems. The general Gauss class number problem was finally solved completely by Goldfeld, Gross, and Zagier in 1985 [Go1; Go2; GZ]. The key idea of the proof is based on the following theorem (see [Go1] (1976), for an essentially equivalent result) which reduced the problem to a finite amount of computation.

THEOREM 1. *Let D be a fundamental discriminant of an imaginary quadratic field. If there exists a modular elliptic curve E (defined over $\mathbb{Q}$) whose associated base change Hasse–Weil L-function $L_{E/\mathbb{Q}(\sqrt{D})}(s)$ has a zero of order ≥ 4 at $s = 1$ then for every $\varepsilon > 0$, there exists an effective computable constant $c_\varepsilon(E) > 0$, depending only on ε, E such that*

$$h(D) > c_\varepsilon(E)(\log|D|)^{1-\varepsilon}.$$

Note that the L-function of $E/\mathbb{Q}$, $L_E(s)$, always divides $L_{E/\mathbb{Q}(\sqrt{D})}(s)$. If an imaginary quadratic field $Q(\sqrt{D})$ has small class number, then many small primes are inert. It is not hard to show that the existence of an elliptic curve whose associated Hasse–Weil L-function has a triple zero at $s = 1$ is enough to usually guarantee that $L_{E/\mathbb{Q}(\sqrt{D})}(s)$ has a fourth order zero. This idea will be clarified in § 3. We also remark, that if $L_{E/\mathbb{Q}(\sqrt{D})}(s)$ had a zero of order $g \geq 4$, then you would get (see [G1]) the lower bound

$$h(D) \gg (\log|D|)^{g-3} e^{-21\sqrt{g(\log\log|D|)}}.$$

Actually, [G1] also gives a similar result for real quadratic fields ($D > 0$),

$$h(D)\log\varepsilon_D \gg (\log|D|)^{g-3} e^{-21\sqrt{g(\log\log|D|)}},$$

where ε_D denotes the fundamental unit. In this case, however, it is required that $L_{E/\mathbb{Q}(\sqrt{D})}(s)$ has a zero of order $g \geq 5$ to get a non–trivial lower bound, because $\log\varepsilon_D \gg \log D$. The term

$$e^{-21\sqrt{g(\log\log|D|)}}$$

(obtained by estimating a certain product of primes dividing D) is far from optimal, because it simultaneously covers the cases of both real and imaginary quadratic fields. If one considers only imaginary quadratic fields, the term can be easily written as a simple product over primes dividing D. This was done by Oesterlé in 1985 [O] who made Theorem 1 explicit. He proved that for

$(D, 5077) = 1$,

$$h(D) > \frac{1}{55} \log |D| \prod_{p|D,\ p \neq |D|} \left(1 - \frac{\lfloor 2\sqrt{p} \rfloor}{p+1}\right),$$

which allowed one to solve the class number 3 problem. More recently, using the above methods, Arno [A] (1992), solved the class number four problem, and subsequently, work with Robinson and Wheeler [ARW] (1998), and work of Wagner [Wag] (1996) gave a solution to Gauss' class number problem for class numbers 5, 6, 7. The most recent advance in this direction is due to Watkins [Wat], who obtained the complete list of all imaginary quadratic fields with class number ≤ 100.

The main aim of this paper is to illustrate the key ideas of the proof of Theorem 1 by giving full details of the proof for the solution of just the class number one problem. The case of class number one is considerably simpler than the general case, but the proof exemplifies the ideas that work in general. We have not tried to compute or optimize constants, but have focused instead on exposition of the key ideas.

2. The Deuring–Heilbronn Phenomenon

Let $\mathbb{Q}(\sqrt{D})$ denote an imaginary quadratic field with class number $h(D) = 1$. If a rational prime p splits completely in $\mathbb{Q}(\sqrt{D})$, then $(p) = \pi \cdot \bar{\pi}$, with

$$\pi = \left(\frac{m + n\sqrt{D}}{2}\right)$$

a principal ideal. It follows that

$$p = \frac{m^2 - n^2 D}{4} \implies p > \frac{1 + |D|}{4}.$$

We have thus shown:

LEMMA 2. *Let $\mathbb{Q}(\sqrt{D})$ be an imaginary quadratic field of class number one. Then all primes less than $\frac{1+|D|}{4}$ must be inert.*

Lemma 2 can be used to write down prime producing polynomials [Ra]

$$x^2 - x + \frac{|D| + 1}{4},$$

(e.g., $x^2 - x + 41$) which takes prime values for $x = 1, 2, \ldots, \frac{|D|-3}{4}$.

Lemma 2 is the simplest example of the more general phenomenon which says that an imaginary quadratic field with small class number has the property that most small rational primes must be inert in that field. It follows that if $h(D) = 1$, then the quadratic character $\chi_D(n) = \left(\frac{D}{n}\right)$ (Kronecker symbol) associated to $\mathbb{Q}(\sqrt{D})$ satisfies $\chi_D(p) = -1$ for most small primes, and thus behaves like the

Liouville function. Consequently, we heuristically expect that as $D \to -\infty$ and s fixed with $\mathrm{Re}(s) > \frac{1}{2}$,

$$\begin{aligned} L(s, \chi_D) &= \prod_p \left(1 - \frac{\chi_D(p)}{p^s}\right)^{-1} \\ &\sim \prod_p \left(1 + \frac{1}{p^s}\right)^{-1} = \frac{\zeta(2s)}{\zeta(s)}, \end{aligned}$$

so that analytically the Dirichlet L-function, $L(s, \chi_D)$, associated to $\mathbb{Q}(\sqrt{D})$ behaves like $\zeta(2s)/\zeta(s)$. By $f(s) \sim g(s)$ in a region $s \in \mathscr{R} \subset \mathbb{C}$ we mean that there exists a small $\varepsilon > 0$ such that $|f(s) - g(s)| < \varepsilon$ in the region $\mathscr{R}$. Here we are appealing to the standard use of approximate functional equations which allow one to replace an L-function by a short (square root of conductor) sum of its early Dirichlet coefficients. This is the basis for the so called zero repelling effects (Deuring–Heilbronn phenomenon) associated to imaginary quadratic fields with small class number. For example, if $h(D) = 1$ and $D \to -\infty$, and D_1 is a fixed discriminant of a quadratic field, then we expect that for $\mathrm{Re}(s) > \frac{1}{2}$,

$$L(s, \chi_{D_1}) L(s, \chi_D \chi_{D_1}) \sim L(2s, \chi_{D_1}),$$

which implies (see [Da]) that $L(s, \chi_{D_1})$ has no zeros $\gamma + i\rho$ with $\gamma > \frac{1}{2}$.

3. Existence of L-functions of Elliptic Curves with Triple Zeros

Let E be an elliptic curve defined over $\mathbb{Q}$ whose associated Hasse–Weil L-function $L_E(s)$ vanishes at $s = 1$. Let $L_E(s, \chi_d)$ denote the L-function twisted by the quadratic character χ_d of conductor d, a fundamental discriminant of an imaginary quadratic field. We shall need the Gross–Zagier formula (see [G–Z])

$$\frac{d}{ds} \big(L_E(s) L_E(s, \chi_d)\big)_{s=1} = c_E \langle P_d, P_d \rangle, \tag{3.1}$$

where $\langle P_d, P_d \rangle$ is the height pairing of a certain Heegner point P_D and c_E is an explicit constant depending on the elliptic curve E. Gross and Zagier showed that if E is an elliptic curve of conductor 37 and $d = -139$, then the Heegner point is torsion and the height pairing $\langle P_d, P_d \rangle$ vanishes. By (3.1), this gives a construction of an L-function with a triple zero at $s = 1$. Actually, their method is quite general, and many other such examples can be constructed.

Henceforth, we fix E to be the above elliptic curve of conductor $N = 37 \cdot 139^2$. Then the Hasse–Weil L-function $L_E(s)$ satisfies the functional equation (see [Shim])

$$\left(\frac{\sqrt{N}}{2\pi}\right)^{1+s} \Gamma(1+s) L_E(1+s) = -\left(\frac{\sqrt{N}}{2\pi}\right)^{1-s} \Gamma(1-s) L_E(1-s),$$

and $L_E(1+s)$ has a MacLaurin expansion of the form

$$L_E(1+s) = c_3 s^3 + c_4 s^5 \ + \ \{\text{higher odd powers of } s\}.$$

Now, let D, with $|D| > 163$, denote a fundamental discriminant of an imaginary quadratic field with class number one. It is not hard to show that $(D,\ 37 \cdot 139) = 1$. Let χ_D denote the quadratic Dirichlet character of conductor D. We define

$$\Lambda_D(s) = \left(\frac{N|D|}{4\pi^2}\right)^s \Gamma(1+s)^2 L_E(s) L_E(s, \chi_D). \tag{3.2}$$

Then it can be shown (see [Shim]) that $\Lambda_D(s)$ satisfies the functional equation

$$\Lambda_D(1+s) = w \cdot \Lambda_D(1-s), \tag{3.3}$$

with root number $w = \chi_D(-37 \cdot 139^2) = \chi_D(-37) = +1$, because the early primes of an imaginary quadratic field $\mathbb{Q}(\sqrt{D})$ with class number one must be inert (Lemma 2). It follows from (3.3) that

$$L_{E/\mathbb{Q}(\sqrt{D})}(s) = L_E(s) L_E(s, \chi_D)$$

has a zero of even order at $s = 1$. Since $L_E(s)$ has a zero of order 3 at $s = 1$, we immediately see that $L_{E/\mathbb{Q}(\sqrt{D})}(s)$ must have a zero of order at least 4 at $s = 1$. This is the main requirement of Theorem 1.

4. Solution of the Class Number One Problem

Assume D is sufficiently large and the class number $h(D)$ of $\mathbb{Q}(\sqrt{D})$ is one. We will get a contradiction using zero-repelling ideas (Deuring–Heilbronn phenomenon) of Section 2. The main idea is to consider the integral I_D defined by

$$I_D = \frac{1}{2\pi i} \int_{2-i\infty}^{2+i\infty} \Lambda_D(1+s) \, \frac{ds}{s^3},$$

where $\Lambda_D(1+s)$ is given in (3.2).

Lemma 3. *We have* $I_D = 0$.

Proof. If we shift the line of integration to $\mathrm{Re}(s) = -2$, the residue at $s = 0$ is zero because $\Lambda_D(1+s)$ has a fourth order zero at $s = 0$. If immediately follows that

$$I_D = \frac{1}{2\pi i} \int_{-2-i\infty}^{-2+i\infty} \Lambda_D(1+s) \, \frac{ds}{s^3} = -\frac{1}{2\pi i} \int_{2-i\infty}^{2+i\infty} \Lambda_D(1+s) \, \frac{ds}{s^3} = -I_D,$$

after applying the functional equation (3.3) and letting $s \to -s$. Consequently, $I_D = 0$. □

We will now show that if $h(D) = 1$ and D is sufficiently large then $I_D \neq 0$. The heuristics for obtaining this contradiction are easily seen. We may write the Euler products

$$L_E(s) = \prod_p \left(1 - \frac{\alpha_p}{p^s}\right)^{-1} \left(1 - \frac{\beta_p}{p^s}\right)^{-1}, \tag{4.1}$$

$$L_E(s, \chi_D) = \prod_p \left(1 - \frac{\alpha_p \chi_D(p)}{p^s}\right)^{-1} \left(1 - \frac{\beta_p \chi_D(p)}{p^s}\right)^{-1}.$$

The assumption that $h(D) = 1$ implies that $\chi_D(p) = -1$ for all primes $p < \frac{1+|D|}{4}$ (Lemma 2). So we expect that analytically the Euler product $L_E(s)L_E(s, \chi_D)$ should behave like

$$\phi(s) := \prod_p \left(1 - \frac{\alpha_p^2}{p^{2s}}\right)^{-1} \left(1 - \frac{\beta_p^2}{p^{2s}}\right)^{-1},$$

where

$$|\alpha_p|^2 = |\beta_p|^2 = \alpha_p \beta_p = p,$$

for all but finitely many primes p. Now, if f is the weight two Hecke eigenform associated to E, we have the symmetric square L-function

$$L(s, \mathrm{sym}^2(f)) := \prod_p \left(1 - \frac{\alpha_p^2}{p^s}\right)^{-1} \left(1 - \frac{\alpha_p\beta_p}{p^s}\right)^{-1} \left(1 - \frac{\beta_p^2}{p^s}\right)^{-1}.$$

Thus $\phi(s)$ is essentially $L\big(2s, \mathrm{sym}^2(f)\big)/\zeta(2s-1)$. It is known that $L\big(s, \mathrm{sym}^2(f)\big)$ is entire which implies that $L\big(2s, \mathrm{sym}^2(f)\big)/\zeta(2s-1)$ vanishes at $s = 1$, a result first proved by [Ogg]. This implies that

$$\phi(1) = 0.$$

In fact, $\phi(1)$ has a simple zero at $s = 1$ which seems to contradict the fact that $L_E(s)L_E(s, \chi_D)$ has a fourth order zero. Although it appears that a contradiction could be obtained if $L_E(s)L_E(s, \chi_D)$ had a double zero at $s = 1$, this, unfortunately is not the case. The contradiction is much more subtle and will be shortly clarified.

We now define

$$I_D^* = \frac{1}{2\pi i} \int_{2-i\infty}^{2+i\infty} \left(\frac{37 \cdot 139^2 |D|}{4\pi^2}\right)^{1+s} \Gamma(1+s)^2 \phi(1+s) \, \frac{ds}{s^3},$$

which allows us to write

$$0 = I_D = I_D^* + \text{ Error}, \tag{4.2}$$

with

$$\text{Error} = \frac{1}{2\pi i}\int\limits_{2-i\infty}^{2+i\infty}\left(\frac{37\cdot 139^2|D|}{4\pi^2}\right)^{1+s}\Gamma(1+s)^2\big(L_E(1+s)L_E(1+s,\chi_D)-\phi(1+s)\big)\,\frac{ds}{s^3}. \tag{4.3}$$

LEMMA 4. *Define Dirichlet coefficients B_n $(n = 1, 2, \ldots)$ by the representation*

$$L_E(1+s)L_E(1+s,\chi_D) - \phi(1+s) \;=\; \sum_{n=1}^{\infty} B_n\, n^{-1-s}.$$

Also define Dirichlet coefficients $\nu_D(n)$ $(n = 1, 2, \ldots)$ by the representation

$$\zeta(s)L(s,\chi_D) = \sum_{n=1}^{\infty} \nu_D(n)\, n^{-s}.$$

Then $B_n = 0$ for $n < \frac{1+|D|}{4}$. In the other cases, we have

$$|B_n| \;\le\; \begin{cases} 2\nu_D(n)\sqrt{n} & \text{if } \frac{1+|D|}{4} \le n < \left(\frac{1+|D|}{4}\right)^2, \\ 2d_4(n)\cdot\sqrt{n} & \text{if } n \ge \left(\frac{1+|D|}{4}\right)^2, \end{cases}$$

where $d_4(n) = \sum_{d_1d_2d_3d_4=n} 1$.

PROOF. That $B_n = 0$ for $n < \frac{1+|D|}{4}$ follows immediately from Lemma 2. The upper bound $|B_n| \le 2d_4(n)\cdot n$ is a consequence of the fact (see (4.1)) that $L_E(1+s)L_E(1+s,\chi_D)$ is an Euler product of degree 4. Thus, the Dirichlet coefficients of $L_E(1+s)L_E(1+s,\chi_D)$ are bounded by the Dirichlet coefficients of the Euler product

$$\prod_p \left(1 - \frac{\sqrt{p}}{p^{1+s}}\right)^{-1} = \sum_{n=1}^{\infty} d_4(n)\sqrt{n}\cdot n^{-1-s}.$$

The extra factor of 2 in the bound for B_n comes from the consideration of the additional Euler product for $\phi(1+s)$.

In the range $\frac{1+|D|}{4} \le n < \left(\frac{1+|D|}{4}\right)^2$, we can only have $B_n \ne 0$ if n is divisible by a prime $q > \frac{1+|D|}{4}$. In this range, it is not possible that q^2 divides n. This implies that $\phi(1+s)$ does not contribute to B_n since $\phi(1+s)$ is a Dirichlet series formed from perfect squares, i.e., of the form

$$\phi(1+s) = \sum_{k=1}^{\infty} \frac{b(k)}{(k^2)^{1+s}}.$$

If we let $n = q\cdot m$, we must have $B_n = a_m \cdot a_q$, where

$$L_E(s) = \sum_{k=1}^{\infty} a_k \cdot k^{-s}.$$

Consequently, $|B_n| \le 2|a_m|\sqrt{q}$. It is easy to see that m must be a perfect square because m can only be divisible by primes less than $\frac{1+|D|}{4}$. Again, by considering the Euler product (4.1), we may conclude that in the range $\frac{1+|D|}{4} \le n < \left(\frac{1+|D|}{4}\right)^2$ the coefficients B_n are bounded by $2\nu_D(n)\sqrt{n}$ where $\nu_D(n)\sqrt{n}$ are the Dirichlet coefficients of the Euler product

$$\prod_p \left(1 - \frac{\sqrt{p}}{p^{1+s}}\right)^{-1} \left(1 - \chi_D(p)\frac{\sqrt{p}}{p^{1+s}}\right)^{-1} = \sum_{n=1}^{\infty} \nu_D(n)\sqrt{n}\cdot n^{-1-s}.$$

Clearly,

$$\zeta(s)L(s,\chi_D) = \sum_{n=1}^{\infty} \nu_D(n)\cdot n^{-s}. \qquad \square$$

LEMMA 5. *Let $x > 1$. Then*

$$\sum_{x\le n\le 2x} \nu_D(n)\sqrt{n} \;\le\; 4e\cdot x^{3/2}L(1,\chi_D) \;+\; O\big(|D|^{3/2}x^{-1/2}\big).$$

If we further assume that $|D| > 4$ and $h(D) = 1$, then

$$\sum_{x\le n\le 2x} \nu_D(n)\sqrt{n} \;\le\; 4\pi e\cdot \frac{x^{3/2}}{|D|^{1/2}} \;+\; O\big(|D|^{3/2}x^{-1/2}\big).$$

PROOF. We shall need the well-known Mellin transform

$$\frac{1}{2\pi i}\int_{2-i\infty}^{2+i\infty} x^s\Gamma(s)\,ds = e^{-1/x}.$$

It follows that

$$\begin{aligned}\sum_{x\le n\le 2x} \nu_D(n)\sqrt{n} &\le \frac{2e}{2\pi i}\int_{2-i\infty}^{2+i\infty} \zeta(s-\tfrac12)L(s-\tfrac12,\chi_D)\big((2x)^s - x^s\big)\,\Gamma(s)\,ds \\ &= 2e\sum_{n=1}^{\infty} \nu_D(n)\sqrt{n}\big(e^{-n/(2x)} - e^{-n/x}\big).\end{aligned}$$

Here we have used the fact that $2e\big(e^{-n/(2x)} - e^{-n/x}\big) > 1$ for $x \le n \le 2x$, and, otherwise, $\nu_D(n) \ge 0$. The above integral can be evaluating by shifting the line of integration to the left to the line $\mathrm{Re}(s) = -\frac12$. There is a pole at $s = \frac32$ coming from the Riemann zeta function. Consequently

$$\begin{aligned}\sum_{x\le n\le 2x} \nu_D(n)\sqrt{n} \le\ & 2eL(1,\chi_D)\big((2x)^{3/2} - x^{3/2}\big) \\ & + \left|\frac{2e}{2\pi i}\int_{-\frac12-i\infty}^{-\frac12+i\infty} \zeta(s-\tfrac12)L(s-\tfrac12,\chi_D)\big((2x)^s - x^s\big)\,\Gamma(s)\,ds\right|. \end{aligned} \tag{4.4}$$

The functional equation

$$\zeta(s)L(s,\chi_D) = \left(\frac{\sqrt{|D|}}{\pi}\right)^{1-2s} \frac{\Gamma\left(\frac{1-s}{2}\right)\Gamma\left(\frac{2-s}{2}\right)}{\Gamma\left(\frac{s}{2}\right)\Gamma\left(\frac{1+s}{2}\right)} \zeta(1-s)L(1-s,\chi_D),$$

together with Stirling's asymptotic formula

$$\lim_{|t|\to\infty} |\Gamma(\sigma+it)|\, e^{(\pi/2)|t|}\, |t|^{(1/2)-\sigma} = \sqrt{2\pi},$$

implies that the shifted integral in (4.4) converges absolutely and is bounded by $O\big(|D|^{3/2}x^{-1}\big)$. This completes the first part of the proof of Lemma 5. For the second part, we simply use Dirichlet's class number formula (see [Da]), $L(1,\chi_D) = \pi h(D)/|D|^{1/2}$, which holds for $|D| > 4$. □

LEMMA 6. *For $y > 0$, define*

$$G(y) := \frac{1}{2\pi i} \int_{2-i\infty}^{2+i\infty} y^{s+1}\Gamma(1+s)^2 \, \frac{ds}{s^3}.$$

Then

$$G(y) \ < \ 2y^2 e^{1/\sqrt{y}}.$$

PROOF. Recall the definition of the Gamma function

$$\Gamma(s) = \int_0^\infty e^{-u}u^s \, \frac{du}{u},$$

which satisfies $\Gamma(s+1) = s\Gamma(s)$. It follows that

$$G(y) = \frac{1}{2\pi i} \int_{2-i\infty}^{2+i\infty} y^{s+1} \int_0^\infty \int_0^\infty e^{-u_1-u_2}(u_1u_2)^s \frac{du_1\, du_2}{u_1u_2} \, \frac{ds}{s}. \tag{4.5}$$

On the other hand, we have the classical integral

$$\frac{1}{2\pi i} \int_{2-i\infty}^{2+i\infty} x^s \, \frac{ds}{s} = \begin{cases} 1 & \text{if } x > 1, \\ \frac{1}{2} & \text{if } x = 1, \\ 0 & \text{if } x < 1. \end{cases}$$

If we now apply the above to (4.5) (after interchanging integrals), we obtain

$$G(y) = y \iint\limits_{u_1u_2 \geq y^{-1}} e^{-u_1-u_2} \, \frac{du_1\, du_2}{u_1u_2}. \tag{4.6}$$

To complete the proof, we use the range of integration, $u_1u_2 \geq y^{-1}$, to show that $1/(u_1u_2) \leq y$, from which it follows from (4.6) that

$$\begin{aligned} G(y) &\leq y^2 \iint\limits_{u_1u_2 \geq y^{-1}} e^{-u_1-u_2}\, du_1\, du_2 \\ &= y^2 \int_0^\infty e^{-1/(u_2y)-u_2}\, du_2 = y^{\frac{3}{2}} \int_0^\infty e^{(-1/\sqrt{y})(u_2+1/u_2)}\, du_2 \\ &\leq 2y^{\frac{3}{2}} \int_1^\infty e^{(-1/\sqrt{y})(u_2+1/u_2)}\, du_2 < 2y^2 e^{-1/\sqrt{y}}. \qquad \square \end{aligned}$$

It now follows from (4.3), Lemma 4, and the definition of $G(y)$ given in Lemma 6 that

$$|\text{Error}| \leq \sum_{n \geq \frac{1+|D|}{4}} |B_n| \cdot G\left(\frac{37 \cdot 139^2 |D|}{4\pi^2 n}\right).$$

The bound for $G(y)$ given in Lemma 6 implies that

$$\begin{aligned} |\text{Error}| \leq & \sum_{\frac{1+|D|}{4} \leq n \leq \left(\frac{1+|D|}{4}\right)^2} 4\nu_D(n)\sqrt{n} \cdot \left(\frac{37 \cdot 139^2|D|}{4\pi^2 n}\right)^2 e^{-\sqrt{\frac{4\pi^2 n}{37 \cdot 139^2 |D|}}} \\ & + \sum_{\left(\frac{1+|D|}{4}\right)^2 < n} 4d_4(n)\sqrt{n} \cdot \left(\frac{37 \cdot 139^2|D|}{4\pi^2 n}\right)^2 e^{-\sqrt{\frac{4\pi^2 n}{37 \cdot 139^2 |D|}}}. \end{aligned}$$

The second sum in the above Error is $O\left(e^{-c_1\sqrt{|D|}}\right)$ for some $c_1 > 0$, so it can be ignored. We can, therefore, estimate the Error by breaking it into smaller sums as follows:

$$\begin{aligned} &|\text{Error}| \leq \\ &4 \sum_{k < \log_2 \frac{1+|D|}{4}} \frac{37^2 \cdot 139^4}{2^{2k-2} \cdot \pi^4} \sum_{\substack{\frac{1+|D|}{4} 2^{k-1} \leq n \\ n \leq \left(\frac{1+|D|}{4}\right) 2^k}} \nu_D(n)\sqrt{n} \cdot \exp\left(-\sqrt{\frac{4\pi^2 n}{37 \cdot 139^2 |D|}}\right) \\ &\qquad + O\left(e^{-\sqrt{|D|}}\right). \end{aligned}$$

For each, k, we can apply Lemma 5 to the inner sum over n in the above. It follows that

$$|\text{Error}| \ll |D| \sum_{k < \log \frac{1+|D|}{4}} 2^{-k/2} \ll |D|.$$

It immediately follows that for D sufficiently large, there exists a fixed, effectively computable constant c such that

$$|\text{Error}| \leq c \cdot |D|$$

as $|D| \to \infty$. Combining this bound with (4.2), we conclude that

$$I_D^* = \frac{1}{2\pi i} \int_{2-i\infty}^{2+i\infty} \left(\frac{37 \cdot 139^2 |D|}{4\pi^2} \right)^{1+s} \Gamma(1+s)^2 \phi(1+s) \, \frac{ds}{s^3} \tag{4.7}$$

satisfies

$$|I_D^*| < c \cdot |D|. \tag{4.8}$$

The integral for I_D^* given in (4.7) can be evaluated by shifting the line of integration to the left. A double pole is encountered at $s = 0$. Actually the term $1/s^3$ contributes a triple pole, but the vanishing of $\phi(1+s)$ at $s = 0$ reduces this to a double pole. Because of the double pole and the known zero–free region for the Riemann zeta function, it is not hard to show that there exists an effectively computable constant $c_1 > 0$ such that

$$|I_D^*| > c_1 D \log D. \tag{4.9}$$

The inequalities (4.8) and (4.9) are contradictory for large D. Consequently, it is not possible that $h(D) = 1$. QED.

Acknowledgment

The author thanks Brian Conrey for several helpful comments.

References

[A] S. Arno, "The imaginary quadratic fields of class number 4", *Acta Arith.* **60** (1992), 321–334.

[ARW] S. Arno, M. Robinson, F. Wheeler, "Imaginary quadratic fields with small odd class number", *Acta Arith.* **83** (1998), 295–330.

[B] A. Baker, "Imaginary quadratic fields with class number 2", *Annals of Math.* (2) **94** (1971), 139–152.

[Da] H. Davenport, *Multiplicative number theory*, second edition, revised by H. Montgomery, *Grad. Texts in Math.* **74**, Springer, 1980.

[G] C.F. Gauss, *Disquisitiones Arithmeticae*, Göttingen (1801); English translation by A. Clarke, revised by W. Waterhouse, New Haven, Yale University Press, 1966; reprinted by Springer, 1986.

[Go1] D. Goldfeld, "The class number of quadratic fields and the conjectures of Birch and Swinnerton-Dyer", *Ann. Scuola Norm. Sup. Pisa Cl. Sci.* (4) **3** (1976), 624–663.

[Go2] D. Goldfeld, "Gauss' class number problem for imaginary quadratic fields", *Bull. Amer. Math. Soc.* **13** (1985), 23–37.

[GZ] B. Gross, D. B. Zagier, "Heegner points and derivatives of L–series", *Invent. Math.* **84** (1986), 225–320.

[Heg] K. Heegner, "Diophantische Analysis und Modulfunktionen", *Math. Z.* **56** (1952), 227–253.

[H] H. Heilbronn, "On the class number in imaginary quadratic fields", *Quarterly J. of Math.* **5** (1934), 150–160.

[O] J. Oesterlé, "Le probléme de Gauss sur le nombre de classes", *Enseign. Math.* **34** (1988), 43–67.

[Ogg] A. Ogg, "On a convolution of L-series", *Invent. Math.* **7** (1969), 297–312.

[Ra] G. Rabinovitch, "Eindeutigkeit der Zerlegung in Primzahlfaktoren in quadratischen Zahlkörpern", pp. 418–421 in *Proc. Fifth Internat. Congress Math.* (Cambridge, 1912), vol. I, edited by E. W. Hobson and A. E. H. Love, Cambridge, University Press, 1913.

[Shim] G. Shimura, *Introduction to the arithmetic theory of automorphic functions*, Tokyo, Iwanami Shoten and Princeton, University Press, 1971.

[St1] H. Stark, "A complete determination of the complex quadratic fields of class-number one", *Mich. Math. J.* **14** (1967), 1–27.

[St2] H. Stark, "A transcendence theorem for class–number problems I, II", *Annals of Math.* (2) **94** (1971), 153–173 and **96** (1972), 174–209.

[Wag] C. Wagner, "Class number 5, 6 and 7", *Math. Comp.* **65** (1996), 785–800.

[Wat] M. Watkins, "Class numbers of imaginary quadratic fields", to appear.

Dorian Goldfeld
Columbia University
Department of Mathematics
New York, NY 10027
United States
goldfeld@columbia.edu

Heegner Points and Rankin L-Series
MSRI Publications
Volume **49**, 2004

Heegner Points and Representation Theory

BENEDICT H. GROSS

ABSTRACT. Our aim in this paper is to present a framework in which the results of Waldspurger and Gross–Zagier can be viewed simultaneously. This framework may also be useful in understanding recent work of Zhang, Xue, Cornut, Vatsal, and Darmon. It involves a blending of techniques from representation theory and automorphic forms with those from the arithmetic of modular curves. I hope readers from one field will be encouraged to pursue the other.

CONTENTS

1. Heegner Points on $X_0(N)$ 38
2. Rankin L-Series and a Height Formula 39
3. Starting from the L-Function 40
4. Local Representation Theory 41
5. Unitary Similitudes 42
6. The L-Group and Its Symplectic Representation 43
7. Inner Forms 43
8. Langlands Parameters 44
9. Local ε-Factors 45
10. Local Linear Forms 46
11. Local Test Vectors 46
12. An Explicit Local Formula 48
13. Adèlic Groups 50
14. A Special Case 51
15. Automorphic Representations 52
16. When $\#S$ Is Even 53
17. Global Test Vectors 53
18. An Explicit Global Formula 55
19. When $\#S$ Is Odd 58
20. Shimura Varieties 58
21. Nearby Quaternion Algebras 60
22. The Global Representation 61
23. The Global Linear Form 62
24. Global Test Vectors 64
Acknowledgment 64
References 64

1. Heegner Points on $X_0(N)$

I first encountered Heegner points in 1978, when I was trying to construct points of infinite order on the elliptic curves $A(p)$ I had introduced in my thesis [G0, page 79]. Barry Mazur gave me a lecture on Bryan Birch's work, and on his amazing computations. I had missed Birch's lectures at Harvard on the subject, as I was in Oxford in 1973-4, bemoaning the fact that no one was there to supervise graduate work in number theory.

By 1978, Birch had found the key definitions and had formulated the central conjectures, relating Heegner points to the arithmetic of elliptic curves (see [B], [B-S]). These concerned certain divisors of degree zero on the modular curves $X_0(N)$, and their images on elliptic factors of the Jacobian. I will review them here; a reference for this material is [G1].

A (noncuspidal) point on the curve $X_0(N)$, over a field k of characteristic prime to N, is given by a pair (E, F) of elliptic curves over k and a cyclic N-isogeny $\phi : E \to F$, also defined over k. We represent the point x by the diagram

$$(E \xrightarrow[\phi]{} F);$$

two diagrams represent the same point if they are isomorphic over a separable closure of k.

The ring $\mathrm{End}(x)$ associated to the point x is the subring of pairs (α, β) in $\mathrm{End}(E)\times\mathrm{End}(F)$ which are defined over k and give a commutative square

$$\begin{array}{ccc} E & \xrightarrow[\phi]{} & F \\ \alpha\downarrow & & \downarrow\beta \\ E & \xrightarrow[\phi]{} & F. \end{array}$$

When $\mathrm{char}(k) = 0$, the ring $\mathrm{End}(x)$ is isomorphic to either $\mathbb{Z}$ or an order $\mathcal{O}$ in an imaginary quadratic field K. In the latter case, we say the point x has complex multiplication.

Assume $\mathrm{End}(x) = \mathcal{O}$, and let $\mathcal{O}_K$ be the full ring of integers of K. The conductor c of $\mathcal{O}$ is defined as the index of $\mathcal{O}$ in $\mathcal{O}_K$. Then $\mathcal{O} = \mathbb{Z} + c\mathcal{O}_K$, and the discriminants of $\mathcal{O}_K$ and $\mathcal{O}$ are d_K and $D = d_K c^2$, respectively. We say x is a Heegner point if $\mathrm{End}(x) = \mathcal{O}$, and the conductor c of $\mathcal{O}$ is *relatively prime to* N. This forces an equality: $\mathrm{End}(x) = \mathrm{End}(E) = \mathrm{End}(F)$.

Heegner points exist (for all conductors c prime to N) precisely when all prime factors p of N are either split or ramified in K, and all factors with $\mathrm{ord}_p(N) \geq 2$ are split in K. The points of conductor c are defined over the ring class field k of conductor c over K, which is an abelian extension of K with Galois group isomorphic to $\mathrm{Pic}(\mathcal{O})$.

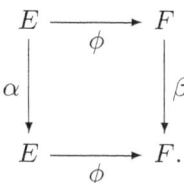

Let x be a Heegner point of conductor c, and let ∞ be the standard cusp on $X_0(N)$, given by the cyclic isogeny $(\mathbb{G}_m/q^{\mathbb{Z}} \underset{N}{\to} \mathbb{G}_m/q^{N\mathbb{Z}})$ of Tate curves over $\mathbb{Q}$. Consider the divisor $(x) - (\infty)$ of degree 0, and let

$$a \equiv (x) - (\infty)$$

be its class in the k-rational points of the Jacobian $J_0(N)$.

The finite dimensional rational vector space

$$W = J_0(N)(k) \otimes \mathbb{Q}$$

is a semi-simple module for the Hecke algebra $\mathbf{T}$, generated over $\mathbb{Q}$ by the operators T_m, for m prime to N, and the involutions w_n, for n dividing N. The Galois group $\mathrm{Gal}(k/K)$ also acts on W, and commutes with $\mathbf{T}$. Note that the complex characters of the commutative $\mathbb{Q}$-algebra $\mathbf{T}[\mathrm{Gal}(k/K)]$ are indexed by pairs (f, χ), where f is a cuspidal eigenform of weight 2 for $\Gamma_0(N)$ and χ is a ring class character of conductor dividing c.

Assume that f is a new form, of level N, and that χ is primitive, of conductor c. Let $a(f, \chi)$ be the projection of the class a in W to the (f, χ)-eigenspace in $W \otimes \mathbb{C}$. The central question on Heegner divisors is to determine *when* $a(f, \chi)$ *is nonzero.*

If a prime p divides both d_K and N, then $(p) = \wp^2$ is ramified in K and $\mathrm{ord}_p(N) = 1$. In this case, $\chi(\wp) = \pm 1$ and $a_p(f) = \pm 1$. When $a_p(f) \cdot \chi(\wp) = 1$, the class $a(f, \chi)$ is zero, for simple reasons [G1]. In what follows, we will assume that $a_p(f) \cdot \chi(\wp) = -1$, for all primes p dividing both d_K and N.

2. Rankin L-Series and a Height Formula

Associated to the pair (f, χ), one has the Rankin L-function

$$L(f, \chi, s) = \prod_p L_p(f, \chi, s),$$

defined by an Euler product in the half plane $\mathrm{Re}(s) > \frac{3}{2}$. The Euler factors have degree ≤ 4 in p^{-s}, and are given explicitly in [G1]. Rankin's method shows that the product

$$\Lambda(f, \chi, s) = \left((2\pi)^{-s}\Gamma(s)\right)^2 L(f, \chi, s)$$

has an analytic continuation to the entire plane, and satisfies the functional equation:

$$\Lambda(f, \chi, 2 - s) = -A^{1-s} \cdot \Lambda(f, \chi, s)$$

with $A = (ND)^2/\gcd(N, D)$. The sign in this functional equation is -1, as the local signs ε_p at finite primes p are all $+1$, and the local sign ε_∞ at the real prime is -1 [G1]. Hence $L(f, \chi, s)$ vanishes to odd order at the central critical point $s = 1$.

Birch considered the projection $a(f,\chi)$ in the special case of *rational* characters (f,χ) of the $\mathbb{Q}$-algebra $\mathbf{T}[\mathrm{Gal}(k/K)]$. In this case, f corresponds to an elliptic curve factor A of $J_0(N)$ over $\mathbb{Q}$, and χ to a factorization $D = d_1 d_2$ into two fundamental discriminants. If A_1 and A_2 are the corresponding quadratic twists of A over $\mathbb{Q}$, then

$$L(f,\chi,s) = L(A_1,s)L(A_2,s).$$

He discovered that $a(f,\chi)$ was nonzero precisely when the order of vanishing of $L(f,\chi,s)$ at $s=1$ was equal to one. In this case, he obtained a wealth of computational evidence in support of a new limit formula, relating the first derivative of the Rankin L-series to the height of the projected Heegner point on the elliptic curve A, over the biquadratic field $\mathbb{Q}(\sqrt{d_1},\sqrt{d_2})$.

The extension of Birch's conjecture to *all* complex characters (f,χ) of the algebra suggested a similar formula for $L'(f,\chi,1)$, of the type

$$L'(f,\chi,1) = \frac{(f,f)}{\sqrt{D}}\,\hat{h}(a(f,\chi)).$$

Here (f,f) is a normalized Petersson inner product, and $\hat{h}$ is the canonical height on $J_0(N)$ over K. Conjecturing the formula in this generality allows one to unwind the various projection operations, and obtain an equivalent identity involving the height paring of a Heegner divisor with a Hecke translate of its Galois conjugate. This is a simpler identity to prove, as one can use Néron's theory of local heights on the original modular curve, where the Heegner points and Hecke operators have a modular interpretation. Another advantage in this formulation is that the Rankin L-series is easier to study than the product of two Hecke L-series, owing to its integral representation.

Working along these lines, Zagier and I obtained such a formula, for D square-free and relatively prime to N, in 1982 [G-Z]. Zhang has recently established a similar formula in great generality [Z1].

3. Starting from the L-Function

In the above formulation, one starts with Heegner divisor classes $a \equiv (x)-(\infty)$ on $J_0(N)$. The Rankin L- function $L(f,\chi,s)$ is introduced in order to study the projection $a(f,\chi)$.

To generalize beyond the modular curves $X_0(N)$, we will *reverse* matters and start with the Rankin L-function $L(f,\chi,s)$. The form f corresponds to an automorphic, cuspidal representation π of the adèlic group $\mathrm{GL}_2(\mathbf{A}_\mathbb{Q})$, with trivial central character and π_∞ in the discrete series of weight 2. The character χ corresponds to a Hecke character of $\mathrm{GL}_1(\mathbf{A}_K)$, with trivial restriction to $\mathrm{GL}_1(\mathbf{A}_\mathbb{Q})$, and $\chi_\infty = 1$. The tensor product $\pi\otimes\chi$ gives an automorphic cuspidal representation of a group G of unitary similitudes over $\mathbb{Q}$, which has L-function $L(f,\chi,s)$ via a four-dimensional, symplectic representation of the L-group of G.

The key idea is to use the local signs in the functional equation of $L(f, \chi, s)$ to define an arithmetic object — either an inner form G' of G, or a Shimura curve $M(G_S)$. We then use the representation theory of G' to study the central critical value $L(f, \chi, 1)$, following Waldspurger [W], and the arithmetic geometry of the Shimura curve and its special points to study the central critical derivative $L'(f, \chi, 1)$. Using the local and global representation theory of GL_2 and its inner forms, we can formulate both cases in a similar manner.

4. Local Representation Theory

We begin with the local theory. Let k be a local field, and let E be an étale quadratic extension of k. Then E is either a field, or is isomorphic to the split k-algebra $k[x]/(x^2 - x) \simeq k + k$. In the latter case, there are two orthogonal idempotents e_1 and e_2 in E, with $e_1 + e_2 = 1$. Let $e \mapsto \bar{e}$ be the nontrivial involution of E fixing k; in the split case $\bar{e}_1 = e_2$. By local class field theory, there is a unique character $\alpha : k^* \to \langle \pm 1 \rangle$ whose kernel is the norm group $\mathbf{N}E^* = \{e\bar{e} : e \in E^*\}$ in k^*.

Let π be an irreducible complex representation of the group $\mathrm{GL}_2(k)$, and let $\omega : k^* \to \mathbb{C}^*$ be the central character of π. We will assume later that π is generic, or equivalently, that π is infinite-dimensional.

Let S be the two-dimensional torus $\mathrm{Res}_{E/k}\mathbb{G}_m$, and let χ be an irreducible complex representation of the group $S(k) = E^*$. Since E has rank 2 over k, we have an embedding of groups:

$$S(k) \simeq \mathrm{Aut}_E(E) \to \mathrm{GL}_2(k) \simeq \mathrm{Aut}_k(E)$$

We will consider the tensor product $\pi \otimes \chi$ as an irreducible representation of the group $\mathrm{GL}_2(k) \times S(k)$, and wish to restrict this representation to the diagonally embedded subgroup $S(k)$.

The central local problem is to compute the space of coinvariants

$$\mathrm{Hom}_{S(k)}(\pi \otimes \chi, \mathbb{C}).$$

If this is nonzero, we must have

$$\omega \cdot \mathrm{Res}(\chi) = 1 \qquad (*)$$

as a character of k^*. Indeed, $\omega \cdot \mathrm{Res}(\chi)$ gives the action of $k^* \subset E^*$ on all vectors in $\pi \otimes \chi$.

We will henceforth assume that $(*)$ holds. Then $\pi \otimes \chi$ is an irreducible representation of $G(k)$, with

$$G = (\mathrm{GL}_2 \times S)/\Delta\mathbb{G}_m,$$

and we wish to restrict it to the subgroup $T(k)$, where T is the diagonally embedded, one-dimensional torus $S/\mathbb{G}_m$.

5. Unitary Similitudes

The group G defined above is a group of unitary similitudes. Indeed, let k be a field, and let $E \subset B$ be an étale, quadratic algebra E over k, contained in a quaternion algebra B over k. The k-algebra B is then graded: $B = B_+ + B_-$ with

$$B_+ = E,$$
$$B_- = \{b \in B : be = \bar{e}b \text{ for all } e \in E\}.$$

Both B_+ and B_- are free E-modules of rank 1. The pairing $\phi : B \times B \to E$, defined by

$$\phi(b_1, b_2) = (b_1\bar{b}_2)_+$$
$$= \text{first component of } b_1\bar{b}_2,$$

is a nondegenerate Hermitian form on the free E module B of rank 2.

The group $\mathrm{GU}(B, \phi)$ of unitary similitudes has k-valued points isomorphic to $B^* \times E^*/\Delta k^*$. To give a specific isomorphism, we let the pair (b, e) in $B^* \times E^*$ act on $x \in B$ by

$$(b, e)(x) = exb^{-1}.$$

Then Δk^* acts trivially on B, and the similitude factor for ϕ is $\mathbf{N}e/\mathbf{N}b$ in k^*.

When E is split, the algebra B is a matrix algebra. If we take $B = \mathrm{End}_k(E) = M_2(k)$, using the basis $\langle e_1, e_2\rangle$ of orthogonal idempotents, then the k-valued points of $\mathrm{GU}(B, \phi)$ are isomorphic to $\mathrm{GL}_2(k) \times k^*$. The pair $\left(\left(\begin{smallmatrix} a & b \\ c & d \end{smallmatrix}\right), \lambda\right)$ acts on $x = \left(\begin{smallmatrix} A & B \\ C & D \end{smallmatrix}\right)$ by

$$\left(\begin{pmatrix} a & b \\ c & d \end{pmatrix}, \lambda\right)(x) = \begin{pmatrix} \lambda & \\ & 1 \end{pmatrix} \cdot x \cdot \begin{pmatrix} a & b \\ c & d \end{pmatrix}^{-1}.$$

The unit element $1 \in B$ satisfies $\phi(1, 1) = 1$, and the subgroup fixing this vector is the diagonally embedded torus $T = S/\mathbb{G}_m$, which acts as the unitary group of the line B_-.

Conversely, if (V, ϕ) is a nondegenerate unitary space of dimension 2 over E, and v is a vector in V satisfying $\phi(v, v) = 1$, we may give V the structure of a quaternion algebra over k, containing the quadratic algebra E. Indeed

$$V = E \cdot v + W, \quad \text{with} \quad W = (Ev)^{\perp}$$

and we define multiplication by

$$(\alpha v + w)(\alpha' v + w') = (\alpha\alpha' - \phi(w, w'))v + (\alpha w' + \bar{\alpha}' w).$$

The group $\mathrm{GU}(V, \phi)(k)$ is then isomorphic to $B^* \times E^*/\Delta k^*$, with B the quaternion algebra so defined.

6. The L-Group and Its Symplectic Representation

The L-group of $G = \mathrm{GU}(V, \phi)$ depends only on E, not on the quaternion algebra B. If E is split, $G \simeq \mathrm{GL}_2 \times \mathbb{G}_m$ and

$$^L G \simeq \mathrm{GL}_2 \times \mathbb{G}_m.$$

If E is a field, $^L G$ is a semidirect product

$$^L G \simeq (\mathrm{GL}_2 \times \mathbb{G}_m) \rtimes \mathrm{Gal}(E/k).$$

The action of the generator τ of $\mathrm{Gal}(E/k)$ is by

$$\tau(\lambda, g) = (\lambda \cdot \det g,\ g \cdot (\det g)^{-1}).$$

In all cases, the L-group $^L G$ has a four-dimensional symplectic representation

$$\rho : {}^L G \to \mathrm{Sp}_4$$

with kernel isomorphic to $\mathbb{G}_m$ and image contained in the normalizer of a Levi factor in a Siegel parabolic. We will encounter this representation in § 8.

7. Inner Forms

We return to the case when k is a local field. If E is split, there is only one possible similitude group, corresponding to the quaternion algebra $M_2(k)$ of 2×2 matrices:

$$G(k) = (\mathrm{GL}_2(k) \times E^*)/\Delta k^* \simeq \mathrm{GL}_2(k) \times k^*.$$

When E is a field, it embeds into $M_2(k)$ as well as into the unique quaternion division algebra B over k (since E is a field, $k \neq \mathbb{C}$). This gives two similitude groups

$$G(k) = (\mathrm{GL}_2(k) \times E^*)/\Delta k^*$$
$$G'(k) = (B^* \times E^*)/\Delta k^*.$$

Both contain the diagonally embedded torus

$$T(k) = E^*/k^*,$$

and the latter is compact modulo its center.

If π is an irreducible, complex representation of $\mathrm{GL}_2(k)$ with central character ω, and χ is a character of E^*, which satisfies $\omega \cdot \mathrm{Res}(\chi) = 1$, then $\pi \otimes \chi$ is an irreducible representation of $G(k)$. If π is square-integrable, it corresponds to a unique finite-dimensional, irreducible, complex representation π' of B^*. The representation π' is described in [J-L]; it is characterized by

$$\mathrm{Tr}(g|\pi) + \mathrm{Tr}(g|\pi') = 0$$

for all elliptic, semi-simple classes g, and has central character ω. If E is a field, then $\pi' \otimes \chi$ gives an irreducible representation of $G'(k)$, which can be restricted to the diagonally embedded torus $T(k)$.

In the next three sections, we will compute the complex vector spaces

$$\mathrm{Hom}_{T(k)}(\pi \otimes \chi, \mathbb{C}) \quad \text{and} \quad \mathrm{Hom}_{T(k)}(\pi' \otimes \chi, \mathbb{C}).$$

8. Langlands Parameters

Let W_k denote the Weil group of k, and normalize the isomorphism $k^* \simeq W_k^{\mathrm{ab}}$ of local class field theory to map a uniformizing parameter to a geometric Frobenius element, in the non-Archimedean case.

The representation π of $\mathrm{GL}_2(k)$ has a local Langlands parameter [G-P]

$$\sigma_\pi : W_k' \to \mathrm{GL}_2(\mathbb{C}).$$

Here W_k' is the Weil–Deligne group, and we normalize the parameter so that $\det(\sigma_\pi) = \omega$.

The character χ of E^* has conjugate χ^τ defined by $\chi^\tau(\alpha) = \chi(\bar{\alpha})$. Since $\chi \cdot \chi^\tau(\alpha) = \chi(\alpha\bar{\alpha}) = \omega^{-1}(\alpha\bar{\alpha})$, we have

$$\chi^\tau = \chi^{-1} \cdot (\omega \circ \mathbf{N})^{-1}.$$

When E is split, so $E^* = k^* \times k^*$, we have $\chi = (\eta_1, \eta_2)$ and $\chi^\tau = (\eta_2, \eta_1)$. In general, the pair (χ, χ^τ) gives a homomorphism, up to conjugacy

$$\sigma_\chi : W_k \to (\mathbb{C}^* \times \mathbb{C}^*) \rtimes \mathrm{Aut}(E/k).$$

The kernel of $\alpha : W_k \to \langle \pm 1 \rangle$ maps to the subgroup $\mathbb{C}^* \times \mathbb{C}^*$ via (χ, χ^τ), and $\mathrm{Aut}(E/k)$ permutes the two factors. The complex group on the right is $GO_2(\mathbb{C})$; in this optic, σ_χ is given by the induced representation $\mathrm{Ind}(\chi) = \mathrm{Ind}(\chi^\tau)$.

We view σ_π as a homomorphism from W_k' to $\mathrm{GSp}(V)$, with $\dim V = 2$, having similitude factor ω. We view σ_χ as a homomorphism from W_k to $GO(V')$, with $\dim V' = 2$, having similitude factor $\mathrm{Res}(\chi) = \mathrm{Res}(\chi^\tau)$. Since $\omega \cdot \mathrm{Res}(\chi) = 1$, the tensor product $V \otimes V'$ is a four-dimensional, symplectic representation of the Weil–Deligne group:

$$\sigma_\pi \otimes \sigma_\chi : W_k' \to \mathrm{Sp}(V \otimes V') = \mathrm{Sp}_4(\mathbb{C}).$$

This is the composition of the Langlands parameter of the representation $\pi \otimes \chi$ of $G(k)$

$$\sigma_{\pi \otimes \chi} : W_k' \to {}^L G(\mathbb{C})$$

with the symplectic representation

$$\rho : {}^L G(\mathbb{C}) \to \mathrm{Sp}_4(\mathbb{C}),$$

mentioned in § 6. The key point is that the image of this symplectic representation lands in the normalizer of a Levi factor.

9. Local ε-Factors

In particular, we have an equality of Langlands L and ε-factors with the Artin–Weil L and ε-factors:

$$L(\pi \otimes \chi, \rho, s) = L(\sigma_\pi \otimes \sigma_\chi, s),$$
$$\varepsilon(\pi \otimes \chi, \rho, \psi, dx, s) = \varepsilon(\sigma_\pi \otimes \sigma_\chi, \psi, dx, s).$$

We normalize the L-function so that the functional equation relates s to $1-s$, and the central point is $s = \frac{1}{2}$. Since this is the only representation ρ of ${}^L G$ we will consider, we suppress it in the notation and write $L(\pi\otimes\chi, s)$ for $L(\pi\otimes\chi, \rho, s)$.

Since $\sigma_\pi \otimes \sigma_\chi$ is symplectic, we can also normalize the local constant following [G4]. If dx is the unique Haar measure on k which is self-dual for Fourier transform with respect to ψ, then

$$\varepsilon(\pi \otimes \chi) = \varepsilon(\sigma_\pi \otimes \sigma_\chi, \psi, dx, \frac{1}{2})$$

depends only on $\sigma_\pi \otimes \sigma_\chi$, and satisfies

$$\varepsilon(\pi \otimes \chi)^2 = 1.$$

Since the representation ρ of ${}^L G$ is not faithful, certain nonisomorphic representations of $G(k)$ have the same local L- and ε-factors. Since $\sigma_\chi = \sigma_{\chi^\tau}$ as representations to $GO_2(\mathbb{C})$, we have

$$L(\pi \otimes \chi, s) = L(\pi \otimes \chi^\tau, s),$$
$$\varepsilon(\pi \otimes \chi) = \varepsilon(\pi \otimes \chi^\tau).$$

Also, if η is any character of k^*, and we define

$$\pi^* = \pi \otimes (\eta \circ \det),$$
$$\chi^* = \chi \otimes (\eta \circ \mathbf{N})^{-1},$$

then $\sigma_{\pi^*} \otimes \sigma_{\chi^*} = \sigma_\pi \otimes \sigma_\chi$, so

$$L(\pi^* \otimes \chi^*, s) = L(\pi \otimes \chi, s),$$
$$\varepsilon(\pi^* \otimes \chi^*) = \varepsilon(\pi \otimes \chi).$$

Note, however, that when $\eta \neq 1$ the representation $\pi \otimes \chi$ and $\pi^* \otimes \chi^*$ of $G(k)$ are *not* isomorphic, as they have distinct central characters. Their restrictions to the subgroup $U_2 \subset \mathrm{GU}_2$ *are* isomorphic, and U_2 contains the torus T.

Finally, since the dual of the representation $\pi \otimes \chi$ of $G(k)$ is isomorphic to

$$(\pi \otimes \chi)^\vee \simeq \pi^\vee \otimes \chi^\vee \simeq (\pi \otimes \omega(\det)^{-1}) \otimes \chi^{-1} \simeq \pi^* \otimes (\chi^\tau)^*,$$

with $\eta = \omega^{-1}$, we have

$$L((\pi \otimes \chi)^\vee, s) = L(\pi \otimes \chi, s),$$
$$\varepsilon((\pi \otimes \chi)^\vee) = \varepsilon(\pi \otimes \chi).$$

10. Local Linear Forms

We can now state the main local result, which is due to Tunnell and Saito (see [T], [S]). Recall that the representation π is generic if it has a nonzero linear functional on which the unipotent radical of a Borel subgroup acts via a nontrivial character. A fundamental result states that this occurs precisely when π is infinite dimensional. Also, recall that π has square-integrable matrix coefficients if and only if it lies in the discrete series for GL_2.

THEOREM. *Assume that π is generic. Then* $\dim \mathrm{Hom}_{T(k)}(\pi \otimes \chi, \mathbb{C}) \leq 1$, *with equality holding precisely when*

$$\varepsilon(\pi \otimes \chi) = \alpha \cdot \omega(-1).$$

If $\dim \mathrm{Hom}_{T(k)}(\pi \otimes \chi, \mathbb{C}) = 0$, *then π is square-integrable and E is a field. In this case, we have* $\dim \mathrm{Hom}_{T(k)}(\pi' \otimes \chi, \mathbb{C}) = 1$.

Informally speaking, this says that the $T(k)$ coinvariants in the representation

$$(\pi \otimes \chi) + (\pi' \otimes \chi)$$

have dimension 1, and the location of the coinvariants is given by the sign:

$$\varepsilon(\pi \otimes \chi)/\alpha \cdot \omega(-1).$$

Note that this result is compatible with the identities

$$\varepsilon(\pi \otimes \chi) = \varepsilon(\pi \otimes \chi^{\tau}) = \varepsilon(\pi^* \otimes \chi^*),$$

where $\pi^* = \pi \otimes (\eta \circ \det)$ and $\chi^* = \chi \otimes (\eta \circ \mathbf{N})^{-1}$.

11. Local Test Vectors

We can refine the result on $T(k)$-invariant linear forms ℓ on $\pi \otimes \chi$ or $\pi' \otimes \chi$ in favorable cases, by giving *test vectors* on which the nonzero invariant linear forms are nonzero. These vectors will lie on a line $\langle v \rangle$ fixed by a compact subgroup M in $G(k)$ or $G'(k)$, and M will be well-defined up to $T(k)$-conjugacy.

For k non-Archimedean, the favorable cases are when either the representation π, or the character χ, is unramified. In this case, $\langle v \rangle$ will be the fixed space of an open, compact subgroup M (see [G-P]). We will give a construction of M from the point of view of Hermitian lattices, first in the case when π is unramified, and then in the case when χ is unramified.

Let A be the ring of integers of k, and let $\mathcal{O}_E$ be the integral closure of A in E. Since either ω or $\mathrm{Res}(\chi)$ is unramified, and $\omega \cdot \mathrm{Res}(\chi) = 1$, both are trivial on the subgroup A^*. Writing $\omega = \eta^{-2}$, where η is an unramified character of k^*, and twisting π by $\eta(\det)$ and χ by $\eta(\mathbf{N})^{-1}$, we may assume that $\omega = \mathrm{Res}(\chi) = 1$.

Let c be the conductor ideal of χ, so χ is trivial when restricted to $(1+c\mathcal{O}_E)$. Since χ is a character of E^*/k^*, c is the extension of an ideal of A, and the order

$$\mathcal{O} = A + c\mathcal{O}_E$$

is stable under conjugation. Since

$$\mathcal{O}^* = A^*(1 + c\mathcal{O}_E),$$

the character χ of E^* is trivial, when restricted to the subgroup $\mathcal{O}^*$.

Now assume that π is unramified. Then, by the theorem in the previous section, we have a $T(k)$-invariant linear form on $\pi \otimes \chi$. The quaternion algebra $B = \mathrm{End}_k(E)$ contains the quadratic algebra $B_+ = \mathrm{End}_E(E) = E$, and $B_- = E\tau$ is the E-submodule of antilinear maps. The associated Hermitian space has

$$\phi(\alpha + \beta\tau, \alpha + \beta\tau) = \mathbf{N}\alpha - \mathbf{N}\beta,$$

and contains the $\mathcal{O}$-lattice of rank 2:

$$L = \mathrm{End}_A(\mathcal{O}) \supset \mathcal{O} + \mathcal{O}\tau.$$

We define the open compact subgroup M of $G(k) = \mathrm{GU}(B,\phi)$ as the stabilizer of L:

$$\begin{aligned} M = \mathrm{GU}(L,\phi) &\simeq \mathrm{Aut}_A(\mathcal{O}) \times \mathcal{O}^*/\Delta A^* \\ &\simeq \mathrm{GL}_2(A) \times \mathcal{O}^*/\Delta A^*. \end{aligned}$$

Note that ϕ, when restricted to L, takes values in $\mathcal{O}^\vee = \mathrm{Hom}(\mathcal{O}, A) \subset E$. Since $L \cap E = \mathcal{O}$, the intersection of M with $T(k) = E^*/k^*$ is $\mathcal{O}^*/A^*$. The proof that M fixes a unique line in $\pi \otimes \chi$, and that nonzero vectors on this line are test vectors for the $T(k)$-invariant linear form, is given in [G-P, § 3].

The construction of $M \subset G(k)$ which we have just given, when π is unramified, appears to be a natural one. But we could also have taken the $\mathcal{O}$-lattice $L' = \mathrm{End}_A(P)$, where $P = \mathcal{O}\alpha$ is any proper $\mathcal{O}$-submodule of E, with $\alpha \in E^*$. The resulting stabilizer $M' = \mathrm{Aut}_A(P) \times \mathcal{O}^*/\Delta A^*$ in $G(k)$ is then the conjugate of M by the image of α in $T(k) = E^*/k^*$. This gives an action of the quotient group $E^*/k^* \cdot \mathcal{O}^*$ on the open compact subgroups M we have defined, and hence on their fixed lines $(\pi \otimes \chi)^M = \langle v \rangle$.

We now turn to the construction of M in the case when the character χ is unramified. The Hermitian lattice

$$L = \mathcal{O}_E + \mathcal{O}_E \cdot w$$

is determined by $\phi(w,w)$ in $A - \{0\}$, up to multiplication by $\mathbf{N}\mathcal{O}_E^*$. If E/k is unramified, $\mathbf{N}\mathcal{O}_E^* = A^*$, and $L = L_n$ is completely determined by $n = \mathrm{ord}(\phi(v,v))$. We let n be the conductor of π, and let $M = M_n$ be the subgroup of $\mathrm{GU}(V,\phi)$ stabilizing L_n.

When E is split, M_n is open and compact in $G(k) \simeq \mathrm{GL}_2(k) \times k^*$, of the form $R_n^* \times A^*$, where R_n is an Eichler order of conductor n. Hence M_n fixes a line in $\pi \otimes \chi$, giving test vectors for the T-invariant form, by [G-P,§ 4]. When E is the unramified quadratic field extension, $\mathrm{GU}(V,\phi)$ is isomorphic to $G(k)$ when n is even, and to $G'(k)$ when n is odd. Since $\varepsilon(\pi\otimes\chi)=(-1)^n$ in this case, we find that M_n contains $T(k)$, so that the line fixed by M_n in $\pi\otimes\chi$ or $\pi'\otimes\chi$ provides test vectors [G-P, prop. 2.6].

Now assume E/k is tamely ramified and that π has conductor $n+1\geq 1$. Then $\mathbf{N}\mathcal{O}_E^*$ has index 2 in A^*, and there are two Hermitian lattices L_n and L'_n of the form $\mathcal{O}_E+\mathcal{O}_E w$, with $\mathrm{ord}(\phi(w,w))=n$. One has $M_n=\mathrm{GU}(L_n,\phi)$ open and compact in $G(k)$, and the other has $M'_n=\mathrm{GU}(L'_n,\phi)$ open and compact in $G'(k)$. Both contain the compact subgroup $T(k)$, and fix a line in $\pi\otimes\chi$ and $\pi'\otimes\chi$, respectively — depending on the sign of $\varepsilon(\pi\otimes\chi)$. This line provides test vectors for the $T(k)$-invariant linear form.

For k real and E complex, the torus $M=T(k)$ is compact. Its fixed space in either $\pi\otimes\chi$ or $\pi'\otimes\chi$ provides test vectors for the invariant form.

The remaining case is when k is Archimedean, and the algebra E is split. Let K be the maximal compact subgroup of $\mathrm{GL}_2(k)=\mathrm{Aut}_k(E)$, which fixes the positive definite form $\mathrm{Tr}(e^2)$, when $k=\mathbf{R}$, and $\mathrm{Tr}(ee')$, when $k=\mathbb{C}$ and $(z,w)'=(\bar{z},\bar{w})$. Let L be the maximal compact subgroup of k^*, so $K\times L$ is a maximal compact subgroup in $G(k)\simeq\mathrm{GL}_2(k)\times k^*$.

Let $W\otimes\chi$ be the minimal $(K\times L)$-type in the representation $\pi\otimes\chi$. By construction, the intersection

$$M=T(k)\cap(K\times L)$$

is isomorphic to L. One shows that the M-invariants in $W\otimes\chi$ have dimension ≤ 1. The favorable situation is when the M-invariants have dimension 1: this line provides test vectors for the $T(k)$-invariant linear form [P].

12. An Explicit Local Formula

We end the local section with a formula for the central critical value of the L-function, which provides a model for the global case.

Assume that k is non-Archimedean, with ring of integers A, and that E is split over k. Let $B=\mathrm{End}_k(E)$, and fix an isomorphism

$$B\simeq M_2(k)=\left\{\begin{pmatrix}a & b\\ c & d\end{pmatrix}:a,b,c,d\in k\right\}$$

such that

$$E=B_+=\left\{\begin{pmatrix}a & 0\\ 0 & d\end{pmatrix}\right\}$$

$$B_-=\left\{\begin{pmatrix}0 & b\\ c & 0\end{pmatrix}\right\}.$$

We assume $\pi\otimes\chi$ is a generic representation of $G(k)=B^*\times E^*/\Delta k^*$, and that the character χ of E^* is unramified.

Let U be the unipotent subgroup of G with

$$U(k) = \left\{ \begin{pmatrix} 1 & b \\ 0 & 1 \end{pmatrix} \times 1 \right\}.$$

We fix an isomorphism $G(k) \simeq \mathrm{GL}_2(k) \times k^*$, so that T maps to the torus with points

$$T(k) = \left\{ g_\lambda = \begin{pmatrix} \lambda & 0 \\ 0 & 1 \end{pmatrix} \times \lambda \right\}.$$

The element g_λ acts by conjugation on $U(k)$, and the isomorphism has been chosen so that this action is given by multiplication by λ. Via the chosen isomorphism, the representation $\pi \otimes \chi$ of $G(k)$ corresponds to a representation $\pi \otimes \eta$ of $\mathrm{GL}_2(k) \times k^*$. The representation $\pi \otimes \chi^\tau$ corresponds to the representation $\pi \otimes \eta'$, with $\eta \cdot \eta' \cdot \omega = 1$, and the contragradient $(\pi \otimes \chi)^\vee$ corresponds to the representation $\pi \cdot \omega^{-1} \otimes \eta^{-1} = \pi \cdot \omega^{-1} \otimes \eta' \omega$. Here we use the notation $\pi \cdot \alpha$, for a character α of k^*, to denote the representation $\pi \otimes \alpha(\det)$ of $\mathrm{GL}_2(k)$. Note that

$$L(\pi \otimes \chi, s) = L(\pi \cdot \eta, s) L(\pi \cdot \eta', s)$$

as the representation σ_χ is the direct sum of the characters η and η'.

Let $\psi : U(k) \to S^1$ be a nontrivial character, with kernel $U(A)$, and let

$$m : \pi \otimes \chi \to \mathbb{C}$$

be a nonzero linear form on which $U(k)$ acts by ψ. This exists, and is unique up to scaling, by the genericity of $\pi \otimes \chi$.

Let $n \geq 0$ be the conductor of π, and

$$K_n = \left\{ \begin{pmatrix} a & b \\ c & d \end{pmatrix} \times \lambda \right\}$$

be the subgroup of $\mathrm{GL}_2(A) \times A^*$ with $c \equiv 0 \pmod{\mathcal{P}_A^n}$. Then K_n fixes a unique line $\langle v \rangle$ in $\pi \otimes \chi$, by results of Casselman [C]. Moreover, one has $m(v) \neq 0$, by [G-P, Lemma 4.1]. Indeed, in the Kirillov model, the linear form m is given by $f \to f(1)$, and the spherical line $\langle v \rangle$ is explicitly determined in the function space. If we take the unique spherical vector on this line with $m(v) = 1$, then

$$\int_{k^*} m(g_\lambda v) \cdot |\lambda|^{s-1/2} \, d^*\lambda = \int_{k^*} m\left(\begin{pmatrix} \lambda & \\ & 1 \end{pmatrix} v_0 \right) \eta(\lambda) \cdot |\lambda|^{s-1/2} \, d^*\lambda,$$

where $v = v_0 \otimes 1$, and $d^*\lambda = \frac{d\lambda}{|\lambda|}$ with $\int_A d\lambda = 1$. But

$$m(gv_0) = W_0(g)$$

is the classical Whittaker function on $\mathrm{GL}_2(k)$, normalized so that $W_0(e) = 1$. Hence, by [J-L],

$$\int_{k^*} m(g_\lambda v_0) \cdot |\lambda|^{s-1/2} d^*\lambda = L(\pi \cdot \eta, s)$$

is the Hecke L-series of the representation $\pi \cdot \eta$.

On the other hand, the map

$$\ell : \pi \otimes \chi \to \mathbb{C}$$

defined on vectors w by the integral

$$\ell(w) = \int_{k^*} m(g_\lambda w)\, d^*\lambda$$

gives (when the indefinite integral is convergent) a $T(k)$-invariant linear form on $\pi \otimes \chi$. Hence the line $\langle v \rangle$ provides test vectors for $\pi \otimes \chi$, whenever $L(\pi \cdot \eta_1, s)$ does *not* have a pole at $s = \frac{1}{2}$, and we have the formula

$$\ell(v) = m(v) \cdot L(\pi \cdot \eta, \tfrac{1}{2})$$

for *any* vector v on that line.

The same considerations apply to the contragradient representation $(\pi \otimes \chi)^\vee$. Denoting the Whittaker linear functional by $m^\vee$, its T-invariant integral by $\ell^\vee$, and the K_n-line of test vectors by $\langle v^\vee \rangle$, we obtain the formula

$$\ell^\vee(v^\vee) = m^\vee(v^\vee) \cdot L(\pi \cdot \eta', \tfrac{1}{2}).$$

Indeed, π is replaced by $\pi \cdot \omega^{-1}$ and η by $\eta^{-1} = \eta'\omega$. Multiplying the two formulas, we obtain

$$\ell(v) \cdot \ell^\vee(v^\vee) = m(v) \cdot m^\vee(v^\vee) \cdot L(\pi \otimes \chi, \tfrac{1}{2}).$$

Since $\langle v \rangle$ is the line of K_n-invariants in $\pi \otimes \chi$, and $\langle v^\vee \rangle$ is the line of K_n-invariants in $(\pi \otimes \chi)^\vee$, we have

$$\langle v, v^\vee \rangle \neq 0$$

under the canonical pairing of $\pi \otimes \chi$ with $(\pi \otimes \chi)^\vee$. Hence, our final formula may be rewritten as

$$\frac{\ell(v)\ell^\vee(v^\vee)}{\langle v, v^\vee \rangle} \cdot \frac{\langle v, v^\vee \rangle}{m(v)m^\vee(v^\vee)} = L(\pi \otimes \chi, \tfrac{1}{2}).$$

It is this form which we will generalize to the global case. Note that $m, m^\vee$, v, and $v^\vee$ are only determined up to scaling, but that the left-hand side of the formula is well defined.

13. Adèlic Groups

We now turn to the global theory. Let k be a global field, with ring of adèles $\mathbf{A}$, and let E be an étale quadratic extension of k. Let

$$\pi \otimes \chi = \hat{\bigotimes_v}(\pi_v \otimes \chi_v)$$

be an admissable, irreducible representation of the adèlic group

$$G(\mathbf{A}) = \mathrm{GL}_2(\mathbf{A}) \times \mathbf{A}_E^*/\Delta\mathbf{A}^*.$$

Then the local components π_v and χ_v are unramified, for almost all finite places v. Consequently, the set

$$S = \{v : \varepsilon(\pi_v \otimes \chi_v) \neq \alpha_v \omega_v(-1)\} \tag{13.1}$$

is finite.

Assume that each local component π_v is infinite-dimensional. Then, by our local results, we may define an admissable, irreducible representation

$$\pi' \otimes \chi = \hat{\bigotimes_v}(\pi'_v \otimes \chi_v)$$

of the locally compact group

$$G_{S,\mathbf{A}} = \prod_v{}' G'(k_v),$$

where $G' = G$ at the places not in S, and G' is the nontrivial local inner form of G at the places in S. Thus $G'(k_v)$ is defined by the split quaternion algebra over k_v, for the places not in S, and by the quaternion division algebra over k_v at the places in S. The representation $\pi' \otimes \chi$ has the property that

$$\mathrm{Hom}_{T(\mathbf{A})}(\pi' \otimes \chi, \mathbb{C})$$

is of dimension 1.

We emphasize that $G_{S,\mathbf{A}}$ need *not* be the adèlic points of a group G' defined over k. This will only be the case when the cardinality of the set S is even. In that case, let B be the quaternion algebra over k ramified at S, which exists by the results of global class field theory. Then, if we define

$$G'(k) = B^* \times E^*/\Delta k^*,$$

we have

$$G'(\mathbf{A}) = G_{S,\mathbf{A}}.$$

In this case, $G_{S,\mathbf{A}}$ not only contains the diagonally embedded adèlic group $T(\mathbf{A})$, but also contains the discrete subgroup $G'(k)$. Their intersection is the subgroup $T(k)$.

14. A Special Case

We now discuss a special case of the above, which is important in arithmetic applications. For this, we assume that k is a number field, and that the set S defined in (13.1) *contains all Archimedean places v of k.* This hypothesis has a number of surprising consequences, and seems essential if we wish to obtain an algebraic theory.

First it implies that the number field k is totally real, and the quadratic extension E of k is totally complex, so is a CM field. Indeed, at each place v of S the algebra $E \otimes k_v$ is a field.

Next, since π_v is square-integrable at each real place of k, and π'_v is finite-dimensional, we find that the representation $(\pi' \otimes \chi)_\infty = \bigotimes_{v|\infty} \pi'_v \otimes \chi_v$ is a finite-dimensional representation of the group $G_{S,\infty}$, which is compact modulo its split center.

We will sometimes *further* assume that the representation $(\pi' \otimes \chi)_\infty$ of $G_{S,\infty}$ is the *trivial* representation. This means that, for each real place v of k, the representation π_v is the discrete series of weight 2 for $\mathrm{PGL}_2(k_v)$, and χ_v is the trivial character of E_v^*/k_v^*. If we also assume that the adèlic representation $\pi \otimes \chi$ of $G(\mathbf{A})$ is automorphic and cuspidal, then the last hypothesis means that π corresponds to a Hilbert modular form, of weight $(2, 2, \ldots, 2)$, with central character ω of finite order, split at infinity, and that χ is a Hecke character of $\mathbf{A}_E^*$, of finite order, with $\chi|\mathbf{A}^* = \omega^{-1}$.

15. Automorphic Representations

We henceforth assume that the adèlic representation $\pi \otimes \chi$ is automorphic and cuspidal, so appears as a submodule in the space of cusp forms

$$\mathcal{F}(G) = \mathcal{F}(G(k)\backslash G(\mathbf{A})).$$

We will also assume that the Hecke character χ of $\mathbf{A}_E^*$ is unitary.

Fixing an embedding:

$$i : \pi \otimes \chi \to \mathcal{F}(G)$$

of $G(\mathbf{A})$-modules, gives a $G(k)$-invariant linear form

$$\ell : \pi \otimes \chi \to \mathbb{C}$$

defined by evaluating the function $i(w)$ on the identity in G:

$$\ell(w) = i(w)(1).$$

From the linear form ℓ, we can recover the embedding i by: $i(w)(g) = \ell(gw)$. Both i and ℓ are well-defined, up to scaling, as the multiplicity of $\pi \otimes \chi$ in $\mathcal{F}(G)$ is equal to 1 (see [J-L]).

Since π is automorphic, its central character ω is an idèle class character. The same is true for the quadratic character α corresponding to E. Hence

$$\prod_v (\alpha\omega)_v(-1) = 1.$$

We therefore find that

$$\varepsilon(\pi \otimes \chi) = \prod_v \varepsilon(\pi_v \otimes \chi_v) = (-1)^{\#S}$$

with S the finite set of places where

$$\varepsilon(\pi_v \otimes \chi_v) \neq \alpha_v \omega_v(-1).$$

This is the global sign in the functional equation of the Rankin L-function $L(\pi \otimes \chi, s)$. In particular, when $\#S$ is even, we will study the central critical value $L(\pi \otimes \chi, \frac{1}{2})$. When $\#S$ is odd, we will study the central critical derivative $L'(\pi \otimes \chi, \frac{1}{2})$, as $L(\pi \otimes \chi, \frac{1}{2}) = 0$.

16. When $\#S$ Is Even

First, assume $\#S$ is even. We then have defined a group G' over k with $G'(\mathbf{A}) = G_{S,\mathbf{A}}$, as well as an irreducible representation $\pi' \otimes \chi$ of $G'(\mathbf{A})$. By a fundamental theorem of Jacquet–Langlands [J-L], the representation $\pi' \otimes \chi$ is also automorphic and cuspidal, and appears with multiplicity 1 in the space $\mathcal{F}(G')$ of cusp forms on G'.

We define a $T(\mathbf{A})$-invariant linear form m on $\mathcal{F}(G')$ by the formula

$$m(f) = \int_{T(k)\backslash T(\mathbf{A})} f(t)\, dt.$$

Here f is a function (with rapid decay) on $G'(\mathbf{A})$ (which is left $G'(k)$-invariant), $f(t)$ denotes its restriction to $T(\mathbf{A})$, and dt is a nonzero invariant measure on the locally compact abelian group $T(\mathbf{A})$ (which is unique up to scaling).

When E is a field, $T(k)\backslash T(\mathbf{A}) = E^* \cdot \mathbf{A}^* \backslash \mathbf{A}_E^*$ is compact. One example of an invariant measure is Tamagawa measure, which gives this quotient volume 2. When E is the split quadratic algebra, $G'(\mathbf{A}) \simeq G(\mathbf{A}) \simeq \mathrm{GL}_2(\mathbf{A}) \times \mathbf{A}^*$, and $T(\mathbf{A})$ is embedded as the subgroup $\begin{pmatrix} t & \\ & 1 \end{pmatrix} \times t$. The integral defining m converges, owing to the rapid decay of f.

If we choose a $G'(\mathbf{A})$-equivariant embedding

$$i : \pi' \otimes \chi \hookrightarrow \mathcal{F}(G')$$

we may restrict m to the image, to obtain an element $m \circ i$ in the one-dimensional vector space $\mathrm{Hom}_{T(\mathbf{A})}(\pi' \otimes \chi, \mathbb{C})$. If ℓ is the $G'(k)$-invariant linear form corresponding to the embedding i, then

$$m \circ i(w) = \int_{T(k)\backslash T(\mathbf{A})} \ell(tw)\, dt = Av_T(\ell)(w).$$

The main global result in this case is due to Waldspurger [W].

Theorem. *The $T(\mathbf{A})$-invariant linear form $m \circ i = Av_T(\ell)$ is nonzero on $\pi' \otimes \chi$ if and only if $L(\pi \otimes \chi, \frac{1}{2}) \neq 0$.*

17. Global Test Vectors

We can refine this result, if we are in the favorable situation where test vectors exist in $\pi'_v \otimes \chi_v$, for *all* places v of k. In this case, we let $\langle w_v \rangle$ be the line of local test vectors, and let

$$w = \bigotimes_v w_v$$

be a basis of the tensor product line in $\pi' \otimes \chi$. Then, by our local results, the linear form $Av_T(\ell)$ is nonzero if and only if

$$Av_T(\ell)(w) \neq 0.$$

Of course, this value depends on the choice of ℓ and w, both of which are only defined up to scalars, as well as the choice of invariant measure dt on $T(\mathbf{A})$ used to define the average. To obtain a number which depends only on $\pi' \otimes \chi$, we choose a $G'(k)$-invariant form $\ell^\vee : (\pi' \otimes \chi)^\vee \to \mathbb{C}$ on the contragradient representation, and a test vector $w^\vee$ for $\ell^\vee$ in $(\pi' \otimes \chi)^\vee$. The space

$$\mathrm{Hom}_{\Delta G'(\mathbf{A})}((\pi' \otimes \chi) \otimes (\pi' \otimes \chi)^\vee, \mathbb{C})$$

has dimension equal to 1, by Schur's lemma. The linear form $Av_{\Delta G'}(\ell \otimes \ell^\vee)$, defined by the integral

$$Av_{\Delta G'}(\ell \otimes \ell^\vee)(u \otimes u^\vee) = \int_{G'_{\mathrm{ad}}(k)\backslash G'_{\mathrm{ad}}(\mathbf{A})} \ell(gu)\, \ell^\vee(gu^\vee)\, dg$$

is a nonzero basis element, where dg is an invariant (positive) measure on the adèlic points of the adjoint group $Z\backslash G' = G'_{\mathrm{ad}}$.

To verify that $Av_{\Delta G'}(\ell \otimes \ell^\vee)$ is nonzero, we observe that if we use ℓ to embed $\pi' \otimes \chi$ as a sub-module of $\mathcal{F}(G')$, so u corresponds to the function f'_u on $G'(k)\backslash G'(\mathbf{A})$ with $\ell(gu) = f'_u(g)$, then the contragradient $(\pi' \otimes \chi)^\vee$ embeds as the functions $g \mapsto \overline{f'_u(g)}$ and the form $\ell^\vee$ is given by evaluation at the identity. Consequently, if $u^\vee$ in $(\pi' \otimes \chi)^\vee$ is the conjugate of f'_u, we find

$$Av_{\Delta G'}(\ell \otimes \ell^\vee)(u \otimes u^\vee) = \int_{Z(\mathbf{A})G'(k)\backslash G'(\mathbf{A})} f'_u(g)\overline{f'_u(g)}\, dg = \langle f'_u, f'_u \rangle.$$

This Petersson product is positive, so is nonzero.

Since the test vector w in $(\pi' \otimes \chi)$ is determined by its M-invariance (and K-type), we find that $w^\vee$ can be taken as the conjugate function of f'_w in $(\pi' \otimes \chi)^\vee$. If we do so, we find that

$$Av_T(\ell)(w) = \int_{T(k)\backslash T(\mathbf{A})} f'_w(t)\, dt,$$

$$Av_T(\ell^\vee)(w^\vee) = \int_{T(k)\backslash T(\mathbf{A})} \overline{f'_w(t)}\, dt = \overline{Av_T(\ell)(w)},$$

$$Av_{\Delta G'}(\ell \otimes \ell^\vee)(w \otimes w^\vee) = \int_{G'_{\mathrm{ad}}(k)\backslash G'_{\mathrm{ad}}(\mathbf{A})} f'_w(g)\overline{f'_w(g)}\, dg > 0.$$

These results hold for any positive, invariant measures dt and dg on the adèlic groups $T(\mathbf{A})$ and $G'_{\mathrm{ad}}(\mathbf{A})$. We now use the product measures

$$dt = \bigotimes_v dt_v, \qquad dg = \bigotimes_v dg_v,$$

which come from our test vector. Namely, at each finite place v of k, we let M_v be the open compact subgroup of $G'(k_v)$ which fixes the test vector w_v, and define dt_v and dg_v by

$$\int_{M_v \cap T(k_v)} dt_v = 1,$$

$$\int_{M_v/M_v \cap Z(k_v)} dg_v = 1.$$

At each Archimedean place,we let dt_v and dg_v be the canonical Haar measure $|\omega_v|$ defined in [G-G]. For example, if $k_v = \mathbf{R}$ and $E_v = \mathbb{C}$, so $T(k_v) = \mathbb{C}^*/\mathbf{R}^*$, we have

$$\int_{T(k_v)} dt_v = 2\pi.$$

If E_v is split, so $T(k_v) \simeq k_v^*$, dt_v is the usual measure on the multiplicative group.

Now the quantities $Av_T(\ell)(w)$, $Av_T(\ell^\vee)(w^\vee)$, and $Av_{\Delta G'}(\ell \otimes \ell^\vee)(w \otimes w^\vee)$ are all defined. The ratio:

$$A(\pi' \otimes \chi) = \frac{Av_T(\ell)(w) \cdot Av_T(\ell^\vee)(w^\vee)}{Av_{\Delta G'}(\ell \otimes \ell^\vee)(w \otimes w^\vee)}$$

is a real number ≥ 0, which is zero *precisely* when $L(\pi \otimes \chi, \frac{1}{2}) = 0$. This is an invariant of $\pi' \otimes \chi$ and $(\pi' \otimes \chi)^\vee$, which is *independent* of the choices of ℓ, $\ell^\vee$, w and $w^\vee$. There should be a simple formula, expressing $L(\pi \otimes \chi, \frac{1}{2})$ as a product $R(\pi \otimes \chi) \cdot A(\pi' \otimes \chi)$, where $R(\pi \otimes \chi)$ is a positive real number, given by periods of $\pi \otimes \chi$. We will make this more precise in the next section.

18. An Explicit Global Formula

To clarify the invariant $A(\pi' \otimes \chi)$, and to prepare for the discussion when $\#S$ is odd, we will obtain an explicit formula for $A(\pi' \otimes \chi)$ under the hypotheses that S has even cardinality and contains all Archimedian places, and that

$$(\pi' \otimes \chi)_\infty \quad \text{is the trivial representation of } G_{S,\infty}.$$

Recall that this implies that k is totally real, that E is a CM field, that π corresponds to a Hilbert modular form of weight $(2, 2, \ldots, 2)$ with central character ω of finite order, split at infinity, and that χ has finite order, with restriction ω^{-1} to $\mathbf{A}^*$.

We also assume that the conductors of π and χ are relatively prime, so a global test vector $w = w_\infty \otimes w_f$ exists in $\pi' \otimes \chi$. We let $M \subset G_{S,f} = G'(\mathbf{A}_f)$ be the open compact subgroup fixing w_f in $(\pi' \otimes \chi)_f$. By hypothesis, $G'(k \otimes \mathbf{R}) = G_{S,\infty}$ fixes w_∞.

Choose a $G'(k)$-invariant linear form (unique up to scaling, by the multiplicity 1 theorem):

$$\ell : \pi' \otimes \chi \to \mathbb{C}.$$

Then, using our test vector w, we get a function

$$f_w(g) = \ell(gw)$$

on the double coset space $G'(k)\backslash G'(\mathbf{A})/G'(k\otimes\mathbf{R})\times M$, which is identified with a function on

$$G'(k)\backslash G'(\mathbf{A}_f)/M,$$

since f_w is constant on $G'(k\otimes\mathbf{R})$. We wish to compute

$$A_T = \int_{T(k)\backslash T(\mathbf{A})} f_w(t)\, dt,$$
$$A_{G'} = \int_{G'_{\mathrm{ad}}(k)\backslash G'_{\mathrm{ad}}(\mathbf{A})} |f_w(g)|^2\, dg,$$

for the Haar measures dt, dg defined in the previous section. Then

$$A(\pi'\otimes\chi) = \frac{A_T\cdot\overline{A}_T}{A_{G'}}.$$

Let $J = M\cap T(\mathbf{A}_f)$, so the restriction of f_w to $T(\mathbf{A}_f)\hookrightarrow G'(\mathbf{A}_f)$ is a function on the finite set $T(k)\backslash T(\mathbf{A}_f)/J$. Recall that A is the ring of integers in k, $\mathcal{O} = A + c\mathcal{O}_E$ the order of conductor c in $\mathcal{O}_E$, and $J = \hat{\mathcal{O}}^*/\hat{A}^*$. By our choice of measures

$$\int_{T(k\otimes\mathbf{R})\times J} dt = (2\pi)^d,$$

where d is the degree of k. Let

$$u = \#(\mathcal{O}^*/A^*) = \#(J\cap T(k)).$$

Then

$$A_T = \int_{T(k)\backslash T(\mathbf{A})} f_w(t)\, dt = (2\pi)^d\cdot\frac{1}{u}\sum_{T(k)\backslash T(\mathbf{A}_f)/J} f_w(t).$$

Similarly, if $M_{\mathrm{ad}} = M/M\cap Z(\mathbf{A}_f)$ is the image of M in $G_{\mathrm{ad}}(\mathbf{A}_f)$, then $|f_w(g)|^2$ is a function on the finite set $G'_{\mathrm{ad}}(k)\backslash G'_{\mathrm{ad}}(\mathbf{A}_f)/M_{\mathrm{ad}}$. For each double coset, we define the integer

$$e_g = \#\big(M_{\mathrm{ad}}\cap g^{-1}G'_{\mathrm{ad}}(k)g\big).$$

We then have the formulas

$$\int_{G'_{\mathrm{ad}}(k\otimes\mathbf{R})\times M_{\mathrm{ad}}} dg = (2\pi)^{2d}$$

$$A_{G'} = (2\pi)^{2d}\cdot\sum_{G'_{\mathrm{ad}}(k)\backslash G'_{\mathrm{ad}}(\mathbf{A}_f)/M_{\mathrm{ad}}} \frac{1}{e_g}\cdot|f_w(g)|^2.$$

The ratio $A(\pi'\otimes\chi) = |A_T|^2/A_{G'}$ has a nice description in the language of "algebraic modular forms" (see [G2]). Consider the finite-dimensional vector space of functions

$$V = \{F : G'(k)\backslash G'(\mathbf{A}_f)/M \to \mathbb{C}\}.$$

The finite abelian group

$$Z(k)\backslash Z(\mathbf{A}_f)/M \cap Z(\mathbf{A}_f) = E^*\backslash \mathbf{A}_{E,f}^*/\hat{\mathcal{O}}^* = \mathrm{Pic}(\mathcal{O})$$

acts on V by the formula

$$zF(g) = F(gz) = F(zg),$$

and V decomposes as a direct sum of eigenspaces $V(\chi)$ for the characters of $\mathrm{Pic}(\mathcal{O})$. Since $\chi \cdot \overline{\chi} = 1$, each eigenspace has a Hermitian inner product

$$\langle F, G\rangle = \sum_{G'_{\mathrm{ad}}(k)\backslash G'_{\mathrm{ad}}(\mathbf{A}_f)/M_{\mathrm{ad}}} \frac{1}{e_g} F(g)\overline{G(g)}.$$

The representation $\pi' \otimes \chi$, and our choice of linear form ℓ and test vector w, give an element $f_w(g)$ in $V(\chi)$. The homomorphism

$$T \to G'$$

of groups over k gives a map of finite sets

$$\phi : T(k)\backslash T(\mathbf{A}_f)/J \to G'(k)\backslash G'(\mathbf{A}_f)/M.$$

This, in turn, gives an element F_T in V, which is defined by

$$F_T(g) = \frac{1}{u} \cdot \#\{t : \phi(t) = g\}.$$

We define the projection

$$F_T(\chi)(g) = \frac{1}{\#\mathrm{Pic}\ \mathcal{O}} \sum_{\mathrm{Pic}(\mathcal{O})} \chi^{-1}(z) F_T(zg),$$

which is an element of $V(\chi)$. We then have the inner product formula

$$\langle f_w, F_T(\chi)\rangle = \sum_g \frac{1}{e_g} f_w(g)\overline{F_T(\chi)(g)} = \frac{1}{u}\sum_t f_w(t).$$

Indeed, $F_T(\chi)$ is supported on $T \cdot Z$, and $T \cap Z = 1$.

Now let $F_T(\pi' \otimes \chi)$ be the projection of the function F_T to the $(\pi' \otimes \chi)$-eigenspace of V, or equivalently the projection of $F_T(\chi)$ to the π'-eigenspace of $V(\chi)$. Since this eigenspace is spanned by f_w, we have the inner product formula

$$F_T(\pi' \otimes \chi) = \frac{\langle F_T(\chi), f_w\rangle}{\langle f_w, f_w\rangle} \cdot f_w.$$

Consequently, we obtain the formula

$$\big\langle F_T(\pi' \otimes \chi), F_T(\pi' \otimes \chi)\big\rangle = \frac{|\langle F_T(\chi), f_w\rangle|^2}{\langle f_w, f_w\rangle} = \frac{|A_T|^2}{A_{G'}} = A(\pi' \otimes \chi).$$

This shows, among other things, that $A(\pi' \otimes \chi)$ is algebraic, and lies in the field of definition of $\pi' \otimes \chi$.

It suggests, by our previous work on special values (see [G3]), that the factor $R(\pi \otimes \chi)$ in the formula

$$L(\pi \otimes \chi, \tfrac{1}{2}) = R(\pi \otimes \chi) \cdot A(\pi' \otimes \chi)$$

has the form

$$R(\pi \otimes \chi) = \frac{1}{\sqrt{\mathbf{N}D}}(f_0, f_0),$$

where f_0 is a new form in π with $a_1(f_0) = 1$, and $(\ ,\ )$ is a normalized Petersson inner product on G_{ad}. There is work of Zhang and Xue in this direction, but the precise result is not yet proved. One may want to renormalize the measures dt and dg so that

$$R(\pi \otimes \chi) = \mathrm{Res}_{s=1} L(\pi \otimes \chi, \mathrm{ad}, s),$$

as this formulation, using the adjoint L-function at $s = 1$, would make sense for more general π and χ.

19. When $\#S$ Is Odd

We now consider the case when $\#S$ is odd. Then

$$\varepsilon(\pi \otimes \chi) = -1 \qquad \text{and} \qquad L(\pi \otimes \chi, \tfrac{1}{2}) = 0.$$

In this case, the adèlic group $G_{S,\mathbf{A}}$ does *not* contain a natural discrete subgroup (like $G'(k)$ in the case when $\#S$ is even), so it is unclear what it means for the representation $\pi' \otimes \chi$ to be automorphic.

To generalize Waldspurger's theorem to a result on $L'(\pi \otimes \chi, \frac{1}{2})$, we need to construct a representation $\mathcal{F}$ of $G_{S,\mathbf{A}}$, analogous to the space of cusp forms on G'. This representation should contain $\pi' \otimes \chi$, and have a naturally defined $T(\mathbf{A})$-invariant linear form. We will construct the representation $\mathcal{F}$ in the special case when k is a number field, the group $G_{S,\infty} = \prod_{v|\infty} G'_v(k_v)$ is compact modulo its center, and the representation $(\pi' \otimes \chi)_\infty$ of $G_{S,\infty}$ is the *trivial* representation.

More generally, one should be able to construct $\mathcal{F}$ whenever the set S contains all Archimedean places of k. In the number field case, this implies that the representation $(\pi' \otimes \chi)_\infty$ is finite-dimensional. When this representation is nontrivial, however, the approach to the first derivative sketched in the next few sections involves the theory of heights on local systems over curves, which is not yet complete. However, there is much preliminary work in the area; see [Br; Z2].

20. Shimura Varieties

We now assume that k is a number field and that the set S has odd cardinality, and contains all Archimedean places. Then k is totally real and E is a CM field.

Let v be a real place of k, and let $B(v)$ be the quaternion algebra over k which is ramified at $S - \{v\}$. Since E_w is a field, for all $w \in S$, E embeds as a subfield of $B(v)$. Let $G' = G(v)$ be the corresponding group of unitary similitudes, with k-points $B(v)^* \times E^*/\Delta k^*$. The torus T with $T(k) = E^*/k^*$ embeds diagonally, as the subgroup of G' fixing a vector u with $\phi(u,u) = 1$. The adèlic pair $T(\mathbf{A}_f) \to G'(\mathbf{A}_f)$ is isomorphic to $T(\mathbf{A}_f) \to G_{S,\mathbf{A}_f}$, independent of the choice of real place v.

Consider the homomorphism

$$h : \mathbb{C}^* \to T(k \otimes \mathbf{R}) = \prod_{w|\infty} (\mathbb{C}^*/\mathbf{R}^*)_w$$

given by

$$h(z) = (z(\text{mod } \mathbf{R}^*), 1, 1, \ldots, 1),$$

where the first coordinate corresponds to the real place v. The inclusion $T \to G'$ then gives rise to a homomorphism

$$h : \mathbb{C}^* \to G'(k \otimes \mathbf{R}).$$

The data (T, h) defines a Shimura variety $M(T, h)$ over $\mathbb{C}$, with an action of $T(\mathbf{A}_f)$. If $J \subset T(\mathbf{A}_f)$ is compact and open, then

$$M(T,h)^J(\mathbb{C}) = T(k)\backslash T(\mathbf{A}_f)/J$$

is a finite set of points.

The data (G', h) define a Shimura variety $M(G', h)$ over $\mathbb{C}$, with an action of $G'(\mathbf{A}_f) = G_{S,\mathbf{A}_f}$. If $K \subset G_{S,\mathbf{A}_f}$ is compact and open, then

$$M(G', h)^K(\mathbb{C}) = G'(k)\backslash(X \times G_{S,\mathbf{A}_f}/K),$$

where X is the Riemann surface of the $G'(k \otimes \mathbf{R})$ conjugacy class of h. We have

$$\begin{aligned} X &\simeq G'(k \otimes \mathbf{R})/Z(h) \\ &= G'(k_v)/Z \cdot T(k_v) \\ &\simeq \mathrm{GL}_2(\mathbf{R})/\mathbb{C}^* \simeq \mathcal{H}^{\pm}. \end{aligned}$$

Hence $M(G', h)^K(\mathbb{C})$ is a disjoint union of a finite number of connected hyperbolic Riemann surfaces with finite volume. They are compact, unless $k = \mathbb{Q}$ and $S = \{\infty\}$, in which case G' is quasi-split and there are finitely many cusps.

The theory of canonical models provides models for $M(T, h)$ and $M(G', h)$ over their reflex fields. In this case, the reflex field is E in both cases, embedded in $(E \otimes k_v) = \mathbb{C}$ by the place v. The actions of $T(\mathbf{A}_f)$ and $G_{S,\mathbf{A}_f}$ are both defined over E, and the morphism $M(T, k) \to M(G', h)$ is defined over E and is $T(\mathbf{A}_f)$-equivariant. The connected components of $M(T, h)$ and $M(G', h)$ are all defined over the maximal abelian extension E^{ab} of E in $\mathbb{C}$, and the action of $\mathrm{Gal}(E^{\mathrm{ab}}/E)$ on these components is given by Shimura's reciprocity law. For proofs of these assertions, see [D] and [Ca].

21. Nearby Quaternion Algebras

In fact, the varieties $M(T,h) \to M(G',h)$ over E do not depend on the choice of a real place v of k, which was used to define G' and the Shimura varieties over $(k_v \otimes E) = \mathbb{C}$. They depend only on S, and we will denote them $M(T) \to M(G_S)$. More precisely, if w is any real place of k, and $G' = G(w)$ is the form of G coming from the quaternion algebra ramified at $S - \{w\}$, we have an isomorphism of Riemann surfaces

$$M(G_S)^K(E_w) = G'(k)\backslash(G'(k_w)/Z \cdot T(k_w) \times G_{S,\mathbf{A}_f}/K).$$

If $J = K \cap T(\mathbf{A}_f)$, then

$$M(T)^J(E_w) = T(k)\backslash(1 \times T(\mathbf{A}_f)/J),$$

and the morphism $M(T) \to M(G_S)$ over E_w is given by the map $T \to G'$.

More generally, if w is a non-Archimedean place in S, and $G' = G(w)$ is the inner form of G made from the quaternion algebra over k which is ramified at $S - \{w\}$, we also have a rigid analytic uniformization of the points of the curve $M(G_S)$ over $E_w = E \otimes k_w$. For simplicity, we describe this in the case where $K \subset G_{S,\mathbf{A}_f}$ has the form $K = K_w \times K^w$, and K_w is the unique maximal compact subgroup (for the general case, see [Dr]). In this case $J_w = T(k_w)$, and all the points of $M(T)^J$ are rational over E_w. We have an isomorphism of rigid analytic spaces:

$$M(G_S)^K(E_w) \simeq G'(k)\backslash\big(G'(k_w)/Z \cdot T(k_w) \times G_{S,\mathbf{A}_f^w}/K^w\big).$$

Here

$$G'(k_w)/Z \cdot T(k_w) \simeq \mathrm{GL}_2(k_w)/E_w^* \simeq \mathbf{P}^1(E_w) - \mathbf{P}^1(k_w)$$

are the E_w-points of Drinfeld's upper half plane. The inclusion of

$$M(T)(E_w) \simeq T(k)\backslash(1 \times T(\mathbf{A}_f^w)/J^w)$$

is again described by group theory. The components of $M(G_S)^K$ containing these "special points" are rational over E_w; the general components are rational over the maximal unramified extension of E_w.

There is also a slightly weaker result for non-Archimedean places w which are *not* in S. Here we get a rigid analytic description of the "supersingular locus" on $M(G_S)$ over E_w, when E_w is the unramified quadratic field extension of k_w. Let $G' = G(w)$ be the inner form of G, corresponding to the quaternion algebra ramified at $S \cup \{w\}$, and assume $K \subset G_{S,\mathbf{A}^f}$ has the form $K = K_w \times K^w$, with K_w the unique hyperspecial maximal compact subgroup of $G(k_v)$ which contains $T(k_v)$. Then $M(G_S)^K$ has a model over the ring of integers $\mathcal{O}_w$ of E_w, with good reduction (mod w). The points reducing to supersingular points (mod w) give a rigid analytic space

$$M(G_S)^K(E_w)_{\mathrm{supersing}} \simeq G'(k)\backslash\big(G'(k_w)/Z \cdot T(k_w) \times G_{S,\mathbf{A}_f^w}/K^w\big).$$

The supersingular points in the residue field $\mathbf{F}_w$ are given by

$$M(G_S)^K(\mathbf{F}_w)_{\text{supersing}} \simeq G'(k)\backslash G_{S,\mathbf{A}_f^w}/K^w,$$

and the reduction map is given by group theory. Note that

$$G'(k_w)/Z \cdot T(k_w) \simeq D_w^*/E_w^*$$

is analytically isomorphic to an open disc over $\mathcal{O}_w$. The special points in $M(T)^J$ are all rational over E_w, and their map to the supersingular locus in $M(G_S)^K$ is given by group theory. Again, not all of the components of $M(G_S)^K$ are rational over E_w, but those containing the special points are.

To recapitulate, the morphism of Shimura varieties $M(T) \to M(G_S)$ over E depends only on the pair of groups $T \to G$ and the odd set S of places, which contains all real places of k. The local study of this morphism over the completion $E_w = E \otimes k_w$ involves the quaternion algebra over k with ramification locus

$$S - \{w\}, \quad \text{when} \quad w \in S,$$
$$S \cup \{w\}, \quad \text{when} \quad w \notin S \quad \text{and} \quad E_w \quad \text{is a field.}$$

These are all quaternion algebras at distance one from the set S.

22. The Global Representation

We are now prepared to construct the representation $\mathcal{F}$ of $G_{S,\mathbf{A}} = G_{S,\infty} \times G_{S,\mathbf{A}_f}$, with a $T(\mathbf{A})$-invariant linear form, under the hypothesis that

$$(\pi' \otimes \chi)_\infty = \mathbb{C} \tag{22.1}$$

is the trivial representation. We will define a natural representation $\mathcal{F}_f$ of $G_{S,\mathbf{A}_f}$, using the Shimura curve $M(G_S)$, as well as a $T(\mathbf{A}_f)$-invariant linear form on it, using the morphism $M(T) \to M(G_S)$. We will then put

$$\mathcal{F} = \mathbb{C} \otimes \mathcal{F}_f$$

as a representation of $G_{S,\infty} \times G_{S,\mathbf{A}_f}$.

Let $\text{Pic}^0(M(G_S)^K)$ be the group of line bundles on the curve $M(G_S)^K$ which have degree zero on each geometric component, and $\text{Pic}^0(M(G_S)^K)(E)$ those line bundles which are rational over E. If $K' \subset K$, we have a finite covering of curves over E

$$\beta : M(G_S)^{K'} \to M(G_S)^K.$$

This induces a homomorphism, by pull-back:

$$\beta^* : \text{Pic}^0(M(G_S)^K)(E) \to \text{Pic}^0(M(G_S)^{K'})(E).$$

The map β^* is an injection, modulo torsion. We define

$$\mathcal{F}_f = \varinjlim \text{Pic}^0(M(G_S)^K)(E) \otimes \mathbb{C},$$

where the direct limit maps are now injective.

The direct limit is a representation of $G_{S,\mathbf{A}_f}$, and the fixed space of any open compact subgroup K

$$(\mathcal{F}_f)^K = \mathrm{Pic}^0(M(G_S)^K)(E) \otimes \mathbb{C}$$

is a finite-dimensional complex vector space, by the Mordell–Weil theorem.

The tangent space to $\mathrm{Pic}^0(M(G_S)^K)$ is the cohomology group

$$H^0(M(G_S)^K, \Omega^1).$$

On the direct limit

$$H_f = \varinjlim H^0(M(G_S)^K, \Omega^1) \otimes \mathbb{C}$$

the group $G_{S,\mathbf{A}_f}$ acts via a direct sum of the representations $(\pi' \otimes \chi)_f$, where $\pi \otimes \chi$ appears in the space of cusp forms for G, is square integrable at all local places in S, and satisfies (22.1).

Since the action of the endomorphisms of abelian varieties in characteristic zero is faithfully represented by their action on the tangent space, these are the *only* representations which *can* appear in the action of $G_{S,\mathbf{A}_f}$ on $\mathcal{F}_f$. Whereas their multiplicity in H_f is equal to 1, their multiplicity in $\mathcal{F}_f$ is predicted by a generalization of the conjecture of Birch and Swinnerton-Dyer.

CONJECTURE. *The multiplicity of* $(\pi' \otimes \chi)_f$ *in* $\mathcal{F}_f$ *is equal to* $\mathrm{ord}_{s=1/2} L(\pi \otimes \chi, s)$.

In particular, since $\#S$ is odd, we have $L(\pi \otimes \chi, \frac{1}{2}) = 0$, and the multiplicity of $(\pi' \otimes \chi)_f$ in $\mathcal{F}_f$ should be positive (and odd). The same holds for the multiplicity of $\pi' \otimes \chi$ in $\mathcal{F} = \mathbb{C} \otimes \mathcal{F}_f$, where $\mathbb{C}$ is the trivial representation of $G_{S,\infty}$.

23. The Global Linear Form

We now use the torus T, and the zero cycle $j : M(T) \to M(G_S)$ over E, to define a $T(\mathbf{A}_f)$-invariant linear form $\ell : \mathcal{F}_f \to \mathbb{C}$.

For each open compact $K \subset G_{S,\mathbf{A}_f}$, we put $J = K \cap T(\mathbf{A}_f)$, which is open, compact in $T(\mathbf{A}_f)$. We define the 0-cycle

$$m(T)^K = j_*(M(T)^J)$$

on the curve $M(G_S)^K$ over E.

Since $M(G_S)^K$ is hyperbolic, there is, for each component c, a class δ_c in $\mathrm{Pic}(M(G_S)^K)(\overline{E}) \otimes \mathbb{Q}$ that has degree 1 on c and degree 0 on all other components. Indeed, if the component c is $X = \mathcal{H}/\Gamma$ over $\mathbb{C}$, then the divisor class

$$d_c = K_X + \sum_{\substack{x \text{ elliptic} \\ \text{or cuspidal}}} \left(1 - \frac{1}{e_x}\right)(x)$$

has positive degree, where K_X is the canonical class, and e_x is the order of the cyclic stabilizer Γ_x of x. This pulls back correctly for coverings given by the subgroups of finite index in Γ. We put

$$\delta_c = d_c / \deg(d_c),$$

and define

$$m_0(T)^K = m(T)^K - \sum_c \deg_c(m(T)^K) \cdot \delta_c.$$

This class lies in $\mathrm{Pic}^0(M(G_S)^K)(E) \otimes \mathbb{Q}$, as $m(T)^K$ has equal degree on conjugate components, and the δ_c are similarly conjugate.

The canonical height pairing of Néron and Tate, on the Jacobian of $M(G_S)^K$ over E, gives a linear form $\ell_K : (\mathcal{F}_f)^K \to \mathbb{C}$, defined by

$$\ell_K(d) = \langle d, m_0(T)^K \rangle.$$

These forms come from a single linear form ℓ on $\mathcal{F}_f$, for when $K' \subset K$ we have a commutative diagram

$$\begin{array}{ccc} M(T)^{J'} & \xrightarrow[j']{} & M(G_S)^{K'} \\ \downarrow & & \downarrow \pi \\ M(T)^{J} & \xrightarrow[j]{} & M(G_S)^{K}. \end{array}$$

Hence $m_0(T)^K = \pi_*(m_0(T)^{K'})$, and if d is in $(\mathcal{F}_f^0)^K$:

$$\ell_{K'}(\pi^* d) = \langle \pi^* d, m_0(T)^{K'} \rangle = \langle d, \pi_* m_0(T)^{K'} \rangle = \langle d, m_0(T)^K \rangle = \ell_K(d).$$

The form ℓ is clearly $T(\mathbf{A}_f)$-invariant, so gives a $T(\mathbf{A})$-invariant form (also denoted ℓ) on $\mathcal{F} = \mathbb{C} \otimes \mathcal{F}_f$. This induces a map

$$\ell_* : \mathrm{Hom}_{G_{S,\mathbf{A}}}(\pi' \otimes \chi, \mathcal{F}) \to \mathrm{Hom}_{T(\mathbf{A})}(\pi' \otimes \chi, \mathbb{C}).$$

We know the space $\mathrm{Hom}_{T(\mathbf{A})}(\pi' \otimes \chi, \mathbb{C})$ is one-dimensional, by the local theory and the construction of $G_{S,\mathbf{A}}$. The space $\mathrm{Hom}_{G_{S,\mathbf{A}}}(\pi' \otimes \chi, \mathcal{F})$ was conjectured to have odd dimension, equal to the order of $L(\pi \otimes \chi, s)$ at $s = \frac{1}{2}$.

The analog of Waldspurger's theorem in this context is the following.

CONJECTURE. *The map ℓ_* is nonzero if and only if $L'(\pi \otimes \chi, \frac{1}{2}) \neq 0$.*

24. Global Test Vectors

We can refine this conjecture, and obtain a statement generalizing [G-Z], in the situation where a global test vector exists (i.e., when the conductors of π and χ are relatively prime). Let $K \subset G_{S,\mathbf{A}_f}$ be the open compact subgroup fixing a line of test vectors in $(\pi' \otimes \chi)_f$. Then one wants a formula relating $L'(\pi \otimes \chi, \frac{1}{2})$ to the height pairing of the $(\pi' \otimes \chi)_f$-eigencomponent of $m_0(T)^K$ with *itself*. This formulation does not require a determination of the entire space $\mathrm{Hom}_{G_{S,\mathbf{A}}}(\pi' \otimes \chi, \mathcal{F})$.

In the spirit of the explicit formula in § 18, I would guess that

$$L'(\pi \otimes \chi, \tfrac{1}{2}) = \frac{(f_0, f_0)}{\sqrt{\mathbf{N}D}} \langle m_0(T)^K(\pi' \otimes \chi)_f, m_0(T)^K(\pi' \otimes \chi)_f \rangle.$$

Zhang [Z1] has done fundamental work in this direction. In the case when $k = \mathbb{Q}$, E is imaginary quadratic, all primes p dividing N (the conductor of π) are split in E, and $c = \text{conductor}(\chi) = 1$, we have $S = \{\infty\}$. Furthermore,

$$K \subset \mathrm{GL}_2(\hat{\mathbb{Z}}) \times \hat{\mathcal{O}}_E^* / \Delta \hat{\mathbb{Z}}^*$$

is the subgroup of those $\left(\begin{smallmatrix} a & b \\ c & d \end{smallmatrix}\right)$ in $\mathrm{GL}_2(\hat{\mathbb{Z}})$ with $c \equiv 0 \pmod{N}$. The desired formula is the one I proved with Zagier, twenty years ago.

Acknowledgment

I want to thank Peter Green for his careful reading of an earlier version of this paper.

References

[B] B. J. Birch, "Heegner points of elliptic curves", pp. 441–445 in *Symposia Mathematica* **15**, Istituto Nazionale di Alta Matematica, New York, Academic Press, 1975.

[B-S] B. J. Birch and N. M. Stephens, "Computation of Heegner points", pp. 13–41 in *Modular forms*, edited by R. A. Rankin, Chichester, Ellis Horwood, 1984.

[Br] J.-L. Brylinski, "Heights for local systems on curves", *Duke Math. J.* **59** (1989), 1–26.

[Ca] H. Carayol, "Sur la mauvaise réduction des courbes de Shimura", *Compositio Math.* **59** (1986), 151–230.

[C] W. Casselman, "On some results of Atkin and Lehner", *Math Ann.* **201** (1973), 301–314.

[D] P. Deligne, "Travaux de Shimura", pp. 123–165 (exposé 389) in *Séminaire Bourbaki*, Lecture Notes in Mathematics **244**, Berlin, Springer, 1971.

[Dr] V. Drinfeld, "Coverings of p-adic symmetric domains", *Funct. Analysis* **10** (1976), 29–40.

[G0] B. H. Gross, *Arithmetic on elliptic curves with complex multiplication*, Lecture Notes in Mathematics **776**, New York, Springer, 1980.

[G1] B. H. Gross, "Heegner points on $X_0(N)$", pp. 87–106 in *Modular forms*, edited by R. A. Rankin, Chichester, Ellis Horwood, 1984.

[G2] B. H. Gross, "Algebraic modular forms", *Israel J. Math.* **113** (1999), 61–93.

[G3] B. H. Gross, "Heights and the special values of L-series", pp. 115–187 in *Number theory* (Montreal, 1985), edited by H. Kisilevsky and J. Labute, CMS Conf. Proc. **7**, Amer. Math. Soc., Providence, RI, 1987.

[G4] B. H. Gross, "L-functions at the central critical point", pp. 527–535 in *Motives*, edited by Uwe Jannsen et al., *Proc. Symp. Pure Math* **55**, Amer. Math. Soc., Providence, RI, 1994.

[G-G] B. H. Gross and W. T. Gan, "Haar measure and the Artin conductor", *Trans. Amer. Math. Soc.* **351** (1999), 1691–1704.

[G-P] B. H. Gross and D. Prasad, "Test vectors for linear forms", *Math. Ann.* **291** (1991), 343–355.

[G-Z] B. H. Gross and D. Zagier, "Heegner points and derivatives of L-series", *Invent. Math.* **84** (1986), 225–320.

[J-L] H. Jacquet, and R. P. Langlands, *Automorphic forms on* GL_2, Lecture Notes in Mathematics **114**, Berlin, Springer, 1970.

[P] A. Popa, "Central values of L-series over real quadratic fields", Ph.D. Thesis, Harvard University (2003).

[S] H. Saito, "On Tunnell's formula for characters of GL_2", *Compositio Math.* **85** (1993), 99–108.

[T] J. Tunnell, "Local ε-factors and characters of GL_2", *Amer. J. Math.* **105** (1983), 1277–1308.

[W] J.-L. Waldspurger, "Sur les valeurs de certaines fonctions L automorphes en leur centre de symétrie", *Compositio Math.* **54** (1985), 173–242.

[Z1] S. Zhang, "Gross–Zagier for GL_2". *Asian J. Math.* **5** (2001), 183–290.

[Z2] S. Zhang, "Heights of Heegner cycles and derivatives of L-series", *Invent. Math.* **130** (1997), 99–152.

BENEDICT H. GROSS
DEPARTMENT OF MATHEMATICS
HARVARD UNIVERSITY
CAMBRIDGE, MA 02138
UNITED STATES
gross@math.harvard.edu

Heegner Points and Rankin L-Series
MSRI Publications
Volume **49**, 2004

Gross–Zagier Revisited

BRIAN CONRAD

WITH AN APPENDIX BY W. R. MANN

CONTENTS

1. Introduction	67
2. Some Properties of Abelian Schemes and Modular Curves	70
3. The Serre–Tate Theorem and the Grothendieck Existence Theorem	76
4. Computing Naive Intersection Numbers	81
5. Intersection Formula Via Hom Groups	85
6. Supersingular Cases with $r_{\mathscr{A}}(m) = 0$	88
7. Application of a Construction of Serre	98
8. Intersection Theory Via Meromorphic Tensors	109
9. Self-Intersection Formula and Application to Global Height Pairings	117
10. Quaternionic Explications	130
Appendix by W. R. Mann: Elimination of Quaternionic Sums	139
References	162

1. Introduction

The aim of this paper is to rework the material in Chapter III of Gross and Zagier's "Heegner points and derivatives of L-series" — see [GZ] in the list of references — based on more systematic deformation-theoretic methods, so as to treat all imaginary quadratic fields, all residue characteristics, and all j-invariants on an equal footing. This leads to more conceptual arguments in several places and interpretations for some quantities which appear to otherwise arise out of thin air in [GZ, Ch. III]. For example, the sum in [GZ, Ch. III, Lemma 8.2] arises for us in (9–6), where it is given a deformation-theoretic meaning. Provided the analytic results in [GZ] are proven for even discriminants, the main results in [GZ] would be valid without parity restriction on the discriminant of the imaginary quadratic field. Our order of development of the basic results follows

This work was partially supported by the NSF and the Alfred P. Sloan Foundation. I thank B. Gross and J.-P. Serre for helpful suggestions, and Mike Roth for helping to resolve some serious confusion.

[GZ, Ch. III], but the methods of proof are usually quite different, making much less use of the "numerology" of modular curves.

Here is a summary of the contents. In Section 2 we consider some background issues related to maps among elliptic curves over various bases and horizontal divisors on relative curves over a discrete valuation ring. In Section 3 we provide a brief survey of the Serre–Tate theorem and the Grothendieck existence theorem, since these form the backbone of the deformation-theoretic methods which underlie all subsequent arguments. For lack of space, some topics (such as intersection theory on arithmetic curves, Gross' paper [Gr1] on quasi-canonical liftings, and p-divisible groups) are not reviewed but are used freely where needed.

In Section 4 we compute an elementary intersection number on a modular curve in terms of cardinalities of isomorphism groups between infinitesimal deformations. This serves as both a warm-up to and key ingredient in Sections 5–6, where we use cardinalities of Hom-groups between infinitesimal deformations to give a formula (in Theorem 5.1) for a local intersection number $(x.T_m(x^\sigma))_v$, where $x \in X_0(N)(H)$ is a Heegner point with CM by the ring of integers of an imaginary quadratic field K (with Hilbert class field H), $\sigma \in \mathrm{Gal}(H/K)$ is an element which corresponds to an ideal class $\mathscr{A}$ in K under the Artin isomorphism, and T_m is a Hecke correspondence with $m \geq 1$ relatively prime to N. An essential hypothesis in Theorem 5.1 is the vanishing of the number $r_{\mathscr{A}}(m)$ of integral ideals of norm m in the ideal class $\mathscr{A}$. This corresponds to the requirement that the divisors x and $T_m(x^\sigma)$ on $X_0(N)$ are disjoint. Retaining the assumption $r_{\mathscr{A}}(m) = 0$, in Section 7 we develop and apply a construction of Serre in order to translate the formula in Theorem 5.1 into the language of quaternion algebras. The resulting quaternionic formulas in Corollary 7.15 are a model for the local intersection number calculation which is required for the computation of global height pairings in [GZ], except the condition $r_{\mathscr{A}}(m) = 0$ in Corollary 7.15 has to be dropped.

To avoid assuming $r_{\mathscr{A}}(m) = 0$, we have to confront the case of divisors which may contain components in common. Recall that global height pairings of degree 0 divisors are defined in terms of local pairings, using a "moving lemma" to reduce to the case in which horizontal divisors intersect properly. We want an explicit formula for the global pairing $\langle c, T_m(d^\sigma)\rangle_H$ where $c = x - \infty$ and $d = x - 0$, so we must avoid the abstract moving lemma and must work directly with improper intersections. Whereas [GZ] deal with this issue by using a technique from [Gr2, § 5] which might be called "intersection theory with a tangent vector", in Section 8 we develop a more systematic method which we call "intersection theory with a meromorphic tensor". This theory is applied in Section 9 to give a formula in Theorem 9.6 which expresses a global height pairing in terms of local intersection numbers whose definition does not require proper intersections.

In Section 10 we use deformation theory to generalize Corollary 7.15 to include the case $r_{\mathscr{A}}(m) > 0$, and the appendix (by W. R. Mann) recovers all three main formulas in [GZ, Ch. III, § 9] as consequences of the quaternionic formulas we obtain via intersection theory with meromorphic tensors. The appendix follows the argument of Gross–Zagier quite closely, but explains some background, elaborates on some points in more detail than in [GZ, Ch. III, § 9], and works uniformly across all negative fundamental discriminants without parity restrictions. This parity issue is the main technical contribution of the appendix, and simply requires being a bit careful.

Some conventions. As in [GZ], we normalize the Artin map of class field theory to associate uniformizers to arithmetic Frobenius elements. Thus, if K is an imaginary quadratic field with Hilbert class field H, then for a prime ideal $\mathfrak{p}$ of K the isomorphism between $\mathrm{Gal}(H/K)$ and the class group Cl_K of K associates the ideal class $[\mathfrak{p}]$ to an arithmetic Frobenius element.

Following [GZ], we only consider Heegner points with CM by the maximal order $\mathscr{O}_K$ in an imaginary quadratic field K. A *Heegner diagram* (for $\mathscr{O}_K$) over an $\mathscr{O}_K$-scheme S is an $\mathscr{O}_K$-linear isogeny $\phi : E \to E'$ between elliptic curves over S which are equipped with $\mathscr{O}_K$-action which is "normalized" in the sense that the induced action on the tangent space at the identity is the same as that obtained through $\mathscr{O}_S$ being a sheaf of $\mathscr{O}_K$-algebras via the $\mathscr{O}_K$-scheme structure on S. For example, when we speak of Heegner diagrams (or Heegner points on modular curves) over $\mathbf{C}$, it is implicitly understood that an embedding $K \hookrightarrow \mathbf{C}$ has been fixed for all time.

Under the action of $\mathrm{Cl}_K \simeq \mathrm{Gal}(H/K)$ on Heegner points

$$X_0(N)(H) \subseteq X_0(N)(\mathbf{C}),$$

with $N > 1$ having all prime factors split in K, the action of $[\mathfrak{a}] \in \mathrm{Cl}_K$ sends the Heegner point

$$([\mathfrak{b}], \mathfrak{n}) \stackrel{\text{def}}{=} (\mathbf{C}/\mathfrak{b} \to \mathbf{C}/\mathfrak{n}^{-1}\mathfrak{b})$$

to $([\mathfrak{b}][\mathfrak{a}]^{-1}, \mathfrak{n})$. The appearance of inversion is due to our decision to send uniformizers to arithmetic (rather than geometric) Frobenius elements. This analytic description of the Galois action on Heegner points will play a crucial role in the proof of Corollary 7.11.

If x is a Heegner point with associated CM field K, we will write u_x to denote the cardinality of the group of roots of unity in K (so $u_x = 2$ unless $K = \mathbf{Q}(\sqrt{-1})$ or $K = \mathbf{Q}(\sqrt{-3})$).

If S is a finite set, we will write $|S|$ or $\#S$ to denote the cardinality of S.

If $(R, \mathfrak{m})$ is a local ring, we write R_n to denote $R/\mathfrak{m}^{n+1}$. If X and Y are R-schemes, we write $\mathrm{Hom}_{R_n}(X, Y)$ to denote the set of morphisms of mod $\mathfrak{m}^{n+1}$ fibers.

If B is a central simple algebra of finite dimension over a field F, we write $\mathrm{N} : B \to F$ and $\mathrm{T} : B \to F$ to denote the reduced norm and reduced trace on B.

It is recommended (but not necessary) that anyone reading these notes should have a copy of [GZ] at hand. One piece of notation we adopt, following [GZ], is that $r_{\mathscr{A}}(m)$ denotes the number of integral ideals of norm $m > 0$ in an ideal class

$\mathscr{A}$ of an imaginary quadratic field K which is fixed throughout the discussion. Since $r_{\mathscr{A}} = r_{\mathscr{A}^{-1}}$, due to complex conjugation inducing inversion on the class group without changing norms, if we change the Artin isomorphism $\mathrm{Gal}(H/K) \simeq \mathrm{Cl}_K$ by a sign then this has no impact on a formula for $\langle c, T_m d^\sigma \rangle_v$ in terms of $r_{\mathscr{A}}$, where $\mathscr{A} \in \mathrm{Cl}_K$ "corresponds" to $\sigma \in \mathrm{Gal}(H/K)$.

2. Some Properties of Abelian Schemes and Modular Curves

We begin by discussing some general facts about elliptic curves. Since the proofs for elliptic curves are the same as for abelian varieties, and more specifically it is not enough for us to work with passage between a number field and $\mathbf{C}$ (e.g., we need to work over discrete valuation rings with positive characteristic residue field, etc.) we state the basic theorem in the more natural setting of abelian varieties and abelian schemes.

THEOREM 2.1. *Let A, B be abelian varieties over a field F.*

(1) *If F is separably closed and F' is an extension field, then $\mathrm{Hom}_F(A, B) \to \mathrm{Hom}_{F'}(A_{/F'}, B_{/F'})$ is an isomorphism. In other words, abelian varieties over a separably closed field never acquire any "new" morphisms over an extension field.*
(2) *If $F = \mathrm{Frac}(R)$ for a discrete valuation ring R and A, B have Néron models $\mathscr{A}$, $\mathscr{B}$ over R which are proper (i.e., A and B have good reduction relative to R), then*
$$\mathrm{Hom}_F(A, B) = \mathrm{Hom}_R(\mathscr{A}, \mathscr{B}) \to \mathrm{Hom}_k(\mathscr{A}_0, \mathscr{B}_0)$$
is injective, where k is the residue field of R and $(\cdot)_0$ denotes the "closed fiber" functor on R-schemes. In other words, $(\cdot)_0$ is a faithful functor from abelian schemes over R to abelian schemes over k. In fact, this latter faithfulness statement holds for abelian schemes over any local ring R whatsoever.

PROOF. The technical details are a bit of a digression from the main aims of this paper, so although we do not know a reference we do not give the details. Instead, we mention the basic idea: for ℓ invertible on the base, ℓ-power torsion is "relatively schematically dense" in an abelian scheme (in the sense of [EGA, IV_3, 11.10]); this ultimately comes down to the classical fact that such torsion is dense on an abelian variety over an algebraically closed field. Such denseness, together with the fact that ℓ^n-torsion is finite étale over the base, provides enough rigidity to descend morphisms for the first part of the theorem, and enough restrictiveness to force injectivity in the second part of the theorem. □

We will later need the faithfulness of $(\cdot)_0$ in Theorem 2.1(2) for cases in which R is an artin local ring, so it is not adequate to work over discrete valuation rings and fields.

Let's now recall the basic setup in [GZ]. We have fixed an imaginary quadratic field $K \subseteq \mathbf{C}$ with discriminant $D < 0$ (and we take $\overline{\mathbf{Q}}$ to be the algebraic closure of $\mathbf{Q}$ inside of $\mathbf{C}$). We write $H \subseteq \overline{\mathbf{Q}}$ for the Hilbert class field of K, and we choose a positive integer $N > 1$ which is relatively prime to D and for which all prime factors of N are split in K. Let $X = X_0(N)_{/\mathbf{Z}}$ be the coarse moduli scheme as in [KM], so X is a proper flat curve over $\mathbf{Z}$ which is smooth over $\mathbf{Z}[1/N]$ but has some rather complicated fibers modulo prime factors of N. We have no need for the assumption that $D < 0$ is odd (i.e, $D \equiv 1 \bmod 4$), whose main purpose in [GZ] is to simplify certain aspects of calculations on the analytic side of the [GZ] paper, so we avoid such conditions (and hence include cases with even discriminant).

Let $x \in X(H) \subseteq X(\mathbf{C})$ be a Heegner point with CM by the maximal order $\mathscr{O}_K$. Because of our explicit knowledge of the action of $\mathrm{Gal}(H/\mathbf{Q})$ on Heegner points, we see that the action of $\mathrm{Gal}(H/\mathbf{Q})$ on $x \in X(H)$ is "free" (i.e., nontrivial elements in $\mathrm{Gal}(H/\mathbf{Q})$ do not fix our Heegner point $x \in X(H)$), so it follows that the map $x : \mathrm{Spec}(H) \to X_{/\mathbf{Q}}$ is a closed immersion. For each closed point v of $\mathrm{Spec}\,\mathscr{O}_H$, by viewing H as the fraction field of the algebraic localization $\mathscr{O}_{H,v}$ we may use the valuative criterion for properness to uniquely extend x to a point in $X(\mathscr{O}_{H,v})$. A simple "smearing out" argument involving denominator-chasing shows that all of these maps arise from a unique morphism $\underline{x} : \mathrm{Spec}(\mathscr{O}_H) \to X$.

It is *not* a priori clear if the map $\underline{x}$ is a closed immersion (in general it isn't). More specifically, if $Z_x \hookrightarrow X$ is the scheme-theoretic closure of x (i.e., the scheme-theoretic image of $\underline{x}$), then $Z_x \to \mathrm{Spec}(\mathbf{Z})$ is proper, flat, and quasi-finite, hence finite flat, so it has the form $Z_x = \mathrm{Spec}(A_x)$ with $\mathbf{Q} \otimes_{\mathbf{Z}} A_x = H$. Thus, A_x is an order in $\mathscr{O}_H$, but it isn't obvious if $A_x = \mathscr{O}_H$. This is an important issue in subsequent intersection theory calculations because the intersection theory will involve closed subschemes of (various base chages on) X. For this reason, we must distinguish $\mathrm{Spec}\,A_x \hookrightarrow X$ and $\underline{x} : \mathrm{Spec}\,\mathscr{O}_H \to X$.

To illustrate what can go wrong, consider the map $\mathbf{Z}_p[T] \to \mathbf{Z}_p[\zeta_{p^2}]$ defined by $T \mapsto p\zeta_{p^2}$. Over $\mathbf{Q}_p$ this defines an immersion

$$\phi_\eta : \mathrm{Spec}(\mathbf{Q}_p(\zeta_{p^2})) \hookrightarrow \mathbf{A}^1_{\mathbf{Z}_p} \subseteq \mathbf{P}^1_{\mathbf{Z}_p}$$

which is a closed immersion since $\mathbf{Q}_p(\zeta_{p^2}) = \mathbf{Q}_p[p\zeta_{p^2}]$, and ϕ_η comes from a map $\phi : \mathrm{Spec}(\mathbf{Z}_p[\zeta_{p^2}]) \to \mathbf{P}^1_{\mathbf{Z}_p}$ which is necessarily the one we would get from applying the valuative criterion for properness to ϕ_η.

The "integral model" map ϕ is *not* a closed immersion because $\mathbf{Z}_p[T] \to \mathbf{Z}_p[\zeta_{p^2}]$ defined by $T \mapsto p\zeta_{p^2}$ is not a surjection. In fact, the generic fiber closed subscheme of $\mathbf{P}^1_{\mathbf{Q}_p}$ defined by ϕ_η has scheme-theoretic closure in $\mathbf{P}^1_{\mathbf{Z}_p}$ given by $\mathrm{Spec}(A)$ where $A = \mathrm{image}(\phi) = \mathbf{Z}_p + p\mathbf{Z}_p[\zeta_{p^2}]$ is a nonmaximal order in $\mathbf{Z}_p[\zeta_{p^2}]$. The moral is that even if we can compute the "field of definition" of a point on a generic fiber smooth curve, it is not true that the closure of this in a particular proper integral model of the curve is cut out by a Dedekind subscheme or that it lies in the relative smooth locus. It is a very fortunate fact in the Gross–Zagier situation that this difficulty usually does not arise for the "horizontal divisors" they need to consider.

Here is the basic result we need concerning Heegner divisors in the relative smooth locus.

LEMMA 2.2. *Let $x \in X(H)$ be a Heegner point and Λ_v denote the valuation ring of the completion H_v at a place v over a prime p. The map* $\mathrm{Spec}(\Lambda_v) \to X_{/\mathbf{Z}_p}$ *corresponding to the pullback $x_v \in X(H_v)$ of x factors through the relative smooth locus, and the induced natural map $\underline{x}_v : \mathrm{Spec}(\Lambda_v) \to X \times_{\mathbf{Z}} \Lambda_v$ arising from x_v is a closed immersion into the relative smooth locus over Λ_v (and hence lies in the regular locus).*

PROOF. Using smoothness of X over $\mathbf{Z}[1/N]$ when p doesn't divide N and [GZ, Ch. III, Prop. 3.1] when $p|N$, we see that the image of the closed point under $\underline{x}_v$ lies in the smooth locus over Λ_v. Since $\underline{x}_v$ is a section to a separated map, it is a closed immersion and necessarily lands inside of the relative smooth locus over Λ_v. □

REMARK 2.3. One reason for the importance of Lemma 2.2 is that the curve $X_v = X \times_{\mathbf{Z}} \Lambda_v$ is usually *not* regular, so to do intersection theory one must resolve singularities. A minimal regular resolution $X^{\mathrm{reg}} = X^{\mathrm{reg}}_{/\mathbf{Z}_{(p)}}$ can be obtained by means of successive normalizations and blow-ups over the nonregular locus (thanks to a deep theorem of Lipman [L]), so the resolution process doesn't do anything over the relative smooth locus.

Thus, even though we can't expect to do intersection theory on X_v, we are assured by Lemma 2.2 that the intersection numbers among properly intersecting Heegner divisors will be computable directly on X_v. We need to keep track of smoothness because regularity is often destroyed by ramified base change, such as $\mathbf{Z}_{(p)} \to \mathscr{O}_{H,v}$ when p is ramified in K (in such cases the relative curve $X^{\mathrm{reg}} \times_{\mathbf{Z}_{(p)}} \Lambda_v$ can be nonregular, but some Heegner divisor intersection numbers may still be computed on this curves, such as happens in [GZ, Ch. III, Prop. 3.3]).

With v fixed as above, let W denote the completion of a maximal unramified extension of Λ_v (if p ramifies in K, this is not a Witt ring). The section $\underline{x}_v \in X_{/W}(W)$ lies in the relative smooth locus. Although $X_{/W}$ is not a fine moduli scheme, the point $\underline{x}_v$ does arise from a $\Gamma_0(N)$-structure over W. This is an important fact. Before we prove it, as a preliminary step recall that from the classical CM theory we can find an algebraic model $\phi : E \to E'$ over H which represents our Heegner point in $X(H)$. It is *not* true in general that E and E' have to admit everywhere good reduction over $\mathscr{O}_H$. But what is true, and suffices for us, is that we can always find such good models over W. This is given by Theorem 2.5, but before proving this we recall a general lemma concerning good reduction for CM elliptic curves.

LEMMA 2.4. *Let R be a complete discrete valuation ring with fraction field F of characteristic* 0, *and let $E_{/F}$ be an CM elliptic curve with CM field $K \subseteq F$ (so the CM-action is automatically defined over F). If $K = \mathbf{Q}(\sqrt{-1})$ or $K = \mathbf{Q}(\sqrt{-3})$, assume moreover that $E_{/F}$ actually has CM by the full ring of integers of K.*

Then there exists a "twist" E' of E over an unramified extension of F (so E' is CM with the same CM ring) such that E' has good reduction.

When E begins life over a number field (which is the only case we really need for treating Heegner points), this lemma follows from Corollary 2 to Theorem 9 in [ST]. Since the argument in [ST] rests on class field theory, we prefer the argument below which uses only general principles concerning abelian varieties. The argument in [ST] has the merit of treating more cases when $K = \mathbf{Q}(\sqrt{-1})$ and $K = \mathbf{Q}(\sqrt{-3})$.

PROOF. Since $\mathrm{End}(E_{/F}) = \mathscr{O}_K$, so $\mathrm{End}(E_{/F_s}) = \mathscr{O}_K$ and hence $\mathrm{Aut}(E_{/F_s}) = \mathscr{O}_K^\times = \mathrm{Aut}(E_{/F})$, twisted forms are described in terms of Galois cohomology $\mathrm{H}^1(\mathrm{Gal}(F_s/F), \mathscr{O}_K^\times)$ with $\mathscr{O}_K^\times$ given a *trivial* action by $\mathrm{Gal}(F_s/F)$. Passing to completions on a henselian discrete valuation ring doesn't affect the Galois group in characteristic 0, so without loss of generality we can assume that R has a separably closed residue field k (i.e., pass to the completion of the maximal unramified extension). Now we seek to find a twist E' of E over F with good reduction over R.

A priori E has potentially good reduction (as does any CM type abelian variety over F), so there exists a finite Galois extension F'/F such that $E_{/F'}$ has good reduction. Let R' be the integral closure of R in F' and let $\mathscr{E}_{/R'}$ denote the Néron model of $E_{/F'}$. Note in particular that the residue field extension k'/k is purely inseparable (since k is separably closed), so $\mathrm{Aut}(k'/k) = \{1\}$. Let $\Gamma = \mathrm{Gal}(F'/F)$, so for each $\sigma \in \Gamma$ there is (by the Néronian property of abelian schemes over R) a unique isomorphism $\phi_\sigma : \mathscr{E} \simeq \sigma^*(\mathscr{E})$ of elliptic curves over R' which extends the evident descent data isomorphism $E_{/F'} \simeq \sigma^*(E_{/F'})$ of elliptic curves over F'. In particular, we have the cocycle property $\sigma^*(\phi_\tau) \circ \phi_\sigma = \phi_{\sigma\tau}$ and also ϕ_σ commutes with the CM actions by K on source and target, as this can all be checked on the generic fibers.

Since σ induces the identity on k' (as $\mathrm{Aut}(k'/k) = \{1\}$!), on the closed fiber we therefore get an *automorphism* $\overline{\phi}_\sigma$ of $\mathscr{E}_{/k'}$ which defines an anti-homomorphism $\rho : \Gamma \to \mathrm{Aut}_{k'}(\mathscr{E}_{/k'})$ landing in the commutator of the K-action. But an imaginary quadratic field acting (in the isogeny category) on an elliptic curve is necessarily its own commutator in the endomorphism ring of the elliptic curve (even in the supersingular case!), so we deduce that $\rho(\Gamma) \subseteq \mathscr{O}_K^\times$ and in particular ρ is actually a homomorphism since $\mathscr{O}_K^\times$ is abelian. Let $\chi(\sigma) \in \mathscr{O}_K^\times$ be the *unique* (see Theorem 2.1(2)) unit inducing the same action as $\rho(\sigma)$ on the closed fiber of $\mathscr{E}_{/R'}$ (recall our initial assumption that when $\mathscr{O}_K^\times$ is larger than $\{\pm 1\}$, then the whole ring $\mathscr{O}_K$ acts on E and hence on the Néron model of $E_{/F'}$). Clearly χ is a homomorphism.

For each $\sigma \in \Gamma$ we see that $\psi(\sigma) \overset{\mathrm{def}}{=} \phi_\sigma \circ \chi(\sigma)^{-1} : \mathscr{E}_{/R'} \simeq \sigma^*(\mathscr{E}_{/R'})$ makes sense and induces the identity on the closed fiber. Since passage to the closed fiber is a faithful functor for abelian schemes (see Theorem 2.1(2)), we conclude that $\sigma \mapsto \psi(\sigma)$ satisfies the Galois cocycle condition on the generic fiber over F'

(indeed, to check this condition we may work over R', and even on closed fibers over R' by faithfulness). Thus, by Galois descent we obtain a twist E' of E over F and an isomorphism $E'_{/F'} \simeq E_{/F'}$ of elliptic curves which carries the canonical descent data on $E'_{/F'}$ over to the "twisted" descent data ψ.

I claim that E' has good reduction over F. To prove this, by Néron–Ogg–Shafarevich it suffices to pick a prime ℓ distinct from the residue characteristic of R and to show that the Γ-action on the ℓ-adic Tate module of $E'_{/F}$ is trivial (note that the Galois action on this Tate module certainly factors through Γ, since at least over F' we know we have good reduction for $E'_{/F'} \simeq E_{/F'}$). But this triviality is equivalent to the natural action of Γ on the closed fiber of the Néron model of E' over R' being the identity. This in turn follows from how E' was constructed. □

Recall that W denotes the completion of the maximal unramified extension of the algebraic localization $\mathscr{O}_{H,v}$ of $\mathscr{O}_H$ at the place v over p.

THEOREM 2.5. *In the above notation, the point in $X(W)$ arising from a Heegner point in $X(H)$ is necessarily induced by a Heegner diagram over W.*

It seems likely that Theorem 2.5 is false if we try to replace W with $\mathscr{O}_{H,v}$ (or its completion Λ_v); we almost surely have to pass to some extension of H unramified over v.

Due to the properness (even finiteness) of the fine moduli scheme of Drinfeld $\Gamma_0(N)$-structures on a given elliptic curve over a base, it follows that a diagram over W as in Theorem 2.5 is automatically a $\Gamma_0(N)$-structure (i.e., a cyclic N-isogeny) since this is true on the generic fiber of characteristic 0.

PROOF. Let's begin with a Heegner diagram $\phi : E \to E'$ over H which induces the given point in $X(H)$. It is generally *not* true that this necessarily admits good reduction over W (e.g., make a ramified quadratic twist), but Lemma 2.4 (whose additional hypothesis for $K = \mathbf{Q}(\sqrt{-1})$ or $K = \mathbf{Q}(\sqrt{-3})$ is satisfied because of our general assumption of CM by the maximal order for our Heegner points) shows that over the fraction field F of W there exists a twist of E with good reduction. More specifically, such a twist is given by an element $\chi \in \mathrm{H}^1(\mathrm{Gal}(F_s/F), \mathscr{O}_K^\times) = \mathrm{Hom}_{\mathrm{cont}}(\mathrm{Gal}(F_s/F), \mathscr{O}_K^\times)$. Since ϕ is $\mathscr{O}_K$-equivariant, we can use the same Galois character χ to twist E', and the resulting $\mathscr{O}_K^\times$-valued twisted descent data automatically has to commute with the $\mathscr{O}_K$-compatible ϕ.

Consequently, we can twist the entire given Heegner diagram relative to χ and thereby obtain a Heegner diagram over F for which the two elliptic curves have good reduction over W and the induced geometric point in $X(F_s)$ coincides with the original Heegner point. Using Néron functoriality we can extend the twisted ϕ over W and thereby get a Heegner diagram over W inducing the point in $X(W)$ arising from our given Heegner point in $X(H)$ (recall that $X(W) \to X(F)$ is bijective). □

In all that follows, we will fix a Heegner diagram $\underline{x}$ over W inducing a given Heegner point in $X(W)$ (or rather, inducing the W-section of $X_{/W}$ arising from an initial choice of Heegner point in $X(H)$). The existence of such a diagram follows from Theorem 2.5. It is important to note that the data of $\underline{x}$ is *determined* up to nonunique isomorphism by the corresponding section in $X(W)$, so this section will also be denoted $\underline{x}$. This uniqueness up to isomorphism follows from the following result, applied to $R = W$:

THEOREM 2.6. *Let R be a complete discrete valuation ring with separably closed residue field k and fraction field L. Let A_1 and A_2 be two abelian schemes over R. Then for any field extension L' of L, the natural map* $\mathrm{Hom}_R(A_1, A_2) = \mathrm{Hom}_L(A_{1/L}, A_{2/L}) \to \mathrm{Hom}_{L'}(A_{1/L'}, A_{2/L'})$ *is an isomorphism.*

In other words, there are no "new" geometric morphisms between A_1 and A_2 over an extension of L which don't already show up over L (or equivalently, over R).

PROOF. By the first part of Theorem 2.1, we are reduced to the case in which L' is a separable closure of L, and by direct limit considerations we can even assume L' is a finite separable extension of L which we may moreover suppose to be Galois. Let $\Gamma = \mathrm{Gal}(L'/L)$. By Galois descent, it suffices to show that $\sigma^*(f) = f$ for any $f : A_{1/L'} \to A_{2/L'}$ and any $\sigma \in \Gamma$. Let A'_j denote the Néron model of $A_{j_{/L'}}$ over the integral closure R' of R in L' (so R' is a Dedekind domain which is finite over R), so $A'_j = A_j \otimes_R R'$. We let $F : A'_1 \to A'_2$ denote the map induced by f on Néron models, so it suffices to prove $\sigma^*(F) = F$ for any $\sigma \in \Gamma$. The crucial role of the hypothesis that R is complete is that it ensures R' is again *local* (i.e., a discrete valuation ring).

By the second part of Theorem 2.1, it suffices to check equality on the closed fiber over R'. But since R' has residue field which is purely inseparable over R (so σ reduces to the identity!), it is immediate that $\sigma^*(F)$ and F coincide on closed fibers over R'. □

Throughout all that follows, the prime p will be fixed and we will write F to denote the fraction field of W. We will also write $\underline{X}$ to denote $X_{/W}$, and π to denote a uniformizer of W.

We now record an important immediate consequence of the above considerations (upon recalling how correspondences act on rational points, via pushforward and pullback).

COROLLARY 2.7. *Fix a positive integer m relatively prime to Np. Consider the Hecke divisor $T_m(\underline{x}_{/F})$ in $X_{/F}$. Its scheme-theoretic closure in $X_{/W}$ is a sum of sections lying inside of the relative smooth locus.*

By "sum of sections" we mean in the sense of relative effective Cartier divisors, which is to say that we take products of ideal sheaves (all of which are invertible, thanks to the sections lying in the relative smooth locus). It is immediate from the corollary, that the closed fiber of this closure is computed exactly by the habitual "Hecke correspondence" formula on the level of geometric points. Thus, this scheme-theoretic closure may be denoted $T_m(\underline{x})$ without risk of confusion.

PROOF. We have to prove that all points in $T_m(\underline{x}_{/F})$ are F-rational with corresponding W-point in the *smooth* locus of $\underline{X}$ over W. It is here that one uses that m is prime to Np, the crux of the matter being that all prime-to-p torsion on elliptic curves over W is finite étale and hence constant (since W has separably closed residue field) and all prime-to-N isogenies induce isomorphisms on level-N structures over any base. Thus, all level structures of interest can be defined over W and we just have to show that if $\underline{z}$ is a $\Gamma_0(N)$-structure over W which is m-isogenous to $\underline{x}$, then the resulting section $\mathrm{Spec}(W) \hookrightarrow \underline{X}$ lands in the smooth locus.

If $p \nmid N$ then $\underline{X}$ is W-smooth, so we're done. If $p|N$, then by [GZ, Ch. III, Prop. 3.1] the closed point $\underline{x}_0 \in \underline{X}(W/\pi)$ is an ordinary point in either the $(0,n)$ or $(n,0)$ component of the closed fiber $\underline{X}_0$ (where $n = \mathrm{ord}_p(N) \geq 1$). Since $\underline{z}_0$ is m-isogenous to $\underline{x}_0$ with $\gcd(m,N) = 1$, we see that the $\Gamma_0(N)$-structure $\underline{z}_0$ is another ordinary point on the same component as $\underline{x}_0$ and in particular is a smooth point. Thus, $\underline{z}$ lies in the relative smooth locus. □

The analogue of this corollary when $p|m$ is much more subtle, and will be treated in Section 6.

3. The Serre–Tate Theorem and the Grothendieck Existence Theorem

The main technical problem in the arithmetic intersection component of [GZ] is to compute various intersection numbers by means of counting morphisms between deformations of elliptic curves. We will have two elliptic curves E and E' over a complete local noetherian ring $(R, \mathfrak{m})$ (often artin local or a discrete valuation ring) and will want to count the size of the finite set $H_n = \mathrm{Hom}_{R_n}(E_n, E'_n)$, where $(\cdot)_n$ denotes reduction modulo $\mathfrak{m}^{n+1}$. By the last line of Theorem 2.1(2), the natural "reduction" map $H_{n+1} \to H_n$ (corresponding to base change by $R_{n+1} \twoheadrightarrow R_n$) is injective for every $n \geq 0$, so we can view the H_n's as a decreasing sequence of subgroups of $H_0 = \mathrm{Hom}_k(E_0, E'_0)$ (where $k = R/\mathfrak{m}$). When k has positive characteristic p, the Serre–Tate theorem (to be stated in Theorem 3.3) will "compute" these Hom-groups H_n between elliptic curves in terms of Hom-groups between p-divisible groups. In fact, the Serre–Tate theorem does much more: it essentially identifies the deformation theory of an elliptic curve with that of its p-divisible group. It will turn out that p-divisible groups are more tractable for our counting purposes, by means of the theory of formal groups (a theory which is closely connected to that of p-divisible groups in the situations of interest).

Once we have a way to work with the H_n's by means of p-divisible groups via the Serre–Tate theorem, it will still be important to understand one further issue:

what is the intersection of the H_n's inside of H_0? For example, the faithfulness at the end of Theorem 2.1(2) yields a natural inclusion

$$\mathrm{Hom}_R(E, E') \subseteq \bigcap H_n \simeq \varprojlim H_n$$

inside of H_0. Using a special case of the Grothendieck existence theorem, to be recorded in Theorem 3.4, this inclusion is an equality.

Let us now set forth the general context in which the Serre–Tate theorem takes place. Since the specificity of elliptic curves leads to no essential simplifications on the argument which is used for abelian schemes, we will work in the more general setting of abelian schemes (although only the case of elliptic curves intervenes in [GZ]). We recommend [Mum] as a basic reference for abelian varieties and [GIT, Ch. 6] as a basic reference for abelian schemes. Let S be a scheme, and A, A' two abelian schemes over S. Assume that a prime number p is locally nilpotent on S (i.e., every point of S has residue field of characteristic p). The case of most interest will be $S = \mathrm{Spec}(W/\pi^{n+1})$ where W is either a ring of Witt vectors or some finite discrete valuation ring extension thereof (with uniformizer π), though more general base rings are certainly needed for deformation theory arguments.

Let $\Gamma = A[p^\infty]$ and $\Gamma' = A'[p^\infty]$ denote the p-divisible groups arising from the p-power torsion schemes on A and A' over S. The problem considered by Serre and Tate was that of lifting morphisms of abelian schemes through an infinitesimal thickening of the base. More specifically, let $S_0 \hookrightarrow S$ be a closed immersion with defining ideal sheaf $\mathscr{I}$ satisfying $\mathscr{I}^N = 0$ for some positive integer N. Consider the problem of lifting an element in $\mathrm{Hom}_{S_0}(A_0, A_0')$ to an element in $\mathrm{Hom}_S(A, A')$. Let's first record that this lifting problem has a unique solution if it has any at all, and that a similar uniqueness holds for p-divisible groups (not necessarily coming from abelian schemes):

LEMMA 3.1. *Let $S_0 \hookrightarrow S$ be as above. Let A and A' be arbitrary abelian schemes over S, and let Γ and Γ' be arbitrary p-divisible groups over S. Then the natural maps $\mathrm{Hom}_S(A, A') \to \mathrm{Hom}_{S_0}(A_0, A_0')$ and $\mathrm{Hom}_S(\Gamma, \Gamma') \to \mathrm{Hom}_{S_0}(\Gamma_0, \Gamma_0')$ are injective.*

PROOF. In both cases we may assume S is local. By the last line of Theorem 2.1(2), base change to the residue field is a faithful functor on abelian schemes over a local ring, so the case of abelian schemes is settled by means of base change to the residue field. For p-divisible groups, use [K, Lemma 1.1.3(2)] (which also handles the abelian scheme case). □

As was mentioned above, one aspect of the Serre–Tate theorem is that it identifies solutions to the infinitesimal lifting problem for morphisms of abelian schemes with solutions to the analogous problem for their p-divisible groups. In order to put ourselves in the right frame of mind, we first record the fact that abelian scheme morphisms can be safely viewed as morphisms between p-divisible groups:

LEMMA 3.2. *If A and A' are abelian schemes over a base S, and Γ and Γ' are the associated p-divisible groups for a prime number p, then the natural map $\mathrm{Hom}_S(A, A') \to \mathrm{Hom}_S(\Gamma, \Gamma')$ is injective.*

Note that this lemma does not impose any requirements on residue characteristics at points in S. We call a map $\Gamma \to \Gamma'$ *algebraic* if it arises from a (necessarily unique) map $A \to A'$.

PROOF. We can assume S is local, and by the last line in Theorem 2.1(2) we can even assume $S = \mathrm{Spec}(k)$ for a field k which we may assume is algebraically closed. If p is distinct from the characteristic of k, then the injectivity can be proven by a relative schematic density argument (this is needed in the proof of Theorem 2.1, and uses results from [EGA, IV_3, 11.10]). If p coincides with the characteristic of k, then under the equivalence of categories between connected p-divisible groups over k and commutative formal groups of finite height over k, the connected component of the p-divisible group of an abelian variety over k is identified with the formal group of the abelian variety. Thus, if $f, g : A \to A'$ satisfy $f[p^\infty] = g[p^\infty]$ then f and g induce the same maps $\widehat{\mathscr{O}}_{A',0} \to \widehat{\mathscr{O}}_{A,0}$ on formal groups at the origin. Thus, $f = g$. □

As an example, if E is an elliptic curve over $\mathbf{Q}_p$ with good reduction over $\mathbf{Z}_p$ and $\mathscr{E}$ is the Néron model of E, then $\mathrm{End}_{\mathbf{Q}_p}(E) = \mathrm{End}_{\mathbf{Z}_p}(\mathscr{E}) \subseteq \mathrm{End}_{\mathbf{Z}_p}(\mathscr{E}[p^\infty])$. We want to describe the image of this inclusion. More generally, suppose we are given a map $f : \Gamma \to \Gamma'$ between the p-divisible groups of two abelian schemes A and A' over S, and assume its reduction f_0 over S_0 is algebraic (in the sense that it comes from a map $A_0 \to A'_0$ which is then necessarily unique). The theorem of Serre and Tate asserts that f is also algebraic (and more):

THEOREM 3.3. (Serre–Tate) *The natural commutative square (with injective arrows)*

$$\begin{array}{ccc} \mathrm{Hom}_S(A, A') & \longrightarrow & \mathrm{Hom}_S(\Gamma, \Gamma') \\ \downarrow & & \downarrow \\ \mathrm{Hom}_{S_0}(A_0, A'_0) & \longrightarrow & \mathrm{Hom}_{S_0}(\Gamma_0, \Gamma'_0) \end{array}$$

is cartesian. In other words, a map $f_0 : A_0 \to A'_0$ lifts to a map $f : A \to A'$ if and only if the induced map $f_0[p^\infty] : \Gamma_0 \to \Gamma'_0$ lifts to a map $h : \Gamma \to \Gamma'$, in which case f and h are unique and $f[p^\infty] = h$.

Moreover, if the ideal sheaf $\mathscr{I}$ of S_0 in S satisfies $\mathscr{I}^N = 0$ for some N and we are just given an abelian scheme A_0 over S_0 and a p-divisible group Γ over S equipped with an isomorphism $\iota_0 : \Gamma_0 \simeq A_0[p^\infty]$, then there exists an abelian scheme A over S and an isomorphism $\iota : \Gamma \simeq A[p^\infty]$ lifting ι_0. The pair (A, ι) is unique up to unique isomorphism.

PROOF. For an exposition of Drinfeld's elegant proof of this theorem, see [K, § 1]. This proof uses the point of view of fppf abelian sheaves. □

In words, Theorem 3.3 says that the infinitesimal deformation theory of an abelian scheme coincides with the infinitesimal deformation theory of its p-divisible group when all points of the base have residue characteristic p. We stress that it is crucial for the Serre–Tate theorem that all points of S have residue characteristic p and that we work with *infinitesimal* deformations. For example, p-divisible groups are étale away from points of residue characteristic p, so there are no obstructions to infinitesimal deformations of p-divisible group maps at such points, whereas there can certainly be obstructions to infinitesimally deforming abelian scheme maps at such points (e.g., supersingular elliptic curves over $k = \overline{\mathbf{F}}_p$ always lift to $W(k)$, and it follows from Corollary 3.5 below that some of the endomorphisms in characteristic p cannot lift through all infinitesimal levels of the deformation to $W(k)$).

One reason for the significance of Theorem 3.3 for our purposes is that if A and A' are abelian schemes over a complete local noetherian ring R of residue characteristic $p > 0$ (the case of elliptic curves being the only one we'll need), then $\mathrm{Hom}_{R_n}(A_n, A'_n)$ is naturally identified with the group of "algebraic" elements in $\mathrm{Hom}_k(\Gamma_0, \Gamma'_0)$ which lift (necessarily uniquely) to $\mathrm{Hom}_{R_n}(\Gamma_n, \Gamma'_n)$. This is particularly interesting when k is algebraically closed and we are working with $A = A'$ equal to a supersingular elliptic curve E_0 over k. In such a case, Γ_0 "is" the unique commutative formal group over k of dimension 1 and height 2, so $\mathrm{End}_k(\Gamma_0)$ is identified with the maximal order in the unique quaternion division algebra over $\mathbf{Q}_p$. Thus, problems in the deformation theory of endomorphisms are transformed into problems in quaternion algebras (and some such problems are solved in [Gr1], as we shall use later on).

The Grothendieck existence theorem addresses the problem of realizing a compatible family of infinitesimal deformations of a proper variety as the reductions of a common noninfinitesimal deformation.

THEOREM 3.4. (Grothendieck) *Let R be a noetherian ring which is separated and complete with respect to the I-adic topology for an ideal I. Let X and Y be proper R-schemes, and R_n, X_n, Y_n the reductions modulo I^{n+1}. The natural map of sets* $\mathrm{Hom}_R(X, Y) \to \varprojlim \mathrm{Hom}_{R_n}(X_n, Y_n)$ *is bijective.*

Moreover, if $\{X_n\}$ is a compatible system of proper schemes over the R_n's and $\mathscr{L}_0$ is an ample line bundle on X_0 which lifts compatibly to a line bundle $\mathscr{L}_n$ on each X_n, then there exists a pair $(X, \mathscr{L})$ consisting of a proper R-scheme and ample line bundle which compatibly reduces to each $(X_n, \mathscr{L}_n)$, and this data over R is unique up to unique isomorphism.

PROOF. See [EGA, III_1, § 5] for the proof and related theory. □

The first part of Grothendieck's theorem identifies the category of proper R-schemes with a full subcategory of the category of compatible systems $\{X_n\}$ of proper schemes over the R_n's, and it is really the second part of the theorem which is usually called the existence theorem (as it asserts the existence of an R-scheme giving rise to specified infinitesimal data over the R_n's).

It is not true that all compatible systems $\{X_n\}$ of proper schemes over the R_n's actually arise from a proper scheme X over R, and those which do are usually described as being *algebraizable.* Since the setup in Grothendieck's theorem is compatible with fiber products, to give compatible group scheme structures on the X_n's is equivalent to giving a group scheme structure on X. Also, various fundamental openness results from [EGA, IV$_3$, § 11–12] ensure that for many important properties **P** of morphisms, a map $X \to Y$ of *proper* R-schemes has property **P** if and only if each $X_n \to Y_n$ does. For example, if each X_n is R_n-smooth of pure relative dimension d, then the same holds for X over R.

Since elliptic curves possess canonical ample line bundles (namely, the inverse ideal sheaf of the identity section), we see that Grothendieck's theorem provides the conceptual explanation (i.e., independent of Weierstrass equations) for why a compatible system of elliptic curve deformations over the R_n's uniquely lifts to an elliptic curve over R. In higher relative dimension, formal abelian schemes need not be algebraizable. However, when given two abelian schemes A and A' over R we conclude from Grothendieck's theorem that

$$\mathrm{Hom}_R(A, A') = \bigcap \mathrm{Hom}_{R_n}(A_n, A'_n) \subseteq \mathrm{Hom}_{R/\mathfrak{m}}(A_0, A'_0).$$

Combining the Grothendieck existence theorem with the Serre–Tate theorem, we arrive at the following crucial result (to be applied in the case of elliptic curves over a complete discrete valuation ring W).

COROLLARY 3.5. *Let $(R, \mathfrak{m})$ be a complete local noetherian ring with residue field $k = R/\mathfrak{m}$ of characteristic $p > 0$, and let A, A' be abelian schemes over R with associated p-divisible groups Γ and Γ'. Then*

$$\mathrm{Hom}_R(A, A') = \mathrm{Hom}_k(A_0, A'_0) \cap \left(\bigcap_n \mathrm{Hom}_{R_n}(\Gamma_n, \Gamma'_n) \right)$$

inside of $\mathrm{Hom}_k(\Gamma_0, \Gamma'_0)$.

Let $(R, \mathfrak{m})$ be as above, with perfect residue field k of characteristic $p > 0$. Let A_0 be an ordinary abelian variety over k. Since k is perfect, the connected-étale sequence $0 \to \Gamma_0^0 \to \Gamma_0 \to \Gamma_0^{\mathrm{et}} \to 0$ of $\Gamma_0 = E_0[p^\infty]$ is uniquely split. One lifting of Γ_0 to a p-divisible group over R is the split form, namely a product of the unique (up to unique isomorphism) liftings of Γ_0^0 and Γ_0^{et} to multiplicative and étale p-divisible groups over R. Such a split lifting involves no ambiguity in the splitting data. More precisely, if the connected-étale sequence of a p-divisible group Γ over R is a split sequence, then the splitting is unique. Indeed, any two splittings differ by a morphism from the étale part to the connected part, so it suffices to show $\mathrm{Hom}_R(\Gamma_1, \Gamma_2) = 0$ for an étale p-divisible group Γ_1 and a connected p-divisible group Γ_2 over R. We may make a local faithfully flat base change on R to get to the case where R has algebraically closed residue field, in which case Γ_1 is a product of $\mathbf{Q}_p/\mathbf{Z}_p$'s, so it suffices to show that $\bigcap p^n \cdot \Gamma_2(R) = 0$. Since Γ_2 is a formal Lie group in finitely many variables, with multiplication by

p given by $T_i \mapsto T_i^p + p(\cdot)$ in terms of formal parameters T_i, $\bigcap p^n \cdot \Gamma_2(R)$ vanishes because $\bigcap \mathfrak{m}^n = 0$ (as follows from Krull's intersection theorem for noetherian local rings). With this uniqueness of splittings in hand, we can make a definition:

DEFINITION 3.6. A *Serre–Tate canonical lift* of A_0 to R is a pair (A, ι) where A is an abelian scheme over R, ι is an isomorphism between $A_{/k}$ and A_0, and the connected-étale sequence of $A[p^\infty]$ is split.

It is immediate from the above theorems of Serre–Tate and Grothendieck that a Serre–Tate canonical lift is characterized by the condition that it is a lift of A_0 whose p-divisible group has a split connected-étale sequence over R, and that it is unique up to unique isomorphism (as a lift of A_0). Moreover, if A_0' is another ordinary abelian variety with Serre–Tate canonical lift (A', ι'), then $\mathrm{Hom}_R(A, A') \to \mathrm{Hom}_k(A_0, A_0')$ is a *bijection.* In general, when R is not artinian one cannot expect a Serre–Tate canonical lift to exist; the best one can do is get a formal abelian scheme. However, in the case of elliptic curves the existence is always satisfied.

4. Computing Naive Intersection Numbers

We are interested in computing the local intersection pairing

$$\langle \underline{x} - (\underline{\infty}), T_m(\underline{x}^{\sigma}) - T_m(\underline{0}) \rangle_v,$$

with $x \in X(H) \subseteq \underline{X}(W)$ a Heegner point and $\sigma \in \mathrm{Gal}(H/K)$. Serious complications will be caused by the possibility that $\underline{x}$ is a component of $T_m(\underline{x}^{\sigma})$. By [GZ, Ch. III, Prop. 4.3], the multiplicity of x as a component in $T_m(x^{\sigma})$ is $r_{\mathscr{A}}(m)$, the number of integral ideals of norm m in the ideal class $\mathscr{A}$ associated to $\sigma \in \mathrm{Gal}(H/K) \simeq \mathrm{Cl}_K$. Thus, we will first concentrate on the case $r_{\mathscr{A}}(m) = 0$, so all divisors of interest on $\underline{X}$ intersect properly. The formula we wish to establish will relate local intersection numbers with Isom-groups and Hom-groups for infinitesimal deformations of Heegner points. This section is devoted to some general preliminaries concerning the simplest situation: intersecting two sections $\underline{y}, \underline{y}'$ in $\underline{X}(W)$ whose generic fibers are distinct and whose closed points $\underline{y}_0$ and $\underline{y}_0'$ are assumed to be disjoint from the cuspidal divisor and supported in the smooth locus. Heegner points will play no role here.

Recall that a section through the regular locus on a proper flat curve over a field or discrete valuation ring is automatically supported in the smooth locus, so we could equivalently be assuming that the closed points of our two sections lie in the (noncuspidal part of the) regular locus of the closed fiber or that the sections lie in the (noncuspidal part of the) regular locus on $\underline{X}$. The most important instance of this setup is given by sections arising from Heegner points or components of prime-to-p Hecke divisors obtained from such points (as in Corollary 2.7).

For such sections $\underline{y}$ and $\underline{y}'$ we need to make an *additional assumption* that there exist $\Gamma_0(N)$-structure diagrams over W which actually induce $\underline{y}$ and $\underline{y}'$. This is an additional assumption because $\underline{X}$ is not a fine moduli scheme (but we'll prove in a moment that in many cases this additional assumption is satisfied). Note in particular (thanks to Theorem 2.6) that such models over W are unique up to nonunique isomorphism, so it is easily checked that subsequent considerations which use such models (e.g., Theorem 4.1 below) are intrinsic to the given sections $\underline{y}$ and $\underline{y}'$ in $\underline{X}(W)$.

In this setting it makes sense to consider the intersection number

$$(\underline{y}.\underline{y}') \stackrel{\text{def}}{=} \text{length}(\underline{y} \cap \underline{y}') \tag{4–1}$$

where $\underline{y} \cap \underline{y}'$ denotes a scheme-theoretic intersection inside of $\underline{X}$. Although general intersection theory on arithmetic curves usually works with degree 0 divisors and requires passing to a regular resolution of $\underline{X}$ and using a moving lemma there, the regular resolution process can be carried out without affecting the relative smooth locus. Thus, the geometric input needed for an intersection calculation on a regular resolution is the length as in (4–1). Taking the point of view of the right side of (4–1) avoids general intersection theory and hence makes sense with *no* assumption of regularity/smoothness. Thus, we could contemplate trying to establish a formula for this length with *no* regularity assumptions at all. However, we'll see that a regularity assumption is crucial if we want to avoid restrictions on automorphism groups at closed points. Here is the formula we wish to prove:

THEOREM 4.1. *Let $\underline{y}, \underline{y}' \in \underline{X}(W)$ be sections which intersect properly and reduce to regular noncuspidal points in the special fiber (and hence are supported in the relative smooth locus over W). Assume moreover that these sections are induced by (necessarily unique) respective $\Gamma_0(N)$-structures denoted y and y' over W. Then* $\text{length}(\underline{y} \cap \underline{y}') = \frac{1}{2}\sum_{n\geq 0} \#\text{Isom}_{W_n}(y', y)$, *where $W_n = W/\pi^{n+1}$.*

Before we prove Theorem 4.1, we should record the following result which explains why the condition of existence of "W-models" for $\underline{y}, \underline{y}' \in \underline{X}(W)$ is usually automatically satisfied.

LEMMA 4.2. *If $\underline{y} \in \underline{X}(W)$ is a section which is disjoint from the cuspidal locus and has* $\text{Aut}(\underline{y}_0) = \{\pm 1\}$, *then there exists a $\Gamma_0(N)$-structure over W which induces $\underline{y}$. Moreover, with no assumptions on* $\text{Aut}(\underline{y}_0)$, *the complete local ring of $\underline{X}$ at the closed point $\underline{y}_0$ of $\underline{y}$ is the subring of* $\text{Aut}(\underline{y}_0)$*-invariants in the universal deformation ring of $\underline{y}_0$, and $\{\pm 1\}$ acts trivially on this ring.*

Since W/π is algebraically closed, every noncuspidal point in $\underline{X}(W/\pi)$ is actually represented by a $\Gamma_0(N)$-structure over W/π (unique up to nonunique isomorphism). Thus, the concept of $\text{Aut}(\underline{y}_0)$ makes sense prior to the proof of the lemma (and the existence of a universal deformation ring follows from the theory of Drinfeld structures on elliptic curves).

Proof of Lemma 4.2. Since formal deformations of $\Gamma_0(N)$-diagrams are algebraizable (thanks to Theorem 3.4), it suffices to prove the assertion concerning universal deformation rings.

If one looks at how the coarse moduli scheme $\underline{X}$ is constructed from a fine moduli scheme $\underline{X}(\iota)$ (away from the cusps) upon adjoining enough prime-to-p level structure ι, and one notes that formation of coarse moduli schemes in our context commutes with flat base change (such as $\mathbf{Z} \to W$), it follows that the complete local ring at $\underline{y}_0$ on $\underline{X}$ is exactly the subring of $\Gamma = \mathrm{Aut}(\underline{y}_0)$-invariants in the complete local ring on $\underline{X}(\iota)$ at a point corresponding to $\underline{y}_0$ with supplementary ι-structure added. The justification of this assertion is a standard argument in deformation theory which we omit.

Since $\underline{X}(\iota)$ is a fine moduli scheme away from the cusps, its complete local rings are formal deformation rings. Combining this with the fact that étale level structure ι is "invisible" when considering formal deformations, we conclude that the complete local ring in question on $\underline{X}$ really is naturally identified with the Γ-invariants in the universal deformation ring of $\underline{y}_0$ (with the *natural* action of Γ!). It therefore remains to check that $-1 \in \Gamma$ acts trivially on this deformation ring. But since inversion uniquely lifts to all deformations of a $\Gamma_0(N)$-structure (other closed fiber automorphisms usually don't lift!), the triviality of the action of -1 on universal deformation rings drops out. □

Proof of Theorem 4.1. If $\underline{y}_0 \neq \underline{y}'_0$, both sides of the formula are 0 and we're done. If these closed points coincide, so $\underline{y}_0 \simeq \underline{y}'_0$ as $\Gamma_0(N)$-structures over W/π, we at least see that the automorphism groups of $\underline{y}_0$ and $\underline{y}'_0$ (by which we mean the automorphism groups of representative $\Gamma_0(N)$-structures over W/π) are abstractly isomorphic.

Let $z \in \underline{X}(W/\pi)$ be the common noncuspidal point arising from $\underline{y}_0$ and $\underline{y}'_0$. For later purposes let's also *fix* choices of isomorphisms of $\Gamma_0(N)$-structures $\underline{y}_0 \simeq z$ and $\underline{y}'_0 \simeq z$ in order to view $\underline{y}$ and $\underline{y}'$ (or more specifically *fixed* choices of W-models for these sections) as W-deformations of the $\Gamma_0(N)$-structure z.

Let A_z denote the universal deformation ring for z. This ring is regular of dimension 2 (as proven in [KM, Ch. 5]) and has a unique structure of W-algebra (compatibly with its residue field identification with W/π), so by commutative algebra the existence of a W-section (such as coming from $\underline{y}$ or $\underline{y}'$) forces A_z to be formally W-smooth, which is to say of the form $W[\![T]\!]$ as an abstract local W-algebra.

By Lemma 4.2, the complete local ring $\widehat{\mathscr{O}}_{\underline{X},z}$ is exactly the subring $A_z^{\overline{\Gamma}}$, where $\overline{\Gamma} = \mathrm{Aut}(z)/\{\pm 1\}$ acts naturally. Let $d = |\overline{\Gamma}|$ denote the size of $\overline{\Gamma}$. A crucial point is that $\overline{\Gamma}$ acts *faithfully* on A_z (i.e., the only elements of $\mathrm{Aut}(z)$ acting trivially on the deformation ring are the elements ± 1). This amounts to the assertion that ± 1 are the only automorphisms of z which lift to all deformations of z. That is, the generic deformation cannot have automorphisms other than

± 1. This follows from the 1-dimensionality of modular curves and the existence of elliptic curves with automorphism group $\{\pm 1\}$.

The finite group $\overline{\Gamma}$ of order d now acts faithfully on the formally smooth W-algebra $A_z \simeq W[\![T]\!]$, and if we define the "norm" $t = \mathrm{Norm}_{\overline{\Gamma}}(T) = \prod_{\gamma \in \overline{\Gamma}} \overline{\gamma}(T)$, then by [KM, p. 508] the subring $\widehat{\mathscr{O}}_{\underline{X},z}$ of $\overline{\Gamma}$-invariants is exactly the formal power series ring $W[\![t]\!]$ and [Mat, Thm. 23.1] ensures that the resulting finite map $W[\![t]\!] \to W[\![T]\!]$ between 2-dimensional regular local rings is necessarily *flat*. Moreover, using Artin's theorem on the subfield of invariants under a faithful action of a finite group on a field (such as $\overline{\Gamma}$ acting on the fraction field of $W[\![T]\!]$) we see that $\widehat{\mathscr{O}}_{\underline{X},z} \to A_z$ is finite flat of degree $d = |\overline{\Gamma}|$.

The two sections $\underline{y}, \underline{y}' : \widehat{\mathscr{O}}_{\underline{X},z} \twoheadrightarrow W$ both arise from specified deformations and hence (uniquely) lift to W-sections of A_z. We can choose T without loss of generality to cut out the section corresponding to the deformation $\underline{y}$, and we let T' correspond likewise to the deformation $\underline{y}'$. Define t' to be the $\overline{\Gamma}$-norm of T' in $\widehat{\mathscr{O}}_{\underline{X},z}$. By *definition*, we have $(\underline{y}.\underline{y}') = \mathrm{length}(\widehat{\mathscr{O}}_{\underline{X},z}/(t,t')) = (1/d)\,\mathrm{length}(A_z/(t,t'))$ since $\widehat{\mathscr{O}}_{\underline{X},z} \to A_z$ is finite *flat* of degree d.

By induction and short exact sequence arguments with lengths, one shows that if A is any 2-dimensional regular local ring and $a, a' \in A$ are two nonzero elements of the maximal ideal with $A/(a,a')$ of finite length and prime factorization $a = \prod p_i$, $a' = \prod q_j$ (recall A is a UFD), then $A/(p_i, q_j)$ has *finite* length for all i, j and $\mathrm{length}(A/(a,a')) = \sum_{i,j} \mathrm{length}(A/(p_i, q_j))$. Thus, in our setup we have

$$\begin{aligned}\mathrm{length}(A_z/(t,t')) &= \sum_{\gamma,\gamma' \in \overline{\Gamma}} \mathrm{length}(A_z/(\gamma(T), \gamma'(T'))) \\ &= d \sum_{\gamma' \in \overline{\Gamma}} \mathrm{length}(A_z/(T, \gamma'(T'))).\end{aligned}$$

Consequently, we get

$$(\underline{y}.\underline{y}') = \sum_{\gamma' \in \overline{\Gamma}} \mathrm{length}(A_z/(T, \gamma'(T'))) \qquad (4\text{–}2)$$

We have $A_z/T = W$ corresponding to the deformation $\underline{y}$, and $A_z/(T, \gamma'(T'))$ is of finite length and hence of the form $W/\pi^{k_{\gamma'}}$ for a unique positive integer $k_{\gamma'}$. If we let $\gamma'(\underline{y}')$ denote the deformation obtained from the $\Gamma_0(N)$-structure $\underline{y}'$ by composing its residual isomorphism with z with a representative automorphism $\widetilde{\gamma}' \in \mathrm{Aut}(z)$ for $\gamma' \in \overline{\Gamma} = \mathrm{Aut}(z)/\{\pm 1\}$, then $\gamma'(\underline{y}')$ corresponds to $A_z/\gamma'(T')$ and hence there exists a (unique!) isomorphism of *deformations* $\underline{y} \bmod \pi^k \simeq \gamma'(\underline{y}') \bmod \pi^k$ if and only if $k \leq k_{\gamma'}$. Equivalently, there exists a (unique) isomorphism $\underline{y} \bmod \pi^k \simeq \underline{y}' \bmod \pi^k$ lifting $\widetilde{\gamma}' \in \mathrm{Aut}(z)$ if and only if $k \leq k_{\gamma'}$.

Any isomorphism $\underline{y} \bmod \pi^k \simeq \underline{y}' \bmod \pi^k$ merely as *abstract* $\Gamma_0(N)$*-structures* (ignoring the deformation structure with respect to z) certainly either lifts a unique $\widetilde{\gamma}'$ or else its negative does (but not both!). Since passage to the mod π

fiber is *faithful* (Theorem 2.1(2)), so any $\Gamma_0(N)$-structure isomorphism between $\underline{y}$ and $\underline{y}'$ modulo π^k is uniquely detected as a compatible system of isomorphisms modulo π^n's for $n \leq k$, we conclude from (4–2) that $(\underline{y}.\underline{y}')$ counts exactly the sum over all $n \geq 1$ of the sizes of the sets $\mathrm{Isom}_{W_n}(\underline{y}, \underline{y}')$ up to identifying isomorphisms with their negatives. This is essentially just a jazzed-up way of interchanging the order of a double summation. Since no isomorphism can have the same reduction as its negative, we have completed the proof of Theorem 4.1 by purely deformation-theoretic means. □

5. Intersection Formula Via Hom Groups

The link between intersection theory and deformation theory is given by:

THEOREM 5.1. *Let $\sigma \in \mathrm{Gal}(H/K)$ correspond to the ideal class $\mathscr{A}$ of K, and assume that the number $r_{\mathscr{A}}(m)$ of integral ideals in $\mathscr{A}$ of norm m vanishes, where $m \geq 1$ is relatively prime to N. Then*

$$(\underline{x}.T_m(\underline{x}^\sigma)) = \frac{1}{2}\sum_{n\geq 0} |\,\mathrm{Hom}_{W_n}(\underline{x}^\sigma, \underline{x})_{\deg m}|. \tag{5–1}$$

REMARK 5.2. Recall that the hypothesis $r_{\mathscr{A}}(m) = 0$ says that $\underline{x}$ and $T_m(\underline{x}^\sigma)$ intersect properly, so Theorem 4.1 is applicable (thanks to the description of $T_m(\underline{x}^\sigma)$ as a sum of Cartier divisors in Corollary 2.7).

In order to prove Theorem 5.1, we have to treat three essentially different cases: p not dividing m, $p|m$ with p split in K, and $p|m$ with p inert or ramified in K. The first case will follow almost immediately from what we have already established, while the second case amounts to a careful study of ordinary elliptic curves and Serre–Tate canonical liftings, and the third case is a subtle variant on the second case where we have to deal with supersingular reduction and must replace Serre–Tate theory with Gross' variant as discussed in [Gr1]. In particular, the phrases "canonical" and "quasi-canonical" liftings will have different meanings depending on whether we are in the ordinary or supersingular cases (for $p|m$).

In this section we take care of the easy case $p \nmid m$ and the less involved case where $p|m$ but p splits in K (i.e., the ordinary case). Our proof will essentially be the one in [GZ], except we alter the reasoning a little bit to avoid needing to use explicit Hecke formulas on divisors. The supersingular case (i.e., p not split in K) will roughly follow the same pattern as the ordinary case, except things are a bit more technical (e.g., it seems unavoidable to make explicit Hecke computations on divisors) and hence we postpone such considerations until the next section.

Let's now take care of the easy case $p \nmid m$. In this case, all m-isogenies are finite étale, so by Corollary 2.7 we see that if $\{C\}$ denotes the (finite) set of order m subgroup schemes of $\underline{x}^\sigma{}_0$, each of these Cs uniquely lifts to any deformation of $\underline{x}^\sigma{}_0$ (thanks to the invariance of the étale site with respect to infinitesimal thickenings). As relative effective Cartier divisors in the relative smooth locus of $\underline{X}$ over W we get the equality

$$T_m \underline{x}^\sigma = \sum_C \underline{x}^\sigma{}_C$$

where $\underline{x}^\sigma{}_C$ denotes the quotient $\Gamma_0(N)$-structure on $\underline{x}^\sigma$ by the unique order m subgroup scheme over W lifting C modulo π (this makes sense since m is relatively prime to N).

By Theorem 4.1 and the symmetry of intersection products, we obtain the formula

$$(\underline{x}.T_m\underline{x}^\sigma) = \sum_C (\underline{x}.\underline{x}^\sigma{}_C) = \sum_{n\geq 0}\sum_C \tfrac{1}{2}|\mathrm{Isom}_{W_n}(\underline{x}^\sigma{}_C, \underline{x})| \qquad (5\text{–}2)$$

with all sums implicitly finite (i.e., all but finitely many terms vanish). Since any nonzero map between elliptic curves over W_n is automatically finite flat (thanks to the fiber-by-fiber criterion for flatness), we conclude that the inner sum over C's is equal to $\frac{1}{2}|\,\mathrm{Hom}_{W_n}(\underline{x}^\sigma, \underline{x})_{\mathrm{deg}m}|$. This gives Theorem 5.1 in case $p \nmid m$.

Now assume $p|m$, so we have $m = p^t r$ with $t, r \geq 1$ and $p \nmid r$. In particular, this forces $p \nmid N$, so the $\Gamma_0(N)$-structures are *étale* and $\underline{X}$ is a *proper smooth curve* over W. Hence, we have a good theory of Cartier divisors and intersection theory on $\underline{X}$ without needing to do any resolutions at all, and we can keep track of compatibilities with $\Gamma_0(N)$-structures merely by checking such compatibility modulo π. Also, we can work with Hecke correspondences directly on the level of "integral model" Cartier divisors (rather than more indirectly in terms of closures from characteristic 0 on abstract regular resolutions). These explications are important because much of our analysis will take place on the level of p-divisible groups and deformations thereof (for which level N structure is invisible), and we will have to do very direct analysis of Hecke correspondences on relative effective Cartier divisors in characteristic 0, characteristic p, and at the infinitesimal level. The a priori knowledge that a deformation of a $\Gamma_0(N)$-compatible map is *automatically* $\Gamma_0(N)$-compatible will be the reason that we can focus most of our attention on p-power torsion and not worry about losing track of morphisms within the category of $\Gamma_0(N)$-structures.

In $\underline{X}$ we have the equality of relative effective Cartier divisors $T_m(\underline{x}^\sigma) = T_{p^t}(T_r(\underline{x}^\sigma))$ with $T_r(\underline{x}^\sigma)$ a sum of various W-sections $\underline{z}$, thanks to Corollary 2.7. Thus,

$$(\underline{x}.T_m(\underline{x}^\sigma)) = \sum_{\underline{z}} (\underline{x}.T_{p^t}(\underline{z})). \qquad (5\text{–}3)$$

Before we focus our attention on the proof of Theorem 5.1 when $p|m$, we make one final observation. Using the connected-étale sequence for finite flat group schemes over W_n (such as the kernels of dual isogenies $\underline{x} \to \underline{x}^\sigma$ of degree m) and the fact that finite flat group schemes of prime-to-p order in residue characteristic p are automatically étale (and therefore constant when the residue field is algebraically closed), we see that both sides of Theorem 5.1 naturally break up into into sums over all $\underline{z}$'s. It suffices to establish the equality

$$(\underline{x}.T_{p^t}(\underline{z})) = \frac{1}{2}\sum_{n\geq 0} |\operatorname{Hom}_{W_n}(\underline{z},\underline{x})_{\deg p^t}| \tag{5–4}$$

for each irreducible component $\underline{z} \in \underline{X}(W)$ of $T_r(\underline{x}^\sigma)$.

We now prove (5–4) when p is split in K, which is to say that all the Heegner points (such as $\underline{x}$ and $\underline{x}^\sigma$) have ordinary reduction. The key observation is that the W-models corresponding to all such Heegner points must be (Serre–Tate) *canonical lifts* of their closed fibers, which is to say that the connected-étale sequences of their p-divisible groups over W are *split* (and noncanonically isomorphic to $\mathbf{Q}_p/\mathbf{Z}_p \times \mathbf{G}_m[p^\infty]$). To prove this, we first note that by Tate's isogeny theorem for p-divisible groups over W it suffices to check that on the generic fiber, the p-adic Tate modules underlying Heegner points over W are isomorphic to $\mathbf{Z}_p \times \mathbf{Z}_p(1)$ as Galois modules.

What we know from ordinary reduction and the fact that W has algebraically closed residue field is that these p-adic Tate modules must be extensions of $\mathbf{Z}_p$ by $\mathbf{Z}_p(1)$. In order to get the splitting, we use the fact that for any elliptic curve E over F with CM by K, the CM-action by K is defined over K and hence also over F, so the Galois action commutes with the *faithful* action of

$$\mathbf{Z}_p \otimes_{\mathbf{Z}} \mathscr{O}_K = \mathbf{Z}_p \otimes_{\mathbf{Z}} \operatorname{Hom}_F(E,E) \hookrightarrow \operatorname{Hom}_F(T_p(E), T_p(E)).$$

Thus, when p is split in K and E has good ordinary reduction then the action of the $\mathbf{Z}_p$-algebra $\mathbf{Z}_p \otimes_{\mathbf{Z}} \mathscr{O}_K \simeq \mathbf{Z}_p \times \mathbf{Z}_p$ gives rise to a decomposition of $T_p(E)$ into a direct sum of two Galois characters which moreover *must* be the trivial and p-adic cyclotomic characters, as desired. This proves that our Heegner points are in fact automatically the Serre–Tate canonical lifts of their ordinary closed fibers. Note that each section $\underline{z} \in \underline{X}(W)$ of $T_r(\underline{x}^\sigma)$ has p-adic Tate module *isomorphic* to that of the Heegner point $\underline{x}^\sigma$ and hence all such $\underline{z}$'s are Serre–Tate canonical lifts of their closed fibers (even though such $\underline{z}$'s generally are not Heegner points).

By the Serre–Tate theorem, we conclude

$$\operatorname{Hom}_W(\underline{x}^\sigma,\underline{x}) \hookrightarrow \operatorname{Hom}_{W/\pi}(\underline{x}^\sigma,\underline{x}) \tag{5–5}$$

is an isomorphism because on the p-divisible groups side it is obvious the endomorphisms of $\mathbf{Q}_p/\mathbf{Z}_p \times \mathbf{G}_m[p^\infty]$ over W/π uniquely lift to W (and recall that for the level-N structure we don't have to check anything once the homomorphism over the residue field respects this data).

The assumption $r_{\mathscr{A}}(m) = 0$ says that there are no degree m isogenies on *generic geometric fibers* between $\underline{x}^\sigma$ and $\underline{x}$, so there are no such isogenies over W thanks to Theorem 2.6. Since (5–5) is an isomorphism, there are no such isogenies modulo π^{n+1} for each $n \geq 0$. This renders the right side of Theorem 5.1 equal to 0. Thus, to prove the theorem we must prove that $(\underline{x}.T_{p^t}(\underline{z})) = 0$

for any point $\underline{z} \in \underline{X}(W)$ which is prime-to-p isogenous to $\underline{x}^\sigma$. In more explicit terms, we need to show that a $\Gamma_0(N)$-structure (over some finite extension F' of F) which is p-power isogenous to $\underline{z}_{/F'}$ (and hence has to have good reduction) must have corresponding closed subscheme in $\underline{X}$ which is *disjoint* from $\underline{x}$.

Assuming the contrary, suppose that $\underline{x}$ actually meets $T_{p^t}(\underline{z})$. Thus, some point in $T_{p^t}(\underline{z})$ (viewed as a $\Gamma_0(N)$-structure over a finite extension W' of W) would have to admit $\underline{x}_{/W'}$ as the canonical lift of its closed fiber. We have just seen that $\underline{x}$ is *not* one of the component points of $T_{p^t}(\underline{z})$, yet $\underline{x}$ as a $\Gamma_0(N)$-structure is the Serre–Tate canonical lift of its (ordinary) closed fiber. To get a contradiction, it therefore suffices to check that for any noncuspidal generic geometric point $\underline{z}$ on $\underline{X}$ which has good reduction and is the Serre–Tate canonical lift of its closed fiber, every point in the geometric generic fiber of of $T_{p^t}(\underline{z})$ has the Serre–Tate canonical lift of its closed fiber (with lifted $\Gamma_0(N)$-structure) as one of the (geometric generic) points of $T_{p^t}(\underline{z})$. Due to the level of generality of this claim (with respect to the hypotheses on $\underline{z}$), we can use induction on t and the recursive formula $T_{p^{t+1}} = T_p T_{p^t} - pT_{p^{t-1}}$ to reduce to the case $t = 1$.

By the very definition of T_p, we break up $T_p(\underline{z})_{/\overline{F}}$ (viewed on the level of Néron models of "elliptic curve moduli" over a sufficiently large extension of W) into two parts: the quotient by the order p connected subgroup and the p quotients by the étale order p subgroups. Since $\underline{z}$ is a Serre–Tate canonical lift, it has two canonical quotients (one with connected kernel and one with étale kernel) which are visibly Serre–Tate canonical lifts of their closed fibers. Moreover, these two quotients yield the two distinct geometric points which we see in $T_p(\underline{z})$ on the closed fiber divisor level. This completes the case when p is split in K.

6. Supersingular Cases with $r_{\mathscr{A}}(m) = 0$

We now treat the hardest case of Theorem 5.1: $p|m$ with p either inert or ramified in K. All elliptic curves in question will have (potentially) supersingular reduction. For as long as possible, we shall simultaneously treat the cases of p inert in K and p ramified in K. We will prove (5–4) for $\underline{z}$ in $T_r(\underline{x}^\sigma)$.

Let $K_p = K \otimes_{\mathbf{Q}} \mathbf{Q}_p$ be the completion of K at the unique place over p, so K_p is a quadratic extension *field* of $\mathbf{Q}_p$, with valuation ring $\mathscr{O} = \mathscr{O}_K \otimes_{\mathbf{Z}} \mathbf{Z}_p$. The p-divisible groups of our Heegner points over W are formal $\mathscr{O}$-modules of height 1 over W. Instead of the Serre–Tate theory of canonical lifts, we will have to use Gross' theory of canonical and quasi-canonical lifts for height 1 formal $\mathscr{O}$-modules. In this theory, developed in [Gr1], the role of canonical lifts is played by the unique (up to noncanonical isomorphism) height 1 formal $\mathscr{O}$-module over W, namely a Lubin–Tate formal group. The formal groups of $\underline{x}$, $\underline{x}^\sigma$, and $\underline{z}$ all admit this additional "module" structure (replacing the property of being Serre–Tate canonical lifts in the ordinary case treated above). The points in the $T_{p^t}(\underline{z})$'s will correspond to various quasi-canonical lifts in the sense of [Gr1], and working out the field/ring of definition of such $\Gamma_0(N)$-structures will play an essential role in the intersection theory calculations in the supersingular case.

Our first order of business is to give a precise description of $T_{p^t}(\underline{z})$ for *any* section $\underline{z} \in \underline{X}(W)$ which is supported away from the cuspidal locus and is represented by a $\Gamma_0(N)$-structure $\phi : E \to E'$ over W for which the "common" p-divisible group of E and E' is endowed with a structure of Lubin–Tate formal $\mathscr{O}$-module over W (note ϕ is an N-isogeny and $p \nmid N$, so ϕ induces an isomorphism on p-divisible groups). The $\underline{z}$'s we care about will be known to have such properties due to their construction from Heegner points, but in order to keep straight what really matters we avoid assuming here that $\underline{z}$ is a CM point or even that $\underline{z}$ has any connection to Heegner points at all.

Since W is the completion of the maximal unramified extension of K, by Lubin–Tate theory (or more specifically local class field theory for K at the unique place over the ramified/inert p) we know that the Galois representation $\chi : G_F \to \mathscr{O}^\times$ on the generic fiber p-divisible group associated to a Lubin–Tate formal $\mathscr{O}$-module of height 1 is surjective and (due to the uniqueness of such formal $\mathscr{O}$-modules up to nonunique isomorphism) this character is independent of the specific choice of such formal group. That is, by thinking in terms of this character χ we are dealing with a canonical concept. One may object that the data of the $\mathscr{O}$-action on the p-divisible group of $\underline{z}$ is extra data which we have imposed, but we'll see that the "counting" conclusions we reach will not depend on this choice. Note also that in local class field theory it is shown that χ is the reciprocal of the reciprocity map for K_p (if one associates local uniformizers to arithmetic Frobenius elements; if one adopts the geometric Frobenius convention of Deligne then χ is the reciprocity map). This will lead us to the connection with local ring class fields over K_p, a connection of paramount importance in the proof of Theorem 6.4 below.

In order to describe $T_{p^t}(\underline{z})$ where $\underline{z} = (\phi : E \to E')_{/W}$, we first determine the geometric generic points of $T_{p^t}(\underline{z})$, which is to say that we consider the situation over $\overline{F}$. By definition as a divisor on $\underline{X}_{/\overline{F}}$,

$$T_{p^t}(\underline{z})_{/\overline{F}} = T_{p^t}(\underline{z}_{/\overline{F}}) = \sum_C (\phi_C : E/C \to E'/\phi(C))$$

where C runs over all order p^t subgroups of E and ϕ_C is the naturally induced map (still an N-isogeny since $p \nmid N$). We break up the collection of C's into collections based on the largest $u \geq 0$ for which C contains $E[p^u]$. We have the natural isomorphism $E/E[p^u] \simeq E$ which carries C over to a cyclic subgroup of order p^{t-2u}, and we note that ϕ carries $E[p^u]$ isomorphically over to $E'[p^u]$. Letting $s = t - 2u \geq 0$, we therefore get

$$T_{p^t}(\underline{z})_{/\overline{F}} = \sum_{\substack{s \equiv t \bmod 2 \\ 0 \leq s \leq t}} \sum_{|C| = p^s} (\phi_C : E/C \to E'/\phi(C))$$

with the inner sum now taken over *cyclic* subgroups of order p^s.

We wish to determine which of the $\overline{F}$-points on $\underline{X}$ in this sum correspond to a common closed point on $\underline{X}_{/F}$. By Galois theory, this amounts to working out the $\mathrm{Gal}(\overline{F}/F)$-orbits on these points, where $\sigma \in \mathrm{Gal}(\overline{F}/F)$ acting through functoriality on $\underline{X}(\overline{F})$ carries a point ϕ_C to the point $\phi_{\sigma(C)}$ where σ acts naturally on $C \subseteq E(\overline{F})$ (and $\sigma(\phi(C)) = \phi(\sigma(C))$ since ϕ is a morphism over F). Fortunately, $\sigma(C)$ is something we can compute because $C \subseteq E[p^\infty](\overline{F})$ and G_F acts on $E[p^\infty](\overline{F})$ through the character χ! Since χ is a surjective map onto $\mathscr{O}^\times$, the problem of working out Galois orbits (for which we may fix s) is exactly the problem of computing orbits for the natural $\mathscr{O}^\times$-action on the $p^{s-1}(p+1)$ cyclic subgroups of order p^s in $\mathscr{O}/p^s \simeq (\mathbf{Z}/p^s)^{\oplus 2}$ when $s \geq 1$ (the case $s = 0$ is trivial: in this case t is even and there is a single orbit corresponding to the subgroup $E[p^{t/2}]$, which is to say the point $\underline{z}$).

First suppose p is inert in K, so $\mathscr{O}/p$ is a field (of order p^2). Every element of additive order p^s in $\mathscr{O}/p^s$ is a multiplicative unit, so there is a unique orbit. This orbit is a closed point whose degree over F is 1 when $s = 0$ and $p^{s-1}(p+1)$ when $s > 0$ (since for $s > 0$ we just saw that the orbit contains $p^{s-1}(p+1)$ distinct geometric points). This yields

$$T_{p^t}(\underline{z}) = \sum_{\substack{0 \leq s \leq t, \\ s \equiv t \bmod 2}} \underline{y}(s) \tag{6–1}$$

with each "horizontal divisor" $\underline{y}(s)$ having its generic fiber equal to a single closed point in $\underline{X}_{/F}$ of degree $p^s + p^{s-1}$ over F for $s > 0$ and of degree 1 over F for $s = 0$.

From the local class field theory interpretation of the preceding calculation, we see that the generic point of $\underline{y}(s)$ is defined over an abelian extension of F which is obtained from the abelian extension of K_p whose norm group has image in $(\mathscr{O}/p^s)^\times$ which is spanned by the stabilizer $(\mathbf{Z}/p^s)^\times$ of a "line". This corresponds to the subgroup $\mathbf{Z}_p^\times \cdot (1 + p^s\mathscr{O}) = (\mathbf{Z}_p + p^s\mathscr{O})^\times$ of index $(p+1)p^{s-1}$ in $\mathscr{O}^\times$. If $F(s)/F$ denotes the corresponding finite extension (abelian over K_p), we see see that the field of definition of $\underline{y}(s)_{/F}$ is exactly $\mathrm{Spec}(F(s))$. We let $W(s)$ denote the corresponding valuation ring and let π_s be a uniformizer, so $\underline{y}(s)$ corresponds to an element in $\underline{X}(W(s))$. It is clear from the construction that the p-divisible group underlying $\underline{y}(s)$ is *exactly* a level s quasi-canonical deformation (in the sense of [Gr1]) of the formal $\mathscr{O}$-module underlying $\underline{z}_0^{(p^t)} \simeq \underline{x}_0$. Moreover, by taking scheme-theoretic closure from the generic fiber (in conjunction with the generic fiber description of quasi-canonical deformations in [Gr1]) we see that *the point $\underline{y}(s) \in \underline{X}(W(s))$ arises from a $\Gamma_0(N)$-structure defined over $W(s)$*. We stress that from the point of view of coarse moduli schemes this is a slightly surprising fact (for which we will be very grateful in the proof of Theorem 6.4). By Theorem 2.6 such a $W(s)$-model is unique up to nonunique isomorphism.

Meanwhile, when p is ramified in K then if $\pi \in \mathscr{O}$ is a uniformizer (recall W is unramified over $\mathscr{O}$, so this is not really an abuse of notation), we see that $\mathscr{O}/p$

is a nonsplit $\mathscr{O}$-module extension of $\mathscr{O}/\pi = \mathbf{F}_p$ by $\pi\mathscr{O}/\pi^2\mathscr{O} \simeq \mathscr{O}/\pi$. For $s > 0$, among the $p^s + p^{s-1}$ cyclic subgroups of order p^s in $\mathscr{O}/p^s$ we have a natural decomposition into the p^{s-1} such subgroups whose generators map to nonzero elements in $\pi\mathscr{O}/\pi^2\mathscr{O} \subseteq \mathscr{O}/p$ and the p^s others whose generators have nonzero image in $\mathscr{O}/\pi$. In the first case the generators are unit multiples of π and in the second case the generators are units. Thus, both such clumps are orbits, so for each $s > 0$ with $s \equiv t \bmod 2$ we get an orbit of size p^s and an orbit of size p^{s-1}. When t is even, the case $s = 0$ also gives rise to another singleton orbit corresponding again to the point $\underline{z}$. For both options of the parity of t, we arrive at $t+1$ distinct closed points $\underline{y}(s)_{/F}$ on $\underline{X}_{/F}$ of degrees p^s over F for $0 \le s \le t$. In other words, we get

$$T_{p^t}(\underline{z}) = \sum_{0 \le s \le t} \underline{y}(s) \tag{6–2}$$

with the generic fiber of $\underline{y}(s)$ of degree p^s over F. These generic fiber closed points of $\underline{X}_{/F}$ have residue field identical to the finite abelian extension $F(s)$ of F corresponding (via local class field theory for K_p) to the subgroup $(\mathbf{Z}_p + \pi^s\mathscr{O})^\times$ in $\mathscr{O}^\times$. Once again we recover level s quasi-canonical deformations in the sense of [Gr1], and the points $\underline{y}(s) \in \underline{X}(W(s))$ are represented by $\Gamma_0(N)$-structures over $W(s)$.

The identities (6–1) and (6–2) are precisely the inert and ramified cases in [GZ, Ch. III, (5.2)]. In both the inert and ramified cases, for $s > 0$ the point $\underline{y}(s) \in \underline{X}(W(s))$ is represented by a $\Gamma_0(N)$-structure over the valuation ring $W(s)$ of the local ring class field of level s, with its p-divisible group endowed with a natural structure of formal module of height 1 over a suitable *nonmaximal* order in $\mathscr{O}$ (depending on s and the $\mathscr{O}$-structure specified on $\underline{z}$). More specifically, the explicit description above shows that for $s > 0$ these formal modules are *quasi-canonical* deformations (in the sense of [Gr1]) which are *not* "canonical" (i.e., not formal modules over $\mathscr{O}$).

When t is even, the point $\underline{y}(0)$ is $\underline{z}$ (use the isogeny $p^{t/2}$), corresponding to a "canonical" deformation of the closed fiber. When $s = 1$ in the inert case (so t is odd), the $\Gamma_0(N)$-structure $\underline{y}(1)_0$ over W/π is the quotient of $\underline{z}_0$ by its unique order p subgroup scheme, which is to say that it is the Frobenius base change $\underline{z}_0^{(p)}$ of the $\Gamma_0(N)$-structure $\underline{z}_0$ over W/π. Finally, when $s = 0$ with odd t in the ramified case, then $\underline{y}(0) \simeq \underline{z}^{\sigma_{\mathfrak{p}}}$ as $\Gamma_0(N)$-structures, where $\sigma_{\mathfrak{p}} \in \mathrm{Gal}(H/K)$ is the arithmetic Frobenius element at the unique prime $\mathfrak{p}$ of K over p. To see this assertion in the ramified case, we just have to construct a p-isogeny between $\underline{z}$ and $\underline{z}^{\sigma_{\mathfrak{p}}}$ (and then compose it with $p^{(t-1)/2}$ to realize $\underline{z}^{\sigma_{\mathfrak{p}}}$ as the unique W-section $\underline{y}(0)$ in $T_{p^t}(\underline{z})$. Since $\sigma_{\mathfrak{p}}$ lies in the decomposition group at $\mathfrak{p}$ and is the restriction to H of the arithmetic Frobenius automorphism of the completed maximal unramified extension F of $K_{\mathfrak{p}}$, the existence of a p-isogeny $\underline{z}^{\sigma_{\mathfrak{p}}} \to \underline{z}$ as $\Gamma_0(N)$-structures follows from Theorem 6.2 below (which does not depend on the preceding discussion).

The next two basic results will help us to construct elements of Hom-groups. This is essential in the proof of (5–4) when $p|m$.

THEOREM 6.1. *Let E and E' be elliptic curves over W with supersingular reduction and CM by $\mathcal{O}_K$, so p is inert or ramified in K. Let $f : E'_{/W_n} \to E_{/W_n}$ be an isogeny, with $n \geq 0$.*

- *If $n \geq 1$ in the inert case and $n \geq 2$ in the ramified case, then the map f has degree divisible by p^2 if and only if $f \bmod \pi^{n+1}/p = [p] \circ g$ for some other isogeny $g : E'_{/W/(\pi^{n+1}/p)} \to E_{/W/(\pi^{n+1}/p)}$.*
- *For any $n \geq 0$ the map $[p] \circ f$ always lifts (uniquely) to an isogeny over $W/p\pi^{n+1}$.*
- *If p is inert in K, $n = 0$, and f has degree pr for $r \geq 1$ not divisible by p, then f does not lift to an isogeny over W/π^2.*

Note that the uniqueness of liftings follows from the second part of Theorem 2.1 over the artinian quotients of W. The last part of the theorem is crucial for success in the inert case with t odd, and it is not needed for any other cases. It should also be noted that Theorem 6.1 rests on the fact that W is the completion of a maximal *unramified* extension of $\mathcal{O} = \mathbf{Z}_p \otimes_{\mathbf{Z}} \mathcal{O}_K$ (i.e., has algebraically closed residue field and *no* ramification over $\mathcal{O}$), as otherwise the theorem is generally false.

PROOF. Since W is the completion of the maximal unramified extension of $\mathcal{O}$, we may (and do) take our uniformizer π of W to come from $\mathcal{O}$.

For the first part of the theorem, we first want to prove that if f has degree divisible by p^2, then $f \bmod \pi^{n+1}/p$ factors through the isogeny $[p]$ on E' mod π^{n+1}/p. Let $\Gamma_{/W}$ denote a fixed Lubin–Tate formal $\mathcal{O}$-module of height 1 and let $R = \mathrm{End}_{W_0}(\Gamma_{/W_0})$, a maximal order in a quaternion algebra over $\mathbf{Q}_p$. We can (and do) fix isomorphisms of formal $\mathcal{O}$-modules $E[p^\infty] \simeq \Gamma$ and $E'[p^\infty] \simeq \Gamma$. By means of these isomorphisms, the map $f[p^\infty]$ induced by f on p-divisible groups is converted into an *endomorphism* of $\Gamma_{/W_n}$ as a formal group (*not* necessarily respecting the $\mathcal{O}$-structure). It is proven in [Gr1] that the endomorphism ring of $\Gamma_{/W_n}$ is exactly the subring $\mathcal{O} + \pi^n R$ inside of the endomorphism ring R of the closed fiber of Γ. In this way, f gives rise to an element $\alpha \in \mathcal{O} + \pi^n R$ with reduced norm $\mathrm{N}(\alpha)$ divisible by p^2. By the Serre–Tate theorem relating the deformation theory of elliptic curves with that of their p-divisible groups, as well as the fact that infinitesimal deformations are automatically compatible with $\Gamma_0(N)$-structures when $p \nmid N$ (provided such compatibility holds over the residue field), the condition that $f \bmod \pi^{n+1}/p$ factor through $[p]$ is exactly the statement that α as an element of $\mathcal{O} + (\pi^n/p)R$ be divisible by p (note that $\pi^n/p \in \mathcal{O}$ in the ramified case since we require $n \geq 2$ in this case). Now we are faced with a problem in quaternion algebras.

For the factorization aspect of the first part of the theorem, we aim to show for any $n \geq 1$ in the inert case and any $n \geq 2$ in the ramified case, $p(\mathcal{O} + (\pi^n/p)R) =$

$p\mathscr{O} + \pi^n R$ contains the elements $\alpha \in \mathscr{O} + \pi^n R$ of reduced norm divisible by p^2. Any element of R lies inside of the valuation ring $\mathscr{O}_L$ of a quadratic extension L of $\mathbf{Q}_p$ which lies inside of $R \otimes_{\mathbf{Z}_p} \mathbf{Q}_p$. Moreover, the reduced norm of such an element coincides with its relative norm from L down to $\mathbf{Q}_p$. An element of $\mathscr{O}_L$ with norm down to $\mathbf{Z}_p$ divisible by p^2 is certainly divisible by p in $\mathscr{O}_L$, so we can definitely write $\alpha = p\beta$ for some $\beta \in R$. Thus, for $\beta \in R$ with $p\beta \in \mathscr{O} + \pi^n R$ we must prove $\beta \in \mathscr{O} + (\pi^n/p)R$. Choose an element $r \in R$ such that $R = \mathscr{O} \oplus \mathscr{O}r$, so if $\beta = a + br$ with $a, b \in \mathscr{O}$ then $pb \in \pi^n\mathscr{O}$ and we want $b \in (\pi^n/p)\mathscr{O}$. This is obvious.

Now consider the assertion that $[p] \circ f$ lifts to $W/p\pi^{n+1}$ for any $n \geq 0$. Using the Serre–Tate theorem and the description in [Gr1] of endomorphism rings of Γ over W_r's as recalled above, we wish to prove that any element $\alpha \in \mathscr{O} + \pi^n R$ has the property that $p\alpha \in \mathscr{O} + p\pi^n R$. This is obvious.

Finally, consider the last part of the theorem. Once again using Serre–Tate and [Gr1], we want to show that if $\alpha \in R$ has reduced norm in $\mathbf{Z}_p$ equal to a unit multiple of p, then α does *not* lie in the subring $\mathscr{O} + \pi R$. Suppose that $\alpha \in \mathscr{O} + \pi R$. Since α is not a unit in R (as its norm down to $\mathbf{Z}_p$ is not a unit), its image in R/pR is not a unit. When p is inert in K, so we may take $\pi = p$, the non-unit image of $\alpha \in \mathscr{O} + pR$ in R/pR lies inside of the field $\mathscr{O}/p$ and hence α has vanishing image in R/pR. We conclude that $\alpha \in pR$, so α has reduced norm divisible by p^2, contrary to hypothesis. □

The case of ramified p and odd t will also require the following theorem.

THEOREM 6.2. *Let E and E' be as in Theorem* 6.1, *but assume p is ramified in K. Let φ denote the Frobenius endomorphism of the completion W of a maximal unramified extension of $\mathscr{O}$. Let $(\cdot)^\varphi$ denote the operation of base change by φ on W-schemes. For any positive integer d and any nonnegative integer n there is a natural isomorphism of groups*

$$\mathrm{Hom}_{W_n}(E', E)_{\mathrm{deg} d} \simeq \mathrm{Hom}_{W_{n+1}}(E'^{\varphi}, E)_{\mathrm{deg} pd}. \tag{6–3}$$

Moreover, if we endow E and E' with $\Gamma_0(N)$-structures and give E'^{φ} the base change $\Gamma_0(N)$-structure, then this bijection carries $\Gamma_0(N)$-compatible maps to $\Gamma_0(N)$-compatible maps.

PROOF. As usual, we may (and do) choose a uniformizer of W to be taken from a uniformizer of $\mathscr{O} = \mathscr{O}_K \otimes_{\mathbf{Z}} \mathbf{Z}_p$. Consider the p-torsion $E'[p]$, a finite flat group scheme over W with order p^2. The action of $\mathscr{O}_K$ on this factors through an action of $\mathscr{O}_K/p = \mathscr{O}_K/\mathfrak{p}^2 \simeq \mathscr{O}/\pi^2$, where $\mathfrak{p}$ is the unique prime of $\mathscr{O}_K$ over p. Thus, the torsion subscheme $E'[\mathfrak{p}]$ coincides with the kernel of the multiplication map by π on the formal $\mathscr{O}$-module $E'[p^\infty]$. In particular, this torsion subscheme is finite flat of order p over W. Let $E'' = E'/E'[\mathfrak{p}]$, so there is a natural degree p isogeny $E' \to E''$ whose reduction is uniquely isomorphic to the relative Frobenius map on the closed fiber of E'. Since $p \nmid N$ there is a unique $\Gamma_0(N)$-structure on

E'' compatible via $E' \to E''$ with a given $\Gamma_0(N)$-structure on E'. Consider the degree p dual isogeny $\psi : E'' \to E'$. Let $\psi_e = \psi \bmod \pi^{e+1}$ for any $e \geq 0$.

We claim that if $f : E'_{/W_n} \to E_{/W_n}$ is an isogeny then $f \circ \psi_n$ lifts to W_{n+1} (with degree clearly divisible by p), and conversely every isogeny over W_{n+1} of degree divisible by p arises via this construction (from a necessarily unique f). This will establish the theorem with E'' in the role of E'^{φ} (with the $\Gamma_0(N)$-aspect easily checked by unwinding the construction process), and we will then just have to construct an isomorphism of elliptic curves $E'' \simeq E'^{\varphi}$ over W compatible with the recipe for $\Gamma_0(N)$-structures induced from E'. Using Serre–Tate, our problem for relating E'' and E is reduced to one on the level of p-divisible groups.

If $\Gamma = E'[p^\infty]$, so Γ is a formal $\mathscr{O}$-module of height 1 over W, and we write R for $\mathrm{End}_{W_0}(\Gamma)$, then multiplication by π makes sense on Γ (whereas it does *not* make sense on the level of elliptic curves in case $\mathfrak{p}$ is not principal). Since all formal $\mathscr{O}$-modules of height 1 over W are (noncanonically) isomorphic, so there exists a W-isomorphism $E[p^\infty] \simeq \Gamma$, by using [Gr1] as in the proof of Theorem 6.1 we can reduce our lifting problem to showing that any element $\alpha \in \mathscr{O} + \pi^n R$ has the property that $\alpha\pi^\vee \in \mathscr{O} + \pi^{n+1}R$ (with $\pi^\vee$ corresponding to the dual isogeny to π with respect to the Cartier–Nishi self-duality of the p-divisible group of any elliptic curve, such as $E'_{/W}$), and conversely that any element of $\mathscr{O} + \pi^{n+1}R$ with reduced norm divisible by p necessarily has the form $\alpha\pi^\vee$ for some $\alpha \in \mathscr{O} + \pi^n R$.

The self-duality of E' induces complex conjugation on $\mathscr{O}_K$, and hence induces the unique nontrivial automorphism on $\mathscr{O}$ as a $\mathbf{Z}_p$-algebra. We conclude that $\pi^\vee$ is a unit multiple of π in $\mathscr{O}$. Thus, we are reduced to the algebra problem of proving that $(\mathscr{O} + \pi^n R)\pi \subseteq R$ is exactly the set of elements in $\mathscr{O} + \pi^{n+1}R$ with reduced norm divisible by p. Since $\pi R = R\pi$ (due to the uniqueness of maximal orders in finite-dimensional division algebras over local fields), we need to prove that $\pi\mathscr{O} + \pi^{n+1}R$ is the set of elements in $\mathscr{O} + \pi^{n+1}R$ with reduced norm divisible by p. Since π has reduced norm p and an element of $\mathscr{O} + \pi^{n+1}R$ not in $\pi\mathscr{O} + \pi^{n+1}R$ is a unit in R and hence has unit reduced norm, we're done.

It remains to construct a W-isomorphism $E'' \simeq E'^{\varphi}$ compatible with $\Gamma_0(N)$-recipes from E'. Over W/π we have a canonical isomorphism

$$\overline{E}'' \simeq \overline{E}' / \ker(\mathrm{Frob}) \simeq \overline{E}'^{(p)} \simeq \overline{E'^{\varphi}}$$

which is visibly "$\Gamma_0(N)$-compatible" (with respect to $\overline{E'}$). Thus, we just have to lift *this* isomorphism to W. By Serre–Tate, it suffices to make a lift on the level of p-divisible groups. For $\Gamma = E'[p^\infty]$ we have $E''[p^\infty] \simeq \Gamma/\Gamma[\pi]$ and $E'^{\varphi}[p^\infty] \simeq \Gamma^{\varphi}$, so we seek a W-isomorphism $\Gamma/\Gamma[\pi] \simeq \Gamma^{\varphi}$ lifting the canonical isomorphism $\overline{\Gamma}/\overline{\Gamma}[\pi] \simeq \overline{\Gamma}^{(p)}$ over W/π. In other words, we seek to construct a p-isogeny $\Gamma \to \Gamma^{\varphi}$ over W which lifts the relative Frobenius over W/π. This assertion is intrinsic to Γ. Since Γ is, up to noncanonical isomorphism over W, the unique formal $\mathscr{O}$-module of height 1 over W, it suffices to make such a p-isogeny lifting the relative Frobenius for *one* formal $\mathscr{O}$-module of height 1 over W. By base change compatibility, it suffices to do the construction for a single formal $\mathscr{O}$-module over $\mathscr{O}$. Since $\mathscr{O}$ has residue field $\mathbf{F}_p$, as p is ramified in $\mathscr{O}_K$, over $\mathscr{O}$ we seek a suitable *endomorphism* of a formal $\mathscr{O}$-module of height 1. Using a Lubin–Tate formal group $\mathscr{F}_{\pi,f}$ for a uniformizer π of $\mathscr{O}$ and the polynomial $f(X) = \pi X + X^p$, the endomorphism $[\pi]$ does the job. □

The nonzero contributions to (5–4) require some control on closed fibers, and this is provided by:

LEMMA 6.3. *For $\underline{z}$ in $T_r(\underline{x}^\sigma)$, the closed fiber of $T_{p^t}(\underline{z})$ is supported at a single point in $\underline{X}(W/\pi)$, corresponding to the $\Gamma_0(N)$-structure $\underline{z}_0$ when t is even and $\underline{z}_0^{(p)}$ when t is odd.*

PROOF. First consider the inert case. When t is even then $\underline{y}(0)$ makes sense and we have seen that its closed fiber is $\underline{z}_0$. When t is odd then $\underline{y}(1)$ makes sense and we have seen that its closed fiber is $\underline{z}_0^{(p)}$. Thus, for the inert case we just have to check that $\underline{y}(s)_0$ in $\underline{X}(W/\pi)$ only depends on s mod 2. From the explicit construction, we get $\underline{y}(s)_0$ from $\underline{z}$ as $\Gamma_0(N)$-structures over W/π by s-fold iteration of the process of passing to quotients by the unique order p subgroup scheme (i.e., the kernel of the relative Frobenius) in our elliptic curves. But over a field (such as W/π) of characteristic p, going through two steps of this process is exactly the same as passing to the quotient by p-torsion, so it brings us back to where we began! Thus, the construction only depends on s mod 2.

Now consider the ramified case. This goes essentially as in the preceding paragraph (where we barely used the property of p being inert). The only difference is that our divisor has contributions from both even and odd values of s. Looking back at where $\underline{y}(s)$ came from, we see that the contribution for $s \equiv t$ mod 2 is $\underline{z}_0$ for t even and $\underline{z}_0^{(p)}$ for t odd. Meanwhile, for $s \not\equiv t$ mod 2 our setup is obtained from an order p^{s+1} quotient on the $\Gamma_0(N)$-structure $\underline{z}_0$, so this again corresponds to $\underline{z}_0$ for t even and to $\underline{z}_0^{(p)}$ for t odd. □

One important consequence of Lemma 6.3 is that the intersection number

$$(\underline{x}.T_{p^t}(\underline{z}))$$

in (5–4) is nonzero if and only if $\underline{x}_0 \simeq \underline{z}_0$ for even t and if and only if $\underline{x}_0 \simeq \underline{z}_0^{(p)}$ for odd t (or in perhaps more uniform style, $\underline{x}_0 \simeq \underline{z}_0^{(p^t)}$). Here, isomorphisms are as $\Gamma_0(N)$-structures.

We need one more result before we can prove (5–4). The following theorem rests on the fact that the points $\underline{y}(s) \in \underline{X}(W(s))$ in the support of $T_{p^t}(\underline{z})$ are quasi-canonical liftings (in the sense of [Gr1]).

THEOREM 6.4. *For a point $\underline{z}$ in $T_r(\underline{x}^\sigma)$ and a point $\underline{y}(s)$ in the divisor $T_{p^t}(\underline{z})$ with $s > 0$, we have*

$$(\underline{x}.\underline{y}(s)) = \tfrac{1}{2}\left|\mathrm{Isom}_{W/\pi}(\underline{y}(s)_0, \underline{x}_0)\right|.$$

This result is [GZ, Ch. III, Prop. 6.1]. Our (very different) proof uses deformation theory.

PROOF. We can assume that there exists an isomorphism $\iota : \underline{x}_0 \simeq \underline{y}(s)_0$ as $\Gamma_0(N)$-structures over W/π, as otherwise both sides of the desired equation vanish. Now *fix such an isomorphism* ι.

Since the p-divisible group underlying $\underline{x}$ is a "canonical" lifting of its closed fiber (i.e., its endomorphism ring is a maximal order in a quadratic extension of $\mathbf{Z}_p$) while the p-divisible group underlying $\underline{y}(s)$ for $s > 0$ is merely a level s quasi-canonical lifting in the sense of [Gr1] (so its endomorphism ring is a nonmaximal order), by [Gr1, Prop. 5.3(3)] we see that for $s > 0$ the p-divisible group of $\underline{y}(s)$ over $W(s)$ is *not* isomorphic modulo π_s^2 to the p-divisible group of $\underline{x}_{/W(s)}$ *as deformations of* $\underline{x}_0$ (i.e., in a manner respecting the map induced by ι on p-divisible groups). Consider the universal formal deformation ring A for $\underline{x}_0$ on the category of complete local noetherian W-algebras with residue field W/π.

Abstractly $A \simeq W[\![t]\!]$, and we get $\operatorname{Spec}(W) \to \operatorname{Spec}(A)$ and $\operatorname{Spec}(W(s)) \to \operatorname{Spec}(A)$ corresponding to $\underline{x}$ and $\underline{y}(s)$ respectively. A key technical point is that both of these maps are *closed immersions* (i.e., the ring maps are surjective). In fact, the corresponding ring maps $A \to W$ and $A \to W(s)$ send t to a uniformizer (using Lubin–Tate theory for $\underline{x}$ and [Gr1, Prop. 5.3(3)] for $\underline{y}(s)$). Since the subscheme of $\operatorname{Spec}(A)$ cut out by $\underline{x}$ corresponds to the principal ideal generated by $t - \pi u$ for some unit $u \in W^\times$ and the map $A \twoheadrightarrow W(s)$ corresponding to $\underline{y}(s)$ sends t to $\pi_s u_s$ for a unit $u_s \in W(s)^\times$, we compute that the scheme-theoretic intersection of our two closed subschemes in $\operatorname{Spec}(A)$ is $\operatorname{Spec}(W(s)/(\pi_s u_s - \pi u)) \simeq \operatorname{Spec}(W(s)/\pi_s)$ since π/π_s lies in the maximal ideal of $W(s)$. In geometric terms, the closed subschemes $\underline{x}$ and $\underline{y}(s)$ in $\underline{X}$ are transverse.

Let $\overline{\Gamma} = \operatorname{Aut}(\underline{x}_0)/\{\pm 1\}$, so we want $d \stackrel{\text{def}}{=} |\overline{\Gamma}|$ to equal $(\underline{x}.\underline{y}(s))$ (assuming this intersection number is nonzero). From Lemma 4.2 and the proof of Theorem 4.1, we see that $\overline{\Gamma}$ acts faithfully on $A \simeq W[\![T]\!]$ and that $\widehat{\mathscr{O}}_{\underline{X},\underline{x}_0}$ is naturally identified with the subring of invariants $A^{\overline{\Gamma}} \simeq W[\![t]\!]$, where $t = \operatorname{Norm}_{\overline{\Gamma}}(T)$. Consider the map $A \to W(s)$. This map depends on ι, and we showed above that it is surjective. The action of $\overline{\Gamma}$ on A is compatible with making permutations on the choices of ι, so the given map $A \to W(s)$ sends $\gamma(T)$ to a uniformizer of $W(s)$ for all $\gamma \in \overline{\Gamma}$. In particular, the map $W[\![t]\!] \simeq \widehat{\mathscr{O}}_{\underline{X},\underline{x}_0} \to W(s)$ corresponding to $\underline{y}(s)$ sends t to an element of normalized order in $W(s)$ equal to $|\overline{\Gamma}| = d$, whence we see that the closed subscheme $\underline{y}(s) \hookrightarrow \underline{X}$ is *not* generally $\operatorname{Spec}(W(s))$ but rather corresponds to the order of index d in $W(s)$. Likewise, the surjective map $W[\![t]\!] \to W$ corresponding to $\underline{x}$ sends t to an element of normalized order d in W. Thus, the scheme-theoretic intersection $\underline{x} \cap \underline{y}(s)$ in $\underline{X}$ is the spectrum of the artinian quotient

$$W(s)/(\pi_s^d(\text{unit}) - \pi^d u). \tag{6–4}$$

Since $W(s)$ is totally ramified over W of degree $p^s + p^{s-1} > 1$ in the inert case and of degree $p^s > 1$ in the ramified case, we see that π_s^d has strictly smaller order than π^d in $W(s)$, so the artinian quotient (6–4) has length d, as desired. □

We are now in position to prove Theorem 5.1 when $p|m$.

When $s = 0$ occurs, we have $\underline{z} \neq \underline{x}$ when t is even. Indeed, since $\underline{z}$ is p^t-isogenous to itself (via $p^{t/2}$) and hence occurs as a point in $T_{p^t}(\underline{z}) \subseteq T_m(\underline{x}^\sigma)$, the vanishing of $r_{\mathscr{A}}(m)$ forces $\underline{z} \neq \underline{x}$. Thus, $\underline{y}(0) = \underline{z}$ in $\underline{X}(W)$ is distinct from $\underline{x}$ for even t. Likewise, when t is odd and the $s = 0$ case occurs (i.e., p is ramified in K), then $\underline{y}(0) = \underline{z}^{\sigma_{\mathfrak{p}}}$ is again distinct from $\underline{x}$. The reason is that $\underline{z}^{\sigma_{\mathfrak{p}}}$ is p-isogenous to $\underline{z}$, whence (by multiplying against $p^{(t-1)/2}$) is p^t-isogenous to $\underline{z}$, so if $\underline{y}(0) = \underline{x}$ then $\underline{x}$ would occur in $T_{p^t}(\underline{z}) \subseteq T_m(\underline{x}^\sigma)$, contradicting that $r_{\mathscr{A}}(m) = 0$. Theorem 4.1 therefore gives us a formula for $(\underline{x}.\underline{y}(0))$ in all cases when $s = 0$ occurs.

Combining this with Lemma 6.3 and Theorem 6.4, we can compute $(\underline{x}.T_{p^t}(\underline{z}))$ as a sum of various terms, with the contribution from $s = 0$ being treated separately. We get: for inert p,

$$(\underline{x}.T_m(\underline{z})) = \begin{cases} \frac{1}{2}\sum\limits_{n\geq 0} \left|\mathrm{Hom}_{W_n}(\underline{z},\underline{x})_{\deg 1}\right| + \frac{t}{2}\cdot\frac{1}{2}\left|\mathrm{Hom}_{W/\pi}(\underline{z},\underline{x})_{\deg 1}\right|, & t \text{ even},\\[2ex] \frac{t+1}{2}\cdot\frac{1}{2}\left|\mathrm{Hom}_{W/\pi}(\underline{z},\underline{x})_{\deg p}\right|, & t \text{ odd},\end{cases} \tag{6–5}$$

and for ramified p,

$$(\underline{x}.T_m(\underline{z})) = \begin{cases} \frac{1}{2}\sum\limits_{n\geq 0} \left|\mathrm{Hom}_{W_n}(\underline{z},\underline{x})_{\deg 1}\right| + t\cdot\frac{1}{2}\left|\mathrm{Hom}_{W/\pi}(\underline{z},\underline{x})_{\deg 1}\right|, & t \text{ even},\\[2ex] \frac{1}{2}\sum\limits_{n\geq 0} \left|\mathrm{Hom}_{W_n}(\underline{z}^{\sigma_{\mathfrak{p}}},\underline{x})_{\deg 1}\right| + t\cdot\frac{1}{2}\left|\mathrm{Hom}_{W/\pi}(\underline{z}^{\sigma_{\mathfrak{p}}},\underline{x})_{\deg 1}\right|, & t \text{ odd},\end{cases} \tag{6–6}$$

with the $\sum_{n\geq 0}(\ldots)$ term corresponding to $s = 0$ contributions. For odd t in the ramified case, note also that $\underline{z}^{\sigma_{\mathfrak{p}}}$ has closed fiber $\underline{z}_0^{(p)}$.

One aspect of (6–5) and (6–6) which perhaps requires some further explanation is in the case of odd t and inert p, for which we need to explain why a degree p map $\underline{z}_0 \to \underline{x}_0$ as $\Gamma_0(N)$-structures is the "same" as an isomorphism $\underline{z}_0^{(p)} \simeq \underline{x}_0$ of $\Gamma_0(N)$-structures; this latter isomorphism data is what one naturally gets when applying Lemma 6.3 and Theorem 6.4 with $\underline{y}(s)_0 \simeq \underline{z}_0^{(p)}$ for odd s. The point is that there is a *unique* order p subgroup scheme in a supersingular elliptic curve over a field of characteristic p, the quotient by which is the Frobenius base change. Thus, there is only one possible kernel for a degree p map $\underline{z}_0 \to \underline{x}_0$, so the passage between degree p maps $\underline{z}_0 \to \underline{x}_0$ and isomorphisms $\underline{z}_0^{(p)} \simeq \underline{x}_0$ (respecting $\Gamma_0(N)$-structures) is immediate.

It remains to identify (6–5) and (6–6) with the right side of (5–4) in all four cases (depending on the parity of t and whether p is inert or ramified in K). For even t and inert p, use multiplication by $p^{t/2}$ to lift isomorphisms to p^t-isogenies over thicker artinian bases by repeated application of the second part of Theorem 6.1, and use the first part of Theorem 6.1 to ensure that iteration of this construction gives the right side of (5–4). A similar argument using multiplication by $p^{(t-1)/2}$ and the third part of Theorem 6.1 takes care of odd t and inert p; the role of the final part of Theorem 6.1 is to ensure that the sum on the right side of (5–4) in such cases has vanishing terms for $n > (t-1)/2$. The case of even t and ramified p goes by an argument as in the case of even t and inert p, upon noting that the second part of Theorem 6.1 causes $[p] \circ f$ to lift from W_n to W_{n+2} since $\mathrm{ord}_K(p) = 2$ in the ramified case. The most subtle case of all is odd t and ramified p, for which Theorem 6.2 (and Theorem 6.1) provides the ability to translate this case of (6–6) into the form on the right side of (5–4).

7. Application of a Construction of Serre

With Theorem 5.1 settled, the next task is to explicate the right side of (5–1) in terms of quaternion algebras (still maintaining the assumption $r_{\mathscr{A}}(m) = 0$; that is, the ideal class $\mathscr{A}$ corresponding to σ under class field theory contains no integral ideals of norm m). The simplest case is when p is split in K, for then $\underline{x}$ and $\underline{x}^\sigma$ are Serre–Tate canonical lifts of their closed fibers, so

$$\mathrm{Hom}_{W_n}(\underline{x}^\sigma, \underline{x}) = \mathrm{Hom}_W(\underline{x}^\sigma, \underline{x}) = \mathrm{Hom}_F(x^\sigma, x)$$

for all $n \geq 0$ (see (5–5)), and therefore the right side of (5–1) vanishes because $r_{\mathscr{A}}(m) = 0$. Thus, for the purpose of explicitly computing the right side of Theorem 5.1 in terms of quaternion algebras, we lose no generality in immediately restricting to the case in which p is not split in K, so p does not divide N, $\mathscr{O} = \mathbf{Z}_p \otimes_{\mathbf{Z}} \mathscr{O}_K$ is the ring of integers of a quadratic extension of $\mathbf{Q}_p$, and $\underline{x}$ has supersingular reduction (as does $\underline{x}^\sigma$).

We need the following classical result (which is noted below [GZ, Ch. III, Prop. 7.1] and for which we give a nonclassical proof).

LEMMA 7.1. *The ring $R = \mathrm{End}_{W/\pi}(\underline{x})$ is an order in a quaternion division algebra B over $\mathbf{Q}$ which is nonsplit at exactly p and ∞. More specifically, $\mathbf{Z}_p \otimes_{\mathbf{Z}} R$ is the maximal order in the division algebra $\mathbf{Q}_p \otimes_{\mathbf{Q}} B$ and for $\ell \neq p$ the order $\mathbf{Z}_\ell \otimes_{\mathbf{Z}} R$ is conjugate to the order*

$$\left\{ \begin{pmatrix} a & b \\ c & d \end{pmatrix} \in M_2(\mathbf{Z}_\ell) \,\middle|\, c \equiv 0 \bmod N \right\}$$

in $\mathbf{Q}_\ell \otimes_{\mathbf{Q}} B \simeq M_2(\mathbf{Q}_\ell)$

PROOF. By the classical theory of supersingular elliptic curves, we know the first part. Thus, the whole point is to work out the local structure, and for this we use Tate's isogeny theorem for abelian varieties over finite fields. Let k_0 be a sufficiently large finite field over which there is a model $\underline{x}_0$ of the Heegner diagram $\underline{x} \bmod \pi$ for which all "geometric" endomorphisms of the underlying elliptic curve (ignoring the étale level structure) are defined, so in particular

$R = \mathrm{End}_{k_0}(\underline{x}_0)$. By Tate's isogeny theorem (with the trivial adaptation to keep track of the level structure), for $\ell \neq p$ the natural map

$$\mathbf{Z}_\ell \otimes_{\mathbf{Z}} R \to \mathrm{End}_{\mathbf{Z}_\ell[G_{k_0}]}(\underline{x}_0[\ell^\infty])) \tag{7–1}$$

is an isomorphism, where the right side denotes the ring of ℓ-adic Tate module endomorphisms which are Galois-equivariant and respect the ℓ-part of the level structure. Since the left side has $\mathbf{Z}_\ell$-rank equal to 4 and the right side is an order inside of $M_2(\mathbf{Z}_\ell)$, we conclude that the Galois group acts through scalars (this only requires injectivity of the Tate map, and this is completely elementary, so we haven't yet used the full force of Tate's theorem). Thus, we can replace $\mathbf{Z}_\ell[G_{k_0}]$ with $\mathbf{Z}_\ell$ on the right side of (7–1), and this yields the desired matrix description for all $\ell \neq p$.

Now consider the situation at p. Since there is no level structure involved, the claim is that if E is a supersingular elliptic curve over a finite field k_0 such that all "geometric" endomorphisms of E are defined over k_0, then $\mathbf{Z}_p \otimes_{\mathbf{Z}} \mathrm{End}_{k_0}(E)$ is the *maximal* order in a quaternion division algebra over $\mathbf{Q}_p$. The crucial point is that the action of a Frobenius element ϕ_{k_0} on E is through an *integer*. Indeed, since $\mathbf{Z}$ is a direct summand of the endomorphism algebra of E it follows that ϕ_{k_0} acts as an integer on E as long as it acts as an ℓ-adic integer on an ℓ-adic Tate module of E. This integrality property follows from the fact that (7–1) is an isomorphism, since $\mathbf{Q}_\ell$ is the center of $\mathbf{Q}_\ell \otimes_{\mathbf{Q}} B$.

It follows that on the height 2 formal group $\Gamma = E[p^\infty]_{/\bar{k}_0}$ of E over an algebraic closure $\bar{k}_0$, every endomorphism commutes with the k_0-Frobenius and hence is actually *defined* over k_0. By the theory of formal groups over *separably closed fields*, the endomorphism ring of Γ is a maximal order in a quaternion division algebra over $\mathbf{Q}_p$. But we have just seen that this endomorphism ring coincides with that of the p-divisible group of E over k_0. Thus, it suffices to prove that the natural map

$$\mathbf{Z}_p \otimes_{\mathbf{Z}} \mathrm{End}_{k_0}(E) \to \mathrm{End}_{k_0}(E[p^\infty]) \tag{7–2}$$

is an isomorphism. Notice that if we replace p with $\ell \neq p$ then this is exactly the usual statement of Tate's isogeny theorem (up to the identification of the category of ℓ-divisible groups over k_0 with a certain category of $\mathbf{Z}_\ell[G_{k_0}]$-modules).

This "$\ell = p$" case of Tate's theorem is proven by essentially the same exact method as Tate's theorem: the only change in the proof is that one has to use Dieudonné modules to replace the use of Tate modules. Nearly every theorem in [Mum] concerning Tate modules also works (usually with the same proof) for Dieudonné modules, and this enables one to extend various results (such as computing the characteristic polynomial of Frobenius over a finite field) to the p-part in characteristic p. □

Using the "closed fiber" functor, we have a natural injection $\mathscr{O}_K = \mathrm{End}_W(\underline{x}) \to \mathrm{End}_{W/\pi}(\underline{x}) = R$ which gives rise to a $\mathbf{Q}$-linear injection $K \to B$. By Skolem–

Noether, there exists $j \in B$ such that $jaj^{-1} = \bar{a}$ for all $a \in K$, where $a \mapsto \bar{a}$ is the nontrivial automorphism of K over $\mathbf{Q}$. In particular, $j^2 \in K^\times$ and any other such element j' of B lies in $K^\times j$ (as K is its own centralizer in B). Thus, $B_- = Kj$ is intrinsic to $K \hookrightarrow B$, so the decomposition $B = K \oplus B_-$ is intrinsic. Observe that $B_- = Kj$ can also be intrinsically described as the set of elements $b \in B$ such that $ba = \bar{a}b$ for all $a \in K \hookrightarrow B$.

If $b \in B$ is written as $b = b_+ + b_-$ according to this decomposition then $b_- = cj$ for some $c \in K$. Since $(cj)^2 = c\sigma(c) \in \mathbf{Q}$, when $b_- \neq 0$ it generates a quadratic field over $\mathbf{Q}$ in which its conjugate is $-b_-$. Thus, we can compute the reduced norm $\mathrm{N}(b)$ as

$$\mathrm{N}(b) = \mathrm{N}(b_+)\mathrm{N}(1+(c/b_+)j) = \mathrm{N}(b_+)(1+(c/b_+)j)(1-(c/b_+)j) = \mathrm{N}(b_+)+\mathrm{N}(b_-)$$

when $b_+ \neq 0$, and the case $b_+ = 0$ is trivial. In other words, the reduced norm is additive with respect to the decomposition $B = K \oplus B_-$.

Just as the decomposition $B = K \oplus B_-$ is intrinsic, for our p which is nonsplit in K (so $\mathrm{Gal}(K/\mathbf{Q}) \simeq \mathrm{Gal}(K_p/\mathbf{Q}_p)$) it follows that tensoring with $\mathbf{Q}_p$ gives the analogous decomposition for the nonsplit $\mathbf{Q}_p \otimes_{\mathbf{Q}} B$. If we define $R_p \stackrel{\mathrm{def}}{=} \mathbf{Z}_p \otimes_{\mathbf{Z}} R \simeq \mathrm{End}_{W/\pi}(\hat{\underline{x}})$ (the isomorphism being Tate's isomorphism (7–2)), then by [Gr1, Prop. 4.3] we see that the subring of p-divisible group endomorphisms lifting to W_n is given by those $b = b_+ + b_- \in R_p$ satisfying $D\,\mathrm{N}(b_-) \equiv 0 \bmod p\mathrm{N}(v)^n$, where $\mathrm{N}(v)$ denotes the ideal-theoretic norm of the unique prime of $\mathscr{O}_K$ over p. Thus,

$$\mathrm{End}_{W_n}(\hat{\underline{x}}) = \{b \in R_p \mid D\mathrm{N}(b_-) \equiv 0 \bmod p\mathrm{N}(v)^n\}.$$

The Serre–Tate lifting theorem ensures that $\mathrm{End}_{W_n}(\underline{x})$ consists of those elements of $R = \mathrm{End}_{W_0}(\underline{x})$ lifting to a W_n-endomorphism of the *p-divisible* group $\hat{\underline{x}}$ of $\underline{x}$. Thus,

$$\mathrm{End}_{W_n}(\underline{x}) = \{b \in R \mid D\mathrm{N}(b_-) \equiv 0 \bmod p\mathrm{N}(v)^n\}. \tag{7–3}$$

Describing $\mathrm{Hom}_{W_n}(\underline{x}^\sigma, \underline{x})$ is more subtle, since it rests on an interesting tensor construction of Serre's which unfortunately appears to not be explained adequately in the literature outside of the context of abelian varieties over a field (which is inadequate for applications such as our present situation where we have to work over artin local rings). There is a more general discussion of Serre's construction in [Gi], but that is also not adequate for our needs. We now develop Serre's tensor construction in a general setting, essentially to make functorial sense of an isomorphism $\underline{x}^\sigma \simeq \mathrm{Hom}(\mathfrak{a}, \underline{x}) \simeq \mathfrak{a}^{-1} \otimes_{\mathscr{O}_K} \underline{x}$ in a relative situation, where $\mathfrak{a}$ represents the ideal class $\mathscr{A}$ associated to σ under the Artin isomorphism. From a more algebraic point of view, the problem is that we do not yet have a recipe *over* W for constructing $\underline{x}^\sigma$ in terms of $\underline{x}$ (unless $\sigma \in \mathrm{Gal}(H/K)$ lies in the decomposition group at v). It is this recipe that we must intrinsically construct; see Corollary 7.11 for the answer. The basic idea is to construct $\underline{x}^\sigma$ as an "$\mathscr{O}_K$-tensor product" of $\underline{x}$ against a suitable fractional ideal of $\mathscr{O}_K$.

To motivate things, if $S = \mathrm{Spec}(\mathbf{C})$ and $E^{\mathrm{an}} \simeq \mathbf{C}/\mathfrak{b}$ for a fractional ideal $\mathfrak{b}$ of $\mathscr{O}_K$, then we expect to have an $\mathscr{O}_K$-linear analytic isomorphism

$$(\mathfrak{a} \otimes_{\mathscr{O}_K} E)^{\mathrm{an}} \overset{?}{\simeq} \mathfrak{a} \otimes_{\mathscr{O}_K} (\mathbf{C}/\mathfrak{b}) \simeq \mathbf{C}/\mathfrak{a}\mathfrak{b}. \tag{7–4}$$

This sort of construction on the analytic side *does* describe Galois actions $x \mapsto x^{\sigma}$ on Heegner points when viewed as $\mathbf{C}$-points of a modular curve (if $\mathfrak{a}$ is a prime ideal and σ is a *geometric* Frobenius at this prime), but if we are to have any hope of working with such Galois twisting on the level of W-points (or *anything* other than $\mathbf{C}$-points), we have to find a nonanalytic mechanism for constructing $\mathbf{C}/\mathfrak{a}\mathfrak{b}$ from $\mathbf{C}/\mathfrak{b}$.

The device for algebraically constructing "$\mathfrak{a} \otimes_{\mathscr{O}_K} E$" is due to Serre, and involves representing a functor. Rather than focus only on $\mathscr{O}_K$-module objects in the construction, we prefer to give a construction valid for more general coefficient rings, since abelian varieties of dimension > 1 (or in positive characteristic) tend to have noncommutative endomorphism algebras and Drinfeld modules are naturally "module schemes" over rings such as $\mathbf{F}_p[t]$. To include such a diversity of phenomena, we will construct a scheme $M \otimes_A \mathfrak{M}$ for an arbitrary associative ring A, a finite projective right A-module M, and an arbitrary left A-module scheme $\mathfrak{M}$ over any base scheme S (i.e., $\mathfrak{M}$ is a commutative S-group scheme endowed with a left A-action).

To save space, we will omit most of the proofs of the assertions we make below concerning Serre's tensor construction; we hope to provide a more detailed development elsewhere. Many proofs are mechanical, though one must carry them out in the correct order to avoid complications, and a few arguments require input from Dieudonné theory. Here is the setup for the basic existence result.

Let S be a scheme and let A be an associative ring with identity. Let M be a finite projective right A-module with dual left A-module $M^{\vee} = \mathrm{Hom}_A(M, A)$ of right A-linear homomorphisms (with $(a.\phi)(m) \overset{\mathrm{def}}{=} a\phi(m)$ for $\phi \in M^{\vee}$, $a \in A$, $m \in M$). By using the analogous duality construction for left modules, there is a natural map $M \to M^{\vee\vee}$ which is an isomorphism of right A-modules (use projectivity of M).

THEOREM 7.2. *Let $\mathfrak{M}$ be a left A-module scheme. The functor*

$$T \rightsquigarrow M \otimes_A \mathfrak{M}(T) \simeq \mathrm{Hom}_A(M^{\vee}, \mathfrak{M}(T))$$

on S-schemes is represented by a commutative group scheme over S, denoted $M \otimes_A \mathfrak{M}$ or $\mathrm{Hom}_A(M^{\vee}, \mathfrak{M})$, and $M \otimes_A (\cdot)$ carries closed immersions to closed immersions, surjections to surjections, and commutes with formation of fiber products. For each of the following properties $\mathbf{P}$ *of S-schemes, if $\mathfrak{M}$ satisfies property* $\mathbf{P}$ *then so does $M \otimes_A \mathfrak{M}$: quasi-compact and quasi-separated, locally finitely presented, finitely presented, locally finite type, separated, proper, finite, locally quasi-finite, quasi-finite and quasi-separated, smooth, étale, flat. In particular, if $\mathfrak{M}$ is finite locally free over S then so is $M \otimes_A \mathfrak{M}$.*

In Theorem 7.7, we'll see some nice applications of this result in the context of finite flat group schemes (beware that without an additional constancy condition on the "A-rank" of M, appropriately defined, the functor $M \otimes_A (\cdot)$ may have a slightly funny effect on the order of a finite locally free group scheme).

By considering left A-modules as right A^{opp}-modules and vice-versa, the property of being a finite projective module is preserved and everything we say below carries over with trivial modifications upon switching the words "left" and "right".

PROOF. First we establish representability, and then will consider flatness; the rest is left as an exercise in functorial criteria, etc. When M is a finite free right A-module, it is clear that an r-fold fiber product of $\mathfrak{M}$ does the job. In general, choose a finite presentation of the finite projective left A-module $M^\vee$:

$$A^{\oplus r} \to A^{\oplus s} \to M^\vee \to 0. \tag{7–5}$$

Applying $\mathrm{Hom}_A(\,\cdot\,, \mathfrak{M}(T))$ yields a left exact sequence, so by representability of scheme-theoretic kernels we deduce representability in general.

I am grateful to Serre for suggesting the following argument for the flatness property; this is much simpler than my original argument. Since M is projective, it is a direct summand of some $A^{\oplus n}$ as a right A-module. Thus, $M \otimes_A \mathfrak{M}$ is a direct factor of the S-flat $\mathfrak{M}^n$ as an S-scheme. It therefore suffices to check that if X and Y are S-schemes with $Y(S)$ nonempty and $X \times_S Y$ is S-flat, then X is S-flat. We may assume S is local and X is affine, and can replace Y by an affine open neighborhood of the closed point of the identity section. We thereby get commutative ring extensions $R \to B$ and $R \to C$ with $B \otimes_R C$ flat over R and $R \to C$ having a section, so $C = R \oplus I$ as A-modules. Thus, B is a direct summand of the R-flat $B \otimes_R C$ as an R-module and hence is R-flat. □

Now we wish to study fibers and exact sequences. This requires us to first define what it means to say that a finite projective right A-module M has constant rank r over A. When A is commutative, the meaning is clear (i.e., the vector bundle $\widetilde{M}$ on $\mathrm{Spec}(A)$ has constant rank) and can be formulated by saying that for any ring map $\phi : A \to k$ to a (commutative) field k, the base change $M \otimes_{A,\phi} k$ is r-dimensional as a k-vector space. Moreover, it suffices to demand this condition for algebraically closed fields k. The correct notion in the noncommutative case is unclear in general. However, rings acting on (reasonable) schemes are special. For example, if $\mathfrak{M}$ is locally finite type over S then A acts on the tangent spaces at geometric points of $\mathfrak{M}$, and hence we get many maps from A to matrix algebras (or more conceptually, central simple algebras of finite dimension) over residue fields at geometric points on S. This example suggests a definition which, while surely "wrong" for finite projective modules over general associative rings, works well for the situations we really care about (i.e., module schemes locally of finite type over a base).

DEFINITION 7.3. We say that a finite projective right A-module M has *constant rank* r if, for every map $\phi : A \to C$ to a finite-dimensional central simple algebra over an algebraically closed field $k = Z(C)$, with $\dim_k C = d_C^2$, the finite right C-module $M \otimes_{A,\phi} C$ has length $d_C r$. In other words, as a right C-module $M \otimes_{A,\phi} C$ is isomorphic to an r-fold direct sum of copies of C.

REMARK 7.4. When A is commutative, this is equivalent to the more familiar notion from commutative algebra (i.e. a rank r vector bundle on $\mathrm{Spec}(A)$) since a finite projective module over a local ring is free and C in Definition 7.3 is isomorphic to a direct sum of d_C copies of its unique simple module I_C. Dropping the commutativity assumption, since $\dim_k I_C = d_C$ in the notation of Definition 7.3, the k-length of a finite C-module is equal to d_C times its C-length. Thus, in Definition 7.3 one could equivalently require $\dim_k M \otimes_{A,\phi} C = r \dim_k(C)$ where k is the center of C. This latter formula provides a definition that works without requiring k to be algebraically closed, shows that the quantifiers in Definition 7.3 involve no set theory or universe issues, and makes it clear that the finite projective dual left A-module $M^\vee$ is of constant rank r (defined in terms of $C \otimes_{\phi,A} M^\vee$) if and only if M is of constant rank r (in the sense of Definition 7.3). One sees this latter compatibility with the help of the natural isomorphism of left C-modules

$$C \otimes_{\phi,A} M^\vee \simeq (M \otimes_{A.\phi} C)^\vee$$

(with the right side a C-linear dual) defined by $c \otimes \ell \mapsto (m \otimes c' \mapsto c\phi(\ell(m))c')$. This is particularly useful when considering (Picard or Cartier) duality for left and right A-module schemes which are abelian schemes or finite locally free.

In order to apply Serre's tensor construction to abelian schemes and their torsion subschemes, we need to record some more properties (with proofs omitted).

THEOREM 7.5. *Let $\mathfrak{M} \to S$ be a locally finite type left A-module scheme, and let M be a finite projective right A-module.*

- *If the S-fibers of $\mathfrak{M}$ are connected, then so are the S-fibers of $M \otimes_A \mathfrak{M}$. In particular, if $\mathfrak{M}$ is an abelian scheme over S then so is $M \otimes_A \mathfrak{M}$.*
- *If $\mathfrak{M}$ has fibers over S of dimension d and M has constant rank r over A then $M \otimes_A \mathfrak{M}$ has fibers of dimension dr.*
- *Let $0 \to \mathfrak{M}' \to \mathfrak{M} \to \mathfrak{M}'' \to 0$ be a short exact sequence of locally finitely presented S-flat left A-module schemes. Then applying $M \otimes_A (\cdot)$ yields another short exact sequence of such A-module schemes, and in particular the map $M \otimes_A \mathfrak{M} \to M \otimes_A \mathfrak{M}''$ is faithfully flat.*

We will be particularly interested in the case that A is the ring of integers of a number field and M is a fractional ideal (corresponding to the case of rank 1), so $M \otimes_A E$ is an elliptic curve for $E \to S$ an elliptic curve (with $S = \mathrm{Spec}(W)$, $\mathrm{Spec}(W/\pi^{n+1})$, etc.).

Before we address the special case of the behavior of Serre's construction on finite locally free commutative group schemes, we digress to study an important example which recovers (7–4).

Suppose that $\mathfrak{M}$ is an abelian variety over $\mathbf{C}$ and A is an associative ring acting on $\mathfrak{M}$ on the left. By the analytic theory, we functorially have $\mathfrak{M}^{\text{an}} \simeq V/\Lambda$ where $V = \text{Tan}_0(\mathfrak{M}^{\text{an}})$ is the universal covering and $\Lambda = \text{H}_1(\mathfrak{M}^{\text{an}}, \mathbf{Z})$. In particular, there is a nature left A-action on V commuting with the $\mathbf{C}$-action and with respect to which the lattice Λ is stable. It then makes sense to form $M \otimes_A V$ and $M \otimes_A \Lambda$, and it follows by using a right A-linear isomorphism $M \oplus N \simeq A^{\oplus r}$ for suitable N that $M \otimes_A V$ is a finite-dimensional $\mathbf{C}$-vector space and $M \otimes_A \Lambda$ is a finitely generated *closed* subgroup which is cocompact and hence a lattice. Thus, the quotient $(M \otimes_A V)/(M \otimes_A \Lambda)$ makes sense as a complex torus, and it is only natural to guess that this must be $(M \otimes_A \mathfrak{M})^{\text{an}}$ (which we know to be a complex torus, since $M \otimes_A \mathfrak{M}$ is already known to be an abelian variety). This guess is of course correct, and is crucial for the usefulness of Serre's construction (see the *proof* of Corollary 7.11). We leave the proof as an exercise.

THEOREM 7.6. *With notation as above, there is an natural A-linear isomorphism of $\mathbf{C}$-analytic Lie groups $(M \otimes_A V)/(M \otimes_A \Lambda) \simeq (M \otimes_A \mathfrak{M})^{\text{an}}$ which is functorial in both $\mathfrak{M}$ and M.*

We will use this theorem in the case $\dim \mathfrak{M} = 1$ (i.e., elliptic curves).

If $\underline{x}$ is a "level N" Heegner diagram over an $\mathscr{O}_K[1/N]$-scheme S (such as W or $\mathbf{C}$), it makes sense to apply Serre's tensor construction $\text{Hom}_{\mathscr{O}_K}(\mathfrak{a}, \cdot)$ to this diagram, provided that this construction carries cyclic isogenies of degree N to cyclic isogenies of degree N. While such a property is easily checked over $\mathbf{C}$ by means of the preceding theorem, in order to work over a base which is not a subfield of $\mathbf{C}$ (such as a non-field ring like W or an artinian quotient of W) we need to do some more work.

First we record a general result, the proof of which requires Dieudonné theory.

THEOREM 7.7. *Let $\mathfrak{M}$ be a left A-module scheme which is finite locally free over S, so its Cartier dual $\mathfrak{M}^\vee$ is a right A-module scheme. Then $M \otimes_A \mathfrak{M}$ is also finite locally free over S, of rank d^r if $\mathfrak{M}$ has constant rank d over S and M has constant rank r over A.*

Quite generally, in terms of Cartier duality there is a natural isomorphism $(M \otimes_A \mathfrak{M})^\vee \simeq \mathfrak{M}^\vee \otimes_A M^\vee$.

Using the above theorems and some additional arguments, one can deduce the following two consequences.

COROLLARY 7.8. *Let S be a henselian local scheme. The functor $M \otimes_A (\cdot)$ respects formation of the connected-étale sequence of a finite locally free commutative group scheme over S.*

Moreover, if S is an arbitrary scheme with all points of positive characteristic p and $\mathfrak{M} \to S$ is a left A-module scheme which is an abelian scheme having ordinary (resp. supersingular) fibers, then $M \otimes_A \mathfrak{M} \to S$ has the same property.

COROLLARY 7.9. *Let $f : \mathfrak{M} \to \mathfrak{N}$ be an isogeny (i.e., finite locally free map) of constant rank d between left A-module schemes which are flat and locally finitely presented schemes over S. Then*

$$1_M \otimes f : M \otimes_A \mathfrak{M} \to M \otimes_A \mathfrak{N}$$

is an isogeny, and the natural map $M \otimes_A \ker(f) \to \ker(1_M \otimes f)$ is an isomorphism.

When f has constant degree d and M has constant rank r over A, then $1 \otimes f$ has constant degree d^r. Likewise, when f is étale then so is $1_M \otimes f$.

The applications to Heegner points require that we keep track of *cyclicity* of kernels of isogenies (in the sense of [KM, Ch. 6]). This amounts to:

COROLLARY 7.10. *Let M be a finite projective right A-module of constant rank 1 and let $\mathfrak{M}$ be a left A-module scheme whose underlying group scheme over S is finite étale and "cyclic" of constant order d. Then the same holds for $M \otimes_A \mathfrak{M}$.*

An important corollary of these considerations is that for a Heegner diagram $\underline{x}$ of level N over any $\mathscr{O}_K[1/N]$-scheme S and any fractional ideal $\mathfrak{a}$ of K, applying $\mathrm{Hom}_{\mathscr{O}_K}(\mathfrak{a}, \cdot)$ to this diagram yields another Heegner diagram. Now we can finally prove the main result:

COROLLARY 7.11. *Let $\underline{x}$ be a Heegner diagram over W arising from a Heegner diagram x over H, and let $\sigma \in \mathrm{Gal}(H/K)$ correspond to the ideal class represented by a fractional ideal $\mathfrak{a}$ of K. There exists an isomorphism $\underline{x}^\sigma \simeq \mathrm{Hom}_{\mathscr{O}_K}(\mathfrak{a}, \underline{x}) \simeq \mathfrak{a}^{-1} \otimes_{\mathscr{O}_K} \underline{x}$ as Heegner diagrams over W.*

PROOF. Note that the assertion only depends on the isomorphism classes of the Heegner diagrams, and by Theorem 2.6 two Heegner diagrams over W are W-isomorphic if and only if they become isomorphic over some extension field of the fraction field of W. Since Serre's tensor construction is of formation compatible with *arbitrary* base change, by using an embedding $W \hookrightarrow \mathbf{C}$ as $\mathscr{O}_H$-algebras (which exists by cardinality considerations on transcendence degrees over H) it suffices to prove the result with W replaced by $\mathbf{C}$. From the analytic description of Galois action on Heegner points, if x over $\mathbf{C}$ is analytically isomorphic to $\mathbf{C}/\mathfrak{b} \to \mathbf{C}/\mathfrak{b}\mathfrak{n}^{-1}$, then x^σ over $\mathbf{C}$ is analytically isomorphic to the analogous diagram with $\mathfrak{b}$ replaced by $\mathfrak{a}^{-1}\mathfrak{b}$. Since $\mathrm{Hom}_{\mathscr{O}_K}(\mathfrak{a}, \mathfrak{M}) \simeq \mathfrak{a}^{-1} \otimes_{\mathscr{O}_K} \mathfrak{M}$ for an A-module scheme $\mathfrak{M}$, if we use functoriality then Theorem 7.6 ensures that applying $\mathrm{Hom}_{\mathscr{O}_K}(\mathfrak{a}, \cdot)$ to x over $\mathbf{C}$ yields a Heegner diagram which is analytically and hence (by GAGA) algebraically isomorphic to the diagram of x^σ over $\mathbf{C}$, using $(\mathfrak{a} \otimes_{\mathscr{O}_K} \mathbf{C})/(\mathfrak{a} \otimes_{\mathscr{O}_K} \Lambda) \simeq \mathbf{C}/\mathfrak{a}\Lambda$ for any $\mathscr{O}_K$-lattice Λ in $\mathbf{C}$ (with this isomorphism functorial in $\Lambda \hookrightarrow \mathbf{C}$). □

Having explored the properties of Serre's tensor construction, we're ready to apply it:

THEOREM 7.12. *Let $\mathfrak{a}$ be an ideal in the ideal class $\mathscr{A}$ and let $\sigma \in \mathrm{Gal}(H/K)$ correspond to this ideal class. Then there is an isomorphism of groups*

$$\mathrm{Hom}_{W_n}(\underline{x}^{\sigma}, \underline{x}) \simeq \mathrm{End}_{W_n}(\underline{x}) \cdot \mathfrak{a} \subseteq B \tag{7–6}$$

under which an isogeny $\phi : \underline{x}^{\sigma} \to \underline{x}$ corresponds to an element $b \in B$ with $\mathrm{N}(b) = \deg(\phi)\mathrm{N}(\mathfrak{a})$ *as ideals in* $\mathbf{Z}$.

In addition to needing the preceding theory to justify the isomorphism in (7–6), particularly over an artin local base such as W_n, we will need to work a little more to keep track of degrees. Postponing the proof of Theorem 7.12 for a short while, note that by using an abstract isomorphism $\underline{x}^{\sigma} \simeq \mathrm{Hom}_{\mathscr{O}_K}(\mathfrak{a}, \underline{x})$ over W (and hence over any W-scheme, such as W_n) as follows from Corollary 7.11, we obtain for any W-scheme S that

$$\mathrm{Hom}_S(\underline{x}^{\sigma}, \underline{x}) \simeq \mathrm{Hom}_S(\mathrm{Hom}_{\mathscr{O}_K}(\mathfrak{a}, \underline{x}), \underline{x}) \simeq \mathfrak{a} \otimes_{\mathscr{O}_K} \mathrm{End}_S(\underline{x}) \tag{7–7}$$

where $\mathscr{O}_K$ acts on $\mathrm{End}_S(\underline{x})$ through "inner composition" and the second isomorphism arises from:

LEMMA 7.13. *Let A be an associative ring and M a finite projective left A-module. For any left A-module scheme $\mathfrak{M}$ over S and any commutative S-group scheme G, view the group $\mathrm{Hom}_S(\mathfrak{M}, G)$ as a right A-module via the A-action on $\mathfrak{M}$. Then the natural map $\xi_M : \mathrm{Hom}_S(\mathfrak{M}, G) \otimes_A M \to \mathrm{Hom}_S(\mathrm{Hom}_A(M, \mathfrak{M}), G)$ defined functorially by $\xi_M(\phi \otimes m) : f \mapsto \phi(f(m))$ (on points in S-schemes) is well-defined and an isomorphism.*

This lemma is a simple exercise in definition-chasing (keep in mind how we *defined* the A-action on $\mathrm{Hom}_S(\mathfrak{M}, \mathfrak{N})$), using functoriality with respect to product decompositions in M and the existence of an isomorphism $M \oplus N \simeq A^{\oplus r}$ to reduce to the trivial case $M = A$.

Strictly speaking, this lemma does not include the case of Hom-groups between the Heegner diagrams $\underline{x}$ and $\underline{x}^{\sigma}$, since these are data of elliptic curves with some additional level structure. Since $\mathfrak{a}$ is typically not principal, we need to be a bit careful to extract what we wish from Lemma 7.13. The simple trick to achieve this is to observe that the $\mathscr{O}_K$-module Hom-groups of $\Gamma_0(N)$-structures which we really want to work with are described as kernels of certain $\mathscr{O}_K$-linear maps of Hom modules to which Lemma 7.13 *does* apply. By exactness of tensoring against a flat module (commuting with formation of kernels), we get the second isomorphism in (7–7) as a canonical isomorphism.

Let's now record an easy mild strengthening of a special case of Lemma 7.13, from which we'll be able to compute $\mathrm{End}_{W_n}(\underline{x}^{\sigma})$ in terms of $\mathrm{End}_{W_n}(\underline{x})$.

LEMMA 7.14. *Let A be an associative ring, M and M' two finite projective left A-modules. Let ${M'}^{\vee}$ denote the right module of left-linear maps from M'*

to A. For any two left A-module schemes $\mathfrak{M}$ and $\mathfrak{M}'$ over a base S, view the group $\mathrm{Hom}_S(\mathfrak{M}, \mathfrak{M}')$ *as a right A-module via the A-action on $\mathfrak{M}$ and as a left A-module via the action on $\mathfrak{M}'$. Then the natural map*

$$\xi_{M',M} : M'^{\vee} \otimes_A \mathrm{Hom}_S(\mathfrak{M}, \mathfrak{M}') \otimes_A M \to \mathrm{Hom}_S(\mathrm{Hom}_A(M, \mathfrak{M}), \mathrm{Hom}_A(M', \mathfrak{M}'))$$

defined functorially by $(\xi_{M',M}(\ell' \otimes \phi \otimes m))(f) : m' \mapsto \ell'(m')\phi(f(m))$ (on the level of points in S-schemes) is well-defined and an isomorphism.

In particular, there is a natural isomorphism $M^{\vee} \otimes_A \mathrm{End}_S(\mathfrak{M}) \otimes_A M \simeq \mathrm{End}_S(\mathrm{Hom}_A(M, \mathfrak{M}))$ *given by $\ell \otimes \phi \otimes m \mapsto (f \mapsto \ell'(\cdot) \cdot \phi(f(m)))$, and this is an isomorphism of associative rings.*

As a consequence of this lemma, we get a natural ring isomorphism

$$\mathrm{End}_S(\underline{x}^{\sigma}) \simeq \mathfrak{a}^{-1} \otimes_{\mathcal{O}_K} \mathrm{End}_S(\underline{x}) \otimes_{\mathcal{O}_K} \mathfrak{a} \tag{7–8}$$

for any W-scheme S, where the ring structure on the right is the obvious one obtained via the pairing $\mathfrak{a}^{-1} \times \mathfrak{a} \to \mathcal{O}_K$. Now we have enough machinery to prove Theorem 7.12.

PROOF OF THEOREM 7.12. Specializing the preceding preliminary considerations to the case $S = W_n$ and identifying $\mathrm{End}_{W_n}(\underline{x})$ with a subring of $\mathrm{End}_{W_0}(\underline{x}) = R$, if we view $\mathcal{O}_K = \mathrm{End}_W(\underline{x})$ as embedded in R via reduction it follows that making $\mathcal{O}_K$ act on $\mathrm{End}_{W_n}(\underline{x})$ through "inner composition" as in Lemma 7.13 corresponds to making it act by right multiplication in R (since multiplication in R is defined as composition of morphisms and all modern algebraists define composition of morphisms to begin on the right). This therefore yields an isomorphism of groups $\mathrm{Hom}_{W_n}(\underline{x}^{\sigma}, \underline{x}) \simeq \mathrm{End}_{W_n}(\underline{x}) \otimes_{\mathcal{O}_K} \mathfrak{a}$ where $\mathcal{O}_K \subseteq R$ acts through right multiplication on $\mathrm{End}_{W_n}(\underline{x}) \subseteq R$. Since $\mathfrak{a}$ is an invertible $\mathcal{O}_K$-module and B is torsion-free, the natural multiplication map $\mathrm{End}_{W_n}(\underline{x}) \otimes_{\mathcal{O}_K} \mathfrak{a} \to B$ is an isomorphism onto $\mathrm{End}_{W_n}(\underline{x}) \cdot \mathfrak{a}$.

It remains to chase degrees of isogenies. Since the degree of an isogeny over W_n can be computed on the W/π-fiber, it suffices to work over the *field* W/π and to show the following: given the data consisting of

- a supersingular elliptic curve E over an algebraically closed field k whose positive characteristic p is a prime which is nonsplit in $K/\mathbf{Q}$,
- an action of $\mathcal{O}_K$ on E (i.e., an injection of $\mathcal{O}_K$ into the quaternion division algebra $B = \mathbf{Q} \otimes_{\mathbf{Z}} \mathrm{End}_k(E)$),
- a fractional ideal $\mathfrak{a}$ of K,

under the isomorphism

$$\mathrm{End}_k(E) \cdot \mathfrak{a} \simeq \mathrm{Hom}_k(\mathrm{Hom}(\mathfrak{a}, E), E) \tag{7–9}$$

defined by $\psi \cdot a \mapsto (f \mapsto \psi(f(a)))$ we claim that an isogeny $\phi : \mathrm{Hom}(\mathfrak{a}, E) \to E$ corresponds to an element $b \in B$ with reduced norm $\deg(\phi)\mathrm{N}(\mathfrak{a})$. The delicate point here is that $\mathfrak{a}$ might not be a principal ideal in $\mathcal{O}_K$. The trick by means of which we will "reduce" to the principal ideal situation is to investigate what happens on ℓ-divisible groups for *every* prime ℓ, since $\mathscr{O}_{K,\ell} \stackrel{\text{def}}{=} \mathbf{Z}_\ell \otimes_{\mathbf{Z}} \mathscr{O}_K$ is semi-local (either a discrete valuation ring or a product of two discrete valuation rings) and hence every invertible ideal in this latter ring is principal.

To make things precise, fix an arbitrary prime ℓ of $\mathbf{Z}$. The ℓ-part of $\deg(\phi)$ is exactly the degree of the isogeny ϕ_ℓ induced by ϕ on ℓ-divisible groups. To keep matters simple, note that everything depends only on the isomorphism class of $\mathfrak{a}$ as an $\mathscr{O}_K$-module, or in other words everything is compatible with replacing this ideal by another representative of the same ideal class. More specifically, if $\phi \mapsto b$ under (7–9) and ϕ_c is the composite of ϕ with the isomorphism $c^{-1} : \mathrm{Hom}(c\mathfrak{a}, E) \simeq \mathrm{Hom}(\mathfrak{a}, E)$ then $\deg(\phi) = \deg(\phi_c)$ and $\phi_c \mapsto cb$, with $\mathrm{N}b = \deg(\phi)\mathrm{N}\mathfrak{a} \iff \mathrm{N}(cb) = \deg(\phi_c)\mathrm{N}(c\cdot\mathfrak{a})$. Thus, we may assume without loss of generality that $\mathfrak{a}$ is an integral ideal and hence elements of this ideal act on E. If we replace k with a sufficiently big finite subfield k_0 down to which E descends and over which all "geometric" morphisms among E and $\mathrm{Hom}_{\mathscr{O}_K}(\mathfrak{a}, E)$ are defined, we can consider the isomorphism $\mathrm{End}_{k_0}(E) \cdot \mathfrak{a} \simeq \mathrm{Hom}_{k_0}(\mathrm{Hom}(\mathfrak{a}, E), E)$. We now apply Tate's isogeny theorem *at any prime* ℓ to both sides. Using the language of ℓ-divisible groups so as to treat all primes on an equal footing, tensoring both sides with $\mathbf{Z}_\ell$ yields an isomorphism

$$\mathrm{End}_{k_0}(E[\ell^\infty]) \cdot \mathfrak{a}_\ell \simeq \mathrm{Hom}_{k_0}(\mathrm{Hom}(\mathfrak{a}, E)[\ell^\infty], E[\ell^\infty])$$

with $\mathfrak{a}_\ell = \mathfrak{a}\mathscr{O}_{K,\ell}$ an invertible ideal of $\mathscr{O}_{K,\ell} = \mathbf{Z}_\ell \otimes_{\mathbf{Z}} \mathscr{O}_K$.

By functoriality, it is clear that

$$\mathrm{Hom}(\mathfrak{a}, E)[\ell^\infty] = \varinjlim(\mathfrak{a}^{-1} \otimes_{\mathscr{O}_K} E[\ell^n]) = \varinjlim(\mathfrak{a}_\ell^{-1} \otimes_{\mathscr{O}_{K,\ell}} E[\ell^n]).$$

Corollary 7.9 ensures these tensored group schemes really do form an ℓ-divisible group (of height 2). If we write $\mathfrak{a}_\ell^{-1} \otimes E[\ell^\infty]$ to denote this ℓ-divisible group, then we have an isomorphism

$$\mathrm{End}_{k_0}(E[\ell^\infty]) \cdot \mathfrak{a}_\ell \simeq \mathrm{Hom}_{k_0}(\mathfrak{a}_\ell^{-1} \otimes E[\ell^\infty], E[\ell^\infty]) \tag{7–10}$$

defined in the obvious manner. If $\pi \in \mathscr{O}_{K,\ell}$ is a generator of the invertible (principal!) ideal $\mathfrak{a}_\ell$ then every element b_ℓ of the left side of (7–10) can be written in the form $b_\ell = \psi_\ell\cdot\pi$ for a unique endomorphism ψ_ℓ of $E[\ell^\infty]$, and b_ℓ corresponds under (7–10) to the composite morphism $\phi_\ell : \mathfrak{a}_\ell^{-1} \otimes E[\ell^\infty] \simeq E[\ell^\infty] \to E[\ell^\infty]$, where the first step is the isomorphism of multiplication by π and the second step is ψ_ℓ. Thus, ϕ_ℓ has degree equal to that of ψ_ℓ, which in turn is just the reduced norm of ψ_ℓ (in the ℓ-adic quaternion algebra $\mathbf{Q}_\ell \otimes_{\mathbf{Q}} B$). But since $b_\ell = \psi_\ell\pi$, this reduced norm is equal to $\mathrm{N}(b_\ell)/\mathrm{N}(\pi) = \mathrm{N}(b_\ell)/\mathrm{N}(\mathfrak{a}_\ell)$, thereby giving the ℓ-part $\deg(\phi_\ell) = \mathrm{N}(b_\ell)/\mathrm{N}(\mathfrak{a}_\ell)$ of the desired "global" equality of $\mathbf{Z}$-ideals $\mathrm{N}b = \deg(\phi)\mathrm{N}\mathfrak{a}$. □

COROLLARY 7.15. *Assume* $r_{\mathscr{A}}(m) = 0$ *and* $\gcd(m, N) = 1$. *Choose a representative ideal* $\mathfrak{a}$ *for* $\mathscr{A}$ *which is prime to* p.

(1) *If p is inert in K and v is a place of H over p, then $q_v = p^2$ and*

$$(\underline{x}.T_m(\underline{x}^\sigma)) = \sum_{b \in R\mathfrak{a}/\pm 1,\, \mathrm{N}b = m\mathrm{N}\mathfrak{a}} \tfrac{1}{2}(1 + \mathrm{ord}_p(\mathrm{N}b_-)). \tag{7–11}$$

(2) *If $p\mathscr{O}_K = \mathfrak{p}^2$ and v is a place of H over p, then $q_v = p^k$ where $k \in \{1, 2\}$ is the order of $[\mathfrak{p}]$ in the class group of K and*

$$(\underline{x}.T_m(\underline{x}^\sigma)) = \sum_{b \in R\mathfrak{a}/\pm 1,\, \mathrm{N}b = m\mathrm{N}\mathfrak{a}} \mathrm{ord}_p(D\mathrm{N}b_-)). \tag{7–12}$$

Proof. This result is [GZ, Ch. III, Cor 7.4], for which the main inputs in the proof are Theorem 5.1, (7–3), and Theorem 7.12. We refer the reader to [GZ] for the short argument to put it all together (which requires that $\mathfrak{a}$ be prime to p). The only aspect on which we offer some further clarification is to explain conceptually why b_- is always nonzero for the b's under consideration and why $\mathrm{ord}_p(\mathrm{N}b_-)$ is odd when $p \nmid D$.

The nonvanishing of b_- is simply because an element $b \in R\mathfrak{a}$ with $b_- = 0$ is an element of which lifts to $\mathrm{Hom}_{W_n}(\underline{x}^\sigma, \underline{x})$ for all $n \geq 0$, and hence corresponds to an element in $\mathrm{Hom}_W(\underline{x}^\sigma, \underline{x})$ by Serre–Tate and Grothendieck's existence theorem. But this latter Hom module vanishes due to the assumption $r_{\mathscr{A}}(m) = 0$. Thus, since $b \neq 0$ (as its reduced norm is $m\mathrm{N}\mathfrak{a} \neq 0$) we have $b_- \neq 0$. Meanwhile, when our nonsplit p satisfies $p \nmid D$, so p is inert K, the choice of j (as in the unramified case of [Gr1]) may be made so that $j^2 \in K^\times$ is a uniformizer at the place over p. Thus, any $b_- = cj \in Kj$ satisfies $\mathrm{N}(b_-) = \mathrm{N}_{K/\mathbf{Q}}(c)\mathrm{N}(j^2)$, where $\mathrm{N}_{K/\mathbf{Q}}(c)$ has even order at p and $\mathrm{N}(j^2)$ is a uniformizer at p (as p is inert in K). Hence, $\mathrm{ord}_p(\mathrm{N}(b_-))$ is odd. □

This completes our consideration of cases with $r_{\mathscr{A}}(m) = 0$.

8. Intersection Theory Via Meromorphic Tensors

In order to go beyond the local cases with $r_{\mathscr{A}}(m) = 0$, we need a stronger geometric technique for computing intersection pairings without using a moving lemma. This section is devoted to developing the necessary generalities in this direction, so modular curves and Heegner points play no role in this section. Unfortunately, the letters K, D, F have already been assigned meaning in our earlier analysis, with F being the fraction field of W. Since W and the discriminant of $K/\mathbf{Q}$ will make no appearance in this section, *for this section only we reserve the letter F to denote a global field and D to denote a divisor on a curve.* We fix a proper smooth geometrically connected curve $X_{/F}$ (having nothing to do with $X_0(N)$). Let D, D' be degree 0 divisors on X whose common support $|D| \cap |D'|$, if nonempty, consists entirely of F-rational points. The example of interest to us is $X_0(N)_{/H}$, $D = (x) - (0)$ (resp. $D = (x) - (\infty)$) for some Heegner point $x \in X_0(N)(H)$ (note the two cusps are H-rational too), and $D' = T_m((x^\sigma) - (\infty))$ (resp. $D' = T_m((x^\sigma) - (0))$) for $\sigma \in \mathrm{Gal}(H/K)$, with

$\gcd(m, N) = 1$. In this case, we see that any overlap of supports between D and D' certainly consists entirely of noncuspidal H-rational points, and moreover such nonempty overlap occurs exactly when $r_{\mathscr{A}}(m) > 0$.

In our general setup, we wish to define a local symbol $\langle D, D' \rangle_v$ for each place v of F with the properties:

- it is bilinear in such pairs $\{D, D'\}$ (note the hypothesis $|D| \cap |D'| \subseteq X(F)$ is preserved under formation of linear combinations);
- it agrees with the canonical local height pairing of D_v and D'_v when $|D|$ and $|D'|$ are disjoint;
- the sum $\sum_v \langle D, D' \rangle_v$ is equal to the canonical global height pairing.

We recall that in the case of disjoint support, the local height pairing is uniquely characterized by abstract properties (including functorial behavior with respect to change of the base field) and for nonarchimedean v is explicitly constructed via intersection theory using any regular proper model over the local ring $\mathscr{O}_{F,v}$ at v. For the generalization allowing $|D| \cap |D'|$ to consist of some F-rational points, the local terms $\langle D, D' \rangle_v$ will *not* be canonical, but rather will depend on a certain noncanonical *global* choice. Happily, the product formula will ensure that this global choice only affects local terms by amounts whose total sum is 0. This will retain the connection to the global height pairings (which is what we really care about).

In order to clarify the local nature of the construction, we will carry it out entirely in the context of a local field, only returning to the global case after the basic local construction has been worked out. Thus, we now write F_v to denote a local (perhaps archimedean) field with normalized absolute value $|\cdot|_v$. Feel free to think of this as arising from completing a global field F at a place v, but such a "global" model will play no role in our local construction. Let $\mathscr{O}_{F_v}$ denote the valuation ring of v in the nonarchimedean case. We fix X_{v/F_v} a proper smooth and geometrically connected curve. The following definition is considered in [Gr2, § 5].

DEFINITION 8.1. If $x \in X_v(F_v)$ is a point, $f \in F_v(X_v)^\times$ is a nonzero rational function, and t_x is a *fixed* uniformizer at x, we define

$$f[x] = f_{t_x}[x] \stackrel{\text{def}}{=} \frac{f}{t_x^{\mathrm{ord}_x(f)}}(x) \in F_v^\times.$$

This definition clearly depends on t_x only through its nonzero image ω_x in the cotangent line $\mathrm{Cot}_x(X_v)$ at x, so we may denote it $f_{\omega_x}[x]$ instead. For nonarchimedean v the absolute value $|f[x]|_v$ only depends on this nonzero cotangent vector up to $\mathscr{O}_{F_v}^\times$-multiple. For technical reasons (e.g., the fact that various Hecke divisors do not have all support consisting of rational points), it is convenient to extend the definition in the absolute value aspect to the case in which $x \in X_v$ is merely a closed point and not necessarily F_v-rational (this mild generalization does not seem to appear in [Gr2]).

For any such x, the residue field $\kappa(x)$ at x is a finite extension of F_v and hence for a cotangent vector ω_x represented by a uniformizer t_x at x, the value

$$f[x] = f_{\omega_x}[x] \stackrel{\text{def}}{=} (f/t_x^{\operatorname{ord}_x(f)})(x) \in \kappa(x)^\times$$

makes sense (depending again only on ω_x), and when $\operatorname{ord}_x(f) = 0$ this is just $f(x)$. Thus, the absolute value $|f_{\omega_x}[x]|_v$ make sense, where we write $|\cdot|_v$ to also denote the unique extension of $|\cdot|_v$ to an absolute value on $\kappa(x)$, and it is just $|f(x)|_v$ when $\operatorname{ord}_x(f) = 0$. Extending by $\mathbf{Z}$-linearity and multiplicativity, we can define the symbol

$$|f_{\underline{\omega}}[D]|_v = \prod_{x \in |D|} |f_{\omega_x}[x]|_v^{\operatorname{ord}_x(D)}$$

for any divisor D on X_v for which we have *fixed* a set $\underline{\omega}$ of nonzero cotangent vectors at each $x \in |D| \cap |\operatorname{div}(f)|$ (keep in mind that $|f_\omega[x]|_v$ does not depend on $\omega \in \operatorname{Cot}_x(X_v)$ when $x \notin |\operatorname{div}(f)|$, so the choice of cotangent vector at such x's really should be made when defining $|f_{\underline{\omega}}[D]|_v$ but actually doesn't matter). We will write $|f[D]|_v$ instead of $|f_{\underline{\omega}}[D]|_v$ when the choice of cotangent vectors is understood from context. We do *not* claim to define $f_{\underline{\omega}}[D]$, as this would make no sense if the support $|D|$ contains several nonrational points (whose residue fields are not canonically identified with each other).

We trivially have $|(fg)[D]|_v = |f[D]|_v \cdot |g[D]|_v$ with both sides depending on choices of cotangent vectors at points of $|D|$ where either f or g has a zero or pole, and we also have $|f[D + D']|_v = |f[D]|_v \cdot |f[D']|_v$ (assuming the same cotangent vector has been chosen relative to D and D' at any common point in their support). Now we are ready for the main definition, which we will justify shortly.

DEFINITION 8.2. Let D, D' be degree 0 divisors on X_{v/F_v} with $|D| \cap |D'| = \{x_1, \ldots, x_n\} \subseteq X_v(F_v)$. *Choose* a set of cotangent vectors $\underline{\omega} = \{\omega_1, \ldots, \omega_n\}$ at each such point. For $f \in F_v(X_v)^\times$ such that $|D + \operatorname{div}(f)| \cap |D'| = \varnothing$, define

$$\langle D, D' \rangle_{v,\underline{\omega}} = \langle D + \operatorname{div}_{X_v}(f), D' \rangle_v - \log |f_{\underline{\omega}}[D']|_v, \tag{8–1}$$

where the first term on the right side is the canonical local height pairing for disjoint divisors of degree 0.

It is clear that this definition does not depend on the auxiliary choice of f, thanks to properties of the canonical local height pairing, so when D and D' have disjoint supports we see via the case $f = 1$ that this definition coincides with the canonical local height pairing for disjoint divisors of degree 0. The bilinearity in D and D' (relative to consistent choices of cotangent vectors) is clear.

A crucial observation is that *if* we begin in a global situation $X_{/F}$ with degree 0 divisors D, D' on X whose supports overlap precisely in global rational points $\{x_1, \ldots, x_n\} \subseteq X(F)$, then by choosing a set of *global* cotangent vectors $\underline{\omega}$ at the x_j's and choosing a *global* $f \in F(X)^\times$ for which $D + \mathrm{div}(f)$ and D' have disjoint support, we get

$$\begin{aligned}\sum_v \langle D_v, D'_v\rangle_{v,\underline{\omega}} &= \sum_v \langle D + \mathrm{div}(f), D'\rangle_v - \sum_v \log|f_{\underline{\omega}}[D']|_v \\ &= \langle D + \mathrm{div}(f), D'\rangle - \sum_{x'\in|D'|} \mathrm{ord}_{x'}(D') \sum_v \sum_j \log|f_{\omega_{x'}}[x'_j]|_{v,j},\end{aligned}$$

where the inner sum is over all factor fields $\kappa(x'_j)$ of the reduced ring $\kappa(x) \otimes_F F_v$ (on which the unique multiplicative norm extending $|\cdot|_v$ corresponds to an absolute value $|\cdot|_{v,j}$ on $\kappa(x'_j)$) and x'_j runs over all points on X_v over $x' \in X$. If $t_{x'}$ is a uniformizer representing $\omega_{x'}$ then

$$\sum_j \log|f_{\omega_{x'}}[x'_j]|_{v,j} = \log|\mathrm{N}_{\kappa(x')/F}(f_{\omega_{x'}}[x'])|_v,$$

so by the product formula for $\mathrm{N}_{\kappa(x')/F}(f_{\omega_{x'}}[x]) \in F^\times$ we get

$$\sum_v \langle D_v, D'_v\rangle_{v,\underline{\omega}} = \langle D + \mathrm{div}(f), D'\rangle = \langle D, D'\rangle,$$

recovering the canonical global height pairing.

It is very important for later purposes to observe that the local construction of $\langle\cdot,\cdot\rangle_{v,\underline{\omega}}$ is unaffected by replacing the choices of the ω_x's in the local theory with $\mathscr{O}_{F_v}^\times$-multiples. Somewhat more generally, if two degree 0 divisors D, D' on X_{v/F_v} satisfy $|D| \cap |D'| = \{x_1, \ldots, x_n\} \subseteq X_v(F_v)$, and we define $\omega'_{x_j} = \alpha_j \omega_{x_j}$, then since $\mathrm{ord}_x(f) = -\mathrm{ord}_x(D)$ for all $x \in |D| \cap |D'|$ in (8–1) we get

$$\langle D, D'\rangle_{v,\underline{\omega}'} = \langle D, D'\rangle_{v,\underline{\omega}} - \sum_j \mathrm{ord}_{x_j}(D)\mathrm{ord}_{x_j}(D')\log|\alpha_j|_v. \qquad (8\text{–}2)$$

This follows immediately from the definitions, using the fact that uniformizers t_x and t'_x representing ω_x and $\omega'_x = \alpha\omega_x$ are related by $t'_x = \alpha t_x + (\ldots)$. Note that (8–2) is exactly in accordance with [GZ, (5.2), p. 250] (which treats the case of a one-point overlap), since our α is the reciprocal of that in [GZ].

An explication of (8–1) which plays an essential role in calculations at complex places is briefly described in [GZ, Ch. II, § 5.1] (in which the *normalized* absolute value $|\cdot|_v$ on $\mathbf{C}$ is denoted $|\cdot|^2$ for the usual reasons). Due to its importance, we want to justify that this alternative description really matches (8–1), and in particular really works for any local field at all (not just $\mathbf{C}$). The starting point for this alternative description is the observation that since X_v is F_v-smooth, for any $x \in X_v(F_v)$ there is a small open neighborhood of x in X_v^{an} which is analytically isomorphic to an open unit disc (so there is a plentiful supply of rational points near x in the sense of the topology on F_v), and the ordinary local height pairing has nice continuity properties with respect to slightly modifying a divisor variable:

THEOREM 8.3. *Let D, D' be degree 0 divisors on X_v with*

$$|D| \cap |D'| = \{x_1, \ldots, x_n\} \subseteq X_v(F_v).$$

Choose $y_j \in X_v(F_v)$ near (but not equal to) x_j in the topological space $X_v(F_v)$. Define $D_{\underline{y}}$ to be the divisor of degree 0 obtained from D by replacing all appearances of x_j with y_j. Then

$$\langle D, D' \rangle_{v,\underline{\omega}} = \lim_{\underline{y} \to \underline{x}} \left(\langle D_{\underline{y}}, D' \rangle_v - \sum_j \mathrm{ord}_{x_j}(D) \mathrm{ord}_{x_j}(D') \log |t_{x_j}(y_j)|_v \right),$$

where t_{x_j} is a uniformizer representing ω_{x_j}. In particular, this limit actually exists.

It is clear that the term in the limit is independent of the choice of uniformizer lifting each ω_{x_j}.

PROOF. Subtracting the left side (as defined in (8–1)) from the right side, we get the limit as $\underline{y} \to \underline{x}$ of

$$\langle \sum_j \mathrm{ord}_{x_j}(D) y_j - \sum_{x \in |\mathrm{div}(f)| - \underline{x}} \mathrm{ord}_x(f) x, D' \rangle_v - \sum_j \mathrm{ord}_{x_j}(D) \mathrm{ord}_{x_j}(D') \log |t_{x_j}(y_j)|_v$$
$$+ \left(\sum_{x' \in |D'| - \underline{x}} \mathrm{ord}_{x'}(D') \log |f(x')|_v + \sum_j \mathrm{ord}_{x_j}(D') \log |\frac{f}{t_{x_j}^{\mathrm{ord}_{x_j}(f)}}(x_j)|_v \right),$$

where the sum of the last two terms is $\log |f_{\underline{\omega}}[D']|_v$. Since $\mathrm{ord}_{x_j}(D) = -\mathrm{ord}_{x_j}(f)$, after some cancellation and noting that

$$\frac{f}{t_{x_j}^{\mathrm{ord}_{x_j}(f)}}(x_j) = \lim_{y_j \to x_j} \frac{f(y_j)}{t_{x_j}(y_j)^{\mathrm{ord}_{x_j}(f)}},$$

we are left with the limiting value of $-\langle \mathrm{div}_{X_v}(f)_{\underline{y}}, D' \rangle_v + \langle D'_{\underline{y}}, \mathrm{div}_{X_v}(f) \rangle_v$ as $\underline{y} \to \underline{x}$. Thus, it suffices to prove the general claim that if D, D' are two degree 0 divisors on X_v with overlap at the set of rational points $\underline{x}$, and $\underline{y}$ is a slight deformation of this set of points, then $\langle D_{\underline{y}}, D' \rangle_v - \langle D, D'_{\underline{y}} \rangle_v \to 0$ as $\underline{y} \to \underline{x}$ (with both terms ordinary canonical local height pairings).

By choosing a rational point P away from these (which we avoid in the limiting process) and expressing D and D' as $\mathbf{Z}$-linear combinations of differences $Q - [\kappa(Q) : F_v]P$ for closed points $Q \in X_v$, we reduce to the following general situation. Let $x_1, x_2 \in X_v$ be two closed points distinct from $P \in X_v(F_v)$, with $\kappa_j = \kappa(x_j)$ the residue field at x_j. Define $\widetilde{x}_j = x_j$ if x_j is not a rational point, and otherwise define $\widetilde{x}_j$ to be a rational point near x_j, with the requirement that $\widetilde{x}_j \neq x_j$ for both j's when $x_1 = x_2$. Note in particular that $\widetilde{x}_j$ has residue field degree over F_v equal to $[\kappa_j : F_v]$ in all cases. The main claim is that

$$\left\langle \widetilde{x}_1 - [\kappa_1 : F_v]P, x_2 - [\kappa_2 : F_v]P \right\rangle_v - \left\langle x_1 - [\kappa_1 : F_v]P, \widetilde{x}_2 - [\kappa_2 : F_v]P \right\rangle_v \to 0$$

as $\underline{\widetilde{x}} \to \underline{x}$ (the condition $\widetilde{x}_j \to x_j$ being a tautology if x_j is not a rational point).

If neither x_1 nor x_2 is a rational point, there's nothing to say. If x_1 is a rational point but x_2 is not (or vice-versa), then by bilinearity we reduce to the continuity of the local height pairing when rational points within a fixed divisor are moved around in $X_v(F_v)$. Thus, we may now assume $x_1, x_2 \in X_v(F_v)$. By treating separately the cases in which we have $\widetilde{x}_j = x_j$ for some j and when this does not hold for either j (and in the latter case, distinguishing $x_1 = x_2$ from the case $x_1 \neq x_2$), one can use related continuity/bilinearity arguments. $\square$

Now we turn to explicating our modified local intersection pairing in the nonarchimedean case. Two natural questions arise for nonarchimedean F_v:

(1) Can we carry out the local construction using a pairing on the level of (suitable) divisors *not* necessarily of degree 0, somewhat generalizing the use of intersection theory for regular proper models in the case of *disjoint* effective divisors?
(2) Are there ways to package the $\mathscr{O}_{F_v}^\times$-multiple class of a nonzero cotangent vector by using higher (meromorphic) tensors?

Thanks to bilinearity, in order to address the first question, the central issue is to properly define $(x.x)_v$ for $x \in X_v(F_v)$ when a nonzero cotangent vector ω_x is chosen at x. More specifically, suppose that $\underline{X}_{v/\mathscr{O}_{F_v}}$ is a regular proper model for X_v, and let $\underline{x} \in \underline{X}_v(\mathscr{O}_{F_v})$ be the section arising from x via scheme-theoretic closure. Since $\underline{X}_v$ is regular, this section $\underline{x}$ lies in the relative smooth locus over $\mathscr{O}_{F_v}$. We have the natural linear isomorphism among cotangent spaces $\mathrm{Cot}_x(X_v) \simeq F_v \otimes_{\mathscr{O}_{F_v}} \mathrm{Cot}_{\underline{x}}(\underline{X}_v)$, so given a nonzero $\omega_x \in \mathrm{Cot}_x(X_v)$ there exists an $\alpha_x \in F_v^\times$ unique up to $\mathscr{O}_{F_v}^\times$-multiple such that $\alpha_x\omega_x \in \mathrm{Cot}_x(X_v)$ is a basis of the rank 1 lattice $\mathrm{Cot}_{\underline{x}}(\underline{X}_v)$. In particular, we see that $\mathrm{ord}_v(\alpha_x)$ depends only on the data of ω_x (not t_x) and $\underline{X}_v$. We'll now see that the choice of $\underline{X}_v$ doesn't matter.

THEOREM 8.4. *Fix a cotangent vector ω_x at every $x \in X_v(F_v)$, and let $\underline{\omega} = \{\omega_x\}_{x\in X_v(F_v)}$ denote the corresponding indexed collection. Define $(x.x)_{v,\underline{\omega}} = \mathrm{ord}_v(\alpha_x)$ for each $x \in X_v(F_v)$ as above. For distinct closed points $x', x'' \in X_v$ define $(x'.x'')_{v,\underline{\omega}} = (x'.x'')_v$ as in (4–1) applied to the regular proper model $\underline{X}_v$. Extending these definitions to define $(D.D')_{v,\underline{\omega}}$ via $\mathbf{Z}$-bilinearity for divisors D, D' on X_v with $|D| \cap |D'| \subseteq X_v(F_v)$, we get*

$$\langle D, D'\rangle_{v,\underline{\omega}} = -(D.D')_{v,\underline{\omega}} \log q_v \ . \tag{8–3}$$

whenever D and D' have degree 0 (defining the left side as in (8–1) via the same choices $\underline{\omega}$) and the "closure" of either D or D' in $\underline{X}_v$ has intersection number 0 with every irreducible component of the closed fiber of $\underline{X}_v$.

REMARK 8.5. Our α_x is the reciprocal of the α in [GZ], so the lack of a minus sign in [GZ, (8.1), p. 263] is consistent with our formula with a minus sign. Beware that $\mathrm{ord}_v(\alpha_x)$ is very dependent on the particular choice of regular model $\underline{X}_v$, so $(x.x)_{v,\underline{\omega}}$ depends on $\underline{X}_v$ (but we omit such dependence from the notation so as to avoid tediousness). Such dependence is reasonable to expect in view of the fact that the ability to *compute* $\langle D, D'\rangle_{v,\underline{\omega}}$ in terms of intersection theory on $\underline{X}_v$ is also heavily dependent on a hypothesis involving $\underline{X}_v$ (namely, that the closure of either D or D' in $\underline{X}_v$ has intersection number 0 with all irreducible components of the closed fiber of $\underline{X}_v$).

REMARK 8.6. It is immediate from the definitions that if $c \in F_v^\times$ then

$$(x.x)_{v,c\underline{\omega}} = (x.x)_{v,\underline{\omega}} - \mathrm{ord}_v(c). \tag{8–4}$$

PROOF. By symmetry, we may assume the closure of D in $\underline{X}_v$ has intersection number 0 with all closed fiber irreducible components. We wish to use the limit formula from Theorem 8.3. For $\underline{y}$ "near" $\underline{x}$ as in that theorem, we first claim that $D_{\underline{y}}$ satisfies the same property which we just imposed on D. This will enable us to compute the term $\langle D_{\underline{y}}, D'\rangle_v$ via intersection theory on $\underline{X}_v$, since $D_{\underline{y}}$ and D' are certainly disjoint divisors on X_v.

When taking the closure of $D_{\underline{y}}$ in $\underline{X}_v$ we get essentially the same divisor as the closure of D in $\underline{X}_v$ except for possibly the contributions of the sections $y_j \in X_v(F_v) = \underline{X}_v(\mathscr{O}_{F_v})$. It suffices to check that if $x \in X_v(F_v)$ is a given point and $y \in X_v(F_v)$ is sufficiently close to x in the topological space $X_v(F_v)$, then for any irreducible component Δ of the closed fiber of $\underline{X}_v$ we have $(\bar{x}.\Delta) = (\bar{y}.\Delta)$, where $\bar{x}$ and $\bar{y}$ are the unique sections of $\underline{X}_v \to \mathrm{Spec}(\mathscr{O}_{F_v})$ extending x and y respectively (which, when viewed as closed subschemes of $\underline{X}_v$, are the scheme-theoretic closures of the respective closed points x, y in the open subscheme $X_v \subseteq \underline{X}_v$). Since $\underline{X}_v$ is regular with smooth generic fiber, the section $\bar{x}$ must factor through the relative smooth locus. Moreover, if x_0 is the closed point of $\bar{x}$ then the preimage $B(x_0)$ of x_0 under the reduction map

$$X_v(F_v) = \underline{X}_v(\mathscr{O}_{F_v}) \to \underline{X}_v(\mathscr{O}_{F_v}/\mathfrak{m}_v)$$

is an open neighborhood of x. Thus, we consider only y in this neighborhood, so $\bar{y}$ has closed point x_0.

If T is a generator of the ideal sheaf of $\bar{x}$ in a neighborhood of x_0, then $\widehat{\mathscr{O}}_{\underline{X}_v,x_0} \simeq \mathscr{O}_{F_v}[\![T]\!]$ and $y \mapsto T(y)$ sets up a topological isomorphism between $B(x_0)$ and the open unit disc in F_v around the origin. Thus, if the ideal sheaf of Δ is generated by the regular element $f \in \mathscr{O}_{F_v}[\![T]\!]$, then f has nonzero constant term (as otherwise the Cartier divisor Δ would contain $\bar{x}$ and so would not be a closed fiber component) and $(\bar{y}.\Delta)_v = \mathrm{ord}(f(T(y)))$. We have to show that $\mathrm{ord}(f(t)) = \mathrm{ord}(f(0))$ for $t \in F_v$ sufficiently near 0. Since $f(0) \neq 0$ and f has integral coefficients, it suffices to take $|t|_v < |f(0)|_v$. This completes the justification that we may compute $\langle D_{\underline{y}}, D'\rangle_v$ using intersection theory on $\underline{X}_v$ when $\underline{y}$ is sufficiently close to $\underline{x}$ and the closure of D in $\underline{X}_v$ has intersection number 0 with all closed fiber irreducible components of $\underline{X}_v$.

We conclude from the limit formula in Theorem 8.3 that it suffices to prove

$$-(x.x)_{v,\omega_x} \log q_v = \lim_{y \to x} \bigl(-(\bar{y}.\bar{x})_v \log q_v - \log |t_x(y)|_v\bigr),$$

where $y \in X_v(F_v) - \{x\}$. In other words, we must prove that this limit not only exists, but is equal to

$$-\mathrm{ord}_v(\alpha_x)\log q_v = \log|\alpha_x|_v$$

(and so in particular only depends on ω_x and not the specific representative uniformizer t_x, which is a priori clear). Since t_x is a regular function in a Zariski neighborhood of x, it can be viewed as an "analytic" function t_x^{an} on a small disc centered at x with parameter T as in the preceding paragraph (the domain of this analytic function might not be the entire open unit disc: t_x might have a pole near x). If $\tau = T(y)$ then $-(\bar{y}.\bar{x})_v \log q_v = -\mathrm{ord}_v(\tau)\log q_v = \log|\tau|_v$ and $t_x(y) = t_x^{\mathrm{an}}(\tau)$. We therefore have to compute

$$\lim_{\tau\to 0}(\log|\tau|_v - \log|t_x^{\mathrm{an}}(\tau)|_v)$$

where $t_x = (1/\alpha_x)T + \cdots$ in $F[\![T]\!]$ with T as defined above serving as a parameter on an open unit disc (so this also gives the expansion of the analytic function t_x^{an}).

Since $t_x^{\mathrm{an}} = (1/\alpha_x)T \cdot h$ where h is an analytic function near 0 with $h(0) = 1$, we conclude that $t_x^{\mathrm{an}}(\tau) = (1/\alpha_x)\tau h(\tau)$, so $\log|\tau|_v - \log|t_x^{\mathrm{an}}(\tau)| = \log|\alpha_x|_v - \log|h(\tau)|_v$, and as $\tau \to 0$ we have $\log|h(\tau)|_v \to \log|1|_v = 0$. Thus, we conclude that $\lim_{y\to x}(-(\bar{y}.\bar{x})_v \log q_v - \log|t_x(y)|_v) = \log|\alpha_x|_v$, as desired. $\square$

As the above proof shows, in order to *compute* $(D.D')_{v,\underline{\omega}}$ we may make a base change to the completion of a maximal unramified extension of F_v, and we note that $\underline{X}_v \times_{\mathscr{O}_{F_v}} \widehat{\mathscr{O}^{\mathrm{sh}}}_{F_v}$ is still regular.

With these preliminary constructions settled, we turn to the question of computing self-intersection numbers with the help of an auxiliary tensor. Fix an integer k (we allow $k \le 0$) and fix a nonzero rational section θ of $(\Omega^1_{X_v/F_v})^{\otimes k}$ with $r_x \stackrel{\mathrm{def}}{=} \mathrm{ord}_x(\theta) \ne -k$. When $k = 0$ we are requiring the nonzero rational function θ to have a zero or pole at x, so the situation is always "abstract". We may write

$$\theta = (C_x t_x^{r_x} + \cdots)(\mathrm{d}t_x)^{\otimes k}, \tag{8–5}$$

where $C_x \in F_v^\times$ and t_x is a uniformizer lifting ω_x (so C_x only depends on ω_x, not t_x). If we use $\omega'_x = c\omega_x$ with $c \in F_v^\times$, then upon using the representative uniformizer $t'_x = ct_x$ we obtain the transformation formula

$$C'_x = \frac{C_x}{c^{r_x+k}}. \tag{8–6}$$

Note that when θ is *given* then ω_x determines C_x up to $\mathscr{O}_{F_v}^\times$-scaling and vice-versa, but when $k + r_x \ne \pm 1$ then because of (8–6) we certainly cannot expect to find ω_x realizing an arbitrary desired $C_x \in F_v^\times$ for a given θ. In general, the best we can do is determine ω_x up to $\mathscr{O}_{F_v}^\times$-multiple in terms of θ_x by the requirement that $0 \le \mathrm{ord}_v(C_x) < |r_x + k|$. When $k = 1$ and $r_x = 0$ this recovers the normalization used for intersection theory via tangent vectors, but in general we cannot expect to make C_x a unit at v. Although this milder normalization does give us a way to use θ to single out a "preferred" ω_x up to $\mathscr{O}_{F_v}^\times$-multiple (since $\mathrm{ord}_x(\theta) \neq -k$), and hence the choice of θ is "all" the input we need to make sense of C_x, it will actually be convenient to not minimize the choices in this way. Rather, we prefer to think of fixing θ in advance and still retaining the freedom to pick whatever ω_x we like at the point x.

Observe that if $\alpha_x\omega_x$ induces a basis of $\mathrm{Cot}_{\underline{x}}(\underline{X}_v)$, there is a generator T_x of the height 1 prime ideal $\mathfrak{p}_{\underline{x}}$ of the section $\underline{x}$ in $\mathscr{O}_{\underline{X}_v,\underline{x}_0}$ such that $\alpha_x t_x \equiv T_x$ in $\mathrm{Cot}_x(X_v)$. Thus,

$$\theta = \left(\frac{C_x}{\alpha_x^{r_x+k}} T_x^{r_x} + \cdots\right)(\mathrm{d}T_x)^{\otimes k}, \tag{8–7}$$

so $(\alpha_x^{r_x+k}/C_x)\theta$ is a basis of the invertible $\mathscr{O}_{F_v}$-module

$$(\Omega^1_{\underline{X}_v/\mathscr{O}_{F_v}})^{\otimes k}(-r_x \cdot \underline{x})_{\underline{x}_0}/\mathfrak{p}_{\underline{x}}.$$

Consequently, we can now recast the entire discussion in terms of a fixed choice of θ.

More specifically, let us now *fix* a choice of nonzero rational section θ of $(\Omega^1_{X_v/F_v})^{\otimes k}$ (in practice this will arise from a global rational tensor). For $v \nmid \infty$ and any $x \in X_v(F_v)$ with $r_x \stackrel{\text{def}}{=} \mathrm{ord}_x(\theta) \neq -k$, we choose a uniformizer t_x at x (in practice arising from a global uniformizer at a global rational point), and let $C_x \in F_v^\times$ be the leading coefficient of the t_x-adic expansion of θ (this coefficient transforms by the reciprocal multiplier when the nonzero cotangent vector $\omega_x \in \mathrm{Cot}_x(X_v)$ attached to t_x is replaced with an $F_v^\times$-multiple). If $\beta_x \in F_v^\times$ is chosen so that $\beta_x\theta$ is a basis of $(\Omega^1_{\underline{X}_v/\mathscr{O}_{F_v}})^{\otimes k}(-r_x \cdot \underline{x})_{\underline{x}_0}/\mathfrak{m}_{\underline{x}}$, then β_x is unique up to $\mathscr{O}_{F_v}^\times$-multiple, β_x *does not depend on* ω_x, and via (8–7) we see

$$\frac{\mathrm{ord}_v(C_x\beta_x)}{r_x + k} = \mathrm{ord}_v(\alpha_x) = (x.x)_{v,\omega_x} \in \mathbf{Z}. \tag{8–8}$$

In this way, we can compute $(x.x)_{v,\omega_x}$ at any x for which $r_x \stackrel{\text{def}}{=} \mathrm{ord}_x(\theta) \neq -k$.

9. Self-Intersection Formula and Application to Global Height Pairings

We wish to apply the theory in Section 8 to generalize Corollary 7.15 to cases with $r_{\mathscr{A}}(m) > 0$. This amounts to finding systematic (noncanonical!) local definitions of self-intersection numbers for rational points in such a way that one still recovers the canonical global height pairing as a sum of local terms. In [GZ] this is carried out with the help of a tangent vector. Unfortunately, the application of this method in the proof of [GZ, Ch. III, Lemma 8.2] uses the 1-form $\eta^4(q)\mathrm{d}q/q$ which doesn't actually live on $X_0(N)$, but only on a degree 6 covering $X' \to X_0(N)$. At points of ramification for this covering (i.e., elliptic points) one gets the zero map on tangent spaces, and elsewhere at points away from the branch locus there is no reason why global H-rational points on $X_0(N)$ have to lift to H-rational points on X'. Without the ability to work with H-rational points, and hence with H-rational tangent vectors, one encounters complications when formulating a global formula over H in terms of local tangent vector calculations: there has got to be *some* link between all of the local tangent vectors (e.g., they all come from a global one) in order for the sum of noncanonical local terms to recover canonical global height pairings in the case of degree 0 divisors with nondisjoint supports. Our alternative approach will also have to confront the issue of what to do at elliptic points, but the merit of using deformation theory is that we will be able to treat all points in a uniform manner without needing to use specialized arguments for the elliptic case.

Our method might be called "intersection theory with meromorphic tensors". The motivation is that although $\eta^4(q)\mathrm{d}q/q$ only lives on a cyclic covering of $X_0(N)$, $\Delta(q)(\mathrm{d}q/q)^{\otimes 6}$ makes sense as a meromorphic tensor on $X_0(N)_{/\mathbf{Q}}$. More importantly, Δ has a functorial interpretation and so makes sense within the context of deformation theory. For our purposes, the special role of Δ is that (via the relative Kodaira–Spencer isomorphism) on universal deformation rings for elliptic curves (*without* level structure), the leading coefficient of Δ as a 6-tensor is always a *unit*. After we have carried out our approach, we will revisit the method used in [GZ] and see how it can be understood as a special case of our approach, at least if one avoids the quadratic fields $K = \mathbf{Q}(\sqrt{-1})$ and $K = \mathbf{Q}(\sqrt{-3})$. An added bonus of our approach via Δ and meromorphic tensors is that we will be able to argue with abstract local deformation theory instead of relying on the global geometry of modular curves. This allows us to avoid complications traditionally caused by small primes and special j-invariants.

The main result of this section is really Theorem 9.6, which gives a rather general formula (9–16) for global height pairings. The proof of this global formula will require a formula for certain self-intersection numbers. In the global formula, the local junk terms in Theorem 9.2 essentially all cancel out, and we are left with something that is computable. In Section 10 we will take up the problem of computing the nonarchimedean terms (denoted $\langle x, T_m(x^\sigma)\rangle_v^{\mathrm{GZ}}$ in (9–16)) in terms of quaternionic data.

We will have to do some hard work in the special case of coarse moduli schemes, whereas the general intersection theory discussion in Section 8 took place on rather general smooth curves over local fields. Let us return to our earlier standard notation, with $\underline{X} = X_0(N)_{/W}$, where W is the completion of a maximal unramified extension of the local ring $\mathscr{O}_{H,v}$ at a place v of H over a prime p of $\mathbf{Q}$. Also, F denotes the fraction field of W and X denotes the generic fiber of $\underline{X}$. We do *not* make any assumptions on the behavior of p in K: it may be inert, split, or ramified.

Let $\underline{x} \in \underline{X}(W)$ be a section disjoint from the cuspidal locus, and assume there exists a $\Gamma_0(N)$-diagram over W which represents $\underline{x}$. We emphasize that $\underline{x}$ need *not* come from a Heegner point, though by Theorem 2.5 our hypotheses are satisfied for sections arising from Heegner points over H, and by Theorem 2.6 such a W-diagram realizing a *specified* F-diagram is unique up to canonical W-isomorphism (such uniqueness having nothing to do with Heegner points). When the F-diagram is not specified (i.e., only the F-point of the coarse moduli scheme is specified) then we lose the canonicalness of the W-diagram. We write $\underline{x}$ to denote a Heegner diagram over W (unique up to noncanonical isomorphism) representing a given section $\underline{x} \in \underline{X}(W)$ for which some such diagram exists. For our present purposes, the connection with Heegner points is not relevant; the isomorphism class of the diagram over W is all that matters. We first focus on a purely local assertion concerning diagrams over W (viewed as certain sections of $\underline{X}_{/W}$).

Let $x \in \underline{X}(F)$ denote the generic fiber rational point corresponding to $\underline{x}$. Fix an arbitrary integer k and a rational section θ of $(\Omega^1_{X_{/F}})^{\otimes k}$ such that $r_x \overset{\text{def}}{=} \mathrm{ord}_x(\theta) \neq -k$. Taking $k = 0$ and θ a nonzero rational function with divisor having a zero or pole at x is one option, for example. Another option of interest is the case $k = 6$ and $\theta = \Delta$, but it doesn't matter what choice we make. In practice the choice we make will have to arise from the global model over H. For conceptual clarity, we avoid specifying a particular choice of k or θ at the outset.

Choose a cotangent vector ω_x at x on $\underline{X}_{/F}$, with t_x a uniformizer representing ω_x. We let C_x denote the leading coefficient of the t_x-adic expansion of the rational section θ (relative to the basis $(\mathrm{d}t_x)^{\otimes k}$), so C_x depends only on ω_x. This data allows us to define $(x.x)_{v,\omega_x}$ in accordance with the recipe of the previous section (see (8–8))

The group $\mathrm{Aut}_F(x)$ coincides with the "geometric" automorphism group of x (by Theorem 2.6), and hence has order $2u_x$ with $u_x \in \{1, 2, 3\}$ (as F has characteristic 0). For example, if $\underline{x}$ came from a global point over H, then by Theorem 2.1(1) the value of u_x would actually be of global nature (i.e., it would depend only on the resulting $\overline{\mathbf{Q}}$-point or $\mathbf{C}$-point) and not at all depend on v. We now introduce an auxiliary quantity which intuitively measures the fact that we are trying to do intersection theory on a (regular model of a) coarse moduli scheme rather than on a Deligne–Mumford stack. To get started, we require a preliminary lemma in deformation theory.

We state the lemma in the specific context we need (i.e., $\Gamma_0(N)$-structures), but the reader will see that the method of proof works much more generally. The basic point is to provide a mechanism for passing between deformation rings of objects in characteristic 0 and objects in characteristic p (in cases for which the universal deformations are algebraizable, so it actually makes sense to base change a formal deformation from residue characteristic p over to residue characteristic 0; e.g., there is a map $\mathrm{Spec}(\mathbf{Q}_p) \to \mathrm{Spec}(\mathbf{Z}_p)$ but there is no map $\mathrm{Spf}(\mathbf{Q}_p) \to \mathrm{Spf}(\mathbf{Z}_p)$).

LEMMA 9.1. *Let $\underline{x}$ be a $\Gamma_0(N)$-structure over W, with generic fiber x over F. Let $\mathscr{R}_0$ be the universal deformation ring of $\underline{x}_0$ on the category of complete local* *noetherian W-algebras with residue field W/π. Let $\mathscr{I}_{\underline{x}} \subseteq \mathscr{R}_0$ be the ideal of the section corresponding to the W-deformation $\underline{x}$. The $F \otimes_W \mathscr{I}_{\underline{x}}$-adic completion of the algebraized universal deformation over $F \otimes_W \mathscr{R}_0$ is the universal deformation of x on the category of complete local noetherian F-algebras with residue field F.*

Let us explicate the meaning of the lemma in the case $N = 1$ (which, strictly speaking, is not a case we have ever been considering. but for which the lemma is true). Let E be an elliptic curve over W with closed fiber E_0 over W/π, and let $\mathscr{R}$ be the universal deformation ring of E_0. Let $\mathscr{E} \to \mathrm{Spec}(\mathscr{R})$ be an elliptic curve lifting E_0 which algebraizes the universal formal deformation. By abstract deformation theory $\mathscr{R}$ is formally smooth over W of relative dimension 1, and the ideal cutting out the W-section of $\mathrm{Spec}(\mathscr{R})$ arising from the special deformation $E_{/W}$ is a height 1 prime. By choosing a generator T of this we can identify $\mathscr{R} \simeq W[\![T]\!]$ such that reduction modulo T recovers $E_{/W}$. The content of the lemma is that the elliptic curve $\mathscr{E}_{/F[\![T]\!]}$ is an (algebraized) universal deformation of the *generic* fiber $E_{/F}$ of our original elliptic curve E over W. In this way, we have a precise link between universal deformation rings of the generic and closed fibers of $E_{/W}$.

To see the general meaning of the lemma (with $N > 1$ allowed), suppose $\underline{x}$ actually occurs in a universal algebraic family over an affine finite type W-scheme $\mathrm{Spec}(R)$. Let $\mathfrak{m}$ denote the maximal ideal associated to $\underline{x}_0$ and let $\mathfrak{p}_{\underline{x}}$ denote the prime associated to $\underline{x}$. Note that the $\mathfrak{p}_{\underline{x}}$-adic completion of R is naturally isomorphic to $\widehat{R}_{\mathfrak{m}}$ with a slightly weaker topology, since $R/\mathfrak{p}_{\underline{x}} \simeq W$ is max-adically complete. For example, if $\widehat{R}_{\mathfrak{m}} \simeq W[\![T]\!]$ with its maximal-adic topology (and T corresponding to the section $\underline{x}$), then the $\mathfrak{p}_{\underline{x}}$-adic completion would be isomorphic to $W[\![T]\!]$ with just the T-adic (rather than (π, T)-adic) topology.

If we let $R_\eta = F \otimes_W R$, then the above lemma is just the obvious commutative algebra assertion that the completion of R_η along the section x is naturally isomorphic to the completed tensor product $F \widehat{\otimes}_W \widehat{R}_{\mathfrak{m}}$ where F and W are given the discrete topology and $\widehat{R}_{\mathfrak{m}}$ is given its topology as $\mathfrak{p}_{\underline{x}}$-adic completion of R. In even more concrete terms, this is a jazzed-up version of the assertion that if we form the T-adic completion of $F \otimes_W (W[\![T]\!])$ then we naturally get $F[\![T]\!]$ (beware that we're not actually assuming formal smoothness over W for the deformation ring $\mathscr{R}_0$ in the lemma, though such smoothness does hold in the cases to which we'll be applying this lemma later on).

PROOF. Although we could give a proof using universal algebraic families, for esthetic reasons we prefer to give a proof entirely within the framework of deformation theory, as this clarifies the general nature of the argument (i.e., the role of elliptic curves and level structures is rather inessential to the argument). Essentially, the same method will apply "whenever" one is studying a moduli problem which is finite over one whose deformation rings are formally smooth, whose Isom-schemes are finite unramified, and whose universal deformations are algebraizable over the base. First, we remove the appearance of $\Gamma_0(N)$-structures (which we view as pairs consisting of an elliptic curve and an auxiliary subgroup scheme with $\Gamma_0(N)$-structure). Let the complete local noetherian W-algebra R_0 denote the universal deformation ring for the elliptic curve underlying $\underline{x}_0$, and let $I_{\underline{x}}$ denote the ideal corresponding to the W-section given by the elliptic curve underlying $\underline{x}$.

The functor of $\Gamma_0(N)$-structures on an elliptic curve is finite, so if R_0' denotes the finite R_0-algebra classifying such structures on deformations of the elliptic curve underlying $\underline{x}_0$, then R_0' is the product of finitely many local rings (as R_0 is local henselian), and the ring $\mathscr{R}_0$ is the unique local factor ring of R_0' corresponding to the $\Gamma_0(N)$-structure $\underline{x}_0$, or equivalently is the unique *local* factor ring of R_0' supporting the W-section arising from $\underline{x}$. It follows that the $F \otimes_W \mathscr{I}_{\underline{x}}$-adic completion of $F \otimes_W \mathscr{R}_0$ is a factor ring of the ring of $\Gamma_0(N)$-structures on the "universal" elliptic curve over the $F \otimes_W I_{\underline{x}}$-adic completion of $F \otimes_W R$. Moreover, this completion of $F \otimes_W \mathscr{R}_0$ also supports the F-section corresponding to x. Since $F \otimes_W (R/I_{\underline{x}}) = F$ and $F \otimes_W (\mathscr{R}_0/\mathscr{I}_{\underline{x}}) = F$ are *local*, so the indicated completions of $F \otimes_W \mathscr{R}_0$ and $F \otimes_W R$ are also local, it follows that the $F \otimes_W \mathscr{I}_{\underline{x}}$-adic completion of $F \otimes_W \mathscr{R}_0$ (along with the $\Gamma_0(N)$-structure over it!) is the desired universal deformation of x *provided* that the $F \otimes_W I_{\underline{x}}$-adic completion of $F \otimes_W R$ (along with the elliptic curve over it!) is the universal deformation of the elliptic curve underlying x. This latter statement is exactly the original problem for elliptic curves without the interference of level structures.

Throwing away the level structures, we begin with an elliptic curve E_0 over W/π and a deformation E of E_0 to W. Let R denote the universal deformation ring of E_0 and let $J \subseteq R$ be the ideal of the W-section corresponding to E. Define $R_\eta = F \otimes_W R$ and $J_\eta = F \otimes_W J$, so J_η is a maximal ideal in R_η (with residue field F). We need to prove that the J_η-adic completion of R_η (along with the elliptic curve over it) is the universal deformation of $E_{/F}$ (on the category of complete local noetherian F-algebras with residue field F). Let $\widetilde{R}_\eta$ be the universal deformation ring of $E_{/F}$, so we have a natural map

$$\widetilde{R}_\eta \to \widehat{(R_\eta)_{J_\eta}} \tag{9–1}$$

which we want to prove to be an isomorphism.

From the deformation theory of elliptic curves, we know that both source and target in (9–1) are formal power series rings over F of the same dimension (which happens to be 1, but we ignore this fact to maintain conceptual understanding of the situation). Thus, it suffices to prove that the map is surjective. More specifically, it suffices to prove that the map

$$\widetilde{R}_\eta \to R_\eta/J_\eta^2 = F \otimes_W (R/J^2) \tag{9–2}$$

is surjective. The whole point is to prove surjectivity onto the maximal ideal $F \otimes_W (J/J^2)$, with J/J^2 a *finite free* W-module. Assuming failure of surjectivity, we can choose an appropriate codimension 1 lattice in J/J^2 that contains the entire image of the maximal ideal of $\widetilde{R}_\eta$ under (9–2) (after tensoring with F). Thus, we can find a W-algebra quotient of $R = W \oplus J$ of the form $W[\varepsilon]$ (with (ε) the image of J) so that the resulting elliptic curve deformation over $F[\varepsilon]$ is trivial. In other words, we will have a nontrivial deformation $\mathscr{E}$ of $E_{/W}$ to $W[\varepsilon]$ which induces a trivial deformation of $E_{/F}$ over $F[\varepsilon]$. We now show that such a deformation cannot exist.

Consider the Isom-scheme $I = \underline{\mathrm{Isom}}(\mathscr{E}, E_{/W[\varepsilon]})$ over $W[\varepsilon]$ which classifies elliptic curve isomorphisms over variable $W[\varepsilon]$-schemes (*not* necessarily respecting structures as deformations of $E_{/W}$!). This is a finite unramified $W[\varepsilon]$-scheme (by the deformation theory of elliptic curves), and we just have to prove that $I(W[\varepsilon]) \to I(F[\varepsilon])$ is surjective, since an isomorphism $\mathscr{E} \simeq E_{/W[\varepsilon]}$ over $W[\varepsilon]$ which induces an isomorphism of $F[\varepsilon]$-*deformations* of $E_{/F}$ is automatically an isomorphism of *deformations* of $E_{/W}$ (as the "generic fiber" functor from flat separated W-schemes to F-schemes is faithful). Since $W[\varepsilon]$ is a henselian local ring, any finite unramified $W[\varepsilon]$-scheme can be realized as a closed subscheme of a finite étale $W[\varepsilon]$-scheme. But since $W[\varepsilon]$ is even strictly henselian, it follows that the only finite étale $W[\varepsilon]$-algebras are finite products of copies of $W[\varepsilon]$. Thus, I is a finite disjoint union of spectra of quotients of $W[\varepsilon]$. But the ideal theory of $W[\varepsilon]$ is sufficiently easy that we see by inspection that the only quotient of $W[\varepsilon]$ which admits a $W[\varepsilon]$-algebra map to $F[\varepsilon]$ is $W[\varepsilon]$ itself. Thus, $I(W[\varepsilon]) \to I(F[\varepsilon])$ is surjective (and even bijective). □

Now we return to our original situation with the universal deformation ring $\mathscr{R}_0$ of $\underline{x}_0$. We have a natural map $\widehat{\mathscr{O}}_{\underline{X},\underline{x}} = \widehat{\mathscr{O}}_{\underline{X},\underline{x}_0} \to \mathscr{R}_0$ which computes the subring of invariants in $\mathscr{R}_0$ under $\mathrm{Aut}_{W/\pi}(\underline{x}_0)$. After inverting π and passing to completions along the sections defining x, the source ring becomes $\widehat{\mathscr{O}}_{X,x}$ while the target ring becomes the universal deformation ring $\mathscr{R}_\eta$ of x, thanks to Lemma 9.1. Moreover, by chasing universal (algebraized) objects in Lemma 9.1 (and projecting to artinian quotients of $\mathscr{R}_\eta$), we see that this induced natural map

$$\widehat{\mathscr{O}}_{X,x} \to \mathscr{R}_\eta \tag{9–3}$$

is indeed the canonical isomorphism of $\widehat{\mathscr{O}}_{X,x}$ onto the subring of $\mathrm{Aut}_F(x)$-invariants in the universal deformation ring $\mathscr{R}_\eta$ of x. Since the group $\mathrm{Aut}_F(x)/\{\pm 1\}$ of order u_x acts faithfully on the universal deformation ring $\mathscr{R}_\eta \simeq F[\![T]\!]$, so the totally (tamely) ramified extension (9–3) has ramification degree u_x, we conclude that the generator θ_x of $(\widehat{\Omega}^1_{\widehat{\mathscr{O}}_{X,x}/F})^{\otimes k}(-r_x)$ maps to a generator of $(\widehat{\Omega}^1_{\mathscr{R}_\eta/F})^{\otimes k}(-(r_x u_x + k(u_x - 1)))$, where the twisting notation "(n)" denotes tensoring with the nth power of the inverse of the maximal ideal; this amounts to nothing more than the calculation that if $T' = \mathrm{unit} \cdot T^u$ then

$$T'^r (\mathrm{d}T')^{\otimes k} = \mathrm{unit}' \cdot T^{ru + k(u-1)} (\mathrm{d}T)^{\otimes k} \tag{9–4}$$

in $\widehat{\Omega}_{F[\![T]\!]/F}$ (since $u \in F^\times$).

If we let $\mathscr{J}$ denote the (invertible!) height 1 prime ideal of the W-section $\underline{x}$ of the formally smooth deformation ring $\mathscr{R}_0$ and let $\mathscr{J}_\eta$ denote the ideal of the F-section x of $\mathscr{R}_\eta$, then for any integer m we have a natural isomorphism of 1-dimensional F-vector spaces

$$(\widehat{\Omega}^1_{\mathscr{R}_\eta/F})^{\otimes k}(m)/\mathscr{J}_\eta \simeq F \otimes_W \left((\widehat{\Omega}^1_{\mathscr{R}_0/W})^{\otimes k}(m)/\mathscr{J}\right),$$

where on the right side we are twisting relative to the invertible ideal $\mathscr{J}$. The left side is a 1-dimensional F-vector space and the "integral" differentials on the right side provide a natural rank 1 lattice in this vector space. In this way, taking $m = -(r_x u_x + k(u_x - 1))$ enables us to define an integer $\mathrm{ord}_{v,x}(\theta)$ which measures the extent to which θ_x mod $\mathscr{J}_\eta$ fails to arise from a generator of the lattice of "integral" differential tensors. In more explicit terms, if we compute the formal expansion of the tensor θ relative to a formal parameter along the section $\underline{x}$ of the "integral" deformation ring $\mathscr{R}_0$ of $\underline{x}_0$, then we get a leading coefficient in $F^\times$ which is well-defined up to unit, and $\mathrm{ord}_{v,x}(\theta)$ is just the ord_v of this coefficient.

With these preliminary considerations settled, we are now in position to state the main local formula:

THEOREM 9.2. *With notation as defined above, including C_x defined as in* (8–5), *we have*

$$(x.x)_{v,\omega_x} = \frac{1}{2}\sum_{n\geq 0}\left(|\operatorname{Aut}_{W_n}(\underline{x})| - |\operatorname{Aut}_W(\underline{x})|\right) + \frac{\mathrm{ord}_v(C_x u_x^k) - \mathrm{ord}_{v,x}(\theta)}{r_x + k} \tag{9–5}$$

REMARK 9.3. Despite possible appearances to the contrary, we will see in Section 10 that (9–5) with $\theta = \Delta$ recovers [GZ, Ch. III, Lemma 8.2] at nonelliptic points.

REMARK 9.4. If x arises from a global point over H and we choose a global θ and a global ω_x, then C_x, u_x, and r_x all have global meaning independent of v (e.g., $2u_x$ is the order of the geometric automorphism group of x, which can be computed over $\overline{\mathbf{Q}}$, $\mathbf{C}$, or an algebraic closure of the fraction field of W). We regard the first term on the right side of (9–5) as the interesting part, and the rest as "junk". The only junk term which is somewhat subtle is $\mathrm{ord}_{v,x}(\theta)$, since it is defined via formal deformation theory at v and hence is not obviously ord_v of some globally defined quantity when x, ω_x, and θ_x are globally defined at the start. Thus, we will need to do a little work to explicate the "global" meaning of $\mathrm{ord}_{v,x}(\theta)$ in such situations. When we take $k = 6$ and $\theta = \Delta$, then typically $r_x = 0$ and $\mathrm{ord}_{v,x}(\theta) = 0$. However, at elliptic points we may have $r_x \neq 0$ and when $v|N$ we may have $\mathrm{ord}_{v,x}(\theta) \neq 0$. The choice $\theta = \Delta$ is the one we will use later on in order to explicate formulas for global height pairings.

For our purposes the extra local "junk" terms will not be problematic because when we form the sum over *all* places v of H (in the context of Heegner points), then by the product formula these terms will essentially add up to zero once the archimedean analogue is adapted to our method of computation and once we have found a more conceptual interpretation of $\mathrm{ord}_{v,x}(\theta)$ for a well-chosen θ (namely, $\theta = \Delta$). It is crucial for such an application of the product formula that the denominator $r_x + k$ in (9–5) does *not* depend on v when $x \in \underline{X}(F)$ arises from a global point over H and θ, ω_x are globally chosen over H.

Now we prove Theorem 9.2.

PROOF. Although $\underline{X}$ need not be regular, as usual it suffices to work with this model for our intersection theory calculations because $\underline{x}$ lies in the W-smooth locus on this model (see (8–8)). We also observe that both sides of (9–5) transform the same way under a change of ω_x. Indeed, if we use $\omega'_x = c\omega_x$ for some $c \in F_v^\times$ and if $\alpha_x\omega_x$ is a basis of $\mathrm{Cot}_{\underline{x}}(\underline{X}_v)$, then $\alpha'_x = \alpha_x/c$ makes $\alpha'_x\omega'_x = \alpha_x\omega_x$ an $\mathscr{O}_{F_v}$-basis of the cotangent line along $\underline{x}$, so

$$(x.x)_{v,\omega'_x} = \mathrm{ord}_v(\alpha'_x) = -\mathrm{ord}_v(c) + \mathrm{ord}_v(\alpha_x) = -\mathrm{ord}_v(c) + (x.x)_{v,\omega_x},$$

while by (8–6) we have $\mathrm{ord}_v(C'_x)/(r_x+k) = -\mathrm{ord}_v(c)+\mathrm{ord}_v(C_x)/(r_x+k)$. Since u_x and $\mathrm{ord}_{v,x}(\theta)$ and the Aut-terms on the right side of (9–5) do not depend on ω_x, we conclude that indeed both sides of (9–5) transform in the same way under change in ω_x. Thus, it suffices to prove the result for one choice of ω_x. We will make the choice later in the proof.

Let $G = \mathrm{Aut}_{W/\pi}(\underline{x}_0)/\{\pm 1\}$, a quotient of the group of automorphisms of $\underline{x}_0$ as a $\Gamma_0(N)$-structure. Let $\mathscr{R}_0$ denote the universal deformation ring of $\underline{x}_0$, so there is a natural *injective* W-algebra action map

$$[\cdot] : G \to \mathrm{Aut}_W(\mathscr{R}_0),$$

an abstract isomorphism $\mathscr{R}_0 \simeq W[\![T_0]\!]$, and an isomorphism $\widehat{\mathscr{O}}_{\underline{X},\underline{x}_0} \simeq \mathscr{R}_0^G$, where T_0 cuts out the deformation $\underline{x}$ over W. The norm $T_{\underline{x}} \overset{\mathrm{def}}{=} \mathrm{Norm}_G(T_0) = \prod_{g\in G}[g](T_0)$ is a formal parameter of $\mathscr{R}_0^G \simeq \widehat{\mathscr{O}}_{\underline{X},\underline{x}_0} \simeq \widehat{\mathscr{O}}_{\underline{X},\underline{x}}$ over W, so $T_{\underline{x}}$ serves as a formal parameter along $\underline{x}$ on $\underline{X}$.

We now give a deformation-theoretic interpretation of

$$\frac{1}{2}\sum_{n\geq 0}\big(|\mathrm{Aut}_{W_n}(\underline{x})| - |\mathrm{Aut}_W(\underline{x})|\big). \tag{9–6}$$

The method we use applies to a rather more general class of deformation problems with formally smooth deformation rings of relative dimension 1 (as the following abstract argument makes clear). Since a representative in $\mathrm{Aut}_{W/\pi}(\underline{x})$ of $g \in G$ lifts to W_n (as an automorphism of $\Gamma_0(N)$-structures) if and only if its negative lifts, it makes to speak of an element of G lifting to W_n. Thus, by including the factor of $1/2$ we see that the nth term in (9–6) counts the number of $g \in G$ which lift to W_n but don't lift to W. But $g \in G$ lifts to W_n if and only if T_0 and $[g](T_0)$ generate the same ideal in the universal *deformation* ring $\mathscr{R}_0/\pi^{n+1}$ of $\underline{x}$ mod π^{n+1} over W_n-algebras (with residue field W/π). Since the quotients $\mathscr{R}_0/(\pi^{n+1}, T_0)$ and $\mathscr{R}_0/(\pi^{n+1}, [g](T_0))$ are abstractly W_n-isomorphic (via $[g]$) and hence have the same *finite* length, a surjection of $\mathscr{R}_0$-algebras $\mathscr{R}_0/(\pi^{n+1}, [g](T_0)) \twoheadrightarrow \mathscr{R}_0/(\pi^{n+1}, T_0)$ is necessarily an isomorphism. Thus, we conclude that g lifts to W_n if and only if $[g](T_0)$ is a multiple of T_0 in $\mathscr{R}_0/\pi^{n+1}$. By viewing $[g](T_0) \in \mathscr{R}_0 \simeq W[\![T_0]\!]$ as a formal power series in T_0, to say that $[g](T_0)$ is a multiple of T_0 modulo π^{n+1} amounts to saying that the constant term $[g](T_0)(0)$ (i.e., the image of $[g](T_0)$ in $\mathscr{R}_0/(T_0) \simeq W$) is divisible by π^{n+1}.

We conclude that $g \in G$ doesn't lift to W if and only if $[g](T_0)$ has nonzero constant term $[g](T_0)(0)$, in which case g lifts to W_n if and only if $n+1 \leq \mathrm{ord}_v([g](T_0)(0))$. It follows that (9–6) is exactly the sum of the ord_v's of the nonzero constant terms among the $[g](T_0)$'s. Equivalently, when we consider the formal parameter $T_{\underline{x}} = \mathrm{Norm}_G(T_0) = \prod_{g \in G}[g](T_0)$ along $\underline{x}$ as an element in the universal deformation ring $\mathscr{R}_0 \simeq W[\![T_0]\!]$, the ord_v of its least degree nonzero coefficient is exactly equal to (9–6).

Since the G-norm $T_{\underline{x}}$ of T_0 formally cuts out $\underline{x} \in \underline{X}(W)$, it also induces a uniformizer in the complete local ring at $x \in X(F)$. Thus, when $T_{\underline{x}}$ is viewed as an element in $W[\![T_0]\!] \subseteq F[\![T_0]\!]$ (with the latter ring naturally identified with the universal deformation ring of x, by Lemma 9.1), it has the form

$$T_{\underline{x}} = b_x T_0^{u_x} + \cdots \tag{9–7}$$

with $b_x \neq 0$ and $u_x = (1/2)|\mathrm{Aut}_F(x)|$ the ramification degree of the universal deformation ring $F[\![T_0]\!]$ over $\widehat{\mathscr{O}}_{X,x}$. We have seen already that when the G-norm $T_{\underline{x}}$ of T_0 is expanded as a product of $[g](T_0)$'s, the product of the nonzero constant terms among the $[g](T_0)$'s (taken with respect to T_0-adic expansions) has ord_v equal to (9–6). This product of constant terms is b_x up to $W^\times$-multiple (arising from the g's lifting to W, for which $[g](T_0)$ is a unit multiple of T_0), so we conclude that (9–6) is exactly $\mathrm{ord}_v(b_x)$.

We will take ω_x to be represented by the formal parameter $t_x = T_{\underline{x}}$ in $\widehat{\mathscr{O}}_{X,x}$ (an inspection of our definition of $(x.x)_{v,\omega_x}$ and the calculation (8–8) makes it clear that we may work with formal uniformizers that don't necessarily arise from rational functions in $\mathscr{O}_{X,x}$ on the algebraic curve $X_{/F}$). In $(\widehat{\Omega}^1_{\widehat{\mathscr{O}}_{X,x}/F})^{\otimes k}(-r_x \cdot x)$, by (9–7) we have

$$\theta = (C_x T_{\underline{x}}^{r_x} + \cdots)(\mathrm{d}T_{\underline{x}})^{\otimes k} = (C_x b_x^{r_x+k} u_x^k T_0^{r_x u_x + k(u_x - 1)} + \cdots)(\mathrm{d}T_0)^{\otimes k}.$$

Thus, by definition we have

$$\mathrm{ord}_{v,x}(\theta) = \mathrm{ord}_v(C_x b_x^{r_x+k} u_x^k) = \mathrm{ord}_v(C_x u_x^k) + (r_x + k)\mathrm{ord}_v(b_x). \tag{9–8}$$

Now let's consider the main identity (9–5). Since θ has its $T_{\underline{x}}$-adic expansion with leading coefficient C_x, where $T_{\underline{x}}$ is a formal parameter along the W-section $\underline{x}$ on $\underline{X}$, we take $\beta_x = 1/C_x$ in (8–8), so $(x.x)_{v,\omega_x} = 0$. Meanwhile, by (9–8) the right side of (9–5) is equal to

$$\mathrm{ord}_v(b_x) + \frac{\mathrm{ord}_v(C_x u_x^k) - \mathrm{ord}_{v,x}(\theta)}{r_x + k} = 0 = (x.x)_{v,\omega_x}. \qquad \square$$

Before we move on to establish formulas for $(\underline{x}.T_m(\underline{x}^\sigma))_v$ — when properly defined — in cases with $r_{\mathscr{A}}(m) > 0$, it is convenient to introduce some shorthand so as to avoid having to carry out the "junk" local terms from (9–5). For a $\Gamma_0(N)$-diagram $\underline{x}$ over W whose associated section in $\underline{X}(W)$ lies in $\underline{X}^{\mathrm{sm}}(W)$ (and is *not* assumed to arise from a Heegner point, though Heegner points *do* satisfy this hypothesis), we define

$$(x.x)_v^{\mathrm{GZ}} = \frac{1}{2}\sum_{n\geq 0}(|\operatorname{Aut}_{W_n}(\underline{x})| - |\operatorname{Aut}_W(\underline{x})|). \tag{9–9}$$

We emphasize that (9–9) is in general not equal to our self-intersection pairing, but we now explain why it does coincide with the one used in [GZ, Ch. III, § 8] away from elliptic points and characteristics dividing N. This requires a definition.

DEFINITION 9.5. For $\underline{x}$ as above which is nonelliptic (i.e., $u_x = 1$, so $\mathrm{ord}_x(\Delta) = (6/u_x)(u_x - 1) = 0$) and factors through the relative smooth locus, let ω denote a 1-form near x lifting a nonzero cotangent vector denoted ω_x at x. Let $g_\omega = \omega^{\otimes 6}/\Delta$, a nonzero rational function with no zero or pole at x. Define

$$(x.x)^{\eta^4} = (x.x)^{\omega_x} + \tfrac{1}{6}\mathrm{ord}_v(g_\omega(x)).$$

This is independent of the choice of ω (and ω_x).

To see the independence of the choice of ω (and ω_x), we note that using $\omega' = h\omega$ with a rational function h requires h to not have a zero or pole at x and hence the equalities $\omega'_x = h(x)\omega_x$ and $g_{\omega'} = h^6 g_\omega$ force

$$(x.x)^{\omega'_x} + \tfrac{1}{6}\mathrm{ord}_v(h(x)^6 g_\omega(x)) = (x.x)^{\omega_x} - \mathrm{ord}_v(h(x)) + \mathrm{ord}_v(h(x)) + \tfrac{1}{6}\mathrm{ord}_v(g_\omega(x)).$$

The reason for the notation $(x.x)^{\eta^4}$ is that if η^4 actually made sense as a meromorphic 1-form on $X_0(N)$ and were nonvanishing and regular at x then we would recover the old definition of "intersection theory with a cotangent vector" (based on η_x^4). To see this, if we pretend η^4 lives on X then we can write $\Delta = (\eta^4)^6$ and hence formally

$$\tfrac{1}{6}\mathrm{ord}_v(g_\omega(x)) = \mathrm{ord}_v\Big(\frac{\omega}{\eta^4}(x)\Big),$$

so

$$(x.x)^{\eta^4} = (x.x)^{\omega_x} + \mathrm{ord}_v((\omega/\eta^4)(x)) \tag{9–10}$$

and since $\omega_x = \eta_x^4 \cdot (\omega/\eta^4)(x)$ we could use (8–4) to thereby "recover" the definition of self-intersection at x using the hypothetical cotangent vector η_x^4.

In any case, to actually compute the intrinsic $(x.x)^{\eta^4}$ we may work formally at x and so can choose $\omega = dT$ with T a formal parameter along the section $\underline{x} \in \underline{X}^{\mathrm{sm}}(W)$. In terms of earlier notation from (8–5) with $\theta = \Delta$ we have $g_\omega(x) = 1/C_x$. Combining this with the equalities $u_x = 1$ and $r_x + k = 6$, we conclude from Theorem 9.2 that

$$(x.x)^{\eta^4} = (x.x)^{\omega_x} - \tfrac{1}{6}\mathrm{ord}_v(C_x) = (x.x)^{\mathrm{GZ}} - \tfrac{1}{6}\mathrm{ord}_{v,x}(\Delta).$$

But we will prove in Lemma 10.1 that under the above hypotheses, $\mathrm{ord}_{v,x}(\Delta)/6 = \mathrm{ord}_v(\bar{\mathfrak{n}})$, where $(N) = \mathfrak{n}\bar{\mathfrak{n}}$ with $\mathfrak{n}$ killing the $\Gamma_0(N)$-structure on x. Thus, as long as $v \nmid N$, we have $(x.x)^{\eta^4} = (x.x)^{\mathrm{GZ}}$. It is the modified intersection pairing based on Definition 9.5 which is used in the intersection theory considerations in [GZ, Ch. III, § 8] (so Theorem 9.2 with $v \nmid N$ really does recover [GZ, Ch. III, Lemma 8.2]), but now that we have seen how to recover the point of view in [GZ] at nonarchimedean places via our perspective if one avoids elliptic points, we won't make any further reference to Definition 9.5 because when $v|N$ or $v|\infty$ our local terms seem different from the ones in [GZ] (but the *global* sums coincide).

We define $\langle x, x\rangle_v^{\mathrm{GZ}} \overset{\mathrm{def}}{=} -(x.x)_v^{\mathrm{GZ}} \log(q_v)$. Since $\mathrm{ord}_x(T_m(x^\sigma)) = r_{\mathscr{A}}(m)$, by Theorem 9.2 this definition then yields

$$(x.T_m(x^\sigma))_{v,\omega_x} = (x.T_m(x^\sigma))_v^{\mathrm{GZ}} + r_{\mathscr{A}}(m)\left(\frac{\mathrm{ord}_v(C_x u_x^k) - \mathrm{ord}_{v,x}(\theta)}{r_x + k}\right),$$

where we define $(\,\cdot\,)_v^{\mathrm{GZ}}$ to be essentially the usual intersection pairing except with (9–9) replacing the self-intersection formula, and

$$\theta = (C_x t_x^{r_x} + \cdots)(\mathrm{d}t_x)^{\otimes k},$$

with t_x a uniformizer representing ω_x (and C_x only depending on ω_x, not t_x). We also define

$$\langle x, T_m(x^\sigma)\rangle_v^{\mathrm{GZ}} = -(x.T_m(x^\sigma))_v^{\mathrm{GZ}} \log(q_v) \tag{9–11}$$

using (9–9) and the canonical local height pairing.

Define $c = x - \infty$, $d = x - 0$. Using Theorem 8.4 and Theorem 9.2, for nonarchimedean places v of H

$$\langle c, T_m(d^\sigma)\rangle_{v,\omega_x} = \langle x, T_m(x^\sigma)\rangle_v^{\mathrm{GZ}} + r_{\mathscr{A}}(m)\frac{\log|C_x u_x^k|_v - \log|\theta|_{x,v}}{r_x + k}, \tag{9–12}$$

where $|\theta|_{x,v} \overset{\mathrm{def}}{=} q_v^{-\mathrm{ord}_{v,x}(\theta)}$ denotes the v-adic absolute value of the leading coefficient of θ when expanded relative to a formal parameter along $\underline{x}$ in the universal deformation ring at $\underline{x}_0$. Here we have used that the closure of $T_m(0)$ is $\sigma_1(m).\underline{0}$ and that $\underline{x}$ and $T_m(\underline{x^\sigma})$ are disjoint from the cuspidal locus since CM points have everywhere potentially good reduction.

We define

$$\mathfrak{a}_x = \prod_{v \nmid \infty} \mathfrak{p}_v^{\mathrm{ord}_{v,x}(\theta)}. \tag{9–13}$$

By Lemma 10.1, when $\theta = \Delta$ and $u_x = 1$ then $\mathfrak{a}_x = \bar{\mathfrak{n}}^6$. We conclude that for every nonarchimedean place v of H,

$$\langle c, T_m(d^\sigma)\rangle_{v,\omega_x} = \langle x, T_m(x^\sigma)\rangle_v^{\mathrm{GZ}} + r_{\mathscr{A}}(m)\frac{\log|C_x u_x^k \mathfrak{a}_x^{-1}|_v}{r_x + k}, \tag{9–14}$$

where the factor $C_x u_x^k$ on the right side is an element in $H^\times$ and $\mathfrak{a}_x$ is a fractional ideal. This puts us in excellent position to use the product formula to check that the junk terms will (almost) globally sum to 0 (and hence we can ignore them). What we need to do is work out the appropriate archimedean analogue of the calculation (9–14).

Fix a complex place v of H and let $\mathbf{C}_v$ denote H_v (with $\mathbf{C}_v$ noncanonically isomorphic to $\mathbf{C}$). Fix a choice of $\sqrt{-1} \in \mathbf{C}_v$ to define an orientation on $\mathbf{C}_v$-manifolds, so we may realize the associated upper half-space $\mathfrak{h}_v$ as the base of a universal $\mathbf{C}_v$-analytic family of elliptic curves with trivialized (oriented) relative homology. We thereby get a canonical $\mathbf{C}_v$-analytic "uniformization" $\pi_v : \mathfrak{h}_v \to X_0(N)^{\mathrm{an}}_{/\mathbf{C}_v} = X_0(N)(\mathbf{C}_v)$ and we may (and do) choose $z_v \in \mathfrak{h}_v$ lifting $x_v \in X_0(N)(\mathbf{C}_v)$. The pullback tensor $\pi_v^*(\theta_v^{\mathrm{an}})$ has order $r_x u_x + k(u_x - 1)$ at z_v because of a calculation such as in (9–4) and the fact that $\mathfrak{h}_v$ is the base of a universal family for an analytic moduli problem which is étale over the $\Gamma_0(N)$-moduli problem in the $\mathbf{C}_v$-analytic category (so π_v computes the same ramification degrees u_x). Let g_{z_v} be a local analytic uniformizer at z_v on $\mathfrak{h}_v$ which enjoys the property that the g_{z_v}-adic analytic expansion

$$\pi_v^*(\theta_v) = (g_{z_v}^{r_x u_x + k(u_x-1)} + \cdots)(\mathrm{d}g_{z_v})^{\otimes k} \tag{9–15}$$

has a leading coefficient of 1. Such a g_{z_v} exists since $\mathbf{C}_v$ is algebraically closed. We will use this g_{z_v} shortly (and we do not care if g_{z_v} extends meromorphically to all of $\mathfrak{h}_v$).

By Theorem 8.3, we have

$$\langle c, T_m(d^\sigma)\rangle_{v,\omega_x} = \lim_{y \to x} (\langle c_y, T_m(d^\sigma)\rangle_v - r_{\mathscr{A}}(m) \log |t_x(y)|_v),$$

where $y \in X_0(N)(\mathbf{C}_v) - \{x\}$ converges to x and c_y is the divisor obtained from c by replacing x with y. Combining this identity with (9–14), we may sum over *all* places v of H and exploit the product formula for $C_x u_x^k \in H^\times$ and the identity

$$\sum_{v \nmid \infty} \log |\mathfrak{a}_x^{-1}|_v = \log |\mathrm{N}_{H/\mathbf{Q}}(\mathfrak{a}_x)|$$

(with this fractional ideal norm viewed as a positive rational number) to obtain a formula for the global height pairing: $\langle c, T_m(d^\sigma)\rangle$ is equal to

$$\sum_{v \nmid \infty} \langle x, T_m(x^\sigma)\rangle_v^{\mathrm{GZ}} + \frac{r_{\mathscr{A}}(m) \log |\mathrm{N}_{H/\mathbf{Q}}(\mathfrak{a}_x)|}{r_x + k}$$
$$+ \sum_{v \mid \infty} \lim_{y \to x} \left(\langle c_y, T_m(d^\sigma)\rangle_v - \frac{r_{\mathscr{A}}(m)}{r_x + k}((r_x + k) \log |t_x(y)|_v + \log |C_x u_x^k|_v) \right).$$

To put this into a more useful form, we need to examine the term

$$(r_x + k) \log |t_x(y)|_v + \log |C_x u_x^k|_v$$

for $v|\infty$. We may write

$$\pi_v^*(t_x^{\mathrm{an}}) = \alpha_{v,x} h_{v,x} g_{z_v}^{u_x}$$

for some $\alpha_{v,x} \in \mathbf{C}_v^\times$ and $h_{v,x}$ analytic near z_v with $h_{v,x}(z_v) = 1$. Since $\theta_v = (C_x t_x^{r_x} + \cdots)(\mathrm{d}t_x^{\otimes k})$, we get

$$\pi_v^*(\theta_v^{\mathrm{an}}) = (C_x \alpha_{v,x}^{r_x+k} u_x^k g_{z_v}^{u_x r_x + k(u_x-1)} + \cdots)(\mathrm{d}g_{z_v})^{\otimes k}.$$

We conclude via (9–15) that $C_x u_x^k \alpha_{v,x}^{r_x+k} = 1$. Thus, $(r_x+k)\log|\alpha_{v,x}|_v = -\log|C_x u_x^k|_v$. Taking y_v in $\mathfrak{h}_v$ near z_v and lying over $y \in X_v(\mathbf{C}_v)$, we obtain

$$\begin{aligned}(r_x+k)\log|t_x(y)|_v + \log|C_x u_x^k|_v \\ &= (r_x+k)(\log|\alpha_{v,x}|_v + \log|h_{v,x}(y_v)|_v + u_x \log|g_{z_v}(y_v)|_v) + \log|C_x u_x^k|_v \\ &= (r_x+k)u_x \log|g_{z_v}(y_v)|_v + (r_x+k)\log|h_{v,x}(y_v)|_v,\end{aligned}$$

where $h_{v,x}(y_v) \to 1$ as $y_v \to z_v$ in $\mathfrak{h}_v$. Thus, when forming the limit as $y \to x$ in $X_v(\mathbf{C}_v)$ we can drop the $h_{v,x}(y)$ term and hence arrive at the main result:

THEOREM 9.6. *Let $x \in X_0(N)(H)$ be a Heegner point, and let u_x be half the size of the geometric automorphism group of x (so $u_x = \frac{1}{2}|\mathscr{O}_K^\times|$). Choose a nonzero rational section θ of $(\Omega^1_{X_0(N)/H})^{\otimes k}$ with $r_x \overset{\mathrm{def}}{=} \mathrm{ord}_x(\theta) \neq -k$. Let $\mathfrak{a}_x$ be the fractional ideal of H constructed via deformation theory as in* (9–13). *Define $c = x - \infty$, $d = x - 0$.*

The global height pairing $\langle c, T_m(d^\sigma)\rangle$ is equal to

$$\sum_{v\nmid\infty} \langle x, T_m(x^\sigma)\rangle_v^{\mathrm{GZ}} + \sum_{v|\infty} \lim_{y_v \to z_v} (\langle c_{y_v}, T_m(d^\sigma)\rangle_v - u_x r_{\mathscr{A}}(m)\log|g_{z_v}(y_v)|_v) + \frac{r_{\mathscr{A}}(m)\log|\mathrm{N}_{H/\mathbf{Q}}\mathfrak{a}_x|}{r_x+k}, \qquad (9\text{–}16)$$

where

- *the local term $\langle x, T_m(x^\sigma)\rangle_v^{\mathrm{GZ}}$ for $v \nmid \infty$ is defined by* (9–9) *and* (9–11),
- *for $v|\infty$, $z_v \in \mathfrak{h}_v$ projects onto x_v^{an} under the analytic uniformization $\pi_v : \mathfrak{h}_v \to X_0(N)^{\mathrm{an}}_{/H_v}$,*
- *g_{z_v} is an analytic uniformizer at z_v with respect to which the local analytic expansion of the meromorphic k-tensor $\pi_v^*(\theta_v^{\mathrm{an}})$ has leading coefficient equal to* 1.
- *for $v|\infty$, the limit runs over $y_v \in \mathfrak{h}_v - \{z_v\}$ converging to z_v,*

As we have seen above, the local factors of $\mathfrak{a}_x$ are determined by the incarnation of θ in local deformation theory on good models for x. There is a particularly nice choice to be made in our situation, namely $k = 6$ and $\theta = \Delta$. Since Δ *over* $\mathbf{C}$ is nowhere vanishing away from the cusps, this choice renders

$$r_x u_x + k(u_x - 1) = 0 \qquad (9\text{–}17)$$

for all x; compare with (9–15). In particular, $r_x + k \neq 0$ for all x (so we may indeed use Δ as the basic tensor in the preceding considerations). Consequently, since $\Delta(q)(\mathrm{d}q/q)^{\otimes 6} = ((2\pi i)\eta^4(z)\mathrm{d}z)^{\otimes 6}$, with this choice it is trivial to check that at *any* (fixed) point $z_v \in \mathfrak{h}_v$ we may take the analytic uniformizer $g_{z_v}(z') = (2\pi i)\eta^4(z_v)(z' - z_v)$ for varying $z' \in \mathfrak{h}_v$. With this choice, the contribution from

$v|\infty$ in (9–16) is an "explicit" limit involving archimedean local heights and certain analytic functions g_{z_v} on the upper half-plane over $H_v \simeq \mathbf{C}$, and is exactly the archimedean height contribution which is computed in [GZ, Ch. II, (5.3), (5.5)]. Since $\log p$'s for distinct rational primes p are $\mathbf{Q}$-linearly independent, it follows that Theorem 9.6 in this case must agree place-by-place (over $\mathbf{Q}$) with the global formula used in [GZ], where the contribution in (9–16) over a prime p comes from the first and third pieces of the formula (upon factoring out a $-\log q_v$, using Theorem 8.4):

$$\langle c, T_m d^\sigma \rangle_p \stackrel{\text{def}}{=} -\sum_{v|p} ((x.T_m(x^\sigma))_v^{\text{GZ}} - \frac{r_{\mathscr{A}}(m)}{r_x + k} \text{ord}_{v,x}(\theta)) \log q_v. \tag{9–18}$$

The aim of Section 10 is to make the v-term on the right side of (9–18) explicit when $\theta = \Delta$; see Theorem 10.5. The appendix uses this to give an explication of (9–18) depending on the splitting behavior of p in K.

The method we have outlined is of fairly general nature for computing global height pairings on coarse moduli schemes. The real problem in any given situation is to make the archimedean height pairings and analytic functions g_{z_v} explicit enough to carry out computations, and to compute the ideal $\mathfrak{a}_x$ (this latter data being the nonarchimedean aspect which is sensitive to the choice of θ). The trick is to find a single well-understood θ that will satisfy $\text{ord}_x(\theta) \neq -k$ at *all* Heegner points. For our purposes, the choice $\theta = \Delta$ works best.

10. Quaternionic Explications

We are now in position to carry out the computation of the local term

$$(x.T_m(x^\sigma))_v^{\text{GZ}} - \frac{r_{\mathscr{A}}(m)}{r_x + k} \text{ord}_{v,x}(\theta) \tag{10–1}$$

from (9–18) for *any* value of $r_{\mathscr{A}}(m)$ but with $\theta = \Delta$. The difference (10–1) is the main geometric local contribution to the nonarchimedean contribution (9–18) to the global height pairing formula in Theorem 9.6, with the intersection pairing $(x.T_m(x^\sigma))_v^{\text{GZ}}$ computed by means of usual local intersection pairings for disjoint divisors and the modified self-pairing $(x.x)_v^{\text{GZ}}$ given by (9–9). In particular, we see that $(x.x)_v^{\text{GZ}} = 0$ whenever $\text{Aut}_W(\underline{x}) = \text{Aut}_{W/\pi}(\underline{x}_0)$.

Since the set $\text{Isom}_W(\underline{x}, \underline{y})$ of W-isomorphisms is empty when the corresponding generic geometric points x, y are distinct (thanks to Theorem 2.6), we conclude trivially that the identity

$$(x.y)_v^{\text{GZ}} = \frac{1}{2} \sum_{n \geq 0} |\,\text{Isom}_{W_n}^{\text{new}}(\underline{x}, \underline{y})| \tag{10–2}$$

holds for *all* (possibly equal) pairs of points $x, y \in X(F)$ whose associated W-points come from (necessarily unique up to isomorphism) $\Gamma_0(N)$-diagrams over W, where we define a "new" isomorphism as one not lifting to W. Keep in mind that when $x \neq y$ we are using Theorem 4.1, and in this formula the local intersection pairing on the left side is computed on the W-scheme $\underline{X} = X_0(N)_{/W}$ (which is often nonregular, but points arising from Heegner data and Hecke correspondences thereof lie in the relative smooth locus, so there is no ambiguity concerning these local intersection numbers).

Consequently, the exact same method which we used to prove Theorem 5.1 in the case $p \nmid m$ carries over to give the (finite) formula

$$(x.T_m(x^\sigma))_v^{\mathrm{GZ}} = \frac{1}{2} \sum_{n \geq 0} \left|\mathrm{Hom}_{W_n}^{\mathrm{new}}(\underline{x}^\sigma, \underline{x})_{\deg(m)}\right| \qquad (10\text{–}3)$$

whenever $v \nmid m$, where a "new" homomorphism is one not lifting to W. Once again, $\underline{x} \in \underline{X}(W)$ can be *any* noncuspidal section which is represented by a $\Gamma_0(N)$-diagram over W (and every W-point coming from a Heegner point over H has been seen to have this property), with x the generic fiber of $\underline{x}$ over W.

In order to make progress toward computing (10–1) explicitly when $\theta = \Delta$, we need to compute $\mathrm{ord}_{v,x}(\Delta)$. This is provided by:

LEMMA 10.1. *Let* $\underline{x} \in \underline{X}(W)$ *be disjoint from the cuspidal locus and represented by a* $\Gamma_0(N)$*-diagram. When* $v \nmid N$ *or* $v|\mathfrak{n}$, *then* $\mathrm{ord}_{v,x}(\Delta) = 0$. *When* $v|\bar{\mathfrak{n}}$ *then* $\mathrm{ord}_{v,x}(\Delta)/(r_x + 6) = u_x \mathrm{ord}_v(N)$. *In other words, for all* $v \nmid \infty$ $\mathrm{ord}_{v,x}(\Delta) = u_x(r_x + k)\mathrm{ord}_v(\bar{\mathfrak{n}}) = k \cdot \mathrm{ord}_v(\bar{\mathfrak{n}})$, *so* $\mathfrak{a}_x = \bar{\mathfrak{n}}^k = \bar{\mathfrak{n}}^6$.

REMARK 10.2. The method of proof is fairly abstract, so the same method should give a formula for $\mathfrak{a}_x$ in terms of $x : \mathrm{Spec}(\mathscr{O}_H) \to \mathscr{M}_{1,1}$ *if* θ merely comes from level 1 (over $\mathscr{O}_H$). Since we have to choose a specific θ eventually, it seems simplest to just prove Lemma 10.1 for $\theta = \Delta$ and to leave more general considerations to the reader's imagination.

PROOF. By (9–17), we have $(r_x + 6)u_x = 6$, so when $v|\mathfrak{n}$ we really want to prove $\mathrm{ord}_{v,x}(\Delta) = \mathrm{ord}_v(N^6)$. The exponent of 6 will arise from the fact that Δ is a 6-tensor. Let R denote the universal deformation ring of the elliptic curve E_0 underlying x_0, equipped with universal (algebraized) elliptic curve $E_{/R}$, and let R' denote the universal deformation ring of the $\Gamma_0(N)$-structure x_0, with ι the corresponding universal structure on the base change $E_{/R'}$ (with closed fiber ι_0 recovering the level structure x_0 enhancing E_0). By the general theory, since the $\Gamma_0(N)$-moduli problem is finite flat over the moduli stack of (smooth) elliptic curves it follows that R' is a finite flat R-algebra: it is simply the factor ring of the finite flat universal $\Gamma_0(N)$-structure algebra over R corresponding to the residual structure x_0 on E_0.

The crucial point is that when $v \nmid N$ or $v|\mathfrak{n}$ then $R \to R'$ is an *isomorphism*. Indeed, in the case $v \nmid N$ the N-torsion on deformations of E_0 is *étale* and hence is invisible from the point of view of deformation theory. For the same reason, when $v|N$ then the prime-to-p part of the $\Gamma_0(N)$-structure lifts uniquely to any deformation of E_0. All the action therefore happens in the p-part. If $v|\mathfrak{n}$ then the residual p-part of the level structure is multiplicative and more specifically is selected out functorially by the connected-étale sequence on the $p^{\mathrm{ord}_p(N)}$-torsion of deformations of E_0. Hence, we once again only have to deform E_0 and the level structure uniquely compatibly deforms for free, so again $R \to R'$ is an isomorphism. The case $v|\bar{\mathfrak{n}}$ is more subtle (the entire $\Gamma_0(N)$-structure is étale, but specifying the p-part on a deformation amounts to *splitting* the connected-étale sequence, so this imposes genuine extra data on deformations, leading to a bigger deformation ring than the one for E_0 alone).

Recall that in general $\mathrm{ord}_{v,x}(\theta)$ is defined as follows: we have an isomorphism $R' \simeq W[\![T]\!]$ with $T = 0$ corresponding to the W-structure x (recall that R' is regular and W-flat of dimension 2 by the general theory as in [KM], and it has a W-section arising from x and hence really does have to have the form $W[\![T]\!]$), and $\mathrm{ord}_{v,x}(\theta)$ is defined to be ord_v of the leading coefficient of the T-adic expansion of the k-tensor θ. It makes a big difference whether R' coincides with the universal deformation ring of E_0 or if it is genuinely bigger.

We first dispose of the cases $v \nmid N$ and $v|\mathfrak{n}$, and then will settle $v|\bar{\mathfrak{n}}$. As long as $v \nmid \bar{\mathfrak{n}}$, we have just seen that $R' = R$. Thus, if we write $W[\![T]\!]$ for the universal deformation ring of E_0 (with $T = 0$ cutting out the elliptic curve over W underlying x) then we must show that the 6-tensor Δ has T-adic expansion with unit leading coefficient. The crux of the matter is that Δ is a generator of the line bundle $\omega^{\otimes 12}$ on the (open) moduli stack $\mathscr{M}_{1,1}$, and the Kodaira–Spencer map $\mathrm{KS}_{1,1} : \omega \to \Omega^1_{\mathscr{M}_{1,1}/\mathbf{Z}}$ is what converts it into a 6-tensor on this stack. This is the usual recipe that converts even weight modular forms into tensors on modular curves. The general theory of the Kodaira–Spencer map ensures that $\mathrm{KS}_{1,1}$ is an *isomorphism* (we are working away from the cuspidal substack) precisely because the deformation theory of an elliptic curve (with the marked identity section!) coincides with the deformation theory of its underlying "bare" curve (as we can always lift the identity section, thanks to smoothness, and can then translate it via the group law to put it in the correct position). Consequently, the image of Δ in the invertible R-module $(\widehat{\Omega}^1_{R/W})^{\otimes 6}$ is a generator and hence it has unit leading coefficient.

There remains the case $v|\bar{\mathfrak{n}}$. Since $v|N$, so p is split in K, the elliptic curve underlying x is the Serre–Tate canonical lift of E_0. Moreover, the deformation theory of E_0 coincides with that of its p-divisible group. Thus, we may identify R with $W[\![T]\!]$ where $q = 1 + T$ is the so-called Serre–Tate parameter. The deformation theory of the $\Gamma_0(N)$-structure only matters through its p-part (as the other primary components deform uniquely). Let p^e be the p-part of N (with $e > 0$). We claim

$$R' = R[T']/((1+T')^{p^e} - (1+T)) \simeq W[\![T']\!]. \tag{10–4}$$

Once this is shown, then $T = p^e(T' + \cdots) + T'^{p^e}$, so $\Delta = \mathrm{unit}'(\mathrm{d}T)^{\otimes 6} = \mathrm{unit}'(p^e)^6(\mathrm{d}T')^{\otimes 6}$. More intrinsically, Δ in the invertible R-module $\widehat{\Omega}^1_{R'/W}$ is

$(p^e)^6$ times a generator, whence $\mathrm{ord}_{v,x}(\Delta) = \mathrm{ord}_v((p^e)^6) = \mathrm{ord}_v(N^6) = \mathrm{ord}_p(N^6)$ as desired.

To establish (10–4) (and thereby complete the proof), recall that the deformation aspect of the level structure only matters through its p-part. More specifically, we are not just imposing an arbitrary $\Gamma_0(p^e)$-structure on deformations of E_0 but rather one which is *étale* (here is where the condition $v|\bar{\mathfrak{n}}$ is used). Hence, our task is really one of deforming *splittings* of the connected-étale sequence on p^e-torsion of deformations of the ordinary E_0. Upon choosing a generator of the étale part of $E_0[p^e]$, we uniquely identify $E_0[p^e]$ with $\mathbf{Z}/p^e \times \mu_{p^e}$ compatibly with the scheme-theoretic Weil pairing. For any infinitesimal deformation E' of E_0 we can uniquely write its connected-étale sequence as $0 \to \mu_{p^e} \to E'[p^e] \to \mathbf{Z}/p^e \to 0$ in a manner lifting the corresponding canonical sequence for $E_0[p^e]$. Let f denote the projection map from $E'[p^e]$ onto its maximal étale quotient. Thus, the deformation problem is equivalently one of choosing a section of the fiber $f^{-1}(1)$ on deformations E' of E_0. In the universal situation over $W[\![T]\!]$ (with $q = 1 + T$ the Serre–Tate canonical parameter and $T = 0$ cutting out the Serre–Tate canonical lift which is the elliptic curve underlying x), the torsion levels are described by rather explicit group schemes considered in [KM, § 8.9ff]. This description rests crucially on the fact that $q = 1 + T$ is the Serre–Tate canonical parameter, and in any case explicates the scheme $f^{-1}(1)$ by means of the equation $X^{p^e} = 1 + T$. Thus, if we define $T' = X - 1$ then we obtain the desired description $R' = R[T']/((1 + T')^{p^e} - (1 + T))$ for the (local) ring R' as a finite R-algebra. □

As an application of (10–3), since maps between *ordinary* elliptic curves over the residue field always uniquely lift to maps between Serre–Tate canonical deformations, we see that if p splits in K and $p \nmid mN$ (so (10–3) can be applied and maps between level-N structures uniquely deform) then the $\mathrm{Hom}^{\mathrm{new}}$-terms on the right side of (10–3) vanish. This yields the following generalization of the vanishing observation which was noted at the beginning of Section 7:

LEMMA 10.3. *If p is split in K, $\gcd(m, N) = 1$, and $p \nmid mN$, then $(\underline{x}.T_m(\underline{x}^\sigma))_v^{\mathrm{GZ}}$ vanishes for any place v of H over p, without restriction on $r_{\mathscr{A}}(m)$.*

How about the case in which $p|N$ (so p splits in K and $p \nmid m$, but the N-torsion schemes need not be étale)? This is answered by:

THEOREM 10.4. *Assume that $v|N$ and $\gcd(m, N) = 1$. Let $\underline{x} \in \underline{X}(W)$ be a Heegner point. Let $\mathfrak{n}|N$ be the annihilator of the $\Gamma_0(N)$-structure. Then*

$$(\underline{x}.T_m(\underline{x}^\sigma))_v^{\mathrm{GZ}} - \frac{r_{\mathscr{A}}(m)}{r_x + k}\mathrm{ord}_{v,x}(\Delta) = \begin{cases} 0 & \text{if } v|\mathfrak{n}, \\ -u_x r_{\mathscr{A}}(m)\mathrm{ord}_p(N) & \text{if } v|\bar{\mathfrak{n}}. \end{cases}$$

PROOF. There are two cases to consider, depending on the factor $\mathfrak{n}$ of N over p which kills the level structure on $\underline{x}$. If $v|\mathfrak{n}$, then the p-part of the level structure on $\underline{x}$ (and hence also on $\underline{x}^\sigma$) is connected and in fact multiplicative. Since infinitesimal deformations of ordinary elliptic curves possess *unique* connected multiplicative subgroups of a specified order and any map between such deformations respects the specification of such a subgroup (thanks to the functoriality of the connected-étale sequence), we can conclude by the same argument as for $p \nmid mN$ split in K that the right side of (10–3) vanishes when $v|\mathfrak{n}$. This gives the $v|\mathfrak{n}$ case of Theorem 10.4, since $\mathrm{ord}_{v,x}(\Delta) = 0$ by Lemma 10.1 for such v.

The case $v|\bar{\mathfrak{n}}$, for which the p-part of the $\Gamma_0(N)$-structure is étale, has the property that the p-part of the $\Gamma_0(N)$-structure on $\underline{x}$ and $\underline{x^\sigma}$ amounts to giving (noncanonical) splittings of connected-étale sequences over W so as to single out the p-part of the étale level structures. General elements in $\mathrm{Hom}_{W/\pi}(\underline{x^\sigma}, \underline{x})$ have no nontrivial compatibility condition imposed on the p-part of the level structure because connected-étale sequences uniquely and canonically split over W/π. However, when such a map is uniquely lifted to a map of the (Serre–Tate canonical) deformed elliptic curves underlying $\underline{x}$ and $\underline{x^\sigma}$ over W it is generally not true that such a lifted map must respect chosen splittings of the connected-étale sequences over W (and hence generally does not give a map of $\Gamma_0(N)$-structures over W). But our situation is special because the p-part of the level structure coincides with the piece of the $\bar{\mathfrak{n}}$-divisible group of order equal to the p-part of N. Since there are no nonzero maps from an étale p-divisible group to a connected one (over a local noetherian base), we conclude that the argument used for $v|\mathfrak{n}$ actually still works for $v|\bar{\mathfrak{n}}$. Thus, we get $(x.T_m(x^\sigma))_v^{\mathrm{GZ}} = 0$ even when $v|\bar{\mathfrak{n}}$, so we must prove

$$\mathrm{ord}_{v,x}(\Delta)/(r_x + k) = u_x \mathrm{ord}_p(N) = u_x \mathrm{ord}_v(N)$$

(the latter equality holding since p is unramified in H). This is provided by Lemma 10.1. □

Now we compute (10–1) when $v \nmid N$ and $\theta = \Delta$. By Lemma 10.1, the $\mathrm{ord}_{v,x}(\Delta)$ term vanishes for such v. Thus, our task comes down to computing $(x.T_m(x^\sigma))_v^{\mathrm{GZ}}$.

THEOREM 10.5. *Assume $v \nmid N$, $\gcd(m, N) = 1$, and $\underline{x} \in \underline{X}(W)$ is a Heegner point as usual. Let $\mathfrak{a}$ represent the ideal class $\mathscr{A}$ and be prime to p, and let $\sigma \in \mathrm{Gal}(H/K)$ correspond to $\mathscr{A}$. Let v be a place of H over p.*

(1) *If p is inert in K, then*

$$(x.T_m(x^\sigma))_v^{\mathrm{GZ}} = \sum_{\substack{b \in R\mathfrak{a}/\pm 1, \\ \mathrm{N}b = m\mathrm{N}\mathfrak{a},\, b_- \neq 0}} \tfrac{1}{2}\big(1 + \mathrm{ord}_p(\mathrm{N}(b_-))\big) + \tfrac{1}{2} u_x r_{\mathscr{A}}(m) \mathrm{ord}_p(m).$$

(2) *If p is ramified in K then*

$$(x.T_m(x^\sigma))_v^{\mathrm{GZ}} = \sum_{\substack{b \in R\mathfrak{a}/\pm 1, \\ \mathrm{N}b = m\mathrm{N}\mathfrak{a},\, b_- \neq 0}} \mathrm{ord}_p(D\mathrm{N}(b_-)) + u_x r_{\mathscr{A}}(m) \mathrm{ord}_p(m).$$

(3) *If* $p = \mathfrak{p} \cdot \bar{\mathfrak{p}}$ *is split in* K *and* v *lies over* $\mathfrak{p}$ *then*

$$(x.T_m(x^\sigma))_v^{\mathrm{GZ}} = u_x \kappa_{\mathfrak{p}}$$

where $\kappa_{\mathfrak{p}}$ *and* $\kappa_{\bar{\mathfrak{p}}}$ *are nonnegative integers which are intrinsic to the prime ideals* $\mathfrak{p}$ *and* $\bar{\mathfrak{p}}$ *(e.g., they are defined without reference to* x*) and satisfy* $\kappa_{\mathfrak{p}} + \kappa_{\bar{\mathfrak{p}}} = r_{\mathscr{A}}(m)\mathrm{ord}_p(m)$.

This is [GZ, Ch. III, Prop. 8.5]. Our proof is longer, but more conceptual.

PROOF. The supersingular cases (i.e., p not split in K) will be treated by essentially the exact same method which we used when $r_{\mathscr{A}}(m) = 0$, but the ordinary case (i.e., p split in K) will require some new work.

It follows from (10–3) that the quaternionic formulas from Corollary 7.15 carry over to the general case (i.e., $r_{\mathscr{A}}(m) > 0$ is allowed) to compute $(x.T_m(x^\sigma))_v^{\mathrm{GZ}}$ *when we also assume* $v \nmid m$, provided one augments the condition on the summation to require $b_- \neq 0$ (as we recall that this is the quaternionic translation of the condition that a morphism not lift to W). This gives the first two cases of the theorem when $\mathrm{ord}_p(m) = 0$. Note that Corollary 7.15 only occurs when $p \nmid N$ (as $p|N$ puts us in the split cases). The more interesting case of Corollary 7.15 is $v|m$, or equivalently $p|m$.

Before considering the modifications needed to get the theorem when $p|m$ (so $p \nmid N$) and $r_{\mathscr{A}}(m) > 0$, we first note that if p is inert in K (so p^2 is the norm of the unique prime $p\mathscr{O}_K$ over p) then the positivity of $r_{\mathscr{A}}(m)$ forces $\mathrm{ord}_p(m)$ to be *even*, with $\frac{1}{2}\mathrm{ord}_v(m) = \frac{1}{2}\mathrm{ord}_p(m)$. Meanwhile, if p is ramified in K then $\mathrm{ord}_v(m)$ is even with $\frac{1}{2}\mathrm{ord}_v(m) = \mathrm{ord}_p(m)$. Thus, the "extra" term on the right side of the inert and ramified cases of the theorem can be uniformly described by the formula

$$\tfrac{1}{2} u r_{\mathscr{A}}(m)\mathrm{ord}_v(m), \tag{10–5}$$

where $u = u(K) = u_x$.

Let $m = p^t r$ with $p \nmid r$ and $t > 0$. The first two cases of the theorem require nothing beyond our earlier work on the analogues when $r_{\mathscr{A}}(m) = 0$. The condition $v \nmid N$ is crucial throughout, since the deformation theory analysis uses quite critically that N-torsion is étale (and hence deformations of elliptic curve maps over any W_r are automatically $\Gamma_0(N)$-compatible when this is true over W_0). Also, the condition $r_{\mathscr{A}}(m) = 0$ played essentially *no* role in our earlier treatment of supersingular cases. The only relevance of this condition was to ensure that when cases with $s = 0$ arise then the various W-structures $\underline{z} = \underline{y}(0)$ which show up in $T_m(\underline{x}^\sigma)$ for $\underline{z}$ in $T_r(\underline{x}^\sigma)$ are never equal to $\underline{x}$. If we allow $r_{\mathscr{A}}(m)$ to perhaps be positive, then the analysis of the $\underline{y}(s)$'s for $s > 0$ goes through completely unchanged (since the value of $r_{\mathscr{A}}(m)$ was never relevant in that analysis) and the associated closed points $\underline{y}(s)_{/F}$ on $\underline{X}_{/F}$ were shown to have residue field $F(s)$ strictly bigger than F. In particular, such points cannot contribute to an appearance of the F-rational point $\underline{x}$ in the divisor $T_m(\underline{x}^\sigma)$. Thus, for $r_{\mathscr{A}}(m) > 0$ we get almost the exact same formulas for $(\underline{x}.T_m(\underline{x}^\sigma))_v^{\mathrm{GZ}}$

as in (6–5) and (6–6) with the two modifications that the case of p inert in K only gives rise to cases with *even* $t = \mathrm{ord}_p(m)$ — as we have seen in the deduction of (10–5) — and the summation terms

$$\frac{1}{2}\sum_{n\geq 0}\left|\mathrm{Hom}_{W_n}(\underline{z}, \underline{x})_{\deg 1}\right|$$

which arise as formulas for $(\underline{z}.\underline{x})$ must be replaced with their "new" counterparts in accordance with (10–2). This latter modification causes the $\sum_{n\geq 0}(\dots)$ terms in the "$r_{\mathscr{A}}(m) = 0$" formula for $(x.T_m(x^\sigma))_v^{\mathrm{GZ}}$ in (6–5) and (6–6) to be replaced with "new" counterparts as well.

If we look back at our argument which used Theorem 6.2 and Theorem 6.1 to translate (6–5) and (6–6) into the language of cardinalities of Hom-groups as in Theorem 5.1, we see that those arguments *never* used any hypothesis concerning the value of $r_{\mathscr{A}}(m)$. Thus, when converting the terms

$$\tfrac{t}{2}\cdot\tfrac{1}{2}\left|\mathrm{Hom}_{W/\pi}(\underline{x}^\sigma, \underline{x})_{\deg r}\right|,\quad t\cdot\tfrac{1}{2}\left|\mathrm{Hom}_{W/\pi}(\underline{x}^\sigma, \underline{x})_{\deg r}\right|,\quad t\cdot\tfrac{1}{2}\left|\mathrm{Hom}_{W/\pi}((\underline{x}^\sigma)^{\sigma_{\mathfrak{p}}}, \underline{x})_{\deg r}\right| \tag{10–6}$$

into various $\frac{1}{2}\left|\mathrm{Hom}_{W_e}(\underline{x}^\sigma, \underline{x})\right|$'s for small e, we just have to account for that fact that we now want to break up such terms into a "new" part and a "non-new" part. As we see from (10–6), there are $\frac{t}{2} = \frac{1}{2}\mathrm{ord}_p(m)$ such terms when p is inert in K and there are $t = \mathrm{ord}_p(m)$ such terms when p is ramified in K. We thereby pick up a "non-new" contribution of

$$\tfrac{t}{2}\cdot\tfrac{1}{2}\left|\mathrm{Hom}_{W}(\underline{x}^\sigma, \underline{x})_{\deg m}\right| = \tfrac{t}{2}ur_{\mathscr{A}}(m) = \frac{1}{2}ur_{\mathscr{A}}(m)\mathrm{ord}_p(m)$$

when p is inert in K and

$$t\cdot\tfrac{1}{2}\left|\mathrm{Hom}_{W}(\underline{x}^\sigma, \underline{x})_{\deg m}\right| = ur_{\mathscr{A}}(m)\mathrm{ord}_p(m)$$

when p is ramified in K, with the remaining "new" part of $(x.T_m(x^\sigma))_v^{\mathrm{GZ}}$ given by the right side of (10–3). We compute the right side of (10–3) in our supersingular situation by the exact same quaternionic method as before, and this yields the summation terms given in the first two parts of the theorem, the extra condition $b_- \neq 0$ simply reflecting the fact that we're only counting the number of "new" homomorphisms (i.e., those not lifting to W) at each infinitesimal level. This completes the justification of the first two parts of the theorem.

Now we turn to the third part of the theorem, with $p = \mathfrak{p}\bar{\mathfrak{p}}$ split in $\mathscr{O}_K$ (and $p \nmid N$), so the Heegner data have underlying elliptic curves with ordinary reduction. We have already seen via (10–3) that when $p \nmid mN$ is split in K then $(x.T_m(x^\sigma))_v^{\mathrm{GZ}} = 0$, exactly in accordance with the asserted formula in the split case. Thus, from now on we may (and do) assume $\mathrm{ord}_p(m) > 0$ (which forces $p \nmid N$ anyway). We cannot use (10–3) in such cases, so we will give a more explicit analysis of the situation. The goal is to get a formula for $(x.T_m(x^\sigma))_v^{\mathrm{GZ}}$ which agrees with the one we want for $p|m$ split in K.

As a first step toward computing $(x.T_m(x^\sigma))_v^{\rm GZ}$ when p is split in K and $v \nmid N$ (but $p|m$), we need to determine which order m subgroup schemes $C \subseteq \underline{x}^\sigma$ have the property that the quotient $\underline{x}^\sigma{}_C$ by C has closed fiber isomorphic to the closed fiber of $\underline{x}$ (with the isomorphism respecting the level structures). That is, we want to know when C_0 is the kernel of an isogeny $\phi_0 : \underline{x}^\sigma{}_0 \to \underline{x}_0$ respecting the $\Gamma_0(N)$-structures. By the Serre–Tate theorem, we have $\mathrm{Hom}_{W/\pi}(\underline{x}^\sigma{}_0, \underline{x}_0) = \mathrm{Hom}_W(\underline{x}^\sigma, \underline{x})$ since $v \nmid N$ (so compatibility with $\Gamma_0(N)$-structures does not impose any additional condition when deforming maps). In particular,

$$(x.x)_v^{\rm GZ} = \frac{1}{2}\sum_{n\geq 0}\left(|\operatorname{Aut}_{W_n}(\underline{x})| - |\operatorname{Aut}_W(\underline{x})|\right) = 0. \tag{10–7}$$

Any map $\underline{x}^\sigma \to \underline{x}$ is automatically $\mathscr{O}_K$-compatible (as can be checked by showing that the $\mathscr{O}_H$-module tangent "lines" of x and x^σ have the same (canonical) $\mathscr{O}_K$-structure through action of $\mathscr{O}_K$ on the elliptic curve, and this is trivial to check since $\sigma \in \mathrm{Gal}(H/K)$ acts as the identity on $K \subseteq H$). It follows that ϕ_0 as above is $\mathscr{O}_K$-compatible, so C_0 is an $\mathscr{O}_K$-submodule scheme of $\underline{x}^\sigma{}_0$.

Let $\mathfrak{a}$ be an integral ideal in the ideal class $\mathscr{A}$ with norm m. Consider the unique degree m isogeny

$$\phi : \mathfrak{a}^{-1} \otimes \underline{x} = \underline{x}^\sigma \to \underline{x}$$

lifting ϕ_0, so $\phi \in \mathfrak{a}$ and $\mathrm{N}(\phi) = \deg(\phi)\mathrm{N}(\mathfrak{a}) = m^2$ (as one checks using the same arguments with ℓ-divisible groups that we employed in our earlier degree calculations during the quaternionic considerations). In other words, in $\mathscr{O}_K$ we have an equality of ideals $\phi\mathscr{O}_K = \mathfrak{a}\cdot\mathfrak{b}$ with $\mathfrak{b}$ an integral ideal of norm m in the ideal class $\mathscr{A}^{-1}$. Conversely, it is clear that when we are given an integral ideal in $\mathscr{A}^{-1}$ with norm m then we get a map ϕ as considered above. Thus, up to composing ϕ (or ϕ_0) with a unit of $\mathscr{O}_K$ we see that the set of order m subgroups $C_0 \subseteq \underline{x}^\sigma{}_0$ of interest to us is of cardinality equal to $r_{\mathscr{A}^{-1}}(m) = r_{\mathscr{A}}(m)$. We fix such a $\mathfrak{b}$ and a representative generator $\phi_{\mathfrak{b}}$ of $\mathfrak{a}\mathfrak{b}$, and we seek to determine all order m subgroups $C \subseteq \underline{x}^\sigma{}_{/\overline{F}}$ for which $C_0 = \ker((\phi_{\mathfrak{b}})_0)$, written more loosely as "$C \equiv \ker((\phi_{\mathfrak{b}})_0)$" (changing $\phi_{\mathfrak{b}}$ by an $\mathscr{O}_K^\times$-multiple doesn't matter).

To find all order m subgroups $C \subseteq x^\sigma_{/\overline{F}}$ such that $C \equiv \ker((\phi_{\mathfrak{b}})_0)$, note that on the prime-to-p part everything is uniquely determined, so we focus our attention on the p-part. Let $t = \mathrm{ord}_p(m) > 0$. There is an evident inclusion $\overline{\mathfrak{b}}\cdot(\mathfrak{a}^{-1}\otimes\underline{x}[p^t]) \subseteq \ker(\phi_{\mathfrak{b}})_p$ inside of the p-part of $\ker(\phi_{\mathfrak{b}})$ and this is an equality for order reasons. Because of the $\mathscr{O}_K$-compatibility of $\phi_{\mathfrak{b}}$ and the canonical splitting of the p-divisible groups of $\underline{x}$ and $\underline{x}^\sigma$ into $\mathfrak{p}$-divisible and $\overline{\mathfrak{p}}$-divisible parts (where $p\mathscr{O}_K = \mathfrak{p}\overline{\mathfrak{p}}$), with the $\mathfrak{p}$ denoting the prime under v, we have a decomposition $\ker(\phi_{\mathfrak{b}})_p = (\overline{\mathfrak{b}}\cdot(\mathfrak{a}^{-1}\otimes\underline{x}[p^t]))_{\overline{\mathfrak{p}}} \times (\overline{\mathfrak{b}}\cdot(\mathfrak{a}^{-1}\otimes\underline{x}[p^t]))_{\mathfrak{p}}$ which expresses $\ker(\phi_{\mathfrak{b}})_p$ as the product of étale and connected pieces over W (the first factor is étale and the second is connected because $\mathfrak{p}$ is the prime under v, and everything is really finite flat because of Raynaud's scheme-theoretic closure trick which is particularly well-behaved in the context of multiplicative and étale group schemes without ramification restrictions). The analogous decomposition of $\mathfrak{a}^{-1} \otimes \underline{x}[p^t] = \underline{x}^\sigma[p^t]$ into a $\overline{\mathfrak{p}}$-part and a $\mathfrak{p}$-part reflects the decomposition arising from the unique splitting of the p-divisible group of $\underline{x}^\sigma$ into étale and connected pieces.

By the Serre–Tate theorem, we conclude (by considering deformations of endomorphisms of the p-divisible groups $\mathbf{Q}_p/\mathbf{Z}_p$ and its dual $\mathbf{G}_m[p^\infty]$ separately) that the condition that a deformation of $(\phi_\mathfrak{b})_0$ to some W'/π'^{n+1} *not* lift to W' (with W' a finite discrete valuation ring extension of W, with uniformizer π') is *exactly* that its kernel not split as a product of connected and étale pieces in accordance with the splitting of the p-divisible group of $\underline{x}^\sigma$ (reduced modulo π^{n+1}). Thus, contributions to the points in the divisor $T_m(x^\sigma)$ (aside from $r_\mathscr{A}(m) \cdot x$) which "reduce to $(\phi_\mathfrak{b})_0$" correspond on the level of *geometric* points to subgroups C inside of the geometric generic fiber of the finite flat W-group scheme $\mathfrak{a}^{-1} \otimes \underline{x}[p^t]$ with $(\overline{\mathfrak{b}} \cdot (\mathfrak{a}^{-1} \otimes \underline{x}[p^t]))_\mathfrak{p} \subseteq C$ and the cokernel of this inclusion projecting isomorphically onto $(\overline{\mathfrak{b}} \cdot (\mathfrak{a}^{-1} \otimes \underline{x}[p^t]))_{\overline{\mathfrak{p}}}$, yet with this quotient map *not* being split compatibly with the natural (Serre–Tate "canonical") splitting of the connected-étale sequence of $\mathfrak{a}^{-1} \otimes \underline{x}[p^t]$. The analysis of such C's (e.g., finding Galois orbits, etc) as $\overline{F}$-points takes places entirely in the p-divisible group of $\underline{x}^\sigma = \mathfrak{a}^{-1} \otimes \underline{x}$, so since this p-divisible group only depends on $\mathfrak{a}$ through the $\mathscr{O}$-module $\mathfrak{a} \otimes_{\mathscr{O}_K} \mathscr{O}$ which is free of rank 1, we can drop the appearance of the functor "$\mathfrak{a}^{-1} \otimes (\cdot)$" without any harm.

As we run over all $r_{\mathscr{A}^{-1}}(m) = r_\mathscr{A}(m)$ possibilities for $\mathfrak{b}$, only the p-part $\mathfrak{p}^i\overline{\mathfrak{p}}^{t-i}$ of $\mathfrak{b}$ matters for the analysis of possible C's (with $\mathfrak{a}$ harmlessly removed). This p-part sequence is $0 \to \underline{x}[\mathfrak{p}^i] \to C \to \underline{x}[\overline{\mathfrak{p}}^{t-i}] \to 0$. Upon trivializing the étale part of the p-divisible group of $\underline{x}$, we may naturally identify $\underline{x}[p^e]$ with $\mu_{p^e} \times \mathbf{Z}/p^e$, where $\mu_{p^e} = \underline{x}[\mathfrak{p}^e]$ and $\mathbf{Z}/p^e = \underline{x}[\overline{\mathfrak{p}}^e]$, all this compatible with change in e. Thus, we really have an exact sequence

$$0 \to \mu_{p^i} \to C \to p^i\mathbf{Z}/p^t\mathbf{Z} \to 0$$

inside of $\mu_{p^t} \times \mathbf{Z}/p^t\mathbf{Z}$. Passing to the quotient by μ_{p^i} on C and taking the part of $\underline{x}[p^t]$ lying over $p^i\mathbf{Z}/p^t\mathbf{Z} \subseteq \mathbf{Z}/p^t\mathbf{Z}$, upon using the canonical isomorphisms $\mu_{p^t}/\mu_{p^i} \simeq \mu_{p^{t-i}}$ and $p^i\mathbf{Z}/p^t\mathbf{Z} \simeq \mathbf{Z}/p^{t-i}\mathbf{Z}$ we see that the specification of C (now collapsed to its étale quotient) amounts to giving a *nontrivial* splitting over $\overline{F}$ of the base change of the connected-étale sequence of $\mu_{p^{t-i}} \times \mathbf{Z}/p^{t-i}$ (the trivial splitting contributes to the point $\underline{x}$ in $T_m(\underline{x}^\sigma)$, but $(x.x)_v^{\mathrm{GZ}} = 0$).

The possible C's are given by subgroups generated by elements $(\zeta, 1) \in (\mu_{p^{t-i}} \times \mathbf{Z}/\mathfrak{p}^{t-i})(\overline{F})$. Up to $\mathrm{Gal}(\overline{F}/F)$-conjugacy such C's are classified by the order p^s of ζ (as $W \simeq W(\overline{\mathbf{F}}_p)$ is absolutely unramified, so its p-adic cyclotomic theory is "as big" as possible), with $1 \leq s \leq t - i$ (the case $s = 0$ corresponds to the point $\underline{x}$ in the support of $T_m(\underline{x}^\sigma)$ which we're not considering). The associated divisor point on $\underline{X}_{/F}$ has residue field $F(s) \stackrel{\mathrm{def}}{=} F(\zeta_{p^s})$, and it is represented by a $\Gamma_0(N)$-structure $\underline{y}(s)$ over the valuation ring $W(s) \stackrel{\mathrm{def}}{=} W[\zeta_{p^s}]$. Let $\pi_s = \zeta_{p^s} - 1$, a uniformizer of $W(s)$. Note that $\underline{y}(s)$ mod π_s^2 is *not* the Serre–Tate lift of its closed fiber $\underline{x}_0$ as a "bare" elliptic curve, since $\zeta_{p^s} \not\equiv 1 \bmod \pi_s^2$ (in other words, $\underline{y}(s)$ is a quasi-canonical lifting in the Serre–Tate sense, as is alluded to at the very end of [Gr1]). Thus, under the canonical map $W[\![T_0]\!] \to W(s)$ from the universal deformation ring of the $\Gamma_0(N)$-structure $\underline{x}_0$ (with $T_0 = 0$ corresponding to $\underline{x}$) we have to have that T_0 maps to a uniformizer of $W(s)$. Beware that this does not imply in general that natural map $\mathrm{Spec}(W(s)) \to \underline{X}$ over W (sending the closed point to $\underline{x}_0$) is a closed immersion, let alone that the corresponding closure of the closed point $\mathrm{Spec}(F(s))$ in $\underline{X}_{/F}$ is transverse to the closed subscheme $\underline{x} = \mathrm{Spec}(W)$. The problem is that the complete local ring at $\underline{x}_0$ on $\underline{X}$ might not be the universal deformation ring (as a $\Gamma_0(N)$-structure).

To keep track of what is really happening on the scheme $\underline{X}$, recall from (9–7) that the natural map

$$W[\![T_{\underline{x}}]\!] \simeq \widehat{\mathscr{O}}_{\underline{X},\underline{x}_0} \to W[\![T_0]\!]$$

is $T_{\underline{x}} \mapsto b_x T_0^{u_x} + \cdots$ with $\mathrm{ord}_v(b_x) = (x.x)_v^{\mathrm{GZ}} = 0$. Thus, $W[\![T_{\underline{x}}]\!] = \widehat{\mathscr{O}}_{\underline{X},\underline{x}_0} \to W(s)$ arising from $\underline{y}(s)$ sends $T_{\underline{x}}$ to the u_xth power of a uniformizer of $W(s)$. Hence, the closure of $y(s) : \mathrm{Spec}(F(s)) \hookrightarrow \underline{X}_{/F}$ in $\underline{X}$ is the order of level u_x in $W(s)$ and $(x.y(s))_v = u_x$ for $1 \le s \le t - i$. We conclude that $\mathfrak{b}$ contributes $u_x(t-i) = u_x \mathrm{ord}_{\bar{\mathfrak{p}}}(\mathfrak{b})$ to $(x.T_m(x^\sigma))_v^{\mathrm{GZ}}$. Consequently, we get

$$(x.T_m(x^\sigma))_v^{\mathrm{GZ}} = u_x \cdot \sum_{\mathfrak{b}} \mathrm{ord}_{\bar{\mathfrak{p}}}(\mathfrak{b}) \tag{10–8}$$

when v lies over $p|m$ which is split in K (with $\mathfrak{p}|p$ the prime of $\mathscr{O}_K$ under v). The sum in (10–8) is taken over all integral ideals of norm m in the ideal class of $\mathscr{A}^{-1}$, and will be called $\kappa_{\mathfrak{p}}$ (it clearly is intrinsic to $\mathfrak{p}$ and in general is not divisible by $r_{\mathscr{A}}(m)$). Since $\mathrm{ord}_{\bar{\mathfrak{p}}}(\mathfrak{b}) + \mathrm{ord}_{\mathfrak{p}}(\mathfrak{b}) = t$ for each of the $r_{\mathscr{A}}(m)$ different $\mathfrak{b}$'s, we conclude that $\kappa_{\mathfrak{p}} + \kappa_{\bar{\mathfrak{p}}} = r_{\mathscr{A}}(m)t = r_{\mathscr{A}}(m)\mathrm{ord}_p(m)$. □

Appendix by W. R. Mann: Elimination of Quaternionic Sums

We wish to explicitly compute $\langle c, T_m(d^\sigma)\rangle_p$ as defined in (9–18) with $\theta = \Delta$ and $x \in X(H)$ a Heegner point (with CM by the ring of integers of K). Recall that K is the imaginary quadratic field we have fixed from the outset, $D < 0$ is its discriminant, and H is the Hilbert class field of K. Also, the level N has prime factors which are split in K.

Our approach will build upon the results in Section 10 and use further arguments with quaternion algebras. The results will be expressed in terms of the arithmetic of K, with the answers separated according to whether p is split, inert, or ramified in K (and keep in mind that we allow the discriminant of K to be even). Combining Theorem 8.4, Lemma 10.1, Theorem 10.4 (if $p|N$), and Theorem 10.5 (if $p \nmid N$), we obtain the split case [GZ, Ch. III, Prop. 9.2]:

THEOREM A.1. *If p is split in K, then*

$$\langle c, T_m(d^\sigma)\rangle_p = -u_x r_{\mathscr{A}}(m) h_K \mathrm{ord}_p(m/N)\log(p).$$

The appearance of $h_K \log p$ in Theorem A.1 is due to the identity $h_K \log p = \sum_{v|\mathfrak{p}} \log q_v$, where q_v is the size of the residue field at the place v of H and $\mathfrak{p}$ is a choice of either prime over p in K.

Now we may assume p is nonsplit in K, so in particular $p \nmid N$. Let $\mathfrak{p}$ be the unique prime of K over p. By Theorem 8.4 and Lemma 10.1, we have

$$\langle c, T_m(x^\sigma)\rangle_p = -\sum_{v|\mathfrak{p}} (x.T_m(x^\sigma))_v^{\mathrm{GZ}} \log q_v$$

where the intersection number for the v-term is given by Theorem 10.5(1) (resp. Theorem 10.5(2)) when p is inert (resp. ramified) in K. Since $\sum_{v|\mathfrak{p}} \log q_v = h_K \log \mathrm{N}\mathfrak{p}$ in the nonsplit case, with $q_v = p^2$ in the inert case since the principal prime $\mathfrak{p} = p\mathscr{O}_K$ is totally split in H (Principal Ideal Theorem), Theorem 10.5(1) implies

$$\langle c, T_m d^\sigma\rangle_p = -ur_{\mathscr{A}}(m) h_K \mathrm{ord}_p(m) \log p - \log p \cdot \sum_{v|\mathfrak{p}} \sum_{\substack{b \in R_v \mathfrak{a}/\pm 1 \\ \mathrm{N}b = m\mathrm{N}\mathfrak{a}, b_- \neq 0}} (1 + \mathrm{ord}_p(\mathrm{N}(b_-))) \tag{A–9}$$

in the inert case, with $R_v = \mathrm{End}_{W_v/\pi_v}(\underline{x}_v)$. Meanwhile, in the ramified case we get

$$\langle c, T_m d^\sigma\rangle_p = -ur_{\mathscr{A}}(m) h_K \mathrm{ord}_p(m) \log p - \log p \cdot \sum_{v|\mathfrak{p}} f_v \sum_{\substack{b \in R_v \mathfrak{a}/\pm 1 \\ \mathrm{N}b = m\mathrm{N}\mathfrak{a}, b_- \neq 0}} \mathrm{ord}_p(D\,\mathrm{N}(b_-)), \tag{A–10}$$

where $q_v = p^{f_v}$ and $R_v = \mathrm{End}_{W_v/\pi_v}(\underline{x}_v)$.

Our aim is to find an expression for the inner quaternionic sums in (A–9) and (A–10) depending solely on the arithmetic of K, and we will see that no small effort is required to combine and manipulate these quaternionic formulae. These local sums are *exactly* the local sums which are analyzed in [GZ, Ch. III, § 9], and are computed in terms of the arithmetic of K in [GZ, Ch. III, Prop. 9.7, Prop. 9.11]. Since the explanations there are sometimes a bit terse, in order to clarify what is happening and moreover to show that the results work for even D, the rest of this appendix is devoted to explaining the analysis of these quaternionic sums and deriving the formulas obtained by Gross–Zagier. The additional burden of treating even discriminants will be a slight nuisance, but will not require any essentially new ideas.

Before getting into the details, let us briefly outline the argument. The sums of interest from Theorem 10.5 involve sums which extend over certain elements inside a quaternionic order isomorphic to the order $\mathrm{End}_{W/\pi}(\underline{x})$. The first task is to find a convenient model for the quaternion algebra in question (this algebra is the unique one over $\mathbf{Q}$ ramified at precisely p and ∞, as we saw in Lemma 7.1), paying careful attention to the specification of an embedding of K into this model. We will then have to find a model for the order of interest within this algebra, *but* it turns out that Lemma 7.1 fails to identify this order up to $\mathscr{O}_K$-algebra isomorphism. Specifically, when one takes into account the embedding of $\mathscr{O}_K$ into the order, there are finitely many nonisomorphic orders which satisfy the conditions of Lemma 7.1. We will see that as $\mathrm{Gal}(H/K)$ acts on the places over p, it also serves to simply transitively permute the isomorphism classes of these orders (as $\mathscr{O}_K$-algebras). This is precisely what enables us to obtain a formula depending only on the arithmetic of K when we sum over all places v of H over the unique prime $\mathfrak{p}$ of K over p in (A–9) and (A–10).

Though we retain the notation K with the same meaning as throughout this paper, in this appendix we will use F to denote an arbitrary field of characteristic 0 (though only characteristic 2 requires extra care). This will pose no risk of conflict with the use of F to denote the fraction field of W in the main paper, since that fraction field will never show up in this appendix.

To get started, we review some terminology in the theory of quaternion algebras. Recall that a *quaternion algebra* B over a field F is a 4-dimensional central simple F-algebra. If E is an extension of F, then $B_E \stackrel{\text{def}}{=} E \otimes_F B$ is a quaternion algebra over E. An important example of a quaternion algebra is the matrix algebra $M_2(F)$. Another family of examples is provided by the following, which we will use (keep in mind we assume F has characteristic 0):

EXAMPLE A.2. Pick $e, f \in F^\times$. There is a unique quaternion algebra B with basis $1, i, j, ij$ satisfying the requirements that $i^2 = e$, $j^2 = f$, and $ij = -ji$ (so $(ij)^2 = -ef$). This algebra is denoted $\left(\frac{e,f}{F}\right)$.

It is a basic fact that a quaternion algebra B over F is either a division algebra or is a matrix algebra over F, and in the latter case the isomorphism $B \simeq M_2(F)$ is unique up to inner automorphism. This latter case is called *split*, and the division algebra case is called *nonsplit*. It is a subtle algebraic problem to determine whether or not the construction in Example A.2 is split (for a given pair $e, f \in F^\times$). An extension field E of F is called a *splitting field* of B if B_E is split. In general there exists a finite Galois extension E of F which splits B. Since the trace and determinant on $M_2(E)$ are invariant under conjugation by $M_2(E)^\times = \mathrm{GL}_2(E)$, if we choose a Galois splitting field E/F for B we can transport the trace and determinant to define $\mathrm{Tr}_\iota : B_E \to E$ and $\mathrm{N}_\iota : B_E \to E$ via an isomorphism $\iota : B_E \simeq M_2(E)$. The $\mathrm{GL}_2(E)$-conjugacy ambiguity in the choice of ι does not affect these constructions. If we extend scalars on ι through $\sigma \in \mathrm{Gal}(E/F)$ on source and target, we get a new isomorphism ι^σ. Thus, $\mathrm{Tr}_{\iota^\sigma} = \mathrm{Tr}_\iota$ and $\mathrm{N}_{\iota^\sigma} = \mathrm{N}_\iota$, yet also $\mathrm{Tr}_{\iota^\sigma}$ is the extension of scalars on Tr_ι by σ, and similarly for the comparison of $\mathrm{N}_{\iota^\sigma}$ and N_ι. It follows that Tr_ι and N_ι are $\mathrm{Gal}(E/F)$-equivariant and independent of ι, so these descend to define the canonical *reduced trace* and *reduced norm*

$$\mathrm{Tr} : B \to F, \quad \mathrm{N} : B \to F.$$

These are F-linear and multiplicative respectively, with $\mathrm{Tr}(bb') = \mathrm{Tr}(b'b)$ for any $b, b' \in B$. For $b \in B$, we define $\bar{b} = b - \mathrm{Tr}(b)$. Extending scalars to a splitting field, $b \mapsto \bar{b}$ becomes $\begin{pmatrix} \alpha & \beta \\ \gamma & \delta \end{pmatrix} \mapsto \begin{pmatrix} \delta & -\beta \\ -\gamma & \alpha \end{pmatrix}$. Using this, one checks that $b\bar{b} = \bar{b}b = \mathrm{N}(b)$ and $b \mapsto \bar{b}$ is an anti-automorphism of B.

Note in particular (by Cayley–Hamilton over a splitting field) that $b \in B$ is a root of $X^2 - \mathrm{Tr}(b)X + \mathrm{N}(b)$, so the reduced trace and norm restrict to the usual trace and norm on any quadratic subfield of B over F. Moreover, for $b \in B$ outside of F, this is the *unique* quadratic polynomial over F satisfied by b. It follows that $F[b]$ is 2-dimensional for any $b \in B - F$, and in Example A.2 we have $\overline{x + yi + zj + wij} = x - yi - zj - wij$ for $x, y, z, w \in F$.

Class field theory leads to an abstract classification:

LEMMA A.3. *If F is a local field other than $\mathbf{C}$, there is a unique quaternion division algebra over F up to isomorphism. Any quadratic extension of F can be embedded into B. If F is a number field, a quaternion algebra B over F is split at all but finitely many places, with the set of nonsplit places of even size. Any even set of places arises as the set of nonsplit places of a unique quaternion algebra over F. In particular, if B is split at all places of F then B is split.*

Although Lemma A.3 is useful for theoretical purposes, we need to build some concrete quaternion algebras. For example, we will need to describe the unique quaternion division algebra over $\mathbf{Q}_p$ in terms of the construction in Example A.2.

DEFINITION A.4. When F is a local field the *Hilbert symbol* $(\cdot, \cdot) : F^\times \times F^\times \to \mu_2(F)$ is defined by the requirement that $(e, f) = 1$ when the quaternion algebra $\left(\frac{e,f}{F}\right)$ is split, and $(e, f) = -1$ when this algebra is a division algebra.

In the case of a number field F, if we write $(\cdot, \cdot)_v$ to denote the associated local Hilbert symbol for F_v at a place v, then the evenness in Lemma A.3 corresponds to the *product formula*: $\prod_v (e, f)_v = 1$ for any $e, f \in F^\times$, the product taken over all places v of F. As one special case, if $(e, f)_v = 1$ for all but possibly one place v_0, then $(e, f)_{v_0} = 1$. We will be interested in the case $F = \mathbf{Q}$, and the first problem we will have to solve is that of building quaternion algebras over $\mathbf{Q}$ with a specified even set of nonsplit places, using the language in Example A.2. Often there will be a particularly problematic place, but if we can check that a quaternion algebra has the desired splitting/nonsplitting away from one place, then the behavior at the missing place is forced by the product formula.

Returning to the original problem of interest, recall that our prime p is either inert or ramified in K, and we want to construct a model for the quaternion division algebra B in Lemma 7.1: this is the unique such algebra over $\mathbf{Q}$ which is nonsplit at precisely p and ∞. Since the sums in Theorem 10.5 involve elements of (an order in) B which are identified with respect to right multiplication by a nonzero ideal $\mathfrak{a}$ of $\mathscr{O}_K$, a model for B needs to encode an embedding of K. We want to make such a model in the simplest manner possible. It makes sense to try to build a model $\left(\frac{e,f}{\mathbf{Q}}\right)$ with $e = i^2 = D$ since we want to keep track of how $K = \mathbf{Q}(\sqrt{D})$ sits in B. Thus, we want to find an $f \in \mathbf{Q}^\times$ for which the algebra $\left(\frac{D,f}{\mathbf{Q}}\right)$ is a division algebra nonsplit at exactly p and ∞. That is, we want $(D,f)_v = -1$ exactly when $v = p, \infty$. To find a suitable f, we need to understand how to compute local Hilbert symbols over places of $\mathbf{Q}$. This is done via:

LEMMA A.5. *Let F be a local field. The Hilbert symbol $(\,\cdot\,,\cdot\,) : F^\times \times F^\times \to \mu_2(F)$ factors through $F^\times/(F^\times)^2 \times F^\times/(F^\times)^2$, yielding a nondegenerate symmetric bilinear form. If $e \in F^\times$ is not a square, then $(e,f) = 1$ if and only if $f \in \mathrm{N}_{F(\sqrt{e})/F}(F(\sqrt{e})^\times)$.*

Before we return to the problem of finding a model for B, for later convenience we recall how Lemma A.5 helps us to compute the local symbols for quaternion algebras over $\mathbf{Q}$.

EXAMPLE A.6. We wish to compute $(a,b)_v$ for $a, b \in \mathbf{Q}^\times$ and v a place of $\mathbf{Q}$. If $v = \infty$ then $\mathbf{Q}_v = \mathbf{R}$, so by Lemma A.5 we see that $(a,b)_\infty = -1$ exactly when a and b are both negative. If $v = \ell$ is an odd prime, the situation is only slightly more difficult. The classes in $\mathbf{Q}_\ell^\times/(\mathbf{Q}_\ell^\times)^2$ are represented by $\{1, \varepsilon, \ell, \varepsilon\ell\}$ where $\varepsilon \in \mathbf{Z}_\ell^\times$ is not a quadratic residue mod ℓ and hence is not a square in $\mathbf{Z}_\ell^\times$. Since $\mathbf{Q}_\ell(\sqrt{\varepsilon})$ is the quadratic unramified extension of $\mathbf{Q}_\ell$, it has norm group generated by $\mathbf{Z}_\ell^\times$ and ℓ^2. Thus, $(\varepsilon,\varepsilon)_\ell = 1$ and $(\varepsilon,\ell)_\ell = -1$. Also, since $-\ell$ is a norm from $\mathbf{Q}_\ell(\sqrt{\ell})$, we have $(\ell,-\ell)_\ell = 1$. Thus, $(\ell,\ell)_\ell = (\ell,-1)_\ell$. By Lemma A.5, this is 1 if and only if -1 is a square mod ℓ, so $(\ell,\ell)_\ell = (-1)^{(\ell-1)/2}$. This generates the other possible pairings by bilinearity (e.g., $(u,v)_\ell = 1$ for $u, v \in \mathbf{Z}_\ell^\times$).

For $v = 2$, there are more cases because $\mathbf{Q}_2^\times/(\mathbf{Q}_2^\times)^2$ has order 8 with generators $-1, 2, 3$. Rather than delve into tedious examination of cases, we use the product formula to compute these symbols. Consider the quaternion algebras over $\mathbf{Q}$ isomorphic to $\left(\frac{a,b}{\mathbf{Q}}\right)$ for some a, b chosen from $\{-1, 2, 3\}$, and recall that an even number of places will be nonsplit. Since these choices of a and b are units for all finite primes $\ell > 3$, our observations above show that $(a,b)_\ell = 1$ for all such ℓ. All that is left are the places $2, 3, \infty$, and each algebra must be nonsplit at either none or two of these three places. For the sake of completeness, we note that an examination of the local symbols at 3 and ∞ yields: $(-1,-1)_2 = (-1,3)_2 = (2,3)_2 = (3,3)_2 = -1$, $(-1,2)_2 = (2,2)_2 = 1$.

Now recall that we wish to find $f \in \mathbf{Q}^\times$ so that $\left(\frac{D,f}{\mathbf{Q}}\right)$ is nonsplit at exactly p and ∞. There will be infinitely many f that work, but we want the simplest possible choice. We must require $f < 0$ to force nonsplitting at ∞. We will have to treat separately the cases when p is inert in K and when p is ramified in K. First consider the inert case. It is necessary that $\mathrm{ord}_p(f)$ be odd. To see this, first note that the completion $K_\mathfrak{p} = \mathbf{Q}_p \otimes_\mathbf{Q} K$ is an unramified quadratic extension inside of B_p, so its norm group in $\mathbf{Q}_p^\times$ consists of elements of even order. Thus, if $\mathrm{ord}_p(f)$ is even then $-f = \mathrm{N}_{K_\mathfrak{p}/\mathbf{Q}_p}(c)$ for some $c \in K_\mathfrak{p}^\times$, so via the decomposition $B = K \oplus Kj$ with $j^2 = f$ (compare the discussion following Lemma 7.1) we'd get

$$\mathrm{N}(c + j) = \mathrm{N}(c) + \mathrm{N}(j) = \mathrm{N}_{K_\mathfrak{p}/\mathbf{Q}_p}(c) + f = 0,$$

a contradiction since B_p is nonsplit and $c + j \neq 0$. One might hope that $\left(\frac{D,-p}{\mathbf{Q}}\right)$ works, as it is clearly nonsplit at p and ∞, but in general this algebra can be nonsplit at some primes dividing D (and possibly 2). For example, if $\ell | D$ is an odd prime then

$$(D, -p)_\ell = (\ell, -p)_\ell (D/\ell, -p)_\ell = (\ell, -p)_\ell = \left(\frac{-p}{\ell}\right), \tag{A–11}$$

and this is generally not equal to 1.

Thus, the simplest possible choice for f in the inert case is $f = -pq$ for some *auxiliary* prime $q \nmid Dp$. For odd primes $\ell \nmid Dq$, we already know $(D, -pq)_\ell = 1$ by Example A.6, since D and $-pq$ are units modulo ℓ. Thus, we're left to handle a finite number of primes, at which we will get congruence conditions on q (for which there will be infinitely many solutions, by Dirichlet's theorem). For odd $\ell | D$, computing as in (A–11) gives the condition $\left(\frac{-pq}{\ell}\right) = 1$, which is a congruence condition (with $(\ell - 1)/2$ solutions) for q modulo each such ℓ. It remains to consider the places 2, p, and q (possibly $p = 2$ or $q = 2$). If $p = 2$ then D and q are odd, so the requirement $(D, -pq)_p = -1$ can be satisfied by a mod 8 condition on q. This makes $\left(\frac{D,-pq}{\mathbf{Q}}\right)$ nonsplit at p and ∞, and split at all odd primes except for possibly q, and splitting is then forced at q by the product formula. Now suppose p is odd. In this case, since p is inert in K, $(D, -pq)_p = (D, -q)_p (D, p)_p = (D, p)_p = \left(\frac{D}{p}\right) = -1$, so $\left(\frac{D,-pq}{\mathbf{Q}}\right)$ is nonsplit at p and ∞. By the product formula, it remains to verify splitting at either 2 or q. This renders the case $q = 2$ trivial, so we may assume q is odd. With pq now odd, we will force splitting at 2 instead, by means of the congruence $-pq \equiv 1 \bmod 8$, which forces $-pq \in \mathbf{Q}_2^\times$ to be a square (so $(D, -pq)_2 = 1$). This settles the construction of our quaternion algebra model in the inert case. Observe that q is automatically split in K. Indeed, for odd q we have $1 = (D, -pq)_q = (D, -p)_q (D, q)_q = (D, q)_q = \left(\frac{D}{q}\right)$, while for $q = 2$ we have $(D, -2p)_2 = 1$, so $D \equiv 1 \bmod 8$ (recall that odd fundamental discriminants are $\equiv 1 \bmod 4$), so 2 splits in K.

REMARK A.7. In all inert cases, q satisfies the condition that for all primes $\ell | D$, $-pq \in \mathbf{Z}_\ell^\times$ is a square. Also, requiring $-pq \equiv 1 \bmod D$ is consistent with the requirements on q in inert cases. In ramified cases, the congruence $-q \equiv 1 \bmod D/p$ (which involves no condition at p, unless $p = 2$) may always be imposed without inconsistency.

We can similarly treat the case when p is ramified in K; i.e., $p | D$. The simplest choice of f we might try is $f = -q$ for some prime $q \nmid D$. If p is odd, then as in (A–11) we get $(D, -q)_p = \left(\frac{-q}{p}\right)$, so requiring this to be -1 is a congruence condition on q mod p. If $p = 2$, we similarly find that $(D, -q)_2 = -1$ follows from a congruence condition on q mod 8. For example, if $8 \nmid D$ then $q \equiv 1 \bmod 4$ is necessary and sufficient, while if $8|D$ then $q \equiv 3 \bmod 8$ is sufficient. For odd $\ell \nmid Dq$, just as above it is clear that $(D, -q)_\ell = 1$. For odd $\ell|D$, we have $(D, -q)_\ell = \left(\frac{-q}{\ell}\right)$, and we want this to be 1, imposing a congruence on q mod ℓ. We're now left with the places 2 and q, which means that we are done if $p = 2$ (as we've considered it already) or if $q = 2$ (product formula). Otherwise q is odd and all congruence conditions have been at odd primes, so we can also require $q \equiv -1 \bmod 8$ to force $(D, -q)_2 = 1$. The product formula now ensures $(D, -q)_q = 1$, so in fact q again splits in K, just as in the inert case. Even if $q = 2$ (so D is odd and $p|D$ is odd), from $(D, -2p)_2 = 1$ we get $D \equiv 1 \bmod 8$, so $q = 2$ is split in K. It is important later that our choice of q is automatically split in K (for p of either inert or ramified type in K).

Having found a model for the quaternion algebra B over $\mathbf{Q}$, we need to understand certain orders R in B, as this is what intervenes in Theorem 10.5. We will begin our discussion of this problem by reviewing some basic properties of orders in quaternion algebras which will be used quite a lot in what follows. Let F be the fraction field of a Dedekind domain A of characteristic 0; we are primarily interested in the case when A is the ring of integers of a local field or number field. Let B be a quaternion algebra over F.

DEFINITION A.8. A *lattice* in B is a finite A-submodule L which spans B over F (so L is locally free of rank 4 over A). An *order* in B is an A-subalgebra R which is also a lattice. We say that R is a *maximal* order if it is not strictly contained in any larger order, and we say that R is an *Eichler order* if it is the intersection of two maximal orders.

Note that if R is an order and $r \in R$, then $A[r]$ is a finite A-module, so r is integral over A. Since $A[r]$ is locally free of rank 2 when $r \notin R \cap F = A$, the minimal polynomial $X^2 - \mathrm{Tr}(r)X + \mathrm{N}(r)$ of r over F must have A coefficients when $r \notin A$. Since Tr and N are A-valued on A, we see that Tr and N are A-valued on any order R in B.

LEMMA A.9. *Suppose A is a complete discrete valuation ring, with maximal ideal $\mathfrak{m}$ and uniformizer π. Let B be a quaternion algebra over F.*

(1) *If B is nonsplit, then it contains a unique maximal order and this contains all other orders.*
(2) *If B is split, then all maximal orders in B are conjugate under $B^\times$ and every order lies in one. In $M_2(F)$ a maximal order is $M_2(A)$, and any Eichler order is conjugate to $\begin{pmatrix} A & A \\ \mathfrak{m}^n & A \end{pmatrix}$ for a unique $n \geq 0$. Any two Eichler orders that are abstractly isomorphic as rings are conjugate in B.*

REMARK A.10. The order $\begin{pmatrix} A & A \\ \mathfrak{m}^n & A \end{pmatrix}$ in (2) is the intersection of the maximal orders $M_2(A)$ and $\gamma_n M_2(A)\gamma_n^{-1} = \begin{pmatrix} A & \mathfrak{m}^{-n} \\ \mathfrak{m}^n & A \end{pmatrix}$, where $\gamma_n = \begin{pmatrix} 0 & 1 \\ \pi^n & 0 \end{pmatrix}$. This example is called a *standard* Eichler order.

PROOF. For the first part, it suffices to show that the set R of elements of B integral over R is an order. That is, we must show R is finite as an A-module and is a subring of B. Note that R is stable under the involution $b \mapsto \bar{b}$. The key to the subring property is that if $b \in B$ has $\mathrm{N}(b) \in A$, then $\mathrm{Tr}(b) \in A$. Indeed, $F[b]$ is a field on which the reduced norm and trace agree with the usual norm and trace (relative to F), and by completeness of A we know that the valuation ring of $F[b]$ is characterized by having integral norm. Thus, to show that R is stable under multiplication we just need that if $x, y \in R$ then $\mathrm{N}(xy) \in A$. But $\mathrm{N}(xy) = \mathrm{N}(x)\mathrm{N}(y)$. Meanwhile, for addition (an issue because noncommutativity does not make it evident that a sum of integral elements is integral), we note that if $x, y \in R$ then

$$\mathrm{N}(x+y) = (x+y)(\bar{x}+\bar{y}) = \mathrm{N}(x) + \mathrm{N}(y) + \mathrm{Tr}(x\bar{y}).$$

But this final reduced trace term lies in A because $x\bar{y} \in R$. Hence, R is a subring of B. In particular, R is an A-submodule since $A \subseteq R$.

To show that R is A-finite, we may pick a model for B as in Example A.2, and may assume $i^2 = e, j^2 = f \in A$. Thus, $i, j \in R$, so $ij \in R$. For any $x \in R$, we have $x, xi, xj, xij \in R$. Taking reduced traces of all four of these elements and writing $x = c + c_1 i + c_2 j + c_3 ij$ for $c, c_1, c_2, c_3 \in F$, we get

$$x \in \frac{1}{2}A + \frac{1}{2e}Ai + \frac{1}{2f}Aj + \frac{1}{2ef}Aij.$$

Thus, R lies inside of a finite A-module and hence is A-finite.

Now we turn to the split case, so we may assume $B = M_2(F)$. To keep the picture clear, we suppose $B = \mathrm{End}_F(V)$ for a 2-dimensional F-vector space V on which we do not impose a basis. If R is any order in B, then for a lattice L in V clearly RL is another finite A-submodule spanning V over F, so $N = RL$ is an R-stable lattice. Hence, $R \subseteq \mathrm{End}_A(N)$. Since any two lattices are conjugate to each other, the assertions concerning maximal orders come down to the claim that if $\mathrm{End}_A(N_0) \subseteq \mathrm{End}_A(N_1)$ for two lattices N_0, N_1 in V, then this inclusion is an equality. We may scale N_1 so $N_1 \subseteq N_0$ and N_1 is not contained in $\mathfrak{m}N_0$, so $N_0/N_1 \simeq A/\mathfrak{m}^n$ for some $n \geq 0$. We can then pick bases so that $N_0 = Ae \oplus Ae'$ and $N_1 = Ae \oplus A\mathfrak{m}^n e'$. If $n > 0$, then it is obvious that $\mathrm{End}_A(N_0)$ does not lie in $\mathrm{End}_A(N_1)$ inside of $\mathrm{End}_F(V) = B$.

Now consider Eichler orders. It suffices to focus attention on those Eichler orders R which are not maximal orders in $B = F \otimes_A R$. Using a suitable conjugation by $B^\times$, given any two distinct nonmaximal Eichler orders we may suppose they have the form $R \cap R_1$ and $R \cap R_2$ for maximal orders R, R_1, R_2 which we may assume to be distinct. Thus, $R = \mathrm{End}_A(N)$ for some lattice N in V and $R_j = \mathrm{End}_A(N_j)$ for a sublattice $N_j \subseteq N$ with $N/N_j \simeq A/\mathfrak{m}^{n_j}$ for some $n_j > 0$.

Relative to a basis $\{e_j, e'_j\}$ of N for which $\{e_j, \pi^{n_j} e'_j\}$ is a basis of N_j, we get an identification of $R \cap R_j$ with $R_j \overset{\text{def}}{=} \begin{pmatrix} A & A \\ \mathfrak{m}^{n_j} & A \end{pmatrix}$ inside of $R \simeq M_2(A)$ (isomorphism using the basis $\{e_j, e'_j\}$), yielding the desired description of Eichler orders. In fact, n_j is intrinsic to R_j because on the quaternion algebra $B = F \otimes_A R_j$ we can apply the involution $b \mapsto \bar{b}$ to form $\overline{R}_j$, and by inspection $R/(R_j \cap \overline{R}_j) \simeq A/\mathfrak{m}^{n_j}$ as an A-module.

Thus, when comparing two abstractly isomorphic Eichler orders $R \cap R_j$ as considered above, necessarily $n_1 = n_2 = n$. But we can then use the two bases on N adapted to the N_j's to define an element in $R^\times = \mathrm{Aut}_A(N)$ which carries N_1 into N_2 and hence conjugates $R \cap R_1$ over to $R \cap R_2$. □

REMARK A.11. One important consequence of the preceding proof is that in the split case, for each nonmaximal Eichler order R in B, there is a *unique* pair of maximal orders S and S' with $S \cap S' = R$. Indeed, by conjugation we may consider the case $B = M_2(F)$ and $R = \begin{pmatrix} A & A \\ \mathfrak{m}^n & A \end{pmatrix}$ a standard Eichler order as in Remark A.10. Thus, a maximal order $S = \mathrm{End}_A(N)$ contains R if and only if $R(N) \subseteq N$. Since we can scale to get $N \subseteq A^2$ with A^2/N a cyclic A-module (necessarily $A/\mathfrak{m}^n$), it is easy to check (using A-module generators of the explicit standard Eichler order R) that N must be the span of e_1 and $\pi^j e_2$ for some $0 \leq j \leq n$. Only the extreme options $j = 0, n$ provide a pair whose intersection is the standard R.

In order to build a model of the global order in Lemma 7.1, there is one further aspect of the local theory of orders in quaternion algebras which we need to address: discriminants. The trace form $(x, y) \mapsto \mathrm{Tr}(xy)$ is a symmetric bilinear form on B, and its restriction to an order R is a symmetric A-valued bilinear form on R. Thus, we can define the *discriminant* $\mathrm{disc}(R)$ of R to be the discriminant of this bilinear form (i.e., the ideal generated by the determinant of its values on pairs from an A-basis of R; note that R is A-free since we are in the local case). Since Tr is invariant under $B^\times$-conjugation, conjugate orders in B have the same discriminant ideal. Also, since the trace form is nondegenerate (as can be checked in the split case upon extending scalars), $\mathrm{disc}(R) \neq 0$. Formation of the discriminant ideal commutes with faithfully flat base change $A \to A'$ to another (complete) discrete valuation ring. If $R' \subseteq R$ is a suborder with the finite-length A-module R/R' of length n, then $\mathrm{disc}(R') = \mathfrak{m}^{2n}\mathrm{disc}(R)$. Since every order lies inside of a maximal order (all of which are conjugate), we see that discriminants of orders in B are off by a square factor from the common discriminant of the maximal orders in B.

EXAMPLE A.12. If $B = M_2(F)$ and R is the Eichler order $\begin{pmatrix} A & A \\ \mathfrak{m}^n & A \end{pmatrix}$, then $\mathrm{disc}(R) = \mathfrak{m}^{2n}$. In particular, in the split case $\mathrm{disc}(R) = A$ if and only if R is a maximal order, and all orders have square discriminant.

In general, B has an unramified splitting field F'/F, so if R is a maximal order in B and A' is the valuation ring of F' then $R' = A' \otimes_A R$ is an order in the split $F' \otimes_F B$. It follows that $\operatorname{disc}(R)A' = \operatorname{disc}(R')$ is a square, hence $\operatorname{disc}(R)$ is a square since $A \to A'$ is unramified. This motivates us to define the *reduced discriminant* $\operatorname{Disc}(R)$ of an order R in B to be the ideal of A whose square is $\operatorname{disc}(R)$.

EXAMPLE A.13. If $R' \subseteq R$ is an inclusion of orders in B and R/R' has A-length n, then $\operatorname{Disc}(R') = \mathfrak{m}^n \operatorname{Disc}(R)$. For example, if R_n is the Eichler order $\left(\begin{smallmatrix} A & A \\ \mathfrak{m}^n & A \end{smallmatrix}\right)$ from Remark A.10, then $\operatorname{Disc}(R_n) = \mathfrak{m}^n$. This clarifies the uniqueness of n in Lemma A.9(2).

We can also compute $\operatorname{Disc}(R)$ which R is maximal and B is nonsplit. Since there is only one such B over F up to isomorphism, and all R's are conjugate in B, it suffices to compute in a single example (over F). Let F' be the unramified quadratic extension of F, with valuation ring A'. Since π is not a norm from F', by Lemma A.5 it follows that $B \stackrel{\text{def}}{=} F' \oplus F'j$ with $j^2 = \pi$ is a nonsplit quaternion algebra over F. One can then check that $R = A' \oplus A'j$ is the set of A-integral elements, so this is the maximal order. Writing down the matrix for the Tr pairing, we get $\operatorname{disc}(R) = \operatorname{disc}(A'/A)^2 (j^2)^2 = \mathfrak{m}^2$, so $\operatorname{Disc}(R) = \mathfrak{m}$. Thus, in the nonsplit case an order R is maximal if and only if $\operatorname{Disc}(R) = \mathfrak{m}$.

The nontriviality of reduced discriminants of maximal orders in the nonsplit case is a reason that nonsplit quaternion algebras in the local case are referred to as "ramified".

Now we are in position to globalize to the case of orders in quaternion algebras over the fraction field F of a Dedekind domain A (the case of a number field, especially $F = \mathbf{Q}$, is the one of most importance for us). Fix a quaternion algebra B over F and an order R in B (of which there are clearly many, for example by intersecting B with an order in a split extension B_E). We can define $\operatorname{disc}(R)$ as in the local case to be the (nonzero) discriminant ideal of the symmetric bilinear nondegenerate Tr pairing. Since R is merely locally free as an A-module, and probably not free, this discriminant ideal is constructed by localizing on A, just as in the definition of the discriminant of a symmetric bilinear form on any locally free module of finite rank over a commutative ring. Note that $\operatorname{disc}(R) = \prod_v (\operatorname{disc}(R_v) \cap A)$, with $\operatorname{disc}(R_v) = A_v$ for all but finitely many v. By passing to the local case, we see that $\operatorname{disc}(R)$ is a square, so we may define the *reduced discriminant* ideal $\operatorname{Disc}(R)$ in A. This ideal is the unit ideal at all but finitely many places, so the preceding discriminant calculations show that R_v is a maximal order and B_v is split for all but finitely many maximal ideals v of A. In contrast with the local case, in the global case it is no longer true that maximal orders have to be conjugate.

We can uniquely construct orders by specifying local data. This will be an essential ingredient in our construction of models for global orders, so let us briefly summarize how this goes. As with lattices in any finite-dimensional F-vector space, if R and R' are two orders in a quaternion algebra B over F, then $R_v = R'_v$ inside of $B_v = F_v \otimes_F B$ for all but finitely many maximal ideals v of A.

Conversely, if we pick an order S_v in B_v for all v with $S_v = R_v$ for all but finitely many v, then there exists a unique order S in B with $A_v \otimes_A S = S_v$ inside of B_v for all v. This is an easy application of weak approximation for Dedekind domains. In particular, an order R in B is maximal if and only if R_v is maximal in B_v for all v, and likewise R is an Eichler order (i.e,, an intersection of two maximal orders) if and only if R_v is an Eichler order for all v.

Note that when R is maximal, $\mathrm{Disc}(R)$ is the product of the maximal ideals of A at precisely those places where B is nonsplit. If R is merely an Eichler order, then $\mathrm{Disc}(R)$ is the product of the maximal ideals at the nonsplit places of B (the ramified primes of B, for which B_v has only one maximal order) and powers of maximal ideals at the split places where R is a nonmaximal Eichler order.

EXAMPLE A.14. The order R in Lemma 7.1 is an Eichler order with reduced discriminant Np. Indeed, the lemma assures us that R is maximal at the nonsplit place p, and the local description at all $\ell \neq p$ (where B is split) is exactly a standard Eichler order in $M_2(\mathbf{Z}_\ell)$ whose index is the ℓ-part of N.

With Example A.14 in hand, specialize to the case $A = \mathbf{Z}$. We seek a *potential* model for the order $R = \mathrm{End}_{W/\pi}(\underline{x})$ inside of B which arises in Theorem 10.5. Inside of our concrete model $\left(\frac{D,?}{\mathbf{Q}}\right)$ for B (in the inert and ramified cases separately), we want to find an Eichler order containing $\mathscr{O}_K$ and having reduced discriminant Np. We will find such an order, but we will not know that this order is $\mathscr{O}_K$-isomorphic, let alone conjugate, to $\mathrm{End}_{W/\pi}(\underline{x})$. This problem will be overcome by the fact that in (A–9) and (A–10) we are summing over all places of H over p. We carry out the construction in two separate cases: p inert in K and p ramified in K.

When p is inert in K, recall that we found an isomorphism $B \simeq \left(\frac{D,-pq}{\mathbf{Q}}\right)$ where $q \nmid pD$ is a prime satisfying certain congruence conditions at the primes $\ell | D$ and possibly at 8 as well. Clearly $R' = \mathscr{O}_K \oplus \mathscr{O}_K j$ is an order, and we can compute its reduced discriminant to be Dpq. Thus, R' is maximal at p and at all primes not dividing Dq. Since a maximal order of B has reduced discriminant p (because B is nonsplit at precisely p and ∞), it follows that R' has index Dq inside of any maximal order containing R'. In particular, an order containing R' with index Dq *must* be a maximal order of B.

To examine how one might find a maximal order in B containing R', we first will find local models for maximal orders at all primes $\ell | Dq$ (exactly the places where R' is not maximal), and then we will globalize as in the discussion preceding Example A.14. First let ℓ be a prime dividing D (so ℓ ramifies in K), and consider the split algebra $B_\ell = M_2(\mathbf{Q}_\ell)$. By Remark A.7, $-pq$ is a unit square in $\mathbf{Z}_\ell$, so $-pq$ is a unit norm from the ramified quadratic extension $\mathscr{O}_{K,\ell}$. Pick $X_\ell \in \mathscr{O}_{K,\ell}$ with $\mathrm{N}(X_\ell) = -pq = j^2$. The element $X_\ell - j \in B_\ell$ has nonzero reduced trace but vanishing reduced norm, so it is a zero divisor in B_ℓ which generates a left B_ℓ-module of dimension 2 over $\mathbf{Q}_\ell$. This module must have $X_\ell - j$ and $\omega(X_\ell - j)$ as a $\mathbf{Z}_\ell$-basis, where $\mathscr{O}_K = \mathbf{Z}[\omega]$ (so $\mathscr{O}_{K,\ell} = \mathbf{Z}_\ell[\omega]$). We may take $\omega = (1+i)/2$ when D is odd and $\omega = i/2$ when D is even, where $i^2 = D$. The natural map from B_ℓ to $\mathrm{End}_{\mathbf{Q}_\ell}(B_\ell(X_\ell - j)) \simeq M_2(\mathbf{Q}_\ell)$ gives a concrete splitting, where we use the basis $X_\ell - j$, $\omega(X_\ell - j)$. Under this isomorphism, $R'_\ell = \mathscr{O}_{K,\ell} \oplus \mathscr{O}_{K,\ell} j$ clearly maps into $M_2(\mathbf{Z}_\ell)$ (compute the action of $\mathbf{Z}_\ell$-generators of R'_ℓ on the $\mathbf{Q}_\ell$-basis $\{X_\ell - j, \omega(X_\ell - j)\}$ of $B_\ell(X_\ell - j)$).

Since $(X_\ell + j)(X_\ell - j) = 0$, we compute (using the identities $jX_\ell = \overline{X}_\ell j$ and $\mathrm{N}(X_\ell) = j^2$):

$$
\begin{aligned}
(X_\ell + j)\omega(X_\ell - j) &= (X_\ell + j)\frac{i}{2}(X_\ell - j) = \frac{i}{2}(X_\ell - j)^2 \\
&= \frac{i}{2}(2X_\ell)(X_\ell - j) \;= (2X_\ell)\Big(\frac{i}{2}(X_\ell - j)\Big).
\end{aligned}
$$

For odd D it follows that $\frac{i}{D}(X_\ell + j)$ lies in $M_2(\mathbf{Z}_\ell)$ and has additive order D_ℓ in B_ℓ/R'_ℓ (where D_ℓ is the ℓ-part of D). Thus, the maximal order $R_\ell = M_2(\mathbf{Z}_\ell)$ is generated by $\frac{i}{D}(X_\ell + j)$ and R'_ℓ (since R'_ℓ has index $(Dq)_\ell = D_\ell$ in a maximal order). Similarly, if $2|D$ then $\frac{i}{D}(X_\ell + j)$ and $\frac{1}{2}(X_\ell + j)$ act as elements of $M_2(\mathbf{Z}_\ell)$ and additively generate a subgroup of order D_ℓ in B_ℓ/R'_ℓ. Thus, these two elements along with R'_ℓ generate R_ℓ when D is even, so the maximal order R_ℓ is always spanned over $\mathbf{Z}_\ell$ by 1, ω, $(i/D)(X_\ell + j)$, and $e \cdot (X_\ell + j)$, where $e = 1$ or $e = \frac{1}{2}$, according as whether $2 \nmid D$ or $2|D$.

A more uniform and succinct way to describe the maximal order $R_\ell \subseteq B_\ell$ at $\ell|D$ is:

$$
\{\alpha + \beta j \,|\, \alpha, \beta \in \mathfrak{d}_\ell^{-1},\ \alpha - X_\ell \beta \in \mathscr{O}_{K,\ell}\}, \tag{A–12}
$$

where $\mathfrak{d}_\ell^{-1}$ is the inverse different of $\mathscr{O}_{K,\ell}$. To see that the lattice (A–12) is R_ℓ, one first checks that it contains the $\mathbf{Z}_\ell$-generators $1, \omega, (i/D)(X_\ell + j), e \cdot (X_\ell + j)$ of R_ℓ, so the problem is to show the reverse containment (direct verification that (A–12) is an order seems quite painful). For any pair (α, β) satisfying the criteria in (A–12), we have $\alpha + \beta j = (\alpha - X_\ell \beta) + \beta(X_\ell + j) \in \mathscr{O}_{K,\ell} + \mathfrak{d}_\ell^{-1}(X_\ell + j)$, so we just need $\mathfrak{d}_\ell^{-1}(X_\ell + j)$ to lie in R_ℓ. Since R_ℓ contains $\mathscr{O}_{K,\ell}$ and is an order, it suffices to pick a generator γ of the fractional ideal $\mathfrak{d}_\ell^{-1}$ and to show $\gamma(X_\ell + j)$ lies in R_ℓ. For odd D or $\ell \neq 2$, we can take $\gamma = i/D$. For even D and $\ell = 2$, we use the fact that i/D and $1/2$ generate $\mathfrak{d}_2^{-1}$ as a $\mathbf{Z}_2$-module in such cases (ultimately because $\mathfrak{d}_2$ is generated by $i = \sqrt{D}$ and $D/2$ over $\mathbf{Z}_2$), by treating separately the cases $D \equiv 4 \bmod 8$ and $8|D$.

The importance of the description (A–12) is that it provides a description which only depends on $X_\ell \bmod \mathfrak{d}_\ell$. To make this point clearer, first observe (for $\ell|D$, being careful if $\ell = 2$) that changing an element of $\mathscr{O}_{K,\ell}$ modulo $\mathfrak{d}_\ell$ does not change its norm (to $\mathbf{Z}_\ell$) modulo D_ℓ. Thus, even though we chose X_ℓ above to satisfy $\mathrm{N}(X_\ell) = -pq$ so as to carry out the preceding calculations, if we replace X_ℓ with some $X'_\ell \in \mathscr{O}_{K,\ell}$ which is the same mod $\mathfrak{d}_\ell$, then $\mathrm{N}(X'_\ell) \equiv -pq \bmod D_\ell$ and using X'_ℓ instead of X_ℓ in (A–12) yields the same lattice (i.e., the maximal order $R_\ell \subseteq B_\ell$ we constructed at ℓ). In fact, the *only* thing which matters about

$X_\ell \in \mathscr{O}_{K,\ell}$ is that it satisfies $\mathrm{N}(X_\ell) \equiv -pq \bmod D_\ell$, since we claim that any such X_ℓ is congruent mod $\mathfrak{d}_\ell$ to an element of $\mathscr{O}_{K,\ell}$ with norm equal to $-pq$. Indeed, since $-pq$ is a unit norm at ℓ, we really just have to show that if $u \in \mathscr{O}_{K,\ell}^\times$ satisfies $\mathrm{N}(u) \equiv 1 \bmod D_\ell$ then $u \equiv u' \bmod \mathfrak{d}_\ell$ with $\mathrm{N}(u') = 1$. The case of odd ℓ (resp. $\ell = 2$ and $D_2 = 8$) is easy since $1 + \ell\mathbf{Z}_\ell$ (resp. $1 + 8\mathbf{Z}_2$) consists of squares, and the case $\ell = 2$ and $D_2 = 4$ (so $D \equiv 4 \bmod 8$) requires checking additionally that some (and thus every) element congruent to 5 mod 8 is a norm, and we see that $\mathrm{N}(1+i) = 1 - D \equiv 5 \bmod 8$. From now on, for each $\ell | D$, rather than require X_ℓ to have exact norm $-pq$, we merely require it to represent an element in $\mathscr{O}_{K,\ell}/\mathfrak{d}_\ell$ whose norm in $\mathbf{Z}_\ell/D_\ell$ is $-pq$.

With the construction of explicit maximal orders at all $\ell | D$ settled, we now turn to the easier task of finding a maximal order at q. This is easier because $q = \mathfrak{q}\bar{\mathfrak{q}}$ splits in K. It is simplest to check directly that if $\alpha \in \mathscr{O}_K$ and $\beta \in \mathfrak{q}^{-1}$, then the resulting elements $\alpha + \beta j$ form an order $R^{(q)}$ in B which is maximal at q. This set is stable under multiplication because if $\beta, \beta' \in \mathfrak{q}^{-1}$ then $(\beta j)(\beta' j) = \beta\overline{\beta}' j^2 = \beta\overline{\beta}' \cdot (-pq) \in \mathscr{O}_K$ since $(q) = \mathfrak{q}\bar{\mathfrak{q}}$. The reduced discriminant of $R^{(q)}$ is a q-unit, so we get maximality of this order at the split place q of B.

A global order that satisfies all of the above local conditions and contains $R' = \mathscr{O}_K \oplus \mathscr{O}_K j$ will have reduced discriminant p, as it would now be maximal at all primes dividing Dq. If we pick $X \in \mathscr{O}_K$ such that $\mathrm{N}(X) \equiv -pq \bmod D$, then X is congruent modulo $\mathfrak{d}_\ell$ to a legitimate choice of X_ℓ for each $\ell | D$. If we let $\mathfrak{d}$ denote the different for K over $\mathbf{Q}$, and $\mathscr{O}_\mathfrak{d}$ the semilocal subring of K consisting of elements which are integral at the places dividing $\mathfrak{d}$, then the lattice $S' = S'_X = \{\alpha + \beta j \in B \,|\, \alpha \in \mathfrak{d}^{-1},\ \beta \in \mathfrak{d}^{-1}\mathfrak{q}^{-1}, \alpha - X\beta \in \mathscr{O}_\mathfrak{d}\}$ contains R' and satisfies the above local conditions at each finite place to make it a maximal order of B (for $\ell \nmid Dpq$, S' has ℓ-unit reduced discriminant). We are looking for an Eichler order containing $\mathscr{O}_K$ and having reduced discriminant Np. Recall that the primes of N are split in K, so we can fix a factorization $N = \mathfrak{n}\bar{\mathfrak{n}}$ as a product of relatively prime conjugate ideals (in fact, $\mathfrak{n}$ is chosen in the main text to correspond to the kernel of an isogeny at a chosen Heegner point, but the choice won't matter here). In this case, $\mathfrak{n}S'\mathfrak{n}^{-1}$ is an order which must also be maximal, as its localization at any place is conjugate to that of S' at the same place (since localizations of $\mathfrak{n}$ are principal). Since $\mathfrak{n}$ is relatively prime to D, this conjugation has no impact at the ramified places, so $\mathfrak{n}S'\mathfrak{n}^{-1}$ and S' have the same congruence conditions at places dividing D. In fact,

$$\mathfrak{n}S'\mathfrak{n}^{-1} = \{\alpha + \beta j \,|\, \alpha \in \mathfrak{d}^{-1},\ \beta \in \mathfrak{d}^{-1}\mathfrak{q}^{-1}\mathfrak{n}\bar{\mathfrak{n}}^{-1},\ \alpha - X\beta \in \mathscr{O}_\mathfrak{d}\};$$

the factor of $\bar{\mathfrak{n}}^{-1}$ is due to the relation $jx = \bar{x}j$ for all $x \in K$.

The intersection $S = S' \cap \mathfrak{n}S'\mathfrak{n}^{-1}$ is an Eichler order containing $\mathscr{O}_K$. Since $\mathfrak{n}$ and $\bar{\mathfrak{n}}$ are relatively prime, S is explicitly described as

$$S = S_X = \{\alpha + \beta j \,|\, \alpha \in \mathfrak{d}^{-1},\ \beta \in \mathfrak{d}^{-1}\mathfrak{q}^{-1}\mathfrak{n},\ \alpha - X\beta \in \mathscr{O}_\mathfrak{d}\}. \tag{A–13}$$

The quotient $\mathscr{O}_K$-module S'/S is $\mathscr{O}_K/\mathfrak{n}$, so S has index N in S'. Thus, S has reduced discriminant Np as desired. If we had also required $-pq \equiv 1 \bmod D$, then we could have taken $X = 1$ (see Remark A.7). This seems to implicitly be the choice made in [GZ]; certainly one does not get an order in (A–13) using $X = 1$ unless $-pq \equiv 1 \bmod D$.

When $p\mathscr{O}_K = \mathfrak{p}^2$ is ramified in K and we take $\left(\frac{D,-q}{\mathbf{Q}}\right)$ as the explicit model for B as discussed earlier, we can similarly compute the reduced discriminant of the order $R' = \mathscr{O}_K + \mathscr{O}_K j$ to be Dq. If p is odd, we need only find $X \in \mathscr{O}_K$ such that $\mathrm{N}(X) \equiv -q \bmod D/p$; in exactly the same way as above, we find that the following is an Eichler order (containing $\mathscr{O}_K$) with reduced discriminant Np:

$$S = S_X = \{\alpha + \beta j \mid \alpha \in \mathfrak{p}\mathfrak{d}^{-1},\ \beta \in \mathfrak{p}\mathfrak{d}^{-1}\mathfrak{q}^{-1}\mathfrak{n},\ \alpha - X\beta \in \mathscr{O}_{\mathfrak{d}}\}. \qquad \text{(A–14)}$$

Note that $\mathfrak{p}\mathfrak{d}^{-1}$ has no factor at $\mathfrak{p}$ (since $p \neq 2$), and thus there is no local congruence condition at $\mathfrak{p}$, as α and β are already integral at $\mathfrak{p}$. Of course, no alteration from R' is required above p, as $\mathrm{ord}_p(Dq) = 1$, so R' is maximal at the nonsplit place p of B.

If $p = 2$ is ramified in K, the situation is somewhat different, but the explicit description of the order is the same as (A–14), where we again pick $X \in \mathscr{O}_K/\mathfrak{d}\mathfrak{p}^{-1}$ with $\mathrm{N}(X) \equiv -q \bmod D/p$. It is only necessary to explain the local conditions above 2, where R' fails to be maximal (since $4|Dq$). Recall that the unique maximal order of a nonsplit quaternion algebra over a local field consists of precisely the elements of integral norm. If α and β are nonintegral at 2, the condition that $\mathrm{N}(\alpha)+q\mathrm{N}(\beta) = \mathrm{N}(\alpha+\beta j)$ be 2-integral is equivalent to $\mathrm{N}(\alpha/\beta) \equiv -q \bmod \mathrm{N}(\beta)^{-1}$. Such a congruence cannot happen $\bmod D_2$, since then q would be a local norm and hence we could find a nonzero $\alpha + \beta j \in B$ with vanishing reduced norm, contrary to the fact that B_2 is a division algebra. There is no obstruction $\bmod D_2/2$, and indeed one can easily check that there is a unique $X_2 \in \mathscr{O}_{K,2}/\mathfrak{p}^{-1}\mathfrak{d}$ satisfying $\mathrm{N}(X_2) \equiv -q \bmod D_2/2$. Thus we conclude that $\alpha - X_2\beta \in \mathscr{O}_{K,2}$ (recall that the valuation ring of a local field is characterized by the property of having integral norm in a local subfield), and $\alpha, \beta \in \mathfrak{p}\mathfrak{d}^{-1}$. Likewise, (A–14) holds. Note that in the ramified case (for $p = 2$ and for $p \neq 2$), Remark A.7 allows us to take $X = 1$ (as in [GZ]) by also requiring $-q \equiv 1 \bmod D/p$ (this does not impose a congruence at p, unless $p = 2$).

What we have done so far is find an Eichler order containing $\mathscr{O}_K$ and having the correct reduced discriminant. This has not yet been shown to have a connection to the specific order $R = \mathrm{End}_{W/\pi}(\underline{x})$ in Theorem 10.5. Fortunately, the data of the embedding of $\mathscr{O}_K$ ensures some connection between such Eichler orders. The following theorem and its corollary are due to Eichler, and we include a proof for the reader's convenience. Corollary A.16 will be applied with $F = \mathbf{Q}$ and $E = K$.

THEOREM A.15. *Let F be a nonarchimedean local field with valuation ring A, maximal ideal $\mathfrak{m}$, and uniformizer π. Let B be a split quaternion algebra over*

F with a fixed embedding of the quadratic field E over F. Let S and S' be Eichler orders in B with the same isomorphism class (i.e., $\mathrm{Disc} S = \mathrm{Disc} S'$*). If $E \cap S = E \cap S' = \mathcal{O}_E$, then there exists a nonzero $x \in \mathcal{O}_E$ such that $xSx^{-1} = S'$.*

PROOF. We will first prove the stament when S and S' are maximal orders, which we can assume are distinct. By Lemma A.9, we can assume $B = \mathrm{End}_F(V) = M_2(F)$ for a vector space $V = F^2$ over F, with $S = \mathrm{End}_A(N) = M_2(A)$ for a lattice $N = A^2$ in V and $S' = \mathrm{End}_A(N')$ for some other lattice N' in V which is not a scalar multiple of N. By scaling N' and changing the basis of N if necessary, we can assume $N = Ae_1 \oplus Ae_2$ and $N' = Ae_1 \oplus \mathfrak{m}^n e_2$ for some $n > 0$, so

$$S' = \mathrm{End}_A(Ae_1 \oplus \mathfrak{m}^n e_2) = \begin{pmatrix} A & \mathfrak{m}^{-n} \\ \mathfrak{m}^n & A \end{pmatrix} = \gamma_n M_2(A) \gamma_n^{-1} = \gamma_n S \gamma_n^{-1},$$

where $\gamma_n = \left(\begin{smallmatrix} 0 & 1 \\ \pi^n & 0 \end{smallmatrix}\right)$. Note that the elements of $B^\times = \mathrm{GL}_2(F)$ which conjugate S into S' are exactly the elements of the coset $\gamma_n S^\times = \gamma_n \mathrm{GL}_2(A)$. Since $S \cap S' = \left(\begin{smallmatrix} A & A \\ \mathfrak{m}^n & A \end{smallmatrix}\right)$ is the standard Eichler order of reduced discriminant $\mathfrak{m}^n$ and $\mathcal{O}_E \subseteq S \cap S'$ by hypothesis, we can find $a, b, c \in A$ so that an A-basis of $\mathcal{O}_E$ is given by the identity 1 and the element $\rho = \left(\begin{smallmatrix} a & b \\ \pi^n c & 0 \end{smallmatrix}\right)$. Since $\rho/\pi \in E$ is not in $\mathcal{O}_E$, yet $S \cap E = S' \cap E = \mathcal{O}_E$, it follows that ρ/π is not contained in either of S or S'. Hence, by inspection of the matrix for ρ we see that either a or bc must be a unit in A. If $bc \in A^\times$ then $\rho \in \gamma_n \mathrm{GL}_2(A)$, while if not then a is a unit and hence we can instead replace ρ with $\rho + \pi^n = \left(\begin{smallmatrix} a+\pi^n & b \\ \pi^n c & \pi^n \end{smallmatrix}\right)$ which does lie in $\gamma_n \mathrm{GL}_2(A)$. That is, we may assume $\rho \in \mathcal{O}_E$ conjugates S into S'. This settles the case when S and S' are maximal.

Now consider the case in which the isomorphic S and S' are nonmaximal, say with reduced discriminant $\mathfrak{m}^n$ for some $n > 0$. By conjugation, we may assume $S = \left(\begin{smallmatrix} A & A \\ \mathfrak{m}^n & A \end{smallmatrix}\right)$ is the standard Eichler order of reduced discriminant $\mathfrak{m}^n$ in $M_2(A)$. Thus, $S = S_1 \cap S_2$ where the S_j's are the maximal orders considered above (i.e., $S_1 = M_2(A)$ and $S_2 = \mathrm{End}_A(Ae_1 \oplus \mathfrak{m}^n e_2)$). We may write $S' = S'_1 \cap S'_2$ where the S'_j's are maximal orders in B. The special case of maximal orders ensures that there exists a nonzero $x \in \mathcal{O}_E$ with $xS_1x^{-1} = S'_1$. If S and $x^{-1}S'x$ are conjugate by a nonzero element of $\mathcal{O}_E$ (as in the statement of the theorem), then we'll be done. Thus, we may assume $S'_1 = M_2(A) = S_1$.

Since the two Eichler orders S and S' are presented as intersections of a *common* maximal order S_1 with other maximal orders S_2 and S'_2, from the *proof* of Lemma A.9 there exists $y \in S_1^\times = \mathrm{GL}_2(A)$ such that $ySy^{-1} = S'$. The given condition $\mathcal{O}_E \subseteq S' = ySy^{-1}$ says exactly that $y^{-1}\rho y \in S_2 = \left(\begin{smallmatrix} A & \mathfrak{m}^{-n} \\ \mathfrak{m}^n & A \end{smallmatrix}\right)$, where $\{1, \rho\}$ is an A-basis of $\mathcal{O}_E$, say with ρ chosen as above. Since $y^{-1}\rho y \in S_1 = M_2(A)$, the condition for membership in S_2 is just the condition that the lower left corner entry of $y^{-1}\rho y$ lie in $\mathfrak{m}^n$. Explicitly, if $y = \left(\begin{smallmatrix} a' & b' \\ c' & d' \end{smallmatrix}\right)$, then this condition is equivalent to saying $c'(aa' + bc') \in \mathfrak{m}^n$. Conversely, any $y = \left(\begin{smallmatrix} a' & b' \\ c' & d' \end{smallmatrix}\right) \in \mathrm{GL}_2(A) = S_1^\times$ satisfying this latter property automatically satisfies $\mathcal{O}_E \subseteq ySy^{-1}$, with the Eichler order ySy^{-1} given by intersecting $S_1 = M_2(A)$ with another maximal order.

Suppose c' is not a unit in A. If $c' \in \mathfrak{m}^n$ (automatic if $n = 1$), then $y \in \Gamma_0(\mathfrak{m}^n) \stackrel{\text{def}}{=} S_2 \cap S_1^\times = S_2 \cap \mathrm{GL}_2(A)$ conjugates S_2 into itself, so $S = S'$ and we are done. If, on the other hand, $n > 1$ and $c' \in \mathfrak{m} - \mathfrak{m}^n$, then since $c'(aa' + bc') \in \mathfrak{m}^n$ we get $aa' + bc' \in \mathfrak{m}$, or in other words $aa' \in \mathfrak{m}$. But $y \in \mathrm{GL}_2(A)$ and $c' \in \mathfrak{m}$, so $a' \notin \mathfrak{m}$ and hence $a \in \mathfrak{m}$. This combination of conditions implies $\rho/\pi \in E$ has integral trace and norm as a matrix, so $\rho/\pi \in E$ is integral over A and therefore lies in $\mathscr{O}_E$. This is a contradiction, so we can assume $c' \in A^\times$, so $aa' + bc' \in \mathfrak{m}^n$.

We must have $a \notin \mathfrak{m}$, as otherwise $a, b \in \mathfrak{m}$, a contradiction (since $\rho/\pi \notin S_1 = M_2(A)$). There is a more useful way to express these conditions: for $t = \left(\begin{smallmatrix} 0 & 1 \\ a & b \end{smallmatrix}\right) \in \mathrm{GL}_2(A)$, the condition on y is precisely that $ty \in \Gamma_0(\mathfrak{m}^n)$. We are now in position to find a nonzero $x \in \mathscr{O}_E$ such that $xSx^{-1} = S'$. Let Z be the group generated by the central element $\pi \in S_1 = M_2(A) \subseteq B$. Note that conjugation by γ_n is an involution of both S and $S^\times = \Gamma_0(\mathfrak{m}^n)$, since it interchanges $S_1 = M_2(A)$ and $S_2 = \gamma_n M_2(A) \gamma_n^{-1}$ (with $S = S_1 \cap S_2$). The group G generated by $\Gamma_0(\mathfrak{m}^n)$, γ_n, and Z is exactly the subgroup of elements of $B^\times = \mathrm{GL}_2(F)$ which conjugate S into itself. To prove this, first note that if $b \in B^\times$ conjugates S into itself, then conjugation by b either preserves or swaps the unique pair of maximal orders S_1 and S_2 whose intersection is the nonmaximal Eichler order S (see Remark A.11). Multiplying against γ_n if necessary permits us to assume that conjugation by b preserves both S_j's. But for a maximal order R in B, the elements of $B^\times$ which conjugate R into R are the elements of $F^\times R^\times = \pi^{\mathbf{Z}} R^\times$ (since $R = \mathrm{End}_A(N)$ forces $bRb^{-1} = \mathrm{End}_A(b(N))$, and $\mathrm{End}_A(b(N)) = \mathrm{End}_A(N)$ if and only if $b(N) = cN$ for some $c \in F^\times$). Thus,

$$b \in \pi^{\mathbf{Z}}(S_1^\times \cap (\pi^{\mathbf{Z}} \cdot S_2^\times)) = \pi^{\mathbf{Z}}(S_1^\times \cap S_2^\times) = \pi^{\mathbf{Z}} S^\times = Z \cdot \Gamma_0(\mathfrak{m}^n).$$

Here, the first equality uses that the reduced norm of a unit in an order is a unit in A.

It suffices to find a nonzero $x \in \mathscr{O}_E$ such that $y^{-1}x \in G$, since then $x = y(y^{-1}x)$ conjugates S into S'. Since $y^{-1} \in \Gamma_0(\mathfrak{m}^n)t$, we need to find $x \in \mathscr{O}_E$ such that $tx \in G$. This is something we can accomplish by observation: letting $x = (-a + \pi^r) + \rho$ for an $r > 0$ to be determined, we compute

$$tx = \begin{pmatrix} 0 & 1 \\ a & b \end{pmatrix} \begin{pmatrix} \pi^r & b \\ c\pi^n & -a + \pi^r \end{pmatrix} = \begin{pmatrix} c\pi^n & -a + \pi^r \\ a\pi^r + bc\pi^n & b\pi^r \end{pmatrix}.$$

If $bc \in \mathfrak{m}$ then $r = n$ makes $tx \in \Gamma_0(\mathfrak{m}^n)$, while if $bc \notin \mathfrak{m}$ then $r = n + 1$ does the same. □

COROLLARY A.16. *Let F be a number field, and B a quaternion algebra over F equipped with a fixed embedding of a quadratic extension E. If S and S' are Eichler orders of B with the same reduced discriminant such that $S \cap E = S' \cap E = \mathscr{O}_E$, then there exists a nonzero ideal I of $\mathscr{O}_E$ such that $ISI^{-1} = S'$.*

PROOF. A priori $S_v = S'_v$ for all but finitely many v. For all such v define $x_v = 1$. At the remaining places, use Theorem A.15 to construct $x_v \in \mathscr{O}_{E_v}$ so that $x_v S x_v^{-1} = S'$. Let I be the ideal with $\mathrm{ord}_v(I) = \mathrm{ord}_v(x_v)$ for all nonarchimedean places v. The equality $ISI^{-1} = S'$ can be checked place by place. □

In the situation given in Corollary A.16, if we have $S' = ISI^{-1}$ with I a *principal* ideal, then picking a generator gives an explicit isomorphism from S to S' which is $\mathscr{O}_E$-linear. Since E is its own centralizer in B and all automorphisms of B are inner (Skolem–Noether), we conclude that the set of $\mathscr{O}_E$-linear isomorphism classes of Eichler orders in B with a fixed reduced discriminant and containing $\mathscr{O}_E$ admits a simply transitive action by the class group of $\mathscr{O}_E$. In our situation of interest (with $F = \mathbf{Q}$ and $E = K$), upon fixing a place v of H over p we have already established (in (7–8)) for $R_v = \mathrm{End}_{W/\pi}(\underline{x})$ that if $\sigma \in \mathrm{Gal}(H/K) \simeq \mathrm{Cl}_K$ corresponds to the ideal class of some nonzero integral ideal $\mathfrak{a}$, then there is an abstract isomorphism $\mathfrak{a}^{-1} R_v \mathfrak{a} \simeq \mathrm{End}_{W/\pi}(\underline{x}^{\sigma})$ as $\mathscr{O}_K$-algebras, where $W = W_v$ is the completion of a maximal unramified extension over v. Thus, if B denotes the unique (up to isomorphism) quaternion algebra over $\mathbf{Q}$ which is ramified at exactly p and ∞ and we fix an embedding of K into B, then we have a bijection between two sets: the set of isomorphism classes of Eichler orders in B containing $\mathscr{O}_K$ and having reduced discriminant Np, and the set of Heeger points $\{x^{\sigma}\}$ in $X_0(N)(H)$, via $\mathrm{End}_{W/\pi}(\underline{x}^{\sigma}) \leftrightarrow \sigma$.

Since p is nonsplit in K, we have a transitive free action of $\mathrm{Gal}(H/K)$ on the v's of H over p. An important consequence is that rather than fixing σ and considering the isomorphism classes $\mathfrak{a}^{-1} R_v \mathfrak{a} \simeq \mathrm{End}_{W_v/\pi}(\underline{x}^{\sigma})$ as the place v varies over p, we can fix one place v_0 and study $\mathfrak{a}^{-1} R_{v_0} \mathfrak{a}$ as the ideal class $\mathscr{A}$ of $\mathfrak{a}$ varies. Both processes exhaust (without repetition) the set of $\mathscr{O}_K$-algebra isomorphism classes of Eichler orders in B with reduced discriminant Np and containing $\mathscr{O}_K$. Thus, we can fix the Eichler order $S = S_X$ from (A–13) and then for (A–9) we may contemplate a sum over the orders $R^{\mathfrak{b}} = \mathfrak{b}^{-1} S \mathfrak{b}$ as $\mathfrak{b}$ runs over integral ideals representing the ideal classes of K. It is this latter point of view (which avoids the language of supersingular elliptic curves) which will dominate all that follows.

We are now ready to prove [GZ, Ch. III, Prop. 9.7, 9.11], allowing D to be even. We need to introduce some notation. We write $[\mathfrak{a}]$ to denote the ideal class of a fractional ideal $\mathfrak{a}$ of K, and for any ideal class $\mathscr{B}$ we define $R_{\mathscr{B}}(m) = \sum_c r_{\mathscr{B}c}(m)$ where c runs over the *squares* in the ideal class group of K. Note that $R_{\mathscr{B}\mathscr{B}'} = R_{\mathscr{B}^{-1}\mathscr{B}'}$ for any two ideal classes $\mathscr{B}$ and $\mathscr{B}'$. For example, if a prime ℓ of $\mathbf{Q}$ is split in K (such as our prime q chosen above), say $\ell = \mathfrak{l}\bar{\mathfrak{l}}$, then $[\mathfrak{l}]$ and $[\bar{\mathfrak{l}}]$ are inverse to each other in Cl_K. Taking $\ell = q$, we conclude that $R_{\mathscr{A}[\mathfrak{q}\mathfrak{n}]}$ is independent of the choice of prime $\mathfrak{q}$ of K over q, and is likewise independent of the choice of factorization $N = \mathfrak{n}\bar{\mathfrak{n}}$. In particular, the formula in Theorem A.17 below does not depend on the choice of $\mathfrak{q}$ (but only through the entire identity do we see that the right side is independent of the choice of q as above):

THEOREM A.17. *Suppose p is inert in K and $\mathfrak{q}$ is chosen over q. For $n \geq 1$, define $\delta(n) = 2^{e(n)}$ where $e(n)$ is the number of prime factors of* $\gcd(D, n)$. *Then*

$$\langle c, T_m d^\sigma \rangle_p = -u r_{\mathscr{A}}(m) h_K \mathrm{ord}_p(m) \log p - \log p \cdot u^2 \cdot \sum_{\substack{0<n<m|D|/N \\ p|n}} \mathrm{ord}_p(pn) r_{\mathscr{A}}(m|D| - nN) \delta(n) R_{\mathscr{A}[\mathfrak{q}\mathfrak{n}]}(n/p).$$

We specify a choice of $\mathfrak{q}$ in the statement of the theorem because of the appearance of $\mathfrak{q}$ as the final term on the product on the right side, but keep in mind that the choice does not affect the value of that term.

PROOF. By (A–9) and the preceding discussion, we have to prove the identity

$$\sum_{\mathfrak{b}} \sum_{\substack{b \in R^{(\mathfrak{b})}\mathfrak{a}/\pm 1 \\ \mathrm{N}(b) = m\mathrm{N}\mathfrak{a},\, b_- \neq 0}} (1 + \mathrm{ord}_p(\mathrm{N}(b_-))) = u^2 \cdot \sum_{\substack{0<n<m|D|/N \\ p|n}} \mathrm{ord}_p(pn) r_{\mathscr{A}}(m|D| - nN) \delta(n) R_{\mathscr{A}[\mathfrak{q}\mathfrak{n}]}(n/p), \quad \text{(A–15)}$$

where $\mathfrak{b}$ runs over representatives of ideal classes $\mathscr{B}$ of K and $R^{(\mathfrak{b})} \stackrel{\text{def}}{=} \mathfrak{b} S \mathfrak{b}^{-1}$ ranges over the $\mathscr{O}_K$-algebra isomorphism classes of Eichler orders in B which contain $\mathscr{O}_K$ and have reduced discriminant Np. Here, $S = S_X$ is chosen as in (A–13), with $\mathrm{N}(X) \equiv -pq \bmod D$. We need to analyze sums over all such $R^{(\mathfrak{b})}$'s, with the sum for each $R = R^{(\mathfrak{b})}$ ranging over $R\mathfrak{a}/\pm 1$, where $\mathfrak{a}$ is a prime-to-p representative of the ideal class $\mathscr{A}$ corresponding to our fixed $\sigma \in \mathrm{Gal}(H/K)$. Multiplying $\mathfrak{a}$ by a principal ideal which is prime to p has no impact on the left side of (A–15), so we may select our choice to satisfy the additional property that $\gcd(\mathfrak{a}, D) = 1$. This will be convenient later.

Taking $\mathfrak{b}$ fixed above (so we write R rather than $R^{(\mathfrak{b})}$), using (A–13) and the fact that j acts on K through the involution in $\mathrm{Gal}(K/\mathbf{Q})$ yields

$$R\mathfrak{a} = \{\alpha + \beta j \,|\, \alpha \in \mathfrak{d}^{-1}\mathfrak{a},\ \beta \in \mathfrak{d}^{-1}\mathfrak{q}^{-1}\mathfrak{n}\mathfrak{b}\bar{\mathfrak{b}}^{-1}\bar{\mathfrak{a}},\ \alpha - X^{\mathfrak{a}\mathfrak{b}}\beta \in \mathscr{O}_{\mathfrak{d}}\}, \quad \text{(A–16)}$$

where $X^{\mathfrak{a}\mathfrak{b}} \in \mathscr{O}_K/\mathfrak{d}$ is an element we need to define in terms of $X \in \mathscr{O}_K/\mathfrak{d}$ from (A–13); recall that only $X \in \mathscr{O}_K/\mathfrak{d}$ (rather than $X \in \mathscr{O}_K$) matters for the construction of the model S in (A–13). Note that the separate conditions on α and β in (A–16) define a lattice and hence are properties which can be checked locally (where all fractional ideals of $\mathscr{O}_K$ such as $\mathfrak{a}$ and $\mathfrak{b}$ become principal).

To define $X^{\mathfrak{a}\mathfrak{b}}$, we define an element X^I more generally for any nonzero integral ideal I of $\mathscr{O}_K$. Picking an element $y \in \mathscr{O}_K$ for which $y\mathscr{O}_{\mathfrak{d}} = I_{\mathfrak{d}}$ (i.e., (y) has the same factors as I at places dividing $\mathfrak{d}$), note that $y\bar{y}^{-1}$ is a unit at $\mathfrak{d}$ since the primes of $\mathfrak{d}$ are ramified over $\mathbf{Q}$. Moreover, the residue fields at such primes coincide with the prime field, so changing y by a $\mathfrak{d}$-unit has no impact on $y\bar{y}^{-1} \bmod \mathfrak{d} \in (\mathscr{O}_K/\mathfrak{d})^\times$. Thus, the quantity $y\bar{y}^{-1}X \bmod \mathfrak{d}$ only depends on $I_{\mathfrak{d}}$ and $X \bmod \mathfrak{d}$. Hence, if we define $X^I \in \mathscr{O}_K$ to be any solution to the congruence $X^I \equiv y\bar{y}^{-1}X \bmod \mathfrak{d}$, then X^I is well-defined modulo $\mathfrak{d}$ and

$\mathrm{N}(X^I) \equiv -pq \bmod D$ since $\mathrm{N}(X) \equiv -pq \bmod D$ and $\mathrm{N}(y\bar{y}^{-1}) = y\bar{y}^{-1}\bar{y}y^{-1} = 1$. Thus, $R = R^{(\mathfrak{b})} = \mathfrak{b}S_X\mathfrak{b}^{-1}$ is equal to $S_{X^{\mathfrak{b}}}$ and we see why $X^{\mathfrak{ab}}$ intervenes in the description (A–16).

The $X \mapsto X^I$ construction provides an "action" of the multiplicative monoid of nonzero integral ideals on the set of solutions in $\mathcal{O}_K/\mathfrak{d}$ to the congruence $\mathrm{N}(X) \equiv -pq \bmod D$. We need to get a handle on the set of such solutions since we are working with an Eichler order S from (A–13) which depends on the specification of such a solution. For each $\ell | D$, we claim that the congruence $\mathrm{N}(X) \equiv -pq \bmod D_\ell$ in $\mathcal{O}_{K,\ell}/\mathfrak{d}_\ell$ has exactly two (necessarily unit) solutions, or equivalently (since the congruence *does* have solutions, due to how we chose q) that $\mathrm{N} : (\mathcal{O}_{K,\ell}/\mathfrak{d}_\ell)^\times \to (\mathbf{Z}_\ell/D_\ell)^\times$ has kernel of order 2. For odd ℓ this map is just squaring on $\mathbf{F}_\ell^\times$. If $\ell = 2$ and $D_2 = 8$ then $\mathcal{O}_{K,\ell} = \mathbf{Z}_2[\sqrt{\pm 2d}]$ with $d \in \mathbf{Z}_2^\times$ mattering mod 8, and one verifies the claim by direct calculation in all four cases. If $\ell = 2$ and $D_2 = 4$ then $\mathcal{O}_{K,\ell} = \mathbf{Z}_2[\sqrt{-1}]$ or $\mathbf{Z}_2[\sqrt{3}]$, and again a direct calculation does the job.

We conclude that if $\mathfrak{d}$ has t prime factors then there are 2^t congruence classes $X \in \mathcal{O}_K/\mathfrak{d}$ that satisfy the norm congruence condition $\mathrm{N}(X) \equiv -pq \bmod D$. At each place, $X \mapsto X^I$ either fixes the two local solutions or swaps them, so we deduce $X^{I^2} \equiv X \bmod \mathfrak{d}$ for any I. On the other hand, the action of a prime factor $\mathfrak{m}$ of $\mathfrak{d}$ is nontrivial on the local congruence condition at $\mathfrak{m}$ and is trivial at the other factors. Indeed, if the residue characteristic ℓ at $\mathfrak{m}$ is odd then we may take y very congruent to $i = \sqrt{D}$ at $\mathfrak{m}$ and very congruent to 1 at the other factors of $\mathfrak{d}$ (so $y\bar{y}^{-1} \equiv -1 \bmod \ell$ but $y\bar{y}^{-1}$ is congruent to 1 along the other primary factors of D). Meanwhile, if ℓ is 2 and $D_2 = 8$ then we can argue similarly using $i/2$. Finally, if $\ell = 2$ and $D_2 = 4$ (so $\mathfrak{d}_2 = (2)$) then we can pick y highly congruent to $1 + i/2$ over 2, so $y\bar{y}^{-1} \equiv i/2 \not\equiv 1 \bmod \mathfrak{d}_2$.

At this point it is helpful to interject a few comments about the class group of quadratic imaginary fields. It is well-known, and easily checked directly, that 2-torsion subgroup of the class group is generated by the ideal classes above the ramified primes. For odd primes $\ell | D$ and $\ell = 2$ if $8 | D$, these are of the form $(\ell, \sqrt{D})$, while for $D_2 = 4$ we have $(2, 1 + \sqrt{D/4})$ above 2. The order of the group generated by these elements is 2^{t-1} if t distinct primes divide D, since the only nonsquare ideal over the ramified places that becomes trivial in the ideal class group is $(\sqrt{D}) = (i)$ (unless $D_2 = 4$, which requires a separate argument). In this way, we see that the above construction really defines a transitive free action of $(\mathbf{Z}/2)^t$ on the set of possible X's, and this $(\mathbf{Z}/2)^t$ may be naturally identified with an extension of $\mathrm{Cl}_K[2]$ by $\mathbf{Z}/2$. Our formula (A–16) is slightly more general than [GZ, Ch. III, (9.3)] in that it allows D to be even, in which case the action $X \mapsto X^I$ affects the local congruence above 2 in a slightly more complicated manner when $D_2 = 4$ (if we avoid even D and take $X = 1$, then $X^{\mathfrak{ab}}$ collapses to a product of local signs, as in [GZ]).

Continuing with the method of Gross–Zagier, note that an element $b = \alpha + \beta j \in R\mathfrak{a}$ has reduced norm which is the sum of the reduced norms of $b_+ = \alpha$

and $b_- = \beta j$, so in the presence of the conditions on the sum on the left side of (A–15) we get $m\mathrm{N}(\mathfrak{a}) = \mathrm{N}(b) = \mathrm{N}(\alpha) + pq\mathrm{N}(\beta)$, where $pq\mathrm{N}(\beta) = \mathrm{N}(b_-) \neq 0$, so necessarily $\beta \neq 0$. If we introduce the ideals

$$\mathfrak{c} = (\alpha)\mathfrak{d}\mathfrak{a}^{-1}, \quad \mathfrak{c}' = (\beta)\mathfrak{d}\mathfrak{q}\mathfrak{n}^{-1}\mathfrak{b}^{-1}\bar{\mathfrak{b}}\bar{\mathfrak{a}}^{-1} \neq 0, \tag{A–17}$$

then

$$\begin{aligned}\mathrm{N}(\mathfrak{c}) + (Np)\mathrm{N}(\mathfrak{c}') &= \mathrm{N}(\alpha)\frac{|D|}{\mathrm{N}(\mathfrak{a})} + \mathrm{N}(\beta)\cdot Np\frac{|D|q}{\mathrm{N}(\mathfrak{a})N} \\ &= (\mathrm{N}(\alpha) + pq\mathrm{N}(\beta))\frac{|D|}{\mathrm{N}(\mathfrak{a})} = m|D|.\end{aligned}$$

In particular, since $\gcd(m, N) = \gcd(N, D) = 1$ and $N > 1$, so $Np \nmid m|D|$, we must have $\mathrm{N}(\mathfrak{c}) \neq 0$, so $\mathfrak{c} \neq 0$ (so $\alpha \neq 0$).

To summarize, the conditions on α and β say that $\mathfrak{c}$ and $\mathfrak{c}'$ are nonzero integral ideals with $\mathrm{N}(\mathfrak{c}) + Np\mathrm{N}(\mathfrak{c}') = m|D|$. Since $\mathfrak{d} = (\sqrt{D})$ is a principal ideal, we see that $\mathfrak{c}$ is in the ideal class $\mathscr{A}^{-1}$ while $\mathfrak{c}'$ is in the ideal class $\mathscr{A}\mathscr{B}^{-2}[\mathfrak{q}\mathfrak{n}^{-1}]$, where $\mathscr{B}$ is the ideal class of $\mathfrak{b}$ (the same $\mathfrak{b}$ which is implicit in our model $R = R^{(\mathfrak{b})}$). Thus, for each $\alpha + \beta j \in R\mathfrak{a}$ with $\alpha, \beta \neq 0$ we have constructed a pair of nonzero integral ideals $\mathfrak{c}$ and $\mathfrak{c}'$ of $\mathscr{O}_K$ which lie in specified ideal classes and satisfy a single norm relation. The key is to prove that this construction exhausts all such pairs of integral ideals, and to determine how badly this construction fails to be injective. Recall that on the left side of (A–15), we are varying $\mathscr{B}$ over all ideal classes of K, each of which we include once. However, there is a complication caused by the fact that the class of $\mathfrak{c}'$ only depends on $\mathscr{B}^2$, not $\mathscr{B}$, so a pair of integral ideals $(\mathfrak{c}, \mathfrak{c}')$ may arise from elements $\alpha + \beta j$ in several Eichler orders in B which contain $\mathscr{O}_K$ and have reduced discriminant Np, but are not conjugate as $\mathscr{O}_K$-algebras.

To work out how much repetition we encounter, suppose that we start with a pair of nonzero integral ideals $\mathfrak{c} \in \mathscr{A}^{-1}$ and $\mathfrak{c}' \in \mathscr{A}\mathscr{B}^{-2}[\mathfrak{q}\mathfrak{n}^{-1}]$ with $\mathrm{N}(\mathfrak{c}) + Np\mathrm{N}(\mathfrak{c}') = m|D|$. Define $n = p\mathrm{N}(\mathfrak{c}')$, so $\mathrm{N}(\mathfrak{c}) = m|D| - nN$. We can reverse the original definitions (A–17) to define nonzero principal ideals

$$\mathfrak{c}\mathfrak{a}\mathfrak{d}^{-1}, \quad \mathfrak{c}'\mathfrak{d}^{-1}\mathfrak{q}^{-1}\mathfrak{n}\mathfrak{b}\bar{\mathfrak{b}}^{-1}\bar{\mathfrak{a}}. \tag{A–18}$$

Since $u = |\mathscr{O}_K^\times|/2$, there are $4u^2 = (2u)^2$ choices of respective generators α and β of these principal ideals. Because of the integrality $\mathfrak{c}$ and $\mathfrak{c}'$, all local conditions in (A–16) necessary to make $\alpha + \beta j \in R\mathfrak{a}$ are automatically satisfied *except* for possibly the congruence conditions $\alpha - X^{\mathfrak{a}\mathfrak{b}}\beta \in \mathscr{O}_{K,\ell}$ at the primes $\ell | D$.

For $\ell | D$, we have

$$\mathrm{N}(\alpha + \beta j) = \mathrm{N}(\alpha) + pq\mathrm{N}(\beta) = \frac{\mathrm{N}(\mathfrak{c}) + Np\cdot \mathrm{N}(\mathfrak{c}')}{|D|}\cdot \mathrm{N}(\mathfrak{a}) = m\mathrm{N}(\mathfrak{a}),$$

so $\mathrm{N}(\alpha) \equiv -pq\mathrm{N}(\beta) \pmod{\mathscr{O}_{K,\ell}}$. In particular, since $-pq$ is a local unit, α and β are either both ℓ-integral or else neither is ℓ-integral and they have the same pole order (so α/β is an ℓ-unit). We claim that there is a squarefree integral ideal $I|\mathfrak{d}$ (possibly more than one) such that $\alpha \equiv X^{\mathfrak{ab}I}\beta \pmod{\mathscr{O}_{\mathfrak{d}}}$. This follows from:

LEMMA A.18. *For each prime $\ell|D$, we have $\alpha - X'_\ell\beta \in \mathscr{O}_{K,\ell}$ for at least one of the two units $X'_\ell \in \mathscr{O}_{K,\ell}/\mathfrak{d}_\ell$ with norm $-pq \in \mathbf{Z}_\ell/D_\ell$. Both such units work if and only if $\ell|\mathrm{N}(\mathfrak{c}')$.*

PROOF. We treat separately the case of odd ℓ and $\ell = 2$. First assume $\ell \neq 2$, and we want to prove that at least one choice works. This is clear when α and β are ℓ-integral (both choices work). When neither is ℓ-integral, $\alpha/\beta \in \mathscr{O}_{K,\ell}^\times$ has norm congruent to $-pq \bmod D_\ell$, so there is a unique $X' \in \mathscr{O}_K/\mathfrak{d}_\ell$ with norm $-pq \bmod D_\ell$ such that $\alpha/\beta \equiv X' \pmod{\mathscr{O}_{K,\ell}}$. But look at the expression $(\beta) = \mathfrak{c}'\mathfrak{d}^{-1}\mathfrak{q}^{-1}\mathfrak{n}\mathfrak{b}\bar{\mathfrak{b}}^{-1}\bar{\mathfrak{a}}$. Since $\mathfrak{d}$ is squarefree at ℓ, $\gcd(D, \mathfrak{a}qN) = 1$, and any ramified prime factor in $\mathfrak{b}$ is canceled against its inverse appearing in $\bar{\mathfrak{b}}^{-1}$, we see that β is ℓ-integral if and only if $\mathfrak{c}'$ is divisible by the prime over ℓ, and that otherwise β (hence α) has a simple pole at that prime.

When α and β have simple poles over ℓ, the congruence $\alpha \equiv X'_\ell\beta \pmod{\mathscr{O}_{K,\ell}}$ is equivalent to $\alpha/\beta \equiv X'_\ell \bmod \mathfrak{d}_\ell$. Thus, for odd $\ell|D$ we see that $\alpha \equiv X'_\ell\beta \pmod{\mathscr{O}_{K,\ell}}$ for at least one of the two possible values of X'_ℓ—and for both if and only if α and β are ℓ-integral, which in turn is equivalent to $\ell|\mathrm{N}(\mathfrak{c}')$.

Now we turn to the more subtle case $\ell = 2|D$. For this case it will be convenient to recall from Remark A.7 that $-pq \equiv 1 \bmod 8$ when D is even in the inert case. Once again, either α and β are both ℓ-integral or neither is, and when both are integral then clearly $2|\mathrm{N}(\mathfrak{c}')$. To keep track of the pole order and the condition of whether $2 \nmid \mathrm{N}(\mathfrak{c}')$ in nonintegral cases, we separately treat the cases $D_2 = 4$ and $D_2 = 8$. First assume $D_2 = 4$, so $D = -4d$ with positive squarefree $d \equiv 1 \bmod 4$. Representatives for $(\mathfrak{d}_2^{-1}/\mathscr{O}_{K,2}) - \{0\}$ are $1/2$, $i/4$, and $1/2 + i/4$ (recall $i = \sqrt{D}$). The norms of these classes are well-defined in $D_2^{-1}\mathbf{Z}_2/\mathbf{Z}_2 = 4^{-1}\mathbf{Z}/\mathbf{Z}$, and are represented by $1/4$, $-D/16 \equiv 1/4$, and $(1+d)/4 \equiv 1/2$. Since $\mathrm{N}(X'_2) \equiv -pq \bmod D_2$ and $-pq \equiv 1 \bmod 8$, the two solutions $X'_2 \in \mathscr{O}_{K,2}/\mathfrak{d}_2$ are 1 and $i/2$ (since $d \equiv 1 \bmod 4$). Since β has at worst a double pole in K_2, so $\mathrm{N}(\beta)$ has at worst a double pole in $\mathbf{Q}_2$, the congruence $\mathrm{N}(\alpha) \equiv -pq\mathrm{N}(\beta) \pmod{\mathbf{Z}_2}$ is equivalent to $\mathrm{N}(\alpha) \equiv \mathrm{N}(\beta) \pmod{\mathbf{Z}_2}$.

Thus, $\alpha, \beta \in \mathfrak{d}_2^{-1}/\mathscr{O}_{K,2}$ are either both in the class of $1/2 + i/4$ or are each in one of the two classes $\{1/2, i/4\}$. Running through both options and recall that $X'_2 \in \{1, i/2\}$, we see that $\alpha \equiv X'_2\beta \pmod{\mathscr{O}_{K,2}}$ can be solved in the nonintegral case, with both X'_2 options working if and only if we are in the case where α and β are in the class of $1/2 + i/4$. This is also the only case in which $(\beta)\mathfrak{d}_2$ lies in the maximal ideal of $\mathscr{O}_{K,2}$ (i.e., β has a simple pole, rather than a double pole), or equivalently $2|\mathrm{N}(\mathfrak{c}')$.

The case $D_2 = 8$ (i.e., $D = -8d$ with odd squarefree $d > 0$) goes similarly, but there are more cases to consider. Now $\mathfrak{d}_2^{-1}/\mathscr{O}_{K,2} \simeq \mathbf{Z}/4 \times \mathbf{Z}/2$ as an abelian group, with generators $i/8$ (of order 4) and $1/2$. Norms on here are well-defined in $D_2^{-1}\mathbf{Z}_2/\mathbf{Z}_2 = 8^{-1}\mathbf{Z}/\mathbf{Z}$. The two solutions in $\mathscr{O}_{K,2}/\mathfrak{d}_2$ to $\mathrm{N}(X'_2) \equiv -pq \bmod D_2 =$

1 mod 8 are $X_2' = \pm 1$, so for nonzero classes $\alpha, \beta \in \mathfrak{d}_2^{-1}/\mathscr{O}_{K,2}$ we want $\mathrm{N}(\alpha) \equiv \mathrm{N}(\beta) \pmod{\mathbf{Z}_2}$ if and only if $\alpha = \pm\beta \pmod{\mathscr{O}_{K,2}}$, and that both such signs work precisely when β is 2-torsion (i.e., does not have a triple pole, which is to say exactly that $(\beta)\mathfrak{d}_2$ lies in the maximal ideal, or in other words $2|\mathrm{N}(\mathfrak{c}'))$. This is straightforward by inspection of the norm of each nonzero class in $\mathfrak{d}_2^{-1}/\mathscr{O}_{K,2}$. □

We conclude that when given $\mathfrak{c}$ and $\mathfrak{c}'$ of the desired type and constructing pairs α and β from these, necessarily $\alpha \equiv X^{\mathfrak{a}\mathfrak{b}I}\beta \pmod{\mathscr{O}_{\mathfrak{d}}}$ for *some* integral squarefree ideal $I|\mathfrak{d}$ (which is visibly 2-torsion in the class group). This corresponds to using $\mathfrak{b}I$ instead of $\mathfrak{b}$ (representing a multiple of $\mathscr{B}$ by a 2-torsion ideal class), so any such choice of α and β satisfies conditions so that $\alpha + \beta j \in R'\mathfrak{a}$ with $R' = IRI^{-1} = R^{(\mathfrak{b}I)}$ for some squarefree integral ideal $I|\mathfrak{d}$. Note that $I\overline{I}^{-1} = (1)$, so replacing $\mathfrak{b}$ with $\mathfrak{b}I$ in the definition of $\mathfrak{c}'$ in terms of (β) (or (β) in terms of $\mathfrak{c}'$) has no impact.

Now we are ready to carry out the calculation of the left side of (A–15). Although the outer sum runs over $\mathscr{O}_K$-algebra isomorphism class of Eichler orders R of reduced discriminant Np (containing $\mathscr{O}_K$) inside of B, rather than considering contributions from suitable elements $b = \alpha + \beta j$ which lie in $R^{(\mathfrak{b})}\mathfrak{a}$ for each of a list of class group representatives $\mathfrak{b}$, it is more convenient to consider a coset ρ in $\mathrm{Cl}_K/\mathrm{Cl}_K[2] \simeq \mathrm{Cl}_K^2$, and determine *only* the subsummation within (A–15) coming from those ideal classes $\mathscr{B}$ in the chosen coset. These classes are precisely those whose square $\mathscr{B}^2$ is some specified square class $\mathscr{C} \in \mathrm{Cl}_K^2$, and if we choose a "base point" representative $\mathfrak{b}$ for one such $\mathscr{B}$, then $\{\mathfrak{b}I\}$ provides a 2-to-1 set of representatives of elements in the coset ρ as I runs over squarefree divisors of the principal ideal $\mathfrak{d}$. Recall that $\mathrm{Cl}_K[2]$ is presented as the free $\mathbf{F}_2$-vector space on the prime factors of $\mathfrak{d}$, subject to one relation: when $D_2 \neq 4$ we require that the sum of these elements is zero, and when $D_2 = 4$ (so $D = -4d$ with positive $d \equiv 1 \bmod 4$) we require the vanishing of the sum over the factors in odd residue characteristic when $d > 1$ (and the class group is trivial for $d = 1$). Though we are interested in the subsummation of (A–15) running over the ideal classes lying in the 2-torsion coset ρ (or having square $\mathscr{C}$), we shall consider instead the analogous sum $\Sigma'_{\mathscr{C}}$ carried out over all concrete orders $R^{(\mathfrak{b}I)}$ as I runs through squarefree factors of $\mathfrak{d}$. This collection of orders indexed by I's collapses 2-to-1 when considering $\mathscr{O}_K$-algebra isomorphism classes, so at the end we will need to divide by 2 (it would be inconvenient to explicitly avoid the double overcount caused by isomorphism class repetition).

Fix data consisting of nonzero integral ideals $\mathfrak{c}$ and $\mathfrak{c}'$ with $\mathfrak{c} \in \mathscr{A}^{-1}$ and $\mathfrak{c}' \in \mathscr{A}\mathscr{C}^{-1}[\mathfrak{q}\mathfrak{n}^{-1}]$ with $\mathrm{N}(\mathfrak{c}) + Np\mathrm{N}(\mathfrak{c}') = m|D|$. The ideals in (A–18) are unaffected by replacing $\mathfrak{b}$ with $\mathfrak{b}I$ for squarefree divisors I of $\mathfrak{d}$. Thus, for $\Sigma'_{\mathscr{C}}$ viewed as a sum of sums labeled by orders $R^{(\mathfrak{b}I)}$, it makes sense to focus on the contribution of elements $b = \alpha + \beta j \in R^{(\mathfrak{b}I)}\mathfrak{a}/\pm 1$ for which α and β are respective generators of the ideals in (A–18). Our collection of 2^t Eichler orders $R^{(\mathfrak{b}I)}$, which is 2-to-1 on the level of $\mathscr{O}_K$-algebra isomorphism classes, is naturally indexed by the 2^t elements $X \in \mathscr{O}_K/\mathfrak{d}$ satisfying $\mathrm{N}(X) \equiv -pq \bmod D$:

$$\{\alpha + \beta j \mid \alpha \in \mathfrak{d}^{-1},\ \beta \in \mathfrak{d}^{-1}\mathfrak{q}^{-1}\mathfrak{n}\mathfrak{b}\bar{\mathfrak{b}}^{-1},\ \alpha - X\beta \in \mathscr{O}_{\mathfrak{d}}\}.$$

Define $n = p\mathrm{N}(\mathfrak{c}')$. Each choice of generators α and β for the principal ideals in (A–18) contributes to $\Sigma'_{\mathscr{C}}$, but may contribute multiple times, via membership of $b = \alpha + \beta j$ in perhaps more than one of the 2^t orders with which we are working. The issue comes down to the fact that for each $\ell|D$, the local congruence $\alpha \overset{?}{\equiv} X'_\ell\beta \bmod \mathscr{O}_{K,\ell}$ might be satisfied for both solutions $X'_\ell \in \mathscr{O}_K/\mathfrak{d}_\ell$ to $\mathrm{N}(X'_\ell) \equiv -pq \bmod D_\ell$. By Lemma A.18, we see that the $\ell|D$ for which both congruences hold are exactly those for which $\ell|\mathrm{N}(\mathfrak{c}') = n/p$, or equivalently $\ell|\gcd(n, D)$. If $e(n)$ denotes the number of prime factors of $\gcd(n, D)$, then we see that such terms $\alpha + \beta j$ contribute $1 + \mathrm{ord}_p(\mathrm{N}(b_-)) = \mathrm{ord}_p(p\mathrm{N}(\beta j))$ exactly $2^{e(n)} = \delta(n)$ times.

Thus, for $\mathfrak{c} \in \mathscr{A}^{-1}$ and $\mathfrak{c}' \in \mathscr{A}\mathscr{C}^{-1}[\mathfrak{q}\mathfrak{n}^{-1}]$ we have $4u^2$ choices of pairs (α, β) generating the principal ideals in (A–18), and each such pair gives rise to an element $b = \alpha + \beta j$ which lies in $R'\mathfrak{a}$ for $\delta(n)$ of the orders R' which arise in $\Sigma'_{\mathscr{C}}$. Since we are really summing over the quotient $R'\mathfrak{a}/\pm 1$, where $\pm b$ are the same, we have to divide by 2. Thus, the amount contributed to $\Sigma'_{\mathscr{C}}$ by the $\delta(n)$ appearances of b is $(1 + \mathrm{ord}_p(\mathrm{N}(\beta j)))/2$. We have $\mathrm{N}(\beta j) = \mathrm{N}(\beta) \cdot (-pq)$, and the p-part of $\mathrm{N}(\beta)$ agrees with that of $n/p = \mathrm{N}(\mathfrak{c}')$, due to (A–17) and the fact that p does not divide $\mathrm{N}(\mathfrak{a}) \cdot qND$. Thus, for each pair $(\mathfrak{c}, \mathfrak{c}')$ we get a total contribution to $\Sigma'_{\mathscr{C}}$ given by $(4u^2)\delta(n) \cdot \mathrm{ord}_p(pn)/2$. Remembering now to divide by an additional factor of 2 from the initial double overcount on isomorphism classes of models (upon fixing $\mathscr{C}$), the pair $(\mathfrak{c}, \mathfrak{c}')$ contributes $u^2\delta(n)\mathrm{ord}_p(pn)$, where $n = p\mathrm{N}(\mathfrak{c}')$ is an integer divisible by p, and $Nn < m|D|$. This only depends on $(\mathfrak{c}, \mathfrak{c}')$ through the value of n. Since $\mathfrak{c}$ has norm $m|D| - nN$, the total number of such pairs $(\mathfrak{c}, \mathfrak{c}')$ is $r_{\mathscr{A}}(m|D| - nN)r_{\mathscr{A}\mathscr{C}^{-1}[\mathfrak{q}\mathfrak{n}^{-1}]}(n/p)$. For the first factor we have used that $r_{\mathscr{A}^{-1}} = r_{\mathscr{A}}$ since conjugation of an ideal inverts the ideal class but fixes the norm.

As we vary the square class $\mathscr{C}$, $\mathscr{A}\mathscr{C}^{-1}[\mathfrak{q}\mathfrak{n}^{-1}]$ varies over all products of $\mathscr{A}[\mathfrak{q}\mathfrak{n}]$ against a square ideal class. The number of integral ideals of norm n/p which lie in such classes is precisely what is counted by $R_{\mathscr{A}[\mathfrak{q}\mathfrak{n}]}(n/p)$. This completes the proof. □

REMARK A.19. Another way to interpret the double overcounting is that we should identify contributions from $\alpha+\beta j$ and $\alpha-\beta j$, since $\alpha-\beta j = i(\alpha+\beta j)i^{-1}$ and it is precisely conjugation by the generator $i = \sqrt{D}$ of $\mathfrak{d}$ which serves to identify repetitions within an $\mathscr{O}_K$-algebra isomorphism class in the collection of orders used in the preceding proof (relative to a chosen coset $\rho \in \mathrm{Cl}_K/\mathrm{Cl}_K[2]$).

In the ramified case, the final formula in [GZ, Ch. III, § 9] is valid without parity restriction on D, but some care is required to handle even D. The essential deviations from the proof of Theorem A.17 are detailed below.

THEOREM A.20. *Suppose $p = \mathfrak{p}^2$ is ramified in K. Pick a prime $\mathfrak{q}$ over q as above, and define $\delta(n) = 2^{e(n)}$ where $e(n)$ is the number of prime factors of* $\gcd(n, D)$. *Then*

$$\langle c, T_m d^\sigma\rangle_p = -r_{\mathscr{A}}(m)h_K u \operatorname{ord}_p(m)\log p \\ - \log p \cdot u^2 \cdot \sum_{\substack{0<n<m|D|/N \\ p|n}} \operatorname{ord}_p(n) r_{\mathscr{A}}(m|D| - nN)\delta(n) R_{\mathscr{A}[\mathfrak{q}\mathfrak{p}\mathfrak{n}]}(n/p).$$

PROOF. The first essential difference from the inert case is that the places $v|\mathfrak{p}$ are generally no longer in a bijective correspondence with the elements of Cl_K. Indeed, if $[\mathfrak{p}]$ has order 2 in Cl_K then $f_v = 2$ for all $v|\mathfrak{p}$. In this case, there are only $h/2$ isomorphism classes of Eichler orders in B with reduced discriminant Np containing $\mathscr{O}_K$: if we fix $S = S_X$ as in (A–14), then $\mathfrak{p}S\mathfrak{p}^{-1} = S$. This is a minor change, as summing over the places v will instead correspond to summing over the orders $R^{(\mathfrak{b})}$ as defined previously, where $\mathfrak{b}$ ranges over a representative set of ideals for $\mathrm{Cl}_K/[\mathfrak{p}]$.

As in the inert case, we will perform the sum over ideal classes by summing on the outside over cosets in Cl_K^2 and on the inside over those ideal classes in the coset *modulo* $[\mathfrak{p}]$—these can be represented by the squarefree divisors I of $\mathfrak{d}$ that are relatively prime to $\mathfrak{p}$, but not necessarily in a 2-to-1 fashion as before. Losing the order 2 action at $\mathfrak{p}$ has cut the cardinality of this representative set in half, but if $[\mathfrak{p}]$ is trivial, the cardinality of the $\mathrm{Cl}_K[2]$-coset is the same as in the inert case. Thus, we can conclude that the representation is 2-to-1 exactly when $f_v = 2$ for all $v|\mathfrak{p}$, and 1-to-1 otherwise.

Respecting these changes, the method of summation is the same, but there is a change (relative to what we saw in the inert case) when counting the number of representative models for a given coset of $\mathrm{Cl}_K[2]/[\mathfrak{p}]$ that will contain a choice of $\alpha + \beta j$ corresponding to ideals $(\mathfrak{c}, \mathfrak{c}')$. Previously this count was $\delta(n)$, but now we always have $p|n$, and the resulting local factor of 2 that p contributes to $\delta(n)$ is superfluous, as there is not a corresponding order-2 action at $\mathfrak{p}$ on the representatives.

Along with the observation that $\operatorname{ord}_p(DN(b_-)) = \operatorname{ord}_p(n)$, the formula in the theorem is assembled in the same manner as in the inert case. □

References

[BLR] S. Bosch, W. Lütkebohmert, and M. Raynaud, *Néron models*, Berlin, Springer, 1990.

[EGA] J. Dieudonné and A. Grothendieck, *Éléments de géométrie algébrique*, Publ. Math. IHES **4**, **8**, **11**, **17**, **20**, **24**, **28**, **32** (1960–67).

[Gi] J. Giraud, "Remarque sur une formule de Shimura-Taniyama", *Inventiones Math.* **5** (1968), 231–236.

[Gr1] B. Gross, "On canonical and quasi-canonical liftings", *Inventiones Math.* **84** (1986), 321–326.

[Gr2] B. Gross, "Local heights on curves", pp. 327–339 in *Arithmetic geometry* (Storrs, CT, 1984), edited by Gary Cornell and Joseph H. Silverman, New York, Springer, 1986.

[GZ] B. Gross, D. Zagier, "Heegner points and derivatives of L-series", *Inventiones Math.* **84** (1986), 225–320.

[K] N. Katz, "Serre–Tate local moduli", pp. 138–202 in *Surfaces algébriques* (Orsay 1976–78), edited par J. Giraud et al., Lecture Notes in Math. **868**, New York, Springer, 1981.

[KM] N. Katz, B. Mazur, *Arithmetic moduli of elliptic curves*, Princeton, University Press, 1985.

[L] J. Lipman, "Desingularization of two-dimensional schemes", *Annals of Math.* **107** (1978), 151–207.

[Mat] H. Matsumura, *Commutative ring theory*, Cambridge, Univ. Press, 1986.

[GIT] D. Mumford, *Geometric invariant theory*, Ergebnisse der Mathematik **34**, New York, Springer, 1965.

[Mum] D. Mumford, *Abelian varieties*, Tata Studies in Mathematics **5**, Tata Institute of Fundamental Research, Bombay, 1970.

[ST] J.-P. Serre and J. Tate, "Good reduction of abelian varieties", *Annals of Math.* **88** (1968), 492–517.

[Tate] J. Tate, "Endomorphisms of abelian varieties over finite fields", *Inventiones Math.* **2** (1966), 134–144.

Brian Conrad
Department of Mathematics
University of Michigan
Ann Arbor, MI 48109
United States
bdconrad@umich.edu

W. R. Mann
Department of Mathematics
Brown University
Providence, RI 02912
United States
wrmann@math.brown.edu

Heegner Points and Rankin L-Series
MSRI Publications
Volume **49**, 2004

Special Value Formulae for Rankin L-Functions

VINAYAK VATSAL

1. Introduction

Let F denote a totally real number field, and let K/F denote a totally imaginary quadratic extension. We fix an automorphic cuspidal representation π of $\mathrm{GL}_2(F)$, and a finite order Hecke character χ of K. Thus χ is a representation of $\mathrm{GL}_1(K)$.

Under certain hypotheses, it is known that the central critical value $L(\pi \otimes \chi, \frac{1}{2})$ is algebraic up to a known transcendental factor. Explicit formulae for this value have been given by a number of authors, notably Gross, Waldspurger, and Zhang. Essentially, the work of Gross and Zhang shows that this value is given by the height of a certain CM divisor on a suitable space, while the work of Waldspurger gives a criterion for nonvanishing of this value in terms of a certain linear functional arising from representation theory, and a formula in terms of torus integrals on a quaternion algebra. Our goal in this article is to explain the connections between these works, and to provide a bridge between the general representation-theoretic framework described by Gross (see his article [Gro] in this volume) and the theorems of Zhang [Zha01a] and Waldspurger [Wal85].

We want to point out that the formula we will discuss has numerous applications to arithmetic and Iwasawa theory (see [BD96] and its various sequels). We will therefore attempt to formulate the representation-theoretic results in terms that are familiar to number theorists. We will *not* however discuss any arithmetic applications directly — the reader will find some of these applications elsewhere in this volume.

Needless to say, the present work is mostly expository. The ideas are largely drawn from [Gro87], [Gro], [Wal85], [Zha01a]. However, the organization here is perhaps novel. Our main contribution is given in Theorem 6.4. While the ingredients in this theorem are all well-known, our formulation seems to be new, and is well-suited for applications to number theory as in [BD96] and [Vat02].

I thank Benedict Gross, Shou-Wu Zhang, and Hui Xue for patiently answering my numerous questions on this subject. The statements in this paper reflect my very incomplete understanding of their work, and the reader interested in the details should consult the original sources. Finally, I would also like to thank Barry Mazur, Christophe Cornut, and the mathematics department at Harvard University for their hospitality in April 2002, during which time most of this article was written.

2. Notation and Hypotheses

We start by briefly recalling the basic notions about modular forms in the adelic setting. For generalities on automorphic forms and representations, se [Bum97, Chapter 3] or [BJ79]. A clear discussion may also be found at the end of [Cas73].

Recall that an automorphic form is a function of moderate growth on

$$\mathrm{GL}_2(F)\backslash \mathrm{GL}_2(\mathbf{A}),$$

fixed under right translations by some open compact subgroup of $\mathrm{GL}_2(\mathbf{A}_f)$, smooth at infinity, and contained in some finite dimensional space which is invariant under both a maximal compact subgroup and the center of the universal enveloping algebra, of the real group $\mathrm{GL}_2(F_{\mathbf{R}})$. There is a map from classical Hilbert modular forms to the space of automorphic forms, as in [BJ79, Section 4.3]. The group $\mathrm{GL}_2(\mathbf{A})$ acts on the space of automorphic forms, with the conventions of [Bum97], section 3.3. In particular, the action of $\mathrm{GL}_2(\mathbf{A}_f)$ is by right translations. The structure at infinity is more complicated, since the space of automorphic forms is not preserved by right translations by $\mathrm{GL}_2(F_{\mathbf{R}})$. For a detailed discussion of the structure at infinity, see [Bum97].

Now let π denote a cuspidal automorphic representation of $\mathrm{GL}_2(\mathbf{A})$. Then π may be viewed in (at least) two different ways. On the one hand, π is by definition an irreducible subquotient of $\mathcal{A}$, where $\mathcal{A}$ is the space of automorphic cuspforms on $\mathrm{GL}_2(\mathbf{A})$. On the other hand, it is a theorem of Flath that any such subquotient is abstractly isomorphic to a restricted tensor product $\pi \cong \otimes\pi_v$, where each π_v is a representation of $\mathrm{GL}_2(F_v)$, at least at the finite places. The components π_v at real places v are not quite representations of $\mathrm{GL}_2(\mathbf{R})$; rather, they are just the $(\mathfrak{g}, K)$-modules of [Bum97], Chapter 2. We will need both descriptions of automorphic representations in the sequel. It is a fact that any cuspidal automorphic representation π occurs with multiplicity one in the space of automorphic forms. To each automorphic representation π is attached a certain integral ideal N of F, called the *conductor* of π. Then N depends only on the local components π_v. We will give the exact definition of N in the next section, following [Cas73].

Next we need to recall some simple facts about Whittaker coefficients of automorphic forms, which are the adelic analogue of Fourier coefficients in the classical theory. We will also need to understand how Whittaker coefficients transform with respect to the Hecke operators. Thus fix a nontrivial additive character η of $\hat{F}/F$. For any automorphic cuspform ϕ and an idele a of F, define a function W_ϕ on $\mathrm{GL}_2(\hat{F})$ by

$$W_\phi(g) = \int_{\hat{F}/F} \phi\left(\begin{pmatrix} 1 & x \\ 0 & 1 \end{pmatrix} g\right) \eta(-x)\, dx.$$

Then ϕ has a 'Fourier expansion'

$$\phi(g) = \sum_{\alpha \in F^\times} W_\phi\left(\begin{pmatrix} \alpha & 0 \\ 0 & 1 \end{pmatrix} g\right).$$

For a proof of the above fact, we refer the reader to [Bum97], Theorem 3.5.5. It follows from the Fourier expansion that ϕ is determined by the Whittaker function W_ϕ. Furthermore, the strong approximation theorem implies that W_ϕ is in turn determined by the numbers

$$W_\phi\left(\begin{pmatrix} a & 0 \\ 0 & 1 \end{pmatrix} g_\infty\right),$$

for $a \in \hat{F}_f^\times$ and $g_\infty \in \mathrm{GL}_2(F_{\mathbf{R}})$. We call these numbers the *Whittaker coefficients* of ϕ.

The Whittaker coefficients satisfy a simple transformation rule under the action of the Hecke operators T_v. To state this, let v denote any finite place of F, relatively prime to the level of ϕ. Then one can define a Hecke operator T_v acting on ϕ. (For the precise definition, we refer to [Bum97], Chapter 4, or equation (7–10) below.) Then $T_v\phi$ is also a cuspform, and we have

$$W_\phi\left(\begin{pmatrix} a\varpi_v & 0 \\ 0 & 1 \end{pmatrix} g_\infty\right) = |\varpi_v| W_{T_v\phi}\left(\begin{pmatrix} a & 0 \\ 0 & 1 \end{pmatrix} g_\infty\right). \tag{2–1}$$

Here ϖ denotes a local uniformizer at v. We will use this formula later.

Finally, we want to point out that the classical Petersson inner product on Hilbert modular forms has an analog from the adelic point of view. Indeed, it can be shown that if ϕ is any automorphic cuspform on $\mathrm{GL}_2(\mathbf{A})$, such that the center $Z(\mathbf{A})$ acts on ϕ via a unitary character, then ϕ is square integrable modulo $Z(\mathbf{A})\,\mathrm{GL}_2(F)$. For this we refer the reader to [BJ79], section 4.4. Thus we may define an inner product pairing on the space of cuspforms by

$$(\phi_1, \phi_2) = \int_{Z(\mathbf{A})\,\mathrm{GL}_2(F)\backslash \mathrm{GL}_2(\mathbf{A})} \phi_1(g)\overline{\phi}_2(g)\, dg,$$

where dg denotes any Haar measure. In practice, one must normalize the measure depending on the application in view. Typically, one requires that some fixed open compact subgroup U gets measure 1. We will attempt to be careful about this in the exact formulae later in this article.

Now we specialize everything to the representations π and χ of interest. Thus recall that π is a representation of $\mathrm{GL}_2(F)$, and the finite order Hecke character χ is a representation of $\mathrm{GL}_1(K)$. We will impose the following basic assumptions and notations which will be in force throughout this paper.

1. If v is any infinite place of F, then the local component π_v of π at v is a weight two discrete series representation.
2. The central character of π is trivial. (This means that the center $Z(\mathbf{A})$ acts trivially.)
3. The conjugate of χ under the action of $\mathrm{Gal}(K/F)$ is equal to χ^{-1}. In particular, χ is trivial on $\mathrm{GL}_1(F)$.
4. If N denotes the conductor of π, and d denotes the relative discriminant of K/F, then $(N, d) = 1$. Here N and d denote integral ideals of the ring of integers O_F of F.
5. If c denotes the conductor of χ, then $(c, Nd) = 1$. Note that it follows from assumption 3 above that the conductor of χ is invariant under $\mathrm{Gal}(K/F)$ and so may be identified with an ideal of F.

More concretely, we assume that π corresponds to a holomorphic Hilbert modular form of weight $(2, \ldots, 2)$, with trivial central character, and that the character χ is anticyclotomic. Some of the hypotheses above may be weakened, but for the sake of clarity, it is convenient to impose the extra conditions. We set $D = dc^2$.

Now consider the representation $\pi \otimes \chi$ of $\mathrm{GL}_2(F) \otimes \mathrm{GL}_1(K)$. Let $L(\pi \otimes \chi, s)$ denote the associated L-function. Then $L(\pi \otimes \chi, s)$ has a functional equation of the form $L(\pi \otimes \chi, s) = \varepsilon(\pi \otimes \chi, s)L(\pi \otimes \chi, 1 - s)$. For a detailed discussion of the representation $\pi \otimes \chi$, we refer to the article of Gross in this volume. The L-function and its functional equation are discussed in Chapter 2 of [Zha01a]. Under the hypotheses on π and χ stated above, it can be shown that

$$\varepsilon(\pi \otimes \chi, \tfrac{1}{2}) = (-1)^{\#\Sigma} \tag{2–2}$$

where Σ denotes the set of infinite places of F, together with the set of finite places v such that $\omega_v(N) = -1$. Here ω denotes the quadratic character of the ideles $\hat{F}^\times$ of F, defined by the extension K/F, and ω_v denotes the local component. In other words, Σ consists of the infinite places together with finite places v such that v is inert in K, and such that $\mathrm{ord}_v(N)$ is odd. We shall say that we are in the definite case if Σ is even, and in the indefinite case if Σ is odd. Observe that, under the present hypotheses, the cardinality of Σ depends only on π and K, and not on the character χ. A general formula for $\varepsilon(\pi \otimes \chi, s)$ may be found in Chapter 3 of [Zha01a], especially equation (3.1.2).

Evidently, we have $L(\pi \otimes \chi, \frac{1}{2}) = 0$ if we are in the indefinite case. This is the case originally treated by Gross and Zagier, and subsequently generalized by Zhang [Zha01a], [Zha01b]. We shall not consider this case here. Rather, we will concentrate on explaining the formulae for the value $L(\pi \otimes \chi, \frac{1}{2})$ given in the definite case by Waldspurger [Wal85], Gross [Gro87], and Zhang [Zha01a]. Thus, from now on, we assume that we are in the definite case, so that the set Σ has even cardinality.

3. Atkin–Lehner Theory on GL_2

Now we want to discuss newforms in the adelic setting. First recall classical Atkin–Lehner theory for modular forms on congruence subgroups of $SL_2(\mathbf{Z})$. The basic result states that each cuspform g of level M which is an eigenform for almost all the Hecke operators is given by $g(z) = \sum_a c_a g_0(az)$, where g_0 is a unique form of some level $N|M$, which is an eigenform for *all* the Hecke operators at level N, and a runs over divisors of M/N. The form g_0 is called the Atkin–Lehner newform; it depends only on the package of eigenvalues attached to g.

We want an analog of this theorem in the adelic set-up. Casselman's beautiful idea is to construct a newform ϕ_v locally in each representation π_v. Then the global newform is just the tensor product $\otimes\phi_v$, as v runs over all places. Note however that this produces the newform as an abstract vector in the restricted tensor product $\pi = \otimes\pi_v$; to obtain a genuine automorphic form, one must embed π in the space $\mathcal{A}$ of automorphic forms as above.

To describe Casselman's construction, let v denote a finite place of F, and let ϖ denote a local uniformizer at v. For a non-negative integer c, we define a group $U_0(\varpi^c)$ by putting

$$U_0(\varpi^c) = \left\{\gamma \in \mathrm{GL}_2(\mathcal{O}_{F,v}) : \gamma \equiv \begin{pmatrix} * & * \\ 0 & * \end{pmatrix} \pmod{\varpi^c}.\right\}$$

Let π_v denote the local component of π at v. Then Casselman proves the following theorem.

THEOREM 3.1. *Let $c = c_v$ denote the smallest non-negative integer such that $U_0(\varpi^c)$ has a nonzero fixed vector in π_v. Then the fixed space of $U_0(\varpi^c)$ in π_v has dimension* 1.

DEFINITION 3.2. The ideal $N_v \subset \mathcal{O}_{F,v}$ generated by $\varpi_v^{c_v}$ is called the conductor of π_v. The conductor of π is the ideal $N = \prod_v N_v$.

It can be shown that $c_v = 0$ for almost every v. A nonzero vector ϕ_v fixed by $U_0(\varpi^{c_v})$ is called a local newform at v. Note that ϕ_v is fixed by $\mathrm{GL}_2(\mathcal{O}_{F,v})$ for almost all v. In this case, we say that π is unramified at v.

There is a corresponding statement at the archimedean places of F. However, matters are somewhat more complicated, since, as we have already remarked, the local factor at infinity is not a local representation. We will not enter into a discussion of this point here. Suffice it to say simply that the nature of the infinite component is given by the weight, which in our case is $(2,\ldots,2)$. Again, there exists a local newform ϕ_v for each $v|\infty$.

DEFINITION 3.3. The Atkin–Lehner newspace is the line in $\pi = \otimes\pi_v$ spanned by $\otimes\phi_v$, where, for each v, ϕ_v is a local newform at v.

Thus an Atkin–Lehner newform for π is a nonzero vector in the newspace. It is fixed by the group

$$U_0(N) = \left\{ \gamma \in \mathrm{GL}_2(\hat{\mathcal{O}}_F) : \gamma \equiv \begin{pmatrix} * & * \\ 0 & * \end{pmatrix} \pmod{N} \right\}.$$

As such, it is defined only up to scalars.

To state the main result of Atkin–Lehner theory, we first need to introduce notation. Let π be given, and fix a newform ϕ for π. Let a denote any ideal of F which is relatively prime to the conductor. We may identify a with an idele of F in the usual way. We let

$$g_a = \begin{pmatrix} a^{-1} & 0 \\ 0 & 1 \end{pmatrix} \in \mathrm{GL}_2(\mathbf{A}_f),$$

and define the function ϕ_a by $\phi_a(x) = \phi(xg_a)$. This is well-defined independent of the idele representing the ideal a, since $(a, N) = 1$, and ϕ_v is fixed by $\mathrm{GL}_2(\mathcal{O}_{F,v})$.

Now let D denote any ideal prime to N, and write $V(\pi, ND)$ for the finite dimensional space of vectors in π fixed by the group $U_0(ND)$. We call $V(\pi, ND)$ the space of vectors of level ND.

THEOREM 3.4 [Cas73]. *The space $V(\pi, ND)$ has a basis consisting of the vectors ϕ_a, as a runs over divisors of D.*

Finally, we need to say a word about normalization of the Atkin–Lehner newform, since it is only defined up to scalars. In the classical theory, it is customary to normalize a newform in terms of the Fourier expansion, by requiring that a normalized newform have first coefficient equal to 1. In the adelic situation, we normalize via Whittaker functions, as in [Bum97], or [Zha01a], section 2.5. The details are somewhat technical and we will not reproduce them here. Essentially, one starts by realizing the newform as function on the adeles. Then the corresponding Whittaker function W_ϕ breaks up as a tensor product of local Whittaker functions $W_{\phi,v}$, and we normalize each $W_{\phi,v}$ so that it takes the value 1 at the identity element, at least at good primes, and if the conductor of the additive character is 1. In general, one can describe the normalization by requiring that the Mellin transform of the local Whittaker function $W_{\phi,v}$ be equal to the local L-function $L(s, \pi_v)$. For discussion of the normalization we refer to Zhang's paper [Zha01a].

4. Quaternion Algebras and the Jacquet–Langlands Correspondence

We want to transfer the representation π of GL_2 to a representation π' of a suitable quaternion algebra B. Indeed, we may take for B the unique quaternion algebra ramified precisely at the set of primes in Σ, which has even cardinality. Furthermore, since Σ contains all the infinite places, we see that the algebra B is totally definite. Since each $v \in \Sigma$ is such that π_v is special or supercuspidal if v is finite, and in the discrete series if v is infinite, we see that π_v is square-integrable for every $v \in \Sigma$. Thus the Jacquet–Langlands correspondence implies that a lift π' of π to B exists. Furthermore, there exists an embedding $K \to B$ (since every prime $v \in \Sigma$ is by definition inert in K). We fix such an embedding once and for all. We remark here that a very readable summary of the Jacquet–Langlands correspondence may be found in the book [Lub94]

Now let $R = \prod R_v$ denote an order of $\hat{B} = B(\hat{F})$ defined by requiring that $R_v \subset B_v$ be an order of reduced discriminant N_v which optimally contains $O_{K,v}$. For a proof that such orders exist, we refer to [Gro88], Proposition 3.6. Here N is the level of π as before, and N_v denotes the local component at v.

We put $U_v = R_v^\times$, so that $U = \prod U_v$ is an open compact subgroup of $\hat{B}^\times$. Then it follows from work of Gross and Prasad [GP91], Propositions 2.3 and 2.6, that the subgroup U_v fixes a unique line L_v in the representation space π_v for each v. We fix a nonzero vector $\psi_v \in L_v$ for each v, and write $\psi = \otimes \psi_v \in \pi$, where we have decomposed π as a restricted tensor product relative to the subgroups U_v. This makes sense because $U_v = R_v^\times$ is a maximal compact subgroup of B_v for almost every v. We may regard ψ as the analog of an Atkin–Lehner newform for the representation π'.

It is known that there is a realization of π' in the space $\mathcal{S}$ of cuspforms on $B(F)^\times \backslash \hat{B}^\times$, where the action of $\hat{B}^\times$ is by right translations. Indeed, π' occurs with multiplicity one in $\mathcal{S}$. We fix an embedding $\pi' \to \mathcal{S}$. Note that such an embedding is defined only up to scaling. From this viewpoint, ψ may be considered as a locally constant function on $\hat{B}^\times$, left invariant under $B(F)^\times$, and right invariant under $\hat{F}^\times \cdot U$, where the invariance under $\hat{F}^\times$ comes from our assumption that π and π' have trivial central character. Note also here that since π has weight $(2, \ldots, 2)$ and B is totally definite at infinity, the Jacquet–Langlands correspondence implies that the infinite component of π' is just the trivial representation of the compact group $B(F_{\mathbf{R}})^\times / F_{\mathbf{R}}^\times$.

5. The Work of Waldspurger

Now we fix an anticyclotomic character χ of $\mathrm{GL}_1(K)$ of conductor c. We retain the hypotheses on N and D made in the previous sections. We identify the quadratic extension K/F with a maximal torus T of $B^\times$, and consider a realization of π' in the space of functions on $\hat{B}^\times$. The character χ of $\mathrm{GL}_1(\hat{K})$ is a homomorphism $\hat{K}^\times / K^\times \to \mathbf{C}^\times$, and we write $\mathbf{C}(\chi)$ to denote the associated representation of $\mathrm{GL}_1(\hat{K})$. If v is any place of F, we will use a subscript v to denote the corresponding local object.

With these notations, the fundamental local result is the following proposition of Waldspurger [Wal85] and Tunnell [Tun83].

PROPOSITION 5.1. *Let v denote any place of F. Put*

$$V(\pi_v, T_v, \chi_v) = \mathrm{Hom}_{T_v}(\pi'_v, \mathbf{C}(\chi_v)).$$

Then $\dim_{\mathbf{C}}(V(\pi_v, T_v, \chi_v)) = 1$.

For a discussion of this result, and connection with local root numbers of π, π', and χ', we refer the reader to Gross' article in this volume.

It follows from the local result above that $V(\pi, T, \chi) = \mathrm{Hom}_T(\pi', \mathbf{C}(\chi))$ has dimension 1. Furthermore, it is easy to exhibit a candidate for an element of this one-dimensional space. Indeed, it is clear that the functional defined by

$$e \mapsto \ell(e) = \int_{T(F)^\times \hat{F}^\times \backslash \hat{T}^\times} e(t)\chi^{-1}(t)\,dt, \tag{5–1}$$

is an element (possibly zero) of $\mathrm{Hom}_T(\pi', \mathbf{C}(\chi))$. Here we may take dt to denote any Haar measure on $\hat{T}$, since any two such differ only by a constant multiple. Note also that the integral converges because the domain of integration is compact.

With this notation, Waldspurger proved the following global result [Wal85, Théorème 2, page 221].

THEOREM 5.2. *The functional* ℓ *is nonzero on* π' *if and only if* $L(\pi \otimes \chi, \frac{1}{2}) \neq 0$.

The above theorem may be viewed as giving a criterion for the nonvanishing of $L(\pi \otimes \chi, \frac{1}{2})$. Namely, to show that $L(\pi \otimes \chi, \frac{1}{2})$ is nonzero, it suffices to exhibit an element $e \in \pi'$ such that $\ell(e)$ is nonzero. Note, however, that the linear form ℓ and the 'test' vector e are defined only up to scalar, and that there is no obvious way (yet) to recover the actual value $L(\pi \otimes \chi, \frac{1}{2})$. We will return to this question later.

6. Test Vectors: The Work of Gross and Prasad

In this section, we will review the basic results of [GP91], where the problem of constructing *local* test vectors is solved. Thus let v denote any finite place of F and let π_v denote the local component of π at v. Recall that we have fixed an embedding $K \to B$ and so an embedding $K_v \to B_v$. According to our hypotheses on N and D, at least one of π and χ is unramified at v. In each case, we wish to construct an explicit vector $\psi_{\chi,v} \in \pi'_v$ such that $\ell_v(\psi_{\chi,v}) \neq 0$, where ℓ_v is any nonzero linear functional in $V(\pi_v, T_v, \chi_v) = \mathrm{Hom}_{T_v}(\pi'_v, \mathbf{C}(\chi_v))$.

First consider the case where π_v is an unramified principal series representation. Then we see that $B_v \cong \mathrm{GL}_2$. Let $c(\chi_v)$ denote the conductor of the character χ_v of K_v, and let $R_v \subset B_v$ denote a maximal order which optimally contains the order of K_v with conductor $c(\chi_v)$. Then Gross and Prasad have shown that the group $R_v^\times$ fixes a unique line in π'_v. In this case, we let $\psi_{\chi,v}$ denote any nonzero vector fixed by $R_v^\times$.

Now suppose that π_v is ramified, so that χ_v is unramified. In this case, let $R_v \subset B_v$ denote an order of reduced discriminant N_v which contains $\mathcal{O}_{K,v}$. (That such orders exist and are unique up to conjugation by K_v is proved in

[Gro88].) Again, the group $R_v^\times$ fixes a unique line in π'_v, and we let $\psi_{\chi,v}$ denote any nonzero vector on this distinguished line.

The following result restates Propositions 2.3 and 2.6 of [GP91].

PROPOSITION 6.1 (GROSS–PRASAD). *Let $\psi_{\chi,v} \in \pi'_v$ be defined as above. Then, if ℓ_v is any nonzero element of $V(\pi_v, T_v, \chi_v)$, we have $\ell_v(\psi_{\chi,v}) \neq 0$.*

Now we want a global test vector. Recall that we have defined a 'newform' ψ in section 4 above by specifying that $\psi = \otimes\psi_v$, where each ψ_v is fixed by $R_v^\times$, for a suitable order $R_v \subset B_v$. We will produce our test vector by modifying ψ at places v dividing $c = c(\chi)$. Indeed, it follows from the definitions that $\psi_v = \psi_{\chi,v}$ for almost all v. Indeed, this equality holds for all $v \nmid c(\chi)$. Thus we may consider the vector $\psi_\chi = \otimes\psi_{\chi,v}$ as an element of the restricted tensor product $\otimes\pi'_v$. Then the following proposition may be extracted from [Wal85], and resolves the question of global test vectors:

PROPOSITION 6.2. *Let ψ_χ be defined as above. Let $\ell \in \mathrm{Hom}_T(\pi', \mathbf{C}(\chi))$ denote the functional defined in equation* (5–1). *Then we have $\ell \neq 0$ if and only if $\ell(\psi_\chi) \neq 0$. In particular, $L(\pi \otimes \chi, \frac{1}{2}) \neq 0$ if and only if $\ell(\psi_\chi) \neq 0$.*

Our next task is to produce a formula for the number $\ell(\psi_\chi)$. According to the definitions, we have

$$\ell(\psi_\chi) = \int_{T(F)^\times \hat{F}^\times \backslash \hat{T}^\times} \psi_\chi(t)\chi^{-1}(t)\, dt. \tag{6–1}$$

Now observe that the function ψ_χ is invariant on the right by the group $\prod R_v^\times$, where R_v is an order of B_v which optimally contains the order of $\mathcal{O}_{K,v}$ with conductor $c_v = c(\chi_v)$. Since χ is invariant under $\hat{\mathcal{O}}_c^\times$, it follows that the integral in (6–1) may be rewritten as a finite sum:

$$\ell(\psi_\chi) = \mu_\chi \cdot \sum_{t \in G_c} \psi_\chi(t)\chi^{-1}(t), \tag{6–2}$$

where G_c denotes the finite set $T(F)^\times \hat{F}^\times \backslash \hat{T}^\times / \hat{\mathcal{O}}_c^\times$, and μ_χ is the volume of the image of $\hat{\mathcal{O}}_c^\times$ in $T(F)^\times \hat{F}^\times \backslash \hat{T}^\times$. Observe that, by class field theory, we may identify G_c with the quotient of $\mathrm{Pic}(\mathcal{O}_c)$ by $\mathrm{Pic}(\mathcal{O}_F)$.

One can even go slightly further, and express the right-hand-side of (6–2) in terms of the newform ψ (which is independent of χ). As we will see, this leads naturally to the appearance of certain CM points of conductor $c(\chi)$.

To begin with, recall that ψ and ψ_χ agree at all places v except those finite places which divide c. If v divides c, then $B_v \cong \mathrm{GL}_2(F_v)$ is split, and ψ_v and $\psi_{\chi,v}$ are fixed by maximal orders R_v and $R_{\chi,v}$ respectively, where $R_v \cap K_v = \mathcal{O}_{K,v}$ and $R_{v,\chi} \cap K_v = \mathcal{O}_{\chi,v}$ is the order of conductor c_v. Since all maximal orders in B_v are conjugate, it follows that $\psi_{\chi,v}$ and ψ_v are related by the equation $\psi_{\chi,v}(z) = \psi_v(zg_v)$, where g_v is such that $g_v^{-1}R_vg_v = R_{\chi,v}$. Thus, if we let $g \in \hat{B}$

denote an element such that $\psi_\chi(z) = \psi(zg)$, the sum in equation (6–2) becomes $\sum_{t \in G_c} \psi(tg)\chi^{-1}(t)$.

Now observe that, for each place v of F, the order $t_v g_v R_v (t_v g_v)^{-1}$ is another local order of B_v which optimally contains $\mathcal{O}_{\chi,v}$. For each t, let R_t denote the global order of B defined by $R_t = B \cap tg\hat{R}(tg)^{-1}$. Then R_t is an order with discriminant N which optimally contains $\mathcal{O}_c \subset \mathcal{O}_K$. Furthermore, it is clear that if R' is any order of B with discriminant N which optimally contains $\mathcal{O}_c$, then $R' = R_t$ for some t. (Here we remind the reader that the embedding $K \to B$ is fixed.)

Note that the function ψ factors through the coset space $\mathrm{Cl}(B) = \mathrm{Cl}(B, N) = B^\times \backslash \hat{B}^\times / \hat{F}^\times \hat{R}^\times$. The set $\mathrm{Cl}(B)$ may be identified with conjugacy classes of oriented orders of discriminant N in B. If R' is any such order, then R' determines an element of $\mathrm{Cl}(B)$, and thus it makes sense to speak of the value $\psi(B)$. From this viewpoint, we see that the sum in (6–2) is just $\sum_t \chi^{-1}(t)\psi(R_t)$ and that the set R_t runs over oriented orders of B of discriminant N that optimally contain $\mathcal{O}_c$.

6.1. CM points. The reader who is familiar with the formalism of [BD96] and [Gro87] will recognize the optimal embeddings $\mathcal{O}_c \to R_t$ occurring in the above as being precisely the points called 'definite' Heegner points, or CM points, in the former and special points in the latter. We now proceed to describe these special points from a more adelic point of view, and rewrite Waldspurger's theorem in terms of an evaluation of ψ on a suitable CM cycle.

Thus let G' denote the algebraic group $B^\times / F^\times$. Let U denote any open compact subgroup of G. Recall that we have fixed an embedding $K \to B$, and let T denote the torus $K^\times / F^\times \subset G$. Then the set of CM points of level U on B associated to the embedding $K \to B$ is defined to be the coset space

$$C = T(F) \backslash G'(\mathbf{A}_f)/U,$$

where $\mathbf{A}_f$ denotes the space of finite adeles. Note that there is an action of $\hat{T}/T(F) = \hat{K}^\times / K^\times \hat{F}^\times$ on the set C. A *CM cycle* is just a compactly supported function on C. In other words, a CM cycle is just a finite linear combination of characteristic functions of cosets in C.

We can make this definition more concrete in the case that U is the image of $\hat{R}^\times$, for some order $R \subset B$ of discriminant N. Indeed, in that case, each element P in C is represented by some $x \in \hat{B}$, and we may form the order $B_x \subset R$ defined by $B_x = B \cap x\hat{R}x^{-1}$. Let $\mathcal{O}_x \subset \mathcal{O}_K$ denote the order given by $\mathcal{O}_x = K \cap B_x$. By definition, the fixed embedding $f : K \to B$ induces an optimal embedding $\mathcal{O}_x \to B_x$. Thus the choice of x yields a pair $(\mathcal{O}_x, R_x)$, where $f : \mathcal{O}_x \to R_x$ is an optimal embedding. One checks furthermore that choice of a different representative x' for the coset defining the CM point P yields a pair $(\mathcal{O}_{x'}, R_{x'})$ which differs from $(\mathcal{O}_x, R_x)$ simply by conjugation by an element of T. The *conductor* of P is defined to be the conductor c of the order $\mathcal{O}_x$, which is obviously independent of the choice of x. If P has conductor c, then the action of $\hat{T}/T(F)$ on P factors through G_c, where $G_c = \hat{K}^\times/\hat{F}^\times K^\times \hat{\mathcal{O}}_c^\times$ as above. Furthermore, it is clear that G_c acts simply transitively on the set of CM points of conductor c. If $\sigma \in G_c$ and P is a CM point of conductor c, we write P^σ to denote the image of P under σ.

REMARK 6.3. Suppose that $F = \mathbf{Q}$. Then a CM point of conductor c is defined in [BD96] to be a pair (f, R) where R is an oriented order in B of discriminant N, and $f : \mathcal{O}_c \to R$ is an oriented embedding, where points (f, R) and (f', R') are identified if they are conjugate under the action of $B^\times$. One checks readily that this notion is equivalent to the one above.

We can now combine Waldspurger's result with those of Gross–Prasad to obtain a simple criterion in terms of CM points for the non-vanishing of $L(\pi \otimes \chi, \frac{1}{2})$.

THEOREM 6.4. *Let the hypotheses and notation be as in section* 2. *Let R denote any order of B of discriminant N which optimally contains $\mathcal{O}_K$. Then, if P is any CM point of conductor c and level $U = \hat{R}^\times$, we have $L(\pi \otimes \chi, \frac{1}{2}) \neq 0$ if and only if $\sum_{\sigma \in G_c} \chi^{-1}(\sigma)\psi(P^\sigma) \neq 0$.*

The theorem above seems to fill a gap in the literature, and is extremely convenient for applications to Iwasawa theory and arithmetic. Indeed, it is freely used in [BD97] and its various sequels, as well as in [Vat02]. (Note also that if $F = \mathbf{Q}$ and K is an imaginary quadratic field, then $G_c = \mathrm{Pic}(\mathcal{O}_c)$ since $\mathbf{Q}$ has class number 1.)

7. The Work of Gross and Zhang

It is natural now to ask for an exact relationship between the numbers $L(\pi \otimes \chi, \frac{1}{2})$ and $\ell(\psi_\chi)$. More generally, one could ask for a relationship between $L(\pi \otimes \chi, \frac{1}{2})$ and the number $\sum_{\sigma \in G_c} \chi^{-1}(\sigma)\psi(P^\sigma)$ appearing in Theorem 6.4. These problems fall in the general category of Gross–Zagier formulae, in the sense that they seek to express L-values in terms of explicit CM cycles. This problem was first taken up by Gross in [Gro87], where the case where $F = \mathbf{Q}$ and N and D are prime was treated. It was subsequently generalized to slightly more general D in [Dag96]. The case of general F and D was finally treated in [Zha01a]. All of these results are more arithmetic in flavour than those of Waldspurger discussed above, being based on the calculation of Fourier coefficients of certain kernel functions for Rankin–Selberg convolutions, and the expression of these Fourier coefficients in terms of a height pairing on the CM cycles.

7.1. Gross' formula. To fix the ideas, we want to discuss the main result of [Gro87], which, as described above, deals with the case where $F = \mathbf{Q}$, and N and D are prime. For the benefit of the number theorists in the audience, we will start by describing the basic idea in classical rather than adelic language.

Thus let N denote a positive rational integer, and let $g(z) = \sum a_n q^n$ denote a cuspform of weight 2 on $\Gamma_0(N)$. We normalize g so that $a_1 = 1$. Let $K/\mathbf{Q}$ denote an imaginary quadratic field of discriminant D. We assume that both N and D are prime, and that N remains inert in K. Let χ denote an unramified Hecke character of K, so that $c = 1$. (Thus the notation for N and D employed here is consistent with the general case set out above.) Then g corresponds to a cuspidal automorphic representation π of $\mathrm{GL}_2(\mathbf{Q})$, and χ is a representation of $\mathrm{GL}_1(K)$. We want to study the value $L(\pi \otimes \chi, \frac{1}{2})$, which in classical notation (see [Maz84] for an overview) is just $L(g \otimes \chi, 1)$. Note that the normalization of the classical L-function $L(g \otimes \chi, s)$ yields a functional equation under $s \mapsto 2 - s$, while the automorphic L-function is symmetric under $s \mapsto 1 - s$. As we have remarked, we will try to keep the classical notation in this section.

The starting point is the expression of $L(g \otimes \chi, s)$ as a Rankin–Selberg convolution. It is not our purpose here to discuss the Rankin–Selberg method in detail, so we will simply extract the one statement that is central to our discussion. The details may be found in [Gro87]. We recall that if Γ is a congruence subgroup of $SL_2(\mathbf{Z})$ and f, g are weight 2 modular forms with respect to Γ, then the Petersson inner product (f, g) relative to Γ is defined by

$$(f, g)_\Gamma = 8\pi^2 \int_{F_\Gamma} f(z)\overline{g}(z)\, dx\, dy,$$

where F_Γ is a fundamental domain for Γ in the upper half-plane, and $z = x + iy$. The integral converges provided that at least one of f and g is cuspidal.

PROPOSITION 7.1. *There exists a kernel function Θ_χ, which is a modular form of level ND, such that*

$$L(g \otimes \chi, 1) = (g, \Theta_\chi)_{ND}.$$

Here the Petersson product $(\cdot, \cdot)_{ND}$ is taken relative to the group $\Gamma_0(ND)$.

The first step in the Gross–Zagier argument is to take the trace of Θ_χ down to level N. Thus put $\Theta = \mathrm{Tr}_N^{ND}(\Theta_\chi)$. Then we clearly have

$$L(g \otimes \chi, 1) = (g, \Theta)_N, \tag{7–1}$$

where this time the inner product is taken at level N. Now, Θ is a modular form of level N, so we may write

$$\Theta = E(z) + \sum_i c_i g_i(z) \tag{7–2}$$

where $E(z)$ is an Eisenstein series, and the sum is taken over a basis of the space of newforms of level N. (Since N is prime and we are in weight 2, there are no oldforms.) We may assume that the numbering is such that $g(z) = g_1$. It can be shown that the coefficients c_i are all algebraic, see [Shi76].

Then, by orthogonality, it is clear that

$$(g, \Theta)_N = (g, E + \sum_i c_i g_i)_N = (g, c_1 g_1) = c_g(g, g), \tag{7–3}$$

where $c_g = c_1$ is the coefficient of g. Given an eigenform g_i, let us put $\Theta_{g_i} = c_i g_i$. We call Θ_{g_i} the g_i-isotypic component of Θ. Thus Θ_{g_i} denotes the projection of Θ to the eigenspace of the Hecke algebra with eigenvalues given by the newform g_i. With this notation, we have

$$L(g \otimes \chi, 1) = (g, \Theta)_N = (g, \Theta_g)_N = c_g(g, g)_N. \tag{7–4}$$

Thus, the evaluation of $L(g \otimes \chi, 1)$ boils down to calculating the coefficient c_g of g in the spectral decomposition (7–2) of the kernel function, which is accomplished by computing the Fourier coefficients of the kernel in terms of CM points on a suitable quaternion algebra.

Thus let B denote the quaternion algebra over $\mathbf{Q}$ ramified precisely at N and infinity, and let $R \subset B$ denote a maximal order. Recall also the notation introduced in Section 6.1, and let X denote the set of CM points of conductor 1 and level $U = \hat{R}^\times$. (These are the 'special points' of discriminant D in [Gro87].) Then Gross shows that X admits an action of the group $\mathrm{Pic}(\mathcal{O}_K)$, and that this action is both simple as well as transitive, as in section 6.1 above. We let P denote any fixed point in X, and define a CM divisor by

$$y = \sum_{\sigma \in \mathrm{Pic}(\mathcal{O}_K)} \chi^{-1}(\sigma) P^\sigma. \tag{7–5}$$

Furthermore, Gross defines an intersection pairing $(\,\cdot\,,\cdot\,)$ on the space of CM divisors as follows. Each point $P \in C$ determines an oriented maximal order R of B as described in section 6.1, and we put

$$(P, P') = \delta(P, P'), \tag{7–6}$$

where $\delta(P, P') = 0$ unless the orders R, R' determined by P, P' are conjugate in B. In the latter case, we put $\delta(P, P') = w$, where w is the order of the finite group $R^\times$. We extend to pairing to CM divisors (which are just finite linear combinations of points) by linearity in the first variable, and skew-linearity in the second. Putting $S(U) = B^\times \backslash \hat{B} / \hat{\mathbf{Q}}^\times \hat{R}^\times$, it is clear that the pairing defined above is in fact a pairing on $S(U) \times S(U)$ (since $S(U)$ is just the set of conjugacy classes of oriented maximal orders) and that the pairing on CM points factors through the evident map $C \to S(U)$.

Finally, it is not hard to see that the space $S(U)$ inherits an action of the Hecke operators T_n, for all integers n.

With this notation, the basic result is the following

PROPOSITION 7.2 (GROSS). *Write $\Theta = \sum b_n q^n$. Then the coefficients b_n are given by*

$$b_n = \frac{(y, T_n y)}{u^2\sqrt{D}},$$

where y and the pairing $(\cdot,\cdot)$ are as described above, and $u = \#O_K^\times/2$.

With all this in hand, it is now easy to prove a Gross–Zagier formula. Indeed, we want to compute the coefficient of the newform g in the spectral decomposition of Θ. In other words, we want to pick off the projection Θ_g of Θ to the space where T_n acts via $a_n = a_n(g)$. It is not hard to see that this projection has coefficients b'_n given by

$$b'_n = \frac{(y_g, T_n y_g)}{u^2\sqrt{D}}$$

where y_g denotes the projection of the CM divisor y to the subspace of $\mathbf{C}[S(U)]$ where the Hecke algebra $\mathbf{T}_B$ (defined by action of the T_i on automorphic forms of level $\hat{R}^\times$ on B) acts via the character given by the Jacquet–Langlands correspondent of g. In particular, the first coefficient of the projection Θ_g is given by

$$b_1(\Theta_g) = \frac{(y_g, y_g)}{u^2\sqrt{D}}. \tag{7–7}$$

Writing $\Theta_g = c_g \cdot g$ as in the discussion following (7–2), we see that $u^2\sqrt{D}\cdot(y_g, y_g) = c_1(\Theta_g) = c_g$, since g was normalized to have first Fourier coefficient 1. Combining this with (7–3), we get the desired formula:

$$\frac{L(g\otimes\chi, 1)}{(g,g)_N} = \frac{(y_g, y_g)}{u^2\sqrt{D}}. \tag{7–8}$$

To compare with Waldspurger's theorem, we can rewrite this in terms of test vectors and torus integrals. To do this, we will need to give an adelic description of the intersection pairing, following [Zha01a], Section 4.1. The argument is in some sense entirely formal, but it is nevertheless instructive to work through the details, since this is in fact what occurs in the general case treated in Zhang's article. And since the construction is completely general, we will revert for the moment to the case of general F and K and π, as described in the introduction.

7.2. The intersection pairing revisited. Recall that the set of CM points of level U is defined to be the coset space $C = T(F)\backslash G'(\mathbf{A}_f)/U$, where $G' = B^\times/F^\times$, and U is some open compact subgroup of $G'(\mathbf{A}_f)$. (In Gross' situation, we will take U to denote the image of $R(\mathbf{A}_f)^\times$, for a fixed maximal order R.) Let $S(U) = G'(F)\backslash\hat{G}'_f/U$, where we have simply written $\hat{G}'_f$ in place of $G'(\mathbf{A}_f)$.

Let m denote the characteristic function of $U \subset \hat{G}'_f$, and consider the kernel function $k(x,y)$ defined on $S(U)\times S(U)$ by

$$k(x,y) = \sum_{\gamma\in G'(F)} m(x^{-1}\gamma y). \tag{7–9}$$

This sum is actually finite. Indeed, if $(x, y) \in \hat{G}'_f \times \hat{G}'_f$, then $m(x^{-1}\gamma y)$ is nonzero only for those γ such that $\gamma \in xUy^{-1}$. But xUy^{-1} is compact in $\hat{G}'_f$, and $G'(F_\infty)$ is compact already, so $G'(F) \cap xUy^{-1}$ injects into the compact set $xUy^{-1} \times G'(F_\infty) \subset G'(\mathbf{A})$. Now $G'(F)$ being discrete in $G'(\mathbf{A})$, it follows that $G'(F) \cap xUy^{-1}$ is finite.

Now let $p_1, p_2 \in S(U)$, and let ξ_1, ξ_2 denote the characteristic functions of the corresponding double cosets. Define the intersection pairing via

$$(p_1, p_2) = (\xi_1, \xi_2) = \int_{S(U)^2} \xi_1(x) k(x, y) \overline{\xi_2(y)} dy dx.$$

Here the measure on $S(U)$ is induced from a left-$G'(F)$-invariant measure on $\hat{G}'_f$, normalized so that U gets volume one. To calculate this pairing, observe that ξ is supported on $G'(F)g_1U$, while ξ_2 is supported on $G'(F)g_2U$. It follows directly from the definitions that the kernel is constant on $G(F)g_1U \times G(F)g_2U$. Indeed, if the elements g_i are in different cosets, the kernel is just zero on this set. If on the other hand the g_i are in the same coset, put $w = G'(F) \cap g_1 U g_1^{-1} = G'(F) \cap g_2 U g_2^{-1}$ as before. Then the kernel takes the value w on $G(F)g_1U \times G(F)g_2U$.

We can transfer the pairing to the set of CM points via the evident map $C \to S(U)$, as follows. Let $k^*(x, y)$ denote the pullback of $k(x, y)$ to $C \times C$. Let P_1, P_2 denote CM points, and let ξ_1 and ξ_2 denote the characteristic functions of the corresponding double cosets in $C = T(F)\backslash G'(\mathbf{A}_f)/U$. The measure on C is understood to be induced from a measure on $\hat{G}'_f$ such that U gets volume 1. Then we can define the pairing by an integral over $C \times C$ as above, and we find that (P_1, P_2) is zero unless the P_i project to the same element in $S(U)$, in which case the pairing has value w/u^2, where u denotes the cardinality of the set $T(F) \cap U$. Indeed, the volume of any coset $T(F)gU$ in C is $1/u$.

EXAMPLE 7.3. Consider the special case of $F = \mathbf{Q}$, and N, D prime. Then we have recovered Gross' pairing, including the fudge factor $1/u^2$ which appears in the final formulae.

For later use, we want to consider the spectral decomposition of the kernel function $k(x, y)$, as in [Zha01a], Lemma 4.4.2. But to state the result, we first need some notation. So recall that G' is compact at infinity by hypothesis. Then $V = U \cdot G'(\mathbf{R})$ is an open compact subgroup of $\hat{G}'$. Recall also that we are interested in representations π of GL_2 of parallel weight $(2, \ldots 2)$, which correspond via Jacquet–Langlands to representations π' on G' whose infinite component is trivial. With this in mind, let $\psi_1, \psi_2, \ldots, \psi_r$ denote an orthonormal basis for the space of automorphic forms on G' that are fixed by the subgroup V. We may assume that the ψ_i are eigenvectors for the good Hecke operators. Since $G'(\mathbf{R}) \subset V$, we may view any such ψ_i as a function on the set $S(U) = G'(F)\backslash\hat{G}'_f/U$. Then the basic result is as follows.

Proposition 7.4 (Zhang). *The kernel function $k(x,y)$ has the spectral decomposition*

$$k(x,y) = \sum_i \psi_i(x)\overline{\psi}_i(y).$$

Proof. This is easy. Indeed, if ψ is any function on $S(U)$, then we have

$$\int_{G'(F)\backslash\hat{G}'_f} \psi(y)k(x,y)\,dy = \int_{\hat{G}'_f} \psi(y)m(x^{-1}y)\,dy,$$

where m denotes the characteristic function of the compact set U. Since U has volume 1, we get $\int_{G'(F)\backslash\hat{G}'_f} \psi(y)k(x,y)\,dy = \psi(x)$. Since $S(U)$ is finite, we have $k(x,y) = \sum c_{ij}\psi_i(x)\overline{\psi}_j(y)$, for some c_{ij}, and so the statement of the proposition follows. □

Remark 7.5. The basis vectors ψ_i above will not, in general, diagonalize the Hecke operators, as there will be some contribution from oldforms. In particular, a given eigenspace for the Hecke operators will contain several vectors ψ_i.

Finally, we want to mention the action of the Hecke operators on the CM cycles. Given a function ϕ on $T(F)\backslash\hat{G}'_f/U$, we define a Hecke operator T_v for each place v such that B_v is split and U_v is maximal as follows:

$$T_v\phi(x) = \int_{H_v} \phi(xh_v)\,dh_v, \tag{7–10}$$

where H_v is the set of matrices h_v in B_v such that $\det(h_v)$ is a uniformizer in F_v. Here the measure on H_v is such that the maximal compact U_v gets measure 1. One may check that the operator T_v is self-adjoint with respect to the L^2 norm, so that $(T_a\phi,\phi') = (\phi, T_a\phi')$, provided that the functions ϕ and ϕ' have compact support so that the L^2 integral makes sense. We can of course make a similar definition for functions on $S(U) = G'(F)\backslash\hat{G}'_f/U$.

7.3. Gross' formula: adelic version. We now go back to Gross' situation, where $F = \mathbf{Q}$, and N, D are prime. The goal is to rewrite the special value formula (7–8) as a torus integral, as in Waldspurger's theorem. Obviously, we must compute the divisor y defined in (7–5), as well as the projection to the appropriate Hecke eigenspace. The first question is rather easy. Indeed, consider the image of $T(\mathbf{Q})\backslash\hat{T} = \mathrm{Pic}(\mathcal{O}_K)$ inside the set $C = T(\mathbf{Q})\backslash G'(\mathbf{A}_f)/U$. Each class $\sigma \in \mathrm{Pic}(\mathcal{O}_K)$ defines a CM point $\xi_\sigma \in C$. We may therefore form the CM divisor

$$y = \sum \chi(\sigma)\xi_\sigma.$$

Next we need to make sense of the projection to the Hecke eigenspace of interest. To do this, it will be convenient to identify the CM point ξ_σ with the characteristic function of the corresponding double coset in C. By definition of the intersection pairing, we have

$$(y,y) = \int_{C^2} \xi_y(u)k^*(u,v)\overline{\xi}_y(v)\,du\,dv.$$

In view of the spectral decomposition of the kernel in Proposition 7.4, the integral above may be rewritten as

$$(y, y) = \sum \int_C \xi_y(u)\psi_i(u)\overline{\psi}_i(v)\overline{\xi}_y(v)\, du\, dv,$$

where the sum is taken over an orthonormal basis for the space of cuspforms on $S(U)$. In the situation where $F = \mathbf{Q}$ and the level N is prime, there are in fact no oldforms, so we may take the basis vectors ψ_i to be Jacquet–Langlands newforms, each belonging to a distinct eigenspace for the Hecke algebra. We may assume by renumbering that our original representation π' corresponds to the vector $\psi = \psi_1$.

Thus the pairing decomposes as

$$(y, y) = \sum_i \left(\int_C \xi_y(u)\psi_i(u)\, du\right)\left(\int_C \overline{\psi}_i(v)\overline{\xi}_y(v)\, dv\right).$$

By self-adjointness of the Hecke operators, we have

$$\int_C T_a\xi_y(u)\psi_i(u)\, du = \int_C \xi_y(u)T_a\psi_i(u)\, du,$$

for any a prime to the level N. Since N is prime, strong multiplicity one implies that, if $y_{\pi'}$ denotes the projection of y to the eigenspace corresponding to π', then we have the formula

$$(y_{\pi'}, y_{\pi'}) = \left(\int_C \xi_y(u)\psi(u)\, du\right)\left(\int_C \overline{\xi}_y(v)\overline{\psi}(v)\, dv\right).$$

But it is easy to see that the integrals above are precisely torus integrals of the test vector ψ, since the function ξ_y is supported on the image of the torus $\hat{T}$. Indeed, we have

$$\begin{aligned}\ell_\chi(\psi) &= \int_C \xi_y(u)\psi(u)\, du\\ &== \frac{1}{u} \sum_{t\in T(F)\backslash\hat{T}/\hat{\mathcal{O}}_K} \chi^{-1}(t)\psi(t) \int_{T(F)\backslash\hat{T}/T(\hat{\mathcal{O}}_K)} \chi^{-1}(t)\psi(T),\end{aligned}$$

where the measure on $\hat{T}$ is such that $T(\hat{\mathcal{O}}_K)$ gets measure 1.

Thus we may rewrite Gross' formula in the form

$$\frac{L(g\otimes\chi, 1)}{(g, g)_N} = \frac{|\ell(\psi)|^2}{\sqrt{D}}, \tag{7–11}$$

where ℓ is the torus integral with respect to a Haar measure giving $T(\hat{\mathcal{O}}_K)$ volume 1, and ψ is a Jacquet–Langlands newform, normalized to have L^2 norm 1, with respect to a Haar measure on $\hat{G}'$. Here the measure on $\hat{G}'$ is normalized so that a maximal compact $U \subset \hat{G}'$ has volume equal 1.

7.4. The work of Zhang. We now want to describe the formula of [Zha01a]. We therefore return to the general situation, keeping the notation fixed in Section 2. Thus we want to study the special value $L(\pi \otimes \chi, \frac{1}{2})$, and the starting point is once again the expression of the L-series as a Rankin L-function. The basic strategy is the same as in the discussion above, but with some important improvements. We will attempt here to at least explain the statement of the final formula, if not all the new ideas introduced in Zhang's proof.

7.5. Quasi-newforms on GL_2. As before, one begins with a theta kernel Θ_χ, which (under the present hypotheses) is an automorphic form of weight $(2,\ldots,2)$ on GL_2, and which satisfies the equation

$$L(\pi \otimes \chi, \tfrac{1}{2}) = (\phi, \Theta_\chi) \tag{7–12}$$

where ϕ denotes the Atkin–Lehner newform of level N associated to π as described in section 3. Here the form Θ_χ is of level ND, where $D = dc^2$, d is the discriminant of K/F, and c is the conductor of χ.

Note that there are various normalizations already implicit in the above formula. Specifically, the formula holds with ϕ normalized as in section 3, and the inner product is taken with respect to the L^2 metric on $\mathrm{GL}_2(F)\hat{F}^\times\backslash\mathrm{GL}_2(\hat{F})$. The Haar measure on the latter is fixed so that the open compact subgroup $U_0(ND)K_\infty$ gets measure 1, where K_∞ is a product of the various maximal compact subgroups at primes dividing infinity. The detailed construction of the kernel Θ_χ is given in Chapter 2 and section 3.1 of [Zha01a].

If we were to follow the classical argument exactly, the next step would be to compute the trace of the theta kernel down to level N. This appears to be rather difficult, since the extra level $D = dc^2$ is large when $c \neq 1$. Thus Zhang works directly with the kernel function at level ND, as follows. Let $\pi_1, \pi_2, \ldots, \pi_r$ be an enumeration of the finitely many non-isomorphic cuspidal automorphic representations of weight $(2,\ldots,2)$ occurring at levels dividing ND. Then $\pi = \pi_j$ for some j, say $j = 1$, and we may decompose Θ_χ in a manner analogous to (7–2), as

$$\Theta_\chi = \text{Eisenstein part} + \sum \phi_i \tag{7–13}$$

where each ϕ_i is a form in the Hecke eigenspace corresponding to the representation π_i. Then one has

$$L(\pi \otimes \chi, \tfrac{1}{2}) = (\phi, \Theta_\chi) = (\phi, \phi_\pi), \tag{7–14}$$

where $\phi_\pi = \phi_1$ is the term in (7–13) corresponding to the fixed representation π.

However, since we are working at level ND, and ϕ has level N, it is no longer true that ϕ_π is simply a scalar multiple of the newform ϕ. Indeed, the best we can say is that ϕ_π is some linear combination

$$\phi_\pi = \sum c_a \phi_a,$$

where the sum is taken over ideals a dividing D, and ϕ_a is as in Casselman's theorem. It is evident from this that, even if one could compute the Fourier coefficients of Θ_χ in terms of CM cycles, a simple argument as in (7–7) cannot be made to work, since one has no evident normalization of the form ϕ_π. Note here that it is crucial to the argument following (7–7) that the newform g was normalized to have first Fourier coefficient 1.

In view of the considerations above, we are led to normalize the vector ϕ_π in some convenient way. In other words, we are looking for some distinguished vector on the line L spanned by the vector ϕ_π.

DEFINITION 7.6. Let $\mathcal{S}$ denote the space of cuspforms of level ND. If L is any line in $\mathcal{S}$, then the *quasi-newform* $\phi^\# = \phi^\#_L$ associated to π on the line L is the orthogonal projection of the Atkin–Lehner newform ϕ on L.

REMARK 7.7. The definition assumes that ϕ has already been normalized. This already occurs in the formula (7–12).

REMARK 7.8. By definition, there is a quasi-newform on each line in the ambient space $\mathcal{S}$. Thus, for the definition to be useful, one needs to specify the line L. Note that the quasi-newform attached to L is zero if L is orthogonal to the Atkin–Lehner newform ϕ.

REMARK 7.9. Let the line L be given. Suppose that L is not orthogonal to ϕ so that $\phi^\#_L \neq 0$. Then if v is any vector on L, we will have $v = c\phi^\#_L$, for some scalar c. For simplicity, we assume that c is real, which will be the case in our applications. One then has

$$(\phi, v) = (\phi, c\phi^\#_L) = c(\phi, \phi^\#_L) = c(\phi^\#_L, \phi^\#_L),$$

since $\phi^\#_L$ is the orthogonal projection of ϕ on L.

Let us apply this to the formula (7–14) with L taken to be the line spanned by ϕ_π. Then (7–14) implies that $L(\pi \otimes \chi, \frac{1}{2}) = 0$ if and only if L is orthogonal to ϕ, or ϕ_π is zero. If this not the case, then we may put $v = \phi_\pi = c_\pi \phi^\#_L$, and this gives the key formula

$$L(\pi \otimes \chi, \tfrac{1}{2}) = (\phi, \phi_\pi) = c_\pi(\phi^\#_L, \phi^\#_L). \tag{7–15}$$

We are therefore led to compute the line L spanned by the vector ϕ_π, and to determine the corresponding quasi-newform $\phi^\#_L$. The former problem has been solved by Zhang, as we now explain.

Thus recall that the form ϕ_π is by definition the projection of the kernel Θ_χ to the eigenspace (for the good Hecke operators) corresponding to the representation π. According to Casselmans' theorem, a basis for this eigenspace is given by the functions ϕ_a, where a runs over the ideals dividing D. Thus we have $\phi_\pi = \sum x_a \phi_a$, and we must compute the coefficient x_a for each a. Note here that the vectors ϕ_a are not orthonormal, or even orthogonal.

In practice, it is much easier to compute the inner products (ϕ_π, ϕ_a), thereby determining the orthogonal complement to our line L, rather than L itself. To state the result, we need some notation. Recall that the character χ is anticyclotomic. For each prime v of F which is ramified in K, one sees therefore that the local character χ_v is quadratic. It follows that χ_v is the composition of some unramified character ν of F_v with the local norm from K_v to F_v.

Now, for any place v dividing ND, define a function $\nu^*(v)$ as follows. If v is ramified in K, we put $\nu^*(v) = \nu(\pi_v)$ where π_v is a uniformizer of F_v, and ν is the local character defined above. If v is unramified in K, we put $\nu^*(v) = 0$. We can extend the function ν^* to the set of ideals dividing ND by multiplicativity, and by putting $\nu^*(1) = 1$.

PROPOSITION 7.10 (ZHANG). *Let $\mathcal{S}_\pi$ denote the linear span of the vectors ϕ_a, as a runs over divisors of ND. Let $\phi_\pi \in \mathcal{S}$ denote the projection of Θ_χ to S_π. Then:*

- *If $L(\pi\otimes\chi, \frac{1}{2}) \neq 0$, then $\phi_\pi \neq 0$, and the line L spanned by ϕ_π is the orthogonal complement in S_π of the hyperplane given by $\sum_a c_a\phi_a$, where*

$$\sum c_a\nu^*(a) = 0.$$

- *If $L(\pi \otimes \chi, \frac{1}{2}) = 0$, then $\phi_\pi = 0$.*

The proof of this is a somewhat elaborate computation, based on the precise definition and normalization of the kernel function Θ_χ, which enables one to calculate (ϕ_a, Θ_χ) for any a. The details may be found in Section 3.1 of [Zha01a]. Note also here that L is not orthogonal to the Atkin–Lehner newform, so the quasi-newform $\phi_L^\#$ is non-zero.

Let L denote the line defined in the first part of Proposition 7.10. Then $\phi^\# = \phi_L^\#$ will denote the (nonzero) quasi-newform associated to this L. We put $c_\pi = 0$ if $L(\pi \otimes \chi) = 0$, and we define it by equation (7–15) if not. Then, with this convention, it is clear that (7–15) holds in general. Indeed, the identity

$$\phi_\pi = c_\pi\phi^\# \tag{7–16}$$

holds in general as well.

7.6. Toric newforms on the quaternion algebra B. The next step in the argument is to calculate the Fourier (or Whittaker) coefficients of the kernel function Θ_χ in terms of CM cycles on the quaternion algebra B, and then to express the final formula in terms of a Waldspurger functional on some suitable test vector. Again, the details are somewhat involved, so we will limit ourselves to explaining the statements and results that are relevant to our purposes. The details may be found in Chapter 4 of [Zha01a]. We point out here that the results in [Zha01a] include a general construction of geometric intersection pairings on CM cycles, and in fact yield a very beautiful local version of the Gross–Zagier formula, neither of which we will attempt to discuss here. We will just describe

Zhang's construction of toric newforms, which are the analog in his set-up of the test vectors described in Section 6 above. The two notions are related, but are not equivalent.

Recall that we have fixed an embedding $K \to B$. Let $\mathcal{O}_K$ denote the ring of integers in K. Then we let $R \subset B$ denote an order defined by

$$R_v = \mathcal{O}_{K,v} + \mathcal{O}_{K,v}\lambda_v c(\chi_v),$$

where $c(\chi_v)$ is the conductor of χ_v and λ_v satisfies two conditions:

1. $\lambda_v x = \overline{x}\lambda_v$, for each $x \in K_v$. Here $\overline{x}$ denotes the conjugate of x over F_v.
2. $\mathrm{ord}(\lambda_v) = \mathrm{ord}(N_v)$.

Note that R_v is maximal for almost every v. Note also that $\lambda = (\lambda_v)$ generates a two-sided ideal in $\hat{R}$, and that there is a map $\hat{R}/\lambda\hat{R} \to \hat{\mathcal{O}}_K/c\hat{\mathcal{O}}_K$. Since χ is a character of conductor c, it defines a character of $(\hat{\mathcal{O}}_K/c\hat{\mathcal{O}}_K)^\times$, and we thereby deduce an extension of χ to $\hat{R}^\times$. By abuse of notation, we will continue to denote the extended character by χ. Note however that, as a character of $\hat{R}^\times$, χ is trivial at all places $v \nmid c$, since χ_v is trivial on the units of K_v at places away from the conductor.

Finally, we define a subgroup $\Delta \subset \hat{B}^\times$ as follows. If v is unramified in K, we put

$$\Delta_v = F_v^\times R_v^\times.$$

If v is ramified in K, we put

$$\Delta_v = K_v^\times R_v^\times.$$

This definition makes sense, since K_v is normalized by R_v, so the product in the definition above is indeed a group. Furthermore, one checks that χ, which is defined on both K_v and R_v, extends in an obvious way to Δ. We write χ_Δ for this extended character.

With this definition, Zhang gives the following definition and existence result.

DEFINITION 7.11 ([Zha01a], SECTION 2.3). Let the group Δ be as above, and let χ_Δ denote the character of Δ constructed above. A vector ϕ_χ in π' is called a toric newform with character χ if

- ϕ_χ is an eigenform for the Hecke operator T_v at all places v away from ND, and
- The action of Δ on ϕ_χ is given by the character χ_Δ.

PROPOSITION 7.12. *There exists a unique line in the representation π' where Δ acts via the character χ_Δ. Thus nonzero toric newforms exist.*

REMARK 7.13. The existence and uniqueness of the toric newvector ψ_χ may be checked locally. As we have already remarked, χ_Δ is trivial at all primes away from D. Thus if $\psi = \otimes\psi_v$ denotes the newform attached to the representation π'

in Section 4, we may simply take $\phi_{\chi,v} = \psi_v$ for all v away from D. This already gives the Hecke eigenvector property.

As for the places v dividing D, one has to replace ψ_v with a local vector $\phi_{\chi,v}$ satisfying the appropriate transformation property under Δ_v. This is similar to the techniques from [GP91]. Note, however, that optimal embeddings do not appear directly in the present set-up. We refer the reader to Chapter 2 of [Zha01a] for the details.

REMARK 7.14. It may be of interest to explicate the connection between the toric newform and the test vector of Gross–Prasad. It is easy to see that the two notions are the same at primes away from the conductor of χ, so the we need only consider primes v where χ is ramified. While the general relationship is complicated, we can make a simple statement at primes $v|c(\chi)$ which are *inert* in K.

Thus, let v denote such a prime. Then the local representation $\pi_v = \pi'_v$ decomposes under the action of the torus T as the sum of one-dimensional invariant subspaces (see [Tun83]). The Gross–Prasad local test vector has the property that it projects nontrivially on to any line where the action of T is via a character of conductor dividing that of χ. Zhang's test vector, on the other hand, lies on the line where T acts via the fixed character χ of interest.

7.7. The final formula. We are now almost ready to state Theorem 1.3.2 of [Zha01a], which gives a formula for $L(\pi \otimes \chi, \frac{1}{2})$. But first we need to specify precisely the groups we work with, and the normalizations of the measures and vectors which will appear.

Let $G = \mathrm{GL}_2(F)/F^\times$. Then we normalize the Haar measure on $G(\mathbf{A})$ by requiring that $U_0(ND)$ get measure 1. On the quaternion side, we let $G' = B^\times/F^\times$, and let $U \subset \hat{R}$ denote the kernel of χ. We fix the measure on $\hat{G}'$ so that Δ gets measure 1. Here R denotes the order constructed in Section 7.6. Finally, we fix a Haar measure on $\hat{T} = \hat{K}^\times/\hat{F}^\times$ by requiring that the maximal compact subgroup gets measure 1. This of course is independent of χ.

REMARK 7.15. The subgroup U giving the level is *not* simply the image of $\hat{R}^\times$. Rather, it is the subgroup of $\hat{R}^\times$ corresponding to the kernel of χ. This will be important in understanding the statements.

THEOREM 7.16 (ZHANG). *We have $L(\pi \otimes \chi, \frac{1}{2}) = 0$ if and only if $\ell(\phi_\chi) \neq 0$. If $L(\pi \otimes \chi, \frac{1}{2}) \neq 0$, then we have the formula*

$$c^\# \cdot L(\pi \otimes \chi, \tfrac{1}{2}) = \frac{2^n|\phi^\#|^2}{\sqrt{D_{K/F}}} \cdot \left(\frac{\int_{F^\times \backslash \hat{T}} \chi^{-1}(t)\phi_\chi(t)\, dt}{|\phi_\chi|} \right)^2 \tag{7–17}$$

Here n denotes the degree of $F/\mathbf{Q}$, $D_{K/F}$ denotes the absolute norm of the discriminant of K/F, and $c^\#$ is the first Whittaker coefficient of the quasi-newform $\phi^\#$ (the precise definition is given below). The norms are understood to be the L^2 norms relative to the measures specified above.

REMARK 7.17. The number $c^{\#}$ is missing from the statement in [Zha01a]. Its appearance is explained below.

Needless to say, we are not in a position to say anything substantial about the proof of this theorem. Broadly, it follows the lines sketched in the discussion of Gross' formula. We have already indicated above how the fact that Zhang works at level ND leads to the appearance of the quasi-newform $\phi^{\#}$, rather than the newform ϕ. Let us therefore briefly indicate how the toric newform makes an appearance on the quaternion algebra side of the question. In the process, we will also explain the factor $c^{\#}$.

As we have remarked several times, the basic point is to relate the Whittaker coefficients of the theta kernel to the values of an intersection pairing. Thus, recall that the theta kernel on GL_2 is determined by its Whittaker coefficients

$$W\left(\begin{pmatrix} a & 0 \\ 0 & 1 \end{pmatrix} \cdot g_\infty\right),$$

where a runs over the finite ideles of F, and $g_\infty \in G_\infty = G(F \otimes \mathbf{R})$. To define the Whittaker functions, we assume fixed a nontrivial additive character η of $\hat{F}/F$. We let δ denote the conductor of η.

We want to express the Whittaker coefficient above, which is a function of a and the parameter g_∞, as the value of a pairing on CM cycles. Let y_χ denote the CM cycle given by the following compactly supported function $y_\chi = \prod y_{\chi,v}$ on C, where $y_{\chi,v}$ is supported on $T(F_v)R_v^\times$ and satisfies

$$y_{\chi,v}(tu) = \chi(t)\chi(u), t \in T(F_v), u \in \Delta_v. \tag{7–18}$$

where $G_c = T\backslash\hat{T}/\hat{O}_c$ as before, and ξ_σ is the corresponding element of C. For each $g_\infty \in G_\infty$, we define an intersection pairing $(\,\cdot\,,\cdot\,)(g_\infty)$ on the space of CM cycles as in Gross' theorem, where the multiplicity function $m = m_{g_\infty}$ depends on g_∞. Indeed, we define

$$m_{g_\infty}(x) = 2^n W_\infty(g_\infty)m(x), x \in G'(F)\backslash\hat{G}'_f$$

where W_∞ is a standard Whittaker function for the weight 2 discrete series on G_∞, and m is the characteristic function of U. (See [Zha01a], Eq. 4.4.2.) With this definition, Zhang proves that the Whittaker coefficients of Θ_χ satisfy

$$W\left(\begin{pmatrix} a\delta^{-1} & 0 \\ 0 & 1 \end{pmatrix} \cdot \varepsilon g_\infty\right) = \frac{|a|(T_a y_\chi, y_\chi)(g_\infty)}{\sqrt{D_{K/F}}}, \tag{7–19}$$

for all ideles a with component 1 at places dividing ND. Here $\varepsilon = \left(\begin{smallmatrix} 1 & 0 \\ 0 & -1 \end{smallmatrix}\right)$.

Let ϕ_π denote the projection of Θ_χ to the π-isotypic part. Then we have

$$\phi_\pi = c_\pi \phi^{\#}$$

for the quasi-newform $\phi^{\#}$, with the conventions of (7–16). We now examine what this means in terms of the intersection pairings, as in the argument following (7–7). According to the *strong* multiplicity one theorem, we can pick off the the

π-isotypic component of Θ_χ by means of the good Hecke operators. In other words, there exists some polynomial t in the good Hecke operators T_a such that if $\Theta_\chi = \sum \phi_i$ as in (7–13), then $t\Theta_\chi = \phi_\pi$ is the π-isotypic part. Thus we can compute the first Whittaker coefficient of ϕ_π, using the formulae (2–1) for the action of Hecke operators on Whittaker coefficients:

$$W_{\phi_\pi}\left(\begin{pmatrix}\delta^{-1} & 0\\ 0 & 1\end{pmatrix}\cdot \varepsilon g_\infty\right) = \frac{(ty_\chi, y_\chi)(g_\infty)}{\sqrt{D_{K/F}}}. \tag{7–20}$$

On the other hand, we have $\phi_\pi = c_\pi\phi^\#$ by definition. So we would like to compare Whittaker coefficients, but we have to be slightly careful, since certain coefficients may *a priori* be zero. Define a number $c^\#(g_\infty)$ by the formula

$$c^\#(g_\infty) = W_{\phi^\#}\left(\begin{pmatrix}\delta^{-1} & 0\\ 0 & 1\end{pmatrix}\varepsilon g_\infty\right). \tag{7–21}$$

LEMMA 7.18. *The function $c^\#(g_\infty)$ is nonzero, and $c^\# = c^\#(1) \neq 0$.*

For a proof of this lemma, we refer the reader to forthcoming work of Zhang and to his article in this volume.

With this is mind, we can write $\phi_\pi = c_\pi\phi^\#$ and compare Whittaker coefficients. Then we get

$$c_\pi = \frac{(ty_\chi, y_\chi)(g_\infty)}{c^\#(g_\infty)\sqrt{D_{K/F}}}.$$

To evaluate c_π, we may take $g_\infty = 1$. In this case, we will simply drop it from the notation. Then comparing with the equation (7–15), we find that

$$\frac{L(\pi\otimes\chi, \frac{1}{2})}{(\phi^\#, \phi^\#)} = \frac{(ty_\chi, y_\chi)}{c^\#\sqrt{D_{K/F}}}.$$

It remains to compute the pairing (ty_χ, y_χ), and this we can compute via the spectral decomposition in Proposition 7.4. When $g_\infty = 1$, it turns out that the kernel for the intersection pairing has the decomposition

$$k^*(x,y) = 2^n \sum_i \phi_i(x)\overline{\phi}_i(y),$$

where n is the degree of $F/\mathbf{Q}$, and the sum is taken over an orthonormal basis for the space of forms of level U. We may assume that these basis vectors are all eigenvectors for the good Hecke operators. Thus the pairing (ty_χ, y_χ) becomes

$$(ty_\chi, y_\chi) = 2^n \sum \left(\int_C ty_\chi(u)\phi_i(u)\,du\right)\left(\int_C \overline{\phi}_i(v)\overline{y}_\chi(v)\,dv\right), \tag{7–22}$$

By the self-adjointness of the Hecke operators and the definition of the projection operator t, the contribution from vectors ψ_i corresponding to eigenspaces other than π' are all wiped out. Here we use the fact the Jacquet–Langlands correspondence preserves the Hecke eigenvalues. The vectors that remain correspond to an orthonormal basis for the π'-isotypic subspace of forms of level U, which will, in general, have dimension strictly greater than 1. Thus suppose that $\phi_1, \dots, \phi_r$ are an orthonormal basis for the space of π'-isotypic forms of level U. We may assume that ϕ_1 is the toric newvector. Then we want to show that the terms for $i = 2, \dots, r$ in (7–22) are all zero as well.

Consider therefore the action of the subgroup Δ on the functions ϕ_i by the usual right translation. Then Δ acts on $\phi_\chi = \phi_1$ by the character χ, by definition of the toric newvector. Since π' is unitary, the action of Δ preserves the orthogonal complement of ϕ_1, namely, the space spanned by $\phi_2, \dots, \phi_r$. Now Δ is a totally disconnected group, so any finite dimensional complex representation of Δ factors through a finite quotient. Then the complement of ϕ_1 breaks up as the sum of irreducible representations, all distinct from χ, since the space of toric newforms is one dimensional. Since Δ acts on the function y_χ by the character χ, it follows that the terms for ϕ_i with $i \geq 2$ are all zero.

Thus we have shown that $(ty_\chi, y_\chi) = 2^n |\int_C y_\chi(u)\phi_\chi(u)|^2$. But it is easy to see that the quantity on the right is just a torus integral of the test vector ϕ_χ.

REMARK 7.19. To conclude, we want to indicate what is not yet proven. For the purposes of Iwasawa theory and p-adic L-functions, it would be desirable to reformulate Zhang's theorem in terms of the fixed level structure N of π. As in [BD96] and [Vat02], the desired formula should have the shape

$$|\ell(\psi)|^2 = C_\chi \frac{L(\pi \otimes \chi, \frac{1}{2})}{(\phi, \phi)_N} \tag{7–23}$$

where ψ is the Gross–Prasad test vector, ϕ denotes the Atkin–Lehner newform for π, and the number C_χ is an explicit constant in $\overline{\mathbf{Q}}$ depending on χ. While it is obvious from Zhang's result that such a formula holds up to some algebraic constant, it does not seem easy to compute the number C_χ. The main problem is determining the length of the quasi-newform in Zhang's theorem. For a discussion of this point we refer the reader to Zhang's article in this volume. We would like to point out, however, that for the purposes of the main results in [Vat02] and [BD96] and its various sequels, it is enough to know that nonvanishing of the L-function is equivalent to nonvanishing of $\ell(\psi)$, and this is true unconditionally in view of the results of Waldspurger and Gross–Prasad: see Theorem 6.4.

References

[BD96] M. Bertolini and H. Darmon, "Heegner points on Mumford–Tate curves", *Invent. Math.* **126**:3 (1996), 413–456.

[BD97] M. Bertolini and H. Darmon, "A rigid analytic Gross–Zagier formula and arithmetic applications", *Ann. of Math.* (2) **146**:1 (1997), 111–147. With an appendix by Bas Edixhoven.

[BJ79] A. Borel and H. Jacquet, "Automorphic forms and automorphic representations", pp. 189–207 in *Automorphic forms, representations and L-functions* (Corvallis, OR, 1977), Proc. Sympos. Pure Math. **33**, Part 1, edited by A. Borel and W. Casselman, Providence, Amer. Math. Soc., 1979. With a supplement "On the notion of an automorphic representation" by R. P. Langlands.

[Bum97] Daniel Bump, *Automorphic forms and representations*, Cambridge, University Press, 1997.

[Cas73] W. Casselman, "On some results of Atkin and Lehner", *Math. Ann.* **201** (1973), 301–314.

[Dag96] H. Daghigh, "Quaternion algebras", Ph.D. thesis, McGill University, 1996.

[GP91] B. Gross and D. Prasad, "Test vectors for linear forms", *Math. Ann.* **291** (1991), 343–355.

[Gro] B. Gross, "Heegner points and representation theory", pp. 37–65 in *Heegner Points and Rankin L-Series*, edited by Henri Darmon and Shou-Wu Zhang, *Math. Sci. Res. Inst. Publications* **49**, Cambridge U. Press, New York, 2004.

[Gro87] B. Gross, "Heights and the special values of L-series", pp. 115–189 in *Number Theory*, edited by H. Kisilevsky and J. Labute, CMS Conference Proceedings **7**, Providence, Amer. Math. Soc., 1987.

[Gro88] B. Gross, "Local orders, root numbers, and modular curves", *Am. J. Math.* **110** (1988), 1153–1182.

[Lub94] A. Lubotzky, *Discrete groups, expanding graphs and invariant measures*, Basel, Birkhäuser, 1994. With an appendix by J. Rogawski.

[Maz84] B. Mazur, "Modular curves and arithmetic", pp. 185–211 in *Proceedings of the International Congress of Mathematicians* (Warsaw, 1983), vol. 1, Warsaw, PWN, 1984.

[Shi76] G. Shimura, "The special values of zeta functions associated with cusp forms", *Comm. Pure Applied Math.* **29** (1976), 783–804.

[Tun83] J. Tunnell, "Local ε-factors and characters of GL_2", *Am. J. Math.* **105** (1983), 1277–1308.

[Vat02] V. Vatsal, "Uniform distribution of Heegner points", *Invent. Math.* **148** (2002), 1–46.

[Wal85] J.-L. Waldspurger, "Sur les valeurs des certaines fonctions L automorphes en leur centre de symmétrie", *Compos. Math.* **54** (1985), 174–242.

[Zha01a] S. Zhang, "Gross–Zagier formula for GL_2", *Asian J. Math.* **5**:2 (2001), 183–290.

[Zha01b] S. Zhang, "Heights of Heegner points on Shimura curves", *Ann. of Math.* (2) **153**:1 (2001), 27–147.

Vinayak Vatsal
University of British Columbia
Department of Mathematics
Vancouver, BC V6T 1Z2
Canada
vatsal@math.ubc.ca

Heegner Points and Rankin L-Series
MSRI Publications
Volume **49**, 2004

Gross–Zagier Formula for GL(2), II

SHOU-WU ZHANG

Contents

1. Introduction and Notation	191
2. Automorphic Forms	195
3. Weights and Levels	197
4. Automorphic L-Series	199
5. Rankin–Selberg L-Series	200
6. The Odd Case	201
7. The Even Case	204
8. The Idea of Gross and Zagier	205
9. Calculus on Arithmetic Surfaces	208
10. Decomposition of Heights	210
11. Construction of the Kernels	212
12. Geometric Pairing	216
13. Local Gross–Zagier Formula	219
14. Gross–Zagier Formula in Level ND	221
15. Green's Functions of Heegner Points	223
16. Spectral Decomposition	226
17. Lowering Levels	228
18. Continuous Spectrum	234
19. Periods of Eisenstein Series	236
Acknowledgments	240
References	240

1. Introduction and Notation

Let A be an abelian variety defined over a number field F and let

$$\rho : \mathrm{Gal}(\bar{F}/F) \longrightarrow \mathrm{GL}_n(\mathbb{C})$$

be a finite dimensional representation of the Galois group of F. Then the Birch and Swinnerton-Dyer conjecture predicts the identity

$$\mathrm{ord}_{s=1} L(s, \rho, A) = \dim(A(\bar{F}) \otimes \rho)^{\mathrm{Gal}(\bar{F}/F)}.$$

Here $L(s, \rho, A)$ denotes an Euler product over all places of F:

$$L(s, \rho, A) := \prod_v L_v(s, \rho, A), \qquad (\mathrm{Re}\, s \gg 0)$$

with good local factors given by

$$L_v(s, \rho, A) = \det(1 - q_v^{-s}\, \mathrm{Frob}_v\, |_{\mathrm{T}_\ell(A)\otimes\rho})^{-1},$$

where ℓ is a prime different than the residue characteristic of v, and $\mathbb{Z}_\ell$ has been embedded into $\mathbb{C}$. More precisely, the Birch and Swinnerton-Dyer conjecture predicts that the leading term of $L(s, \rho, A)$ in the Taylor expansion in $(s-1)$ is given in terms of periods, Tate–Shafarevich groups, and Mordell–Weil group. We refer to Tate's Bourbaki talk [9] for the details of the formulation.

In this paper, we will restrict ourself to the following very special situation:

- A/F is an abelian variety associated to a Hilbert newform ϕ over a totally real field F with trivial central character;
- ρ is a representation induced from a ring class character χ of $\mathrm{Gal}(\bar{K}/K)$ where K/F is a totally imaginary quadratic extension;
- the conductor N of ϕ, the conductor c of χ, and the discriminant $d_{K/F}$ of K/F are coprime to each other.

In this case, $L(s+\frac{1}{2}, \rho, A)$ is a product of the Rankin L-series $L(s, \chi^\sigma, \phi^\sigma)$, where χ^σ and ϕ^σ are the Galois conjugates of χ and ϕ. Moreover, $L(s, \chi, \phi)$ has a *symmetric functional equation*:

$$L(s, \chi, \phi) = \varepsilon(\chi, \phi) \cdot \mathrm{N}_{F/\mathbb{Q}}(ND)^{1-2s} \cdot L(1-s, \chi, \phi)$$

where

$$\varepsilon(\chi, \phi) = \pm 1, \qquad D = c^2 d_{K/F}.$$

The main result in our *Asian Journal* and *Annals* papers [16; 17] is to express $L'(1, \chi, \phi)$ when $\varepsilon(\chi, \phi) = -1$ and $L(1, \chi, \phi)$ when $\varepsilon(\chi, \phi) = +1$ in terms of Heegner cycles in certain Shimura varieties of dimension 1 and 0, respectively, of level ND. This result is a generalization of the landmark work of Gross and Zagier in their *Inventiones* paper [6] on Heegner points on $X_0(N)/\mathbb{Q}$ with squarefree discriminant D.

The aim of this paper is to review the proofs in our previous papers [16; 17]. We also take this opportunity to deduce a new formula for Shimura varieties of level N. In odd case, the formula reads as

$$L'(\tfrac{1}{2}, \chi, \phi) = \frac{2^{g+1}}{\sqrt{\mathrm{N}(D)}} \|\phi\|^2 \|x_\phi\|^2$$

where x_ϕ is a Heegner point in the Jacobian of a Shimura curve. See Theorem 6.1 for details. In even case, the formula reads as

$$L(\tfrac{1}{2}, \chi, \phi) = \frac{2^{g}}{\sqrt{\mathrm{N}(D)}} \|\phi\|^2 |(\widetilde{\phi}, P_\chi)|^2$$

where $(\widetilde{\phi}, P_\chi)$ is the evaluation of certain test form $\widetilde{\phi}$ on a CM-cycle P_χ on a Shimura variety of dimension 0. See Theorem 7.1 for details. These results have more direct applications to the Birch and Swinnerton-Dyer conjecture and p-adic L-series and Iwasawa theory. See the papers [1; 11] of Bertolini–Darmon and Vatsal for details.

To do so, we need to compute various constants arising in the comparisons of normalizations of newforms or test vectors. This will follow from a comparison of two different ways to compute the periods of Eisenstein series. One is an extension of the method for cusp form in our *Asian Journal* paper [16], and another one is a direct evaluation by unfolding the integrals. Notice that the residue and constant term of Dedekind zeta function can be computed by the periods formula for Eisenstein series. Thus, the Gross–Zagier formula can be considered as an extension of the class number formula and the Kronecker limit formula not only in its statement but also in its method of proof.

Notice that Waldspurger has obtained a formula (when χ is trivial [12]) and a criterion (when χ is non trivial [13]) in the general situation where

- K/F is any quadratic extension of number fields, and
- ϕ is any cusp form for $\mathrm{GL}_2(\mathbb{A}_F)$, and
- χ is any automorphic character of $\mathrm{GL}_1(\mathbb{A}_K)$ such that the central character is reciprocal to $\chi|_{\mathbb{A}_F^\times}$.

We refer to the papers of Gross and Vatsal in this volume [5; 10] for the explanation of connections between our formula and his work. There seems to be a lot of rooms left to generalize our formula to the case considered by Waldspurger. In this direction, Hui Xue in his thesis [15] has obtained a formula for the central values for L-series attached to a holomorphic Hilbert modular form of parallel weight $2k$.

This paper is organized as follows. In the first part (Sections 2–7), we will give the basic definitions of forms, L-series, Shimura varieties, CM-points, and state our main formula (Theorem 6.1 and Theorem 7.1) in level N. The definitions here are more or less standard and can be found from our previous work as well as the work of Jacquet, Langlands, Waldspurger, Deligne, Carayol, Gross, and Prasad. Forms have been normalized as *newforms or test vectors* according to the action of unipotent or torus subgroup.

In the second part (Sections 8–10), we will review the original ideas of Gross–Zagier in their *Inventiones* paper [6] on $X_0(N)$ with squarefree D and its generalization to Shimura curves of (N, K)-type in our *Annals* paper [17]. The central idea is to compare the Fourier coefficients of certain *natural* kernel functions of level N with certain *natural* CM-points on Shimura curves $X(N, K)$ of (N, K)-type. This idea only works perfectly when D is squarefree and when $X(N, D)$ has regular integral model but has essential difficulty for the general case.

In the third part (Sections 11–16), we review the basic construction and the proof in our *Asian Journal* paper [16] for formulas in level ND. The kernel function and CM-points we pick are good for computation but have level ND. Their correspondence is given by the local Gross–Zagier formula, which is of course the key of the whole proof. The final formulas involve the notion of *quasi-newforms or toric newforms* as variations of newforms or test vectors.

In the last part (Sections 17–19) which is our new contribution in addition to our previous papers, we will deduce the formula in level N from level ND. The plan of proof is stated in the beginning of Section 17 in three steps. The central idea is to use Eisenstein series to compute certain local constants. This is one more example in number theory of the possibility of solving local questions by a global method, as in the early development of local class field theory and in the current work of Harris and Taylor on the local Langlands conjecture.

The first three parts (Sections 2–18) simply review ideas used in our previous papers. For details one may need to go to the original papers. For an elementary exposition of the Gross–Zagier formula (with variants) and its applications to Birch and Swinnerton–Dyer conjecture, see our paper in *Current Developments in Mathematics* [18].

Notation. The notations of this note are mainly adopted from our *Asian Journal* paper [16] with some simplifications.

1. Let F denote a totally real field of degree g with ring of integers $\mathcal{O}_F$, and adeles $\mathbb{A}$. For each place v of F, let F_v denote the completion of of F at v. When v is finite, let $\mathcal{O}_v$ denote the ring of integers and let π_v denote a uniformizer of $\mathcal{O}_v$. We write $\widehat{\mathcal{O}}_F$ for the product of $\mathcal{O}_v$ in $\mathbb{A}$.

2. Let ψ denote a fixed nontrivial additive character of $F\backslash\mathbb{A}$. For each place v, let ψ_v denote the component of ψ and let $\delta_v \in F_v^\times$ denote the conductor of ψ_v. When v is finite, $\delta_v^{-1}\mathcal{O}_v$ is the maximal fractional ideal of F_v over which ψ_v is trivial. When v is infinite, $\psi_v(x) = e^{2\pi\delta_v x}$. Let δ denote $\prod \delta_v \in \mathbb{A}^\times$. Then the norm $|\delta|^{-1} = d_F$ is the discriminant of F.

3. Let dx denote a Haar measure on $\mathbb{A}$ such that the volume of $F\backslash\mathbb{A}$ is one. This measure has a decomposition $dx = \bigotimes dx_v$ into local measures dx_v on F_v which are self-dual with respect to characters ψ_v. Let $d^\times x$ denote a Haar measure on $\mathbb{A}^\times$ which has a decomposition $d^\times x = \bigotimes d^\times x_v$ such that $d^\times x_v = dx_v/x_v$ on $F_v^\times = \mathbb{R}^\times$ when v is infinite, and such that the volume of $\mathcal{O}_v^\times$ is one when v is finite. Notice that our choice of multiplicative measures is different than that in Tate's thesis, where the volume of $\mathcal{O}_v^\times$ is $|\delta_v|^{1/2}$.

4. Let K denote a totally imaginary quadratic extension of F and T denote the algebraic group $K^\times/F^\times$ over F. We will fix a Haar measure dt and its decomposition $dt = \bigotimes dt_v$ such that $T(F_v)$ has volume 1 when v is infinite.

5. Let B denote a quaternion algebra over F and let G denote the algebraic group $B^\times/F^\times$ over F. We will fix a Haar measure dg on $G(\mathbb{A})$ and a decomposition $dg = \bigotimes dg_v$ such that at an infinite place $G(F_v)$ has volume one if it is

compact, and that when $G(F_v) \simeq \mathrm{PGL}_2(\mathbb{R})$,

$$dg_v = \frac{|dx\,dy|}{2\pi y^2} d\theta,$$

with respect to the decomposition

$$g_v = z \begin{pmatrix} y & x \\ 0 & 1 \end{pmatrix} \begin{pmatrix} \cos\theta & \sin\theta \\ -\sin\theta & \cos\theta \end{pmatrix}.$$

In this way, the volume $|U|$ of the compact open subgroup U of $G(\mathbb{A}_f)$ (or $G(F_v)$ for some $v \nmid \infty$) is well defined. We write $(f_1, f_2)_U$ for the hermitian product

$$(f_1, f_2)_U = |U|^{-1} \int_{G(\mathbb{A})} f_1 \bar{f}_2 \, dg$$

for functions f_1, f_2 on $G(\mathbb{A})$ (or $G(\mathbb{A}_f)$, or $G(F_v)$). This product depends only on the choice of U but not on dg.

2. Automorphic Forms

Let F be a totally real field of degree g, with ring of adeles $\mathbb{A}$, and discriminant d_F. Let ω be a (unitary) character of $F^\times \backslash \mathbb{A}^\times$. By an *automorphic form* on $\mathrm{GL}_2(\mathbb{A})$ with central character ω we mean a continuous function ϕ on $\mathrm{GL}_2(\mathbb{A})$ such that the following properties hold:

- $\phi(z\gamma g) = \omega(z)\phi(g)$ for $z \in Z(\mathbb{A}), \gamma \in \mathrm{GL}_2(F)$;
- ϕ is invariant under right action of some open subgroup of $\mathrm{GL}_2(\mathbb{A}_f)$;
- for a place $v \mid \infty$, ϕ is smooth in $g_v \in \mathrm{GL}_2(F_v)$, and the vector space generated by

  $$\phi(gr_v), \quad r_v \in \mathrm{SO}_2(F_v) \subset \mathrm{GL}_2(\mathbb{A})$$

  is finite dimensional;
- for any compact subset Ω there are positive numbers C, t such that

  $$\left|\phi\left(\begin{pmatrix} a & 0 \\ 0 & 1 \end{pmatrix} g\right)\right| \le C(|a| + |a^{-1}|)^t$$

  for all $g \in \Omega$.

Let $\mathcal{A}(\omega)$ denote the space of automorphic forms with central character ω. Then $\mathcal{A}(\omega)$ admits an *admissible representation* ρ by $\mathrm{GL}_2(\mathbb{A})$. This is a combination of a representation ρ_f of $\mathrm{GL}_2(\mathbb{A}_f)$ via right action:

$$\rho_f(h)\phi(g) = \phi(gh), \qquad h \in \mathrm{GL}_2(\mathbb{A}_f),\ \phi \in \mathbb{A}(\omega),\ g \in \mathrm{GL}_2(\mathbb{A}),$$

and an action ρ_∞ by pairs

$$(M_2(F_v), O_2(F_v)), \qquad v \mid \infty.$$

Here the action of $O_2(F_v)$ is the same as above while the action of $M_2(F_v)$ is given by

$$\rho_\infty(x)\phi(g) = \frac{d\phi}{dt}(ge^{tx})|_{t=0}, \qquad x \in M_2(F_v),\ \phi \in \mathcal{A}(\omega),\ g \in \mathrm{GL}_2(\mathbb{A}),\ v \mid \infty.$$

An admissible and irreducible representation Π of $\mathrm{GL}_2(\mathbb{A})$ is called *automorphic* if it is isomorphic to a sub-representation of $\mathcal{A}(\omega)$. It is well-known that the multiplicity of any irreducible representation in $\mathcal{A}(\omega)$ is at most 1. Moreover, if we decompose such a representation into local representations $\Pi = \bigotimes \Pi_v$ then *the strong multiplicity one* says that Π is determined by all but finitely many Π_v.

Fix an additive character ψ on $F\backslash\mathbb{A}$. Then any automorphic form will have a Fourier expansion:

$$\phi(g) = C_\phi(g) + \sum_{\alpha \in F^\times} W_\phi\left(\begin{pmatrix} \alpha & 0 \\ 0 & 1 \end{pmatrix} g\right), \tag{2.1}$$

where C_ϕ is the constant term:

$$C_\phi(g) := \int_{F\backslash\mathbb{A}} \phi\left(\begin{pmatrix} 1 & x \\ 0 & 1 \end{pmatrix} g\right) dx, \tag{2.2}$$

and $W_\phi(g)$ is the Whittaker function:

$$W_\phi(g) = \int_{F\backslash\mathbb{A}} \phi\left(\begin{pmatrix} 1 & x \\ 0 & 1 \end{pmatrix} g\right) \psi(-x)\, dx. \tag{2.3}$$

It is not difficult to show that a form with vanishing Whittaker function will have the form $\alpha(\det g)$ where α is a function on $F^\times\backslash\mathbb{A}^\times$. Every automorphic representation of dimension 1 appears in this space and corresponds to a characters μ of $F^\times\backslash\mathbb{A}^\times$ such that $\mu^2 = \omega$.

We say that an automorphic form ϕ is *cuspidal* if the constant term $C_\phi(g) = 0$. The space of cuspidal forms is denoted by $\mathcal{A}_0(\omega)$. We call an automorphic representation *cuspidal* if it appears in $\mathcal{A}_0(\omega)$.

An irreducible automorphic representation which is neither one dimensional nor cuspidal must be isomorphic to the space $\Pi(\mu_1, \mu_2)$ of Eisenstein series associated to two quasi characters μ_1, μ_2 of $F^\times\backslash\mathbb{A}^\times$ such that $\mu_1\mu_2 = \omega$. To construct an Eisenstein series, let Φ be a Schwartz–Bruhat function on $\mathbb{A}^2$. For s a complex number, define

$$f_\Phi(s, g) := \mu_1(\det g)|\det g|^{s+1/2} \int_{\mathbb{A}^\times} \Phi[(0,t)g]\mu_1\mu_2^{-1}(t)|t|^{1+2s} d^\times t. \tag{2.4}$$

Then $f_\Phi(s, g)$ belongs to the space $\mathcal{B}(\mu_1 \cdot |\cdot|^s, \mu_2 \cdot |\cdot|^{-s})$ of functions on $\mathrm{GL}_2(\mathbb{A})$ satisfying

$$f_\Phi\left(s, \begin{pmatrix} a & x \\ 0 & b \end{pmatrix} g\right) = \mu_1(a)\mu_2(b)\left|\frac{a}{b}\right|^{1/2+s} f_\Phi(s, g). \tag{2.5}$$

The Eisenstein series $E(s,g,\Phi)$ is defined as follows:

$$E(s,g,\Phi) = \sum_{\gamma\in P(F)\backslash \mathrm{GL}_2(F)} f_\Phi(s,\gamma g). \tag{2.6}$$

One can show that $E(s,g,\Phi)$ is absolutely convergent when $\mathrm{Re}\, s$ is sufficiently large, and has a meromorphic continuation to the whole complex plane. The so-defined meromorphic function $E(s,g,\Phi)$ has at most simple poles with constant residue. The space $\Pi(\mu_1,\mu_2)$ consists of the Eisenstein series

$$E(g,\Phi) := \lim_{s\longrightarrow 0}\left(E(s,g,\Phi) - (\text{residue})s^{-1}\right). \tag{2.7}$$

3. Weights and Levels

Let N be an ideal of $\mathcal{O}_F$ and let $U_0(N)$ and $U_1(N)$ be the following subgroups of $\mathrm{GL}_2(\mathbb{A}_f)$:

$$U_0(N) := \left\{\begin{pmatrix} a & b\\ c & d\end{pmatrix} \in \mathrm{GL}_2(\widehat{\mathcal{O}}_F) : c\equiv 0 \pmod N\right\}, \tag{3.1}$$

$$U_1(N) := \left\{\begin{pmatrix} a & b\\ c & d\end{pmatrix} \in U_0(N) : \quad d\equiv 1 \pmod N\right\}. \tag{3.2}$$

For each infinite place v of F, let k_v be an integer such that $\omega_v(-1) = (-1)^{k_v}$.

An automorphic form $\phi \in \mathcal{A}(\omega)$ is said to have *level* N *and weight* $k = (k_v : v \mid \infty)$ if the following conditions are satisfied:

- $\phi(gu) = \phi(g)$ for $u\in U_1(N)$;
- for a place $v\mid\infty$,

$$\phi(gr_v(\theta)) = \phi(g)e^{2\pi i k_v\theta}$$

  where $r_v(\theta)$ is an element in $\mathrm{SO}_2(F_v)\subset \mathrm{GL}_2(\mathbb{A})$ of the form

$$r_v(\theta) = \begin{pmatrix} \cos\theta & \sin\theta\\ -\sin\theta & \cos\theta\end{pmatrix}.$$

Let $\mathcal{A}_k(N,\omega)$ denote the space of forms of weight k, level N, and central character ω. For any level $N'\mid N$ of N and weight $k'\le k$ by which we mean that $k-k'$ has nonnegative components, we may define embeddings

$$\mathcal{A}_{k'}(N',\omega) \longrightarrow \mathcal{A}_k(N,\omega)$$

by applying some of the operators

$$\phi \mapsto \rho_v\begin{pmatrix} \pi_v^{-1} & 0\\ 0 & 1\end{pmatrix}\phi \qquad (v\nmid\infty),$$

$$\phi \mapsto \rho_v\begin{pmatrix} 1 & i\\ i & -1\end{pmatrix}\phi \qquad (v\mid\infty).$$

The first operator increases level by order 1 at a finite place v; while the second operator increases weight by 2 at an infinite place v. Let $\mathcal{A}_k^{\mathrm{old}}(N,\omega)$ denote the subspace of forms obtained from lower level N' or lower weight k' by applying *at least one of the above operators.*

For any ideal a prime to N, the *Hecke operator* T_a on $\mathcal{A}_k(N,\omega)$ is defined as follows:

$$\mathrm{T}_a\phi(g) = \sum_{\substack{\alpha\beta=a \\ x \mod \alpha}} \phi\left(g\begin{pmatrix}\alpha & x\\ 0 & \beta\end{pmatrix}\right) \tag{3.3}$$

where α and β runs through representatives of integral ideles modulo $\widehat{\mathcal{O}}_F^\times$ with trivial component at the place dividing N such that $\alpha\beta$ generates a. One has for the Whittaker function the formula

$$W_\phi\left(g\begin{pmatrix}a\delta^{-1} & 0\\ 0 & 1\end{pmatrix}\right) = |a|W_{\mathrm{T}_a\phi}(g), \tag{3.4}$$

where $g \in \mathrm{GL}_2(\mathbb{A})$ with component 1 at places $v \nmid N\cdot\infty$.

We say that ϕ is an *eigenform* if for any ideal a prime to N, ϕ is an eigenform under the Hecke operator T_a. We say an eigenform ϕ is *new* if all $k_v \geq 0$, and if there is no old eigenform with the same eigenvalues as ϕ. One can show that two new eigenforms are proportional if and only if they share the same eigenvalues for all but finitely many T_v.

For $\phi \in \mathcal{A}_k(N,\omega)$, let's write $\Pi(\phi)$ for the space of forms in

$$\mathcal{A}(\omega) = \cup_{k,N}\mathcal{A}_k(N,\omega)$$

generated by ϕ by right action of $\mathrm{GL}_2(\mathbb{A})$. Then one can show that $\Pi(\phi)$ is irreducible if and only if ϕ is an eigenform. Conversely, any irreducible representation Π of $\mathrm{GL}_2(\mathbb{A})$ in $\mathcal{A}(\omega)$ contains a unique line of new eigenform. An eigenform ϕ with $\dim\Pi(\phi) < \infty$ will have vanishing Whittaker function and is a multiple of a character.

It can be shown that an eigen newform ϕ with $\dim\Pi(\phi) = \infty$ will have Whittaker function nonvanishing and decomposable:

$$W_\phi(g) = \bigotimes W_v(g_v) \tag{3.5}$$

where $W_v(g_v)$ at finite places can be normalized such that

$$W_v\begin{pmatrix}\delta_v^{-1} & 0\\ 0 & 1\end{pmatrix} = 1. \tag{3.6}$$

Each local component Π_v is realized in the subspace

$$\mathcal{W}(\Pi_v,\psi_v) = \Pi(W_v)$$

generated by W_v under the right action of $\mathrm{GL}_2(F_v)$ (or $(M_2(F_v), O_2(F_v))$ when v is infinite.)

4. Automorphic L-Series

For an automorphic form ϕ, we define its L-series by

$$\begin{aligned} L(s,\phi) :&= d_F^{1/2-s}\int_{F^\times\backslash\mathbb{A}^\times}(\phi - C_\phi)\begin{pmatrix} a & 0\\ 0 & 1\end{pmatrix}|a|^{s-1/2}d^\times a \\ &= d_F^{1/2-s}\int_{\mathbb{A}^\times}W_\phi\begin{pmatrix} a & 0\\ 0 & 1\end{pmatrix}|a|^{s-1/2}d^\times a, \end{aligned} \tag{4.1}$$

which is absolutely convergent for $\operatorname{Re} s \gg 0$ and has a meromorphic continuation to the entire complex plane, and satisfies a functional equation.

Assume that ϕ is an eigen newform. Then its Whittaker function is decomposable. The L-series $L(s,\phi)$ is then an Euler product

$$L(s,\phi) = \prod_v L_v(s,\phi) \tag{4.2}$$

where

$$L_v(s,\phi) = |\delta_v|^{s-1/2}\int_{F_v^\times}W_v\begin{pmatrix} a & 0\\ 0 & 1\end{pmatrix}|a|^{s-1/2}d^\times a. \tag{4.3}$$

For a finite place v, the L-factor has the usual expression:

$$L_v(s,\phi) = \begin{cases} (1-\lambda_v|\pi_v|^s + \omega(\pi_v)|\pi_v|^{2s})^{-1}, & \text{if } v \nmid N,\\ (1-\lambda_v|\pi_v|^s)^{-1}, & \text{if } v \mid N, \end{cases} \tag{4.4}$$

where $\lambda_v \in \mathbb{C}$ is such that $\lambda_v|\pi_v|^{-1/2}$ is the eigenvalue of T_v if $v \nmid N$.

For an archimedean place v, the local factor $L_v(s,\phi)$ is a certain product of gamma functions and is determined by analytic properties of ϕ at v. For the purposes of this paper, we will only consider newforms that at an infinite place are either *holomorphic* or *even of weight* 0, that is, invariant under $O_2(F_v)$ rather than $\mathrm{SO}_2(F_v)$. More precisely, at an infinite place v let's consider the function on $\mathcal{H}\times \mathrm{GL}_2(\mathbb{A}^v)$ defined by

$$f(z,g^v) := |y|^{-(k_v+w_v)/2}\phi\left(\begin{pmatrix} y & x\\ 0 & 1\end{pmatrix}, g^v\right), \quad z = x+yi, \tag{4.5}$$

where $w_v = 0$ or 1 is such that $\omega_v(-1) = (-1)^{w_v}$. Then we require that $f(z,g^v)$ is holomorphic in z if $k_v \geq 1$, and that $f(z,g^v) = f(-\bar{z},g^v)$ if $k_v = 0$. If ϕ is of weight 0, then ϕ is an eigenform for the Laplacian

$$\Delta = -y^2\left(\frac{\partial^2}{\partial x^2}+\frac{\partial^2}{\partial y^2}\right). \tag{4.6}$$

We write eigenvalues as $\frac{1}{4}+t_v^2$ and call t_v the *parameter* of ϕ at v. Let's define *the standard Whittaker function* at archimedean places v of weight k_v in the following way: if $k_v > 0$,

$$W_v\begin{pmatrix} a & 0\\ 0 & 1\end{pmatrix} = \begin{cases} 2a^{(w_v+k_v)/2}e^{-2\pi a} & \text{if } a > 0,\\ 0 & \text{if } a < 0, \end{cases} \tag{4.7}$$

and if $k_v = 0$,

$$W_v\begin{pmatrix} a & 0 \\ 0 & 1 \end{pmatrix} = |a|^{1/2}\int_0^\infty e^{-\pi|a|(y+y^{-1})}y^{it_v}d^\times y. \tag{4.8}$$

In this manner (up to a constant $c \neq 0$) ϕ will have a Whittaker function decomposable as in (3.5) with local function W_v normalized as in (3.6), (4.7), (4.8). We say that ϕ is a *newform* if $c = 1$. Equivalently, ϕ is a newform if and only if $L_v(s,\phi)$ has decomposition (4.2) with local factors given by (4.4) when $v \nmid \infty$, and the following when $v \mid \infty$:

$$L_v(s,\phi) = \begin{cases} G_2(s+k_v+w_v), & \text{if } k_v > 0, \\ G_1(s+it_v)G_1(s-it_v), & \text{if } k_v = 0, \end{cases} \tag{4.9}$$

where

$$\begin{aligned} G_1(s) &= \pi^{-s/2}\Gamma(s/2), \\ G_2(s) &= 2(2\pi)^{-s}\Gamma(s) = G_1(s)G_1(s+1). \end{aligned} \tag{4.10}$$

If Π is an automorphic representation generated by a newform ϕ, we write $L(s,\Pi)$ and $L(s,\Pi_v)$ for $L(s,\phi)$ and $L_v(s,\phi)$, respectively.

5. Rankin–Selberg L-Series

Let K be a totally imaginary quadratic extension of F, and let ω be the nontrivial quadratic character of $\mathbb{A}^\times/F^\times \mathrm{N}\mathbb{A}_K^\times$. The conductor $c(\omega)$ is the relative discriminant of K/F. Let χ be a character of finite order of $\mathbb{A}_K^\times/K^\times\mathbb{A}^\times$. The conductor $c(\chi)$ is an ideal of $\mathcal{O}_F$ which is maximal such that χ is factorized through

$$\mathbb{A}_K^\times/K^\times\mathbb{A}^\times\widehat{\mathcal{O}}_{c(\chi)}^\times K_\infty^\times = \mathrm{Gal}(H_{c(\chi)}/K),$$

where $\mathcal{O}_c = \mathcal{O}_F + c(\chi)\mathcal{O}_K$ and H_c is the ring class filed of conductor $c(\chi)$. We define the ideal $D = c(\chi)^2c(\omega)$, and call χ a *ring class character of conductor* $c(\chi)$.

Let ϕ be a newform with trivial central character and of level N. The Rankin–Selberg convolution L-function $L(s,\chi,\phi)$ is defined by an Euler product over primes v of F:

$$L(s,\chi,\phi) := \prod_v L_v(s,\chi,\phi) \tag{5.1}$$

where the factors have degree ≤ 4 in $|\pi_v|^s$. This function has an analytic continuation to the entire complex plane, and satisfies a functional equation. We will assume that the ideals $c(\omega)$, $c(\chi)$, N are coprime each other. Then the local factors can be defined explicitly as follows.

For v a finite place, let's write

$$\begin{aligned} L_v(s,\phi) &= (1-\alpha_1|\pi_v|^s)^{-1}(1-\alpha_2|\pi_v|^s)^{-1}, \\ \prod_{w|v} L(s,\chi_w) &= (1-\beta_1|\pi_v|^s)^{-1}(1-\beta_2|\pi_v|^s)^{-1}. \end{aligned}$$

Then

$$L_v(s,\chi,\phi) = \prod_{i,j}(1-\alpha_i\beta_j|\pi_v|^s)^{-1}. \tag{5.2}$$

Here, for a place w of K, the local factor $L(s,\chi_w)$ is defined by

$$L(s,\chi_w) = \begin{cases} (1-\chi(\pi_w)|\pi_w|^s)^{-1}, & \text{if } w \nmid c(\chi)\cdot\infty, \\ G_2(s), & \text{if } v \mid \infty, \\ 1, & \text{if } v \mid c(\chi). \end{cases} \tag{5.3}$$

At an infinite place v, using formula $G_2(s) = G_1(s)G_1(s+1)$ we may write

$$L_v(s,\phi) = G_1(s+\sigma_1)G_1(s+\sigma_2),$$
$$L_v(s,\chi) = G_1(s+\tau_1)G_1(s+\tau_2).$$

Then the L-factor $L_v(s,\chi,\phi)$ is defined by

$$\begin{aligned} L_v(s,\chi,\phi) &= \prod_{i,j} G_1(s+\sigma_i+\tau_j) \\ &= \begin{cases} G_2(s+\frac{1}{2}(k_v-1))^2, & \text{if } k_v \geq 2, \\ G_2(s+it_v)G_2(s-it_v), & \text{if } k_v = 0, \end{cases} \end{aligned} \tag{5.4}$$

where t_v is the parameter associated to ϕ at a place v where the weight is 0.

The functional equation is then

$$L(1-s,\chi,\phi) = (-1)^{\#\Sigma}\mathrm{N}_{F/\mathbb{Q}}(ND)^{1-2s}L(s,\chi,\phi), \tag{5.5}$$

where $\Sigma = \Sigma(N,K)$ is the following set of places of F:

$$\Sigma(N,K) = \left\{ v \,\middle|\, \begin{array}{l} v \text{ is infinite and } \phi \text{ has weight } k_v > 0 \text{ at } v, \\ \text{or } v \text{ is finite and } \omega_v(N) = -1. \end{array} \right\} \tag{5.6}$$

6. The Odd Case

Now we assume that all $k_v = 2$ and that the sign of the functional equation (5.5) is -1, so $\#\Sigma$ is odd. Our main formula expresses the central derivative $L'(\frac{1}{2},\chi,\phi)$ in terms of the heights of CM-points on a Shimura curve. Let τ be any real place of F, and let B be the quaternion algebra over F ramified exactly at the places in $\Sigma - \{\tau\}$. Let G be the algebraic group over F, which is an inner form of PGL_2, and has $G(F) = B^\times/F^\times$.

The group $G(F_\tau) \simeq \mathrm{PGL}_2(\mathbb{R})$ acts on $\mathcal{H}^\pm = \mathbb{C} - \mathbb{R}$. If $U \subset G(\mathbb{A}_f)$ is open and compact, we get an analytic space

$$M_U(\mathbb{C}) = G(F)_+\backslash\mathcal{H} \times G(\mathbb{A}_f)/U \tag{6.1}$$

where $G(F)_+$ denote the subgroup of elements of $G(F)$ with totally positive determinants. Shimura proved these were the complex points of an algebraic curve M_U, which descends canonically to F (embedded in $\mathbb{C}$, by the place τ). The curve M_U over F is independent of the choice of τ in Σ.

To specify M_U, we must define $U \subset G(\mathbb{A}_f)$. To do this, we fix an embedding $K \longrightarrow B$, which exists, as all places in Σ are either inert or ramified in K. One can show that there is an order R of B containing $\mathcal{O}_K$ with reduced discriminant N. For an explicit description of such an order, we fix a maximal ideal $\mathcal{O}_B$ of B containing $\mathcal{O}_K$ and an ideal $\mathcal{N}$ of $\mathcal{O}_K$ such that

$$\mathrm{N}_{K/F}\mathcal{N} \cdot \mathrm{disc}_{B/F} = N, \tag{6.2}$$

where $\mathrm{disc}_{B/F}$ is the reduced discriminant of $\mathcal{O}_B$ over $\mathcal{O}_F$. Then we take

$$R = \mathcal{O}_K + \mathcal{N} \cdot \mathcal{O}_B. \tag{6.3}$$

We call R *an order of* (N, K)*-type*. Define an open compact subgroup U_v of $G(F_v)$ by

$$U_v = R_v^\times / \mathcal{O}_v^\times. \tag{6.4}$$

Let $U = \prod_v U_v$. This defines the curve M_U up to F-isomorphism. Let X be its compactification over F, so $X = M_U$ unless $F = \mathbb{Q}$ and $\Sigma = \{\infty\}$, where X is obtained by adding many cusps. We call X *a Shimura curve of* (N, K)*-type*. We write $R(N, K)$, $U(N, K)$, $X(N, K)$ when types need to be specified.

We will now construct points in $\mathrm{Jac}(X)$, the connected component of $\mathrm{Pic}(X)$, from CM-points on the curve X. The CM-points corresponding to K on $M_U(\mathbb{C})$ form a set

$$G(F)_+ \backslash G(F)_+ \cdot h_0 \times G(\mathbb{A}_f)/U = T(F) \backslash G(\mathbb{A}_f)/U, \tag{6.5}$$

where $h_0 \in \mathcal{H}$ is the unique fixed point of the torus points $T(F) = K^\times / F^\times$. Let P_c denote a point in X represented by (h_0, i_c) where $i_c \in G(\mathbb{A}_f)$ such that

$$U_T := i_c U i_c^{-1} \cap T(\mathbb{A}_f) \simeq \widehat{\mathcal{O}}_c^\times / \widehat{\mathcal{O}}_F^\times. \tag{6.6}$$

By Shimura's theory, P_c is defined over the ring class field H_c of conductor c corresponding to the Artin map

$$\mathrm{Gal}(H_c/K) \simeq T(F) \backslash T(\mathbb{A}_f)/T(F_\infty)U_T.$$

Let P_χ be a divisor on X with complex coefficients defined by

$$P_\chi = \sum_{\sigma \in \mathrm{Gal}(H_c/K)} \chi^{-1}(\sigma)[P_c^\sigma]. \tag{6.7}$$

If χ is not of form $\chi = \nu \cdot \mathrm{N}_{K/F}$ with ν a quadratic character of $F^\times \backslash \mathbb{A}^\times$, then P_χ has degree 0 on each connected component of X. Thus P_χ defines a class x in $\mathrm{Jac}(X) \otimes \mathbb{C}$. Otherwise we need a reference divisor to send P_χ to $\mathrm{Jac}(X)$. In the modular curve case, one uses cusps. In the general case, we use the *Hodge class* $\xi \in \mathrm{Pic}(X) \otimes \mathbb{Q}$: the unique class whose degree is 1 on each connected component and such that

$$\mathrm{T}_m \xi = \deg(\mathrm{T}_m)\xi$$

for all integral nonzero ideals m of $\mathcal{O}_F$ prime to ND. The Heegner class we want now is the class difference

$$x := [P_\chi - \deg(P_\chi)\xi] \in \mathrm{Jac}(X)(H_c) \otimes \mathbb{C}, \tag{6.8}$$

where $\deg(P_\chi)$ is the multi-degree of P_χ on geometric components.

Notice that the curve X and its Jacobian have an action by the ring of good Hecke operators. Thus x is a sum of eigen vectors of the Hecke operators.

THEOREM 6.1. *Let x_ϕ denote the ϕ-typical component of x. Then*

$$L'(\tfrac{1}{2}, \chi, \phi) = \frac{2^{g+1}}{\sqrt{\mathrm{N}(D)}} \cdot \|\phi\|^2 \cdot \|x_\phi\|^2,$$

where $\|\phi\|^2$ is computed using the invariant measure on

$$\mathrm{PGL}_2(F)\backslash\mathcal{H}^g \times \mathrm{PGL}_2(\mathbb{A}_f)/U_0(N)$$

induced by $dx\,dy/y^2$ on $\mathcal{H}$, and $\|x_\phi\|^2$ is the Néron–Tate pairing of x_ϕ with itself.

To see the application to the Birch and Swinnerton-Dyer conjecture, we just notice that x_ϕ actually lives in a unique abelian subvariety A_ϕ of the Jacobian $\mathrm{Jac}(X)$ such that

$$L(s, A_\phi) = \prod_{\sigma:\mathbb{Z}[\phi]\to\mathbb{C}} L(s, \phi^\sigma). \tag{6.9}$$

Y. Tian [14] has recently generalized the work of Kolyvagin and Bertolini–Darmon to our setting and showed that the rank conjecture of Birch and Swinnerton-Dyer for A in the case $\mathrm{ord}_{s=1/2}L(s, \chi, \phi) \le 1$.

Notice that $\|\phi\|^2$ is not exactly the periods of A_ϕ appearing in the Birch and Swinnerton-Dyer conjecture, but it has an expression in L-series:

$$\|\phi\|^2 = 2\mathrm{N}(N) \cdot d_F \cdot L(1, \mathrm{Sym}^2\phi) \tag{6.10}$$

where $L(s, \mathrm{Sym}^2\phi)$ is the L-series defined by an Euler product with local factors $L_v(s, \mathrm{Sym}^2\phi)$ given by

$$L_v(s, \mathrm{Sym}^2\phi) = G_2(s + \tfrac{1}{2})^2 G_1(s)^{-1}, \tag{6.11}$$

if $v \mid \infty$, and by

$$L_v(s, \mathrm{Sym}^2\phi) = (1 - \alpha^2|\pi_v|^s)^{-1}(1 - \beta^2|\pi_v|^s)^{-1}(1 - \alpha\beta|\pi_v|^s)^{-1}, \tag{6.12}$$

if $v \nmid \infty$, and α and β are given as follows:

$$L_v(s, \phi) = (1 - \alpha|\pi_v|^s)^{-1}(1 - \beta|\pi_v|^s)^{-1}.$$

It will be an interesting question to see how this relates the periods in A_ϕ.

7. The Even Case

We now return to the case where ϕ has possible nonholomorphic components, but we assume that all weights be either 0 or 2 and that the sign of the functional equation of $L(s,\chi,\phi)$ is $+1$, or equivalently, Σ is even. In this case, we have an explicit formula for $L(\frac{1}{2},\chi,\phi)$ in terms of CM-points on locally symmetric varieties covered by $\mathcal{H}^n$ where n is the number of real places of F where ϕ has weight 0.

More precisely, let B be the quaternion algebra over F ramified at Σ, and G the algebraic group associated to $B^\times/F^\times$. Then

$$G(F\otimes\mathbb{R})\simeq \mathrm{PGL}_2(\mathbb{R})^n\times \mathrm{SO}_3^{g-n} \tag{7.1}$$

acts on $(\mathcal{H}^\pm)^n$. The locally symmetric variety we will consider is

$$M_U=G(F)_+\backslash\mathcal{H}^n\times G(\mathbb{A}_f)/U, \tag{7.2}$$

where $U=\prod U_v$ was defined in the previous section. Again we call M_U or its compactification X a *quaternion Shimura variety of* (N,K)*-type*. We will also have a CM-point P_c and a CM-cycle P_χ defined as in (6.6) and (6.7) but with $\mathrm{Gal}(H_c/K)$ replaced by $T(F)\backslash T(\mathbb{A}_f)/T(F_\infty)U_T$.

By results of Waldspurger, Tunnel, and Gross–Prasad [17, Theorem 3.2.2], there is a unique line of cuspidal functions $\widetilde{\phi}$ on M_U such that for each finite place v not dividing $N\cdot D$, $\widetilde{\phi}$ is the eigenform for Hecke operators T_v with the same eigenvalues as ϕ. We call any such a form a *test form* of (N,K)-type.

THEOREM 7.1. *Let $\widetilde{\phi}$ be a test form of norm 1 with respect to the measure on X induced by $dx\,dy/y^2$ on $\mathcal{H}$. Then*

$$L(\tfrac{1}{2},\chi,\phi)=\frac{2^{g+n}}{\sqrt{\mathrm{N}(D)}}\cdot\|\phi\|^2\cdot|(\widetilde{\phi},P_\chi)|^2.$$

Here

$$(\widetilde{\phi},P_\chi)=\sum_{t\in T(F)\backslash T(\mathbb{A}_f)/U_T}\chi^{-1}(t)\widetilde{\phi}(tP_c).$$

Remark. There is a naive analogue between even and odd cases via Hodge theory which is actually a starting point to believe that there will be a simultaneous proof for both cases. To see this, let's consider the space $Z(\Omega^1_X)$ of closed smooth 1-forms on a Shimura curve X of (N,K)-type with hermitian product defined by

$$(\alpha,\beta)=\frac{i}{2}\int\alpha\bar{\beta}.$$

The Hodge theory gives a decomposition of this space into a direct sum

$$Z(\Omega^1_X)=\left(\bigoplus\nolimits_\alpha\mathbb{C}\alpha\right)\oplus(\text{continuous spectrum})$$

where α runs through eigenforms under the Hecke operators and the Laplacian. Each α is either holomorphic, anti-holomorphic or exact. In either case, α corresponds to a test form $\widetilde{\phi}$ of weight 2, -2, or 0 on X, in the sense that

$$\alpha = \begin{cases} \widetilde{\phi}\, dz, & \text{if } \alpha \text{ is holomorphic,} \\ \widetilde{\phi}\, d\bar{z}, & \text{if } \alpha \text{ is anti-holomorphic,} \\ d\widetilde{\phi}, & \text{if } \alpha \text{ is exact.} \end{cases}$$

We may take integration $c \mapsto \int_{\widetilde{c}} \alpha$ to define a map

$$\pi_\alpha : \mathrm{Div}^0(X) \otimes \mathbb{C} \longrightarrow \begin{cases} A_\phi \otimes \mathbb{C}, & \text{if } \alpha \text{ is holomorphic,} \\ \mathbb{C}, & \text{if } \alpha \text{ is exact.} \end{cases}$$

Here $\widetilde{c}$ is an 1-cycle on X with boundary c. In this manner, we have

$$\pi_\alpha(P_{\bar{\chi}}) = \begin{cases} z_\phi, & \text{if } \alpha \text{ is holomorphic,} \\ (\phi, P_\chi), & \text{if } \alpha \text{ is exact.} \end{cases}$$

Thus we can think of $\mathbb{C}$ as an abelian variety corresponding to ϕ in the even case with Néron–Tate heights given by absolute value. This gives a complete analogue of the right-hand side of the Gross–Zagier formulas in the even and odd case.

On the other hand, in the even case, one can define L-series $L(s, \chi, \partial\phi/\partial z_v)$ by (4.1) where v is the only archimedean place where $k_v = 0$. It is not difficult to see that this L-series is essentially $(s - \frac{1}{2})L(s, \chi, \phi)$. Its derivative is given by $L(\frac{1}{2}, \chi, \phi)$. Thus we have an analogue of the left-hand side as well!

8. The Idea of Gross and Zagier

We now describe the original idea of Gross and Zagier in the proof of a central derivative formula (for Heegner points on $X_0(N)/\mathbb{Q}$ with squarefree discriminant D) in their famous *Inventiones* paper. For simplicity, we fix N, χ and assume that $\Sigma(N, K)$ is odd. For a holomorphic form ϕ of weight 2, we define its *Fourier coefficient* $\widehat{\phi}(a)$ *at an integral idele* a by the equation

$$W_\phi \begin{pmatrix} a y_\infty \delta^{-1} & 0 \\ 0 & 1 \end{pmatrix} = \widehat{\phi}(a) W_\infty \begin{pmatrix} y_\infty & 0 \\ 0 & 1 \end{pmatrix}, \tag{8.1}$$

where $W_\infty = \prod_{v \nmid \infty} W_v$ is the standard Whittaker function for weight 2 defined in (4.7).

With the notation of Section 6, there is a cusp form Ψ of level N whose Fourier coefficient is given by

$$\widehat{\Psi}(a) = |a| \langle x, \mathrm{T}_a x \rangle. \tag{8.2}$$

This follows from two facts:

- the subalgebra $\mathbb{T}'$ of $\mathrm{Jac}(X)\otimes\mathbb{C}$ generated by Hecke operators T_a is a quotient of the subalgebra $\mathbb{T}$ in $\mathrm{End}(S_2(N))$ generated by Hecke operators T_a. Here $S_2(N)$ is the space of holomorphic cusp forms of weight $(2,\dots,2)$, level N, with trivial central character;
- any linear functional ℓ of $\mathbb{T}$ is represented by a cusp form $f \in S_2(N)$ in the sense that $|a|\ell(\mathrm{T}_a) = \widehat{f}(a)$.

(This form Ψ is not unique in general. But it is if we can normalize it to be a sum of newforms.)

It is then easy to see that

$$(\phi, \Psi) = \langle x_\phi, x_\phi\rangle(\phi,\phi). \tag{8.3}$$

Here (ϕ, Ψ) denotes the inner product as in Theorem 6.1, which is the same as $(\phi, \Psi)_{U_0(N)}$ in our notations in Introduction.

Thus, the question is reduced to showing that

$$L'(\tfrac{1}{2}, \chi, \phi) = \frac{2^{g+1}}{\sqrt{\mathrm{N}(D)}}(\phi, \Psi). \tag{8.4}$$

On the other hand, one can express $L(s, \chi, \phi)$ using a method of Rankin and Selberg:

$$L(s,\chi,\phi) = \frac{d_F^{1/2-s}}{|U_0(ND)|}\int_{\mathrm{PGL}_2(F)\backslash\mathrm{PGL}_2(\mathbb{A})} \phi(g)\theta(g)E(s,g)\,dg. \tag{8.5}$$

We need to explain various term in this integration.

First of all, θ is a theta series associated to χ. More precisely, θ is an eigen form of weight $(-1,\dots,-1)$, level D, and central character ω such that its local Whittaker functions $W_v(g)$ produces the local L-functions for χ:

$$|\delta_v|^{s-1/2}\int_{F_v^\times} W_v\begin{pmatrix} -a & 0\\ 0 & 1\end{pmatrix}|a|^{s-1/2}d^\times a = \prod_{w|v} L(s,\chi_w). \tag{8.6}$$

Here $L(s, \chi_w)$ is defined in (5.3). It follows that the automorphic representation $\Pi(\chi) := \Pi(\theta)$ generated by θ is irreducible with newform $\theta_\chi(g) = \theta(g\varepsilon)$, where $\varepsilon = \left(\begin{smallmatrix} -1 & \\ -0 & 1\end{smallmatrix}\right)$.

Secondly, $E(s,g) = E(s,g,\mathcal{F})$ is an Eisenstein series (2.6) for the quasi-characters $|\cdot|^{s-1/2}$ and $|\cdot|^{1/2-s}\omega$ with decomposition $\mathcal{F} = \bigotimes \mathcal{F}_v \in \mathcal{S}(\mathbb{A}^2)$. If v is a finite place, then

$$\mathcal{F}_v(x,y) = \begin{cases} 1, & \text{if } v\nmid c(\omega),\ |x|\le |ND|_v,\ |y|\le 1,\\ \omega_v^{-1}(y), & \text{if } v\mid c(\omega),\ |x|\le |ND|_v,\ |y| = 1,\\ 0, & \text{otherwise.}\end{cases} \tag{8.7}$$

If v is an infinite place, then

$$\mathcal{F}_v(x,y) = (\pm ix + y)e^{-\pi(x^2+y^2)} \tag{8.8}$$

where we take the + sign if $k_v = 2$ and the − sign if $k_v = 0$.

Taking a trace, we obtain a form of level N:

$$\Phi_s(g) = \mathrm{tr}_D \Phi_s(g) = \sum_{\gamma \in U_0(D)/U_0(ND)} \overline{d_F^{1/2-s} \theta(g\gamma) E(s, g\gamma)}. \tag{8.9}$$

This form has the property:

$$L(s, \chi, \phi) = (\phi, \Phi_s)_{U_0(N)}. \tag{8.10}$$

The idea of Gross and Zagier (in the odd case) is to compute the derivative $\Phi'_{1/2}$ of Φ_s with respect to s at $s = \frac{1}{2}$ and take a holomorphic projection to obtain a holomorphic form Φ so that

$$L'(\tfrac{1}{2}, \chi, \phi) = (\phi, \Phi)_{U_0(N)}. \tag{8.11}$$

(See [4] for a direct construction of the kernel using Poincaré series instead of the Rankin–Selberg method and holomorphic projection.) Now the problem is reduced to proving that

$$\Phi - \frac{2^{g+1}}{\sqrt{\mathrm{N}(D)}} \Psi$$

is an old form. In other words, we need to show that the Fourier coefficients of Φ are given by height pairings of Heegner points on $\mathrm{Jac}(X)$:

$$\widehat{\Phi}(a) = |a| \langle x, \mathrm{T}_a x \rangle \tag{8.12}$$

for any finite integral ideles a prime to ND.

One expects to prove the above equality by explicit computations for both sides respectively. These computations have been successfully carried out by Gross and Zagier [6] when $F = \mathbb{Q}$, D is squarefree, and $X(N, K) = X_0(N)$. The computation of Fourier coefficients of Φ is essentially straightforward and has been carried out for totally real fields [17]. For the computation of $\langle x, \mathrm{T}_a x \rangle$, Gross and Zagier represented the Hodge class ξ by cusps 0 and ∞ on $X_0(N)$:

$$\langle x, \mathrm{T}_a x \rangle = \big\langle P_\chi - h_\chi 0,\, \mathrm{T}_a(P_\chi) - h_{\chi,a}\infty \big\rangle \tag{8.13}$$

where h_χ and $h_{\chi,a}$ are integers that make both divisors to have degree 0. The right-hand side can be further decomposed into local height pairings by deforming self-intersections using Dedekind η-functions. These local height pairings can be finally computed by a modular interpretation in terms of deformation of formal groups.

When F is arbitrary (even when D is squarefree), the computation of heights has a lot of problems as there is no canonical representatives for the Hodge class, and no canonical modular form for self-intersections. In Sections 9–10, we will see how Arakelov theory been used to compute the heights.

When D is arbitrary (even when $F = \mathbb{Q}$), the computations of both kernels and heights for Theorem 6.1 seem impossible to carry out directly because of *singularities* in both analysis and geometry. Alternatively, we will actually prove a Gross–Zagier formula for level ND (Sections 11–16) and try to reduce the level by using continuous spectrum (Sections 17–19).

9. Calculus on Arithmetic Surfaces

The new idea in our *Annals* paper [17] is to use Arakelov theory to decompose the heights of Heegner points as locally as possible, and to show that the contribution of those terms that we don't know how to compute is *negligible.*

Let F be a number field. By an arithmetic surface over $\mathrm{Spec}\mathcal{O}_F$, we mean a projective and flat morphism $\mathcal{X} \longrightarrow \mathrm{Spec}\mathcal{O}_F$ such that that $\mathcal{X}$ is a regular scheme of dimension 2. Let $\widehat{\mathrm{Div}}(\mathcal{X})$ denote the group of *arithmetic divisors* on $\mathcal{X}$. Recall that an arithmetic divisor on $\mathcal{X}$ is a pair $\widehat{D} := (D, g)$ where D is a divisor on $\mathcal{X}$ and g is a function on

$$X(\mathbb{C}) = \coprod X_\tau(\mathbb{C})$$

with some logarithmic singularities on $|D|$. The form $-(\partial\bar{\partial}/\pi i)g$ on $X(\mathbb{C}) - |D|$ can be extended to a smooth form $c_1(\widehat{D})$ on $X(\mathbb{C})$ which is called the *curvature* of the divisor $\widehat{D}$. If f is a nonzero rational function on $\mathcal{X}$ then we can define the corresponding *principal arithmetic divisor* by

$$\widehat{\mathrm{div}} f = (\mathrm{div} f, -\log|f|). \tag{9.1}$$

An arithmetic divisor (D, g) is called *vertical* if D is supported in the special fibers, and *horizontal* if D does not have component supported in the special fiber).

The group of arithmetic divisors is denoted by $\widehat{\mathrm{Div}}(\mathcal{X})$ while the subgroup of principal divisor is denoted by $\widehat{\mathrm{Pr}}(\mathcal{X})$. The quotient $\widehat{\mathrm{Cl}}(\mathcal{X})$ of these two groups is called the *arithmetic divisor class group* which is actually isomorphic to the group $\widehat{\mathrm{Pic}}(\mathcal{X})$ of hermitian line bundles on $\mathcal{X}$. Recall that a hermitian line bundle on $\mathcal{X}$ is a pair $\bar{\mathcal{L}} = (\mathcal{L}, \|\cdot\|)$, where $\mathcal{L}$ is a line bundle on $\mathcal{X}$ and $\|\cdot\|$ is hermitian metric on $\mathcal{L}(\mathbb{C})$ over $X(\mathbb{C})$. For a rational section ℓ of $\mathcal{L}$, we can define the corresponding divisor by

$$\widehat{\mathrm{div}}(\ell) = (\mathrm{div}\ell, -\log\|\ell\|). \tag{9.2}$$

It is easy to see that the divisor class of $\widehat{\mathrm{div}}(\ell)$ does not depend on the choice of ℓ. Thus one has a well defined map from $\widehat{\mathrm{Pic}}(\mathcal{X})$ to $\widehat{\mathrm{Cl}}(\mathcal{X})$. This map is actually an isomorphism.

Let $\widehat{D}_i = (D_i, g_i)$ $(i = 1, 2)$ be two arithmetic divisors on $\mathcal{X}$ with disjoint support in the generic fiber:

$$|D_{1F}| \cap |D_{2F}| = \varnothing.$$

Then one can define an *arithmetic intersection pairing*

$$\widehat{D}_1 \cdot \widehat{D}_2 = \sum_v (\widehat{D}_1 \cdot \widehat{D}_2)_v, \tag{9.3}$$

where v runs through the set of places of F. The intersection pairing only depends on the divisor class. It follows that we have a well defined pairing on $\widehat{\mathrm{Pic}}(\mathcal{X})$:

$$(\bar{\mathcal{L}}, \overline{\mathcal{M}}) \longrightarrow \widehat{c}_1(\bar{\mathcal{L}}) \cdot \widehat{c}_1(\overline{\mathcal{M}}) \in \mathbb{R}. \tag{9.4}$$

Let $V(\mathcal{X})$ be the group of *vertical metrized line bundles*: namely $\bar{\mathcal{L}} \in \widehat{\mathrm{Pic}}(\mathcal{X})$ with $\mathcal{L} \simeq \mathcal{O}_X$. Then we have an exact sequence

$$0 \longrightarrow V(\mathcal{X}) \longrightarrow \widehat{\mathrm{Pic}}(\mathcal{X}) \longrightarrow \mathrm{Pic}(\mathcal{X}_F) \longrightarrow 0.$$

Define the group of *flat* bundles $\widehat{\mathrm{Pic}}^0(\mathcal{X})$ as the orthogonal complement of $V(\mathcal{X})$. Then we have an exact sequence

$$0 \longrightarrow \widehat{\mathrm{Pic}}(\mathcal{O}_F) \longrightarrow \widehat{\mathrm{Pic}}^0(\mathcal{X}) \longrightarrow \mathrm{Pic}^0(X_F) \longrightarrow 0.$$

The following formula, called the *Hodge index theorem*, gives a relation between intersection pairing and height pairing: for $\bar{\mathcal{L}}, \overline{\mathcal{M}} \in \widehat{\mathrm{Pic}}^0(\mathcal{X})$,

$$\langle \mathcal{L}_F, \mathcal{M}_F \rangle = -\widehat{c}_1(\bar{\mathcal{L}}) \cdot \widehat{c}_1(\overline{\mathcal{M}}), \tag{9.5}$$

where the left-hand side denotes the Néron–Tate height pairing on $\mathrm{Pic}^0(X) = \mathrm{Jac}(X)(F)$.

For X a curve over F, let $\widehat{\mathrm{Pic}}(X)$ denote the direct limit of $\widehat{\mathrm{Pic}}(\mathcal{X})$ over all models over X. Then the intersection pairing can be extended to $\widehat{\mathrm{Pic}}(X)$. Let $\bar{F}$ be an algebraic closure of F and let $\widehat{\mathrm{Pic}}(X_{\bar{F}})$ be the direct limit of $\mathrm{Pic}(X_L)$ for all finite extensions L of F, then the intersection pairing on $\widehat{\mathrm{Pic}}(X_L)$ times $[L : F]^{-1}$ can be extended to an intersection pairing on $\widehat{\mathrm{Pic}}(X_{\bar{F}})$.

Let $\bar{\mathcal{L}} \in \widehat{\mathrm{Pic}}(\mathcal{X})_{\mathbb{Q}}$ be a fixed class with degree 1 at the generic fiber. Let $x \in X(F)$ be a rational point and let $\bar{x}$ be the corresponding section $\mathcal{X}(\mathcal{O}_F)$. Then $\bar{x}$ can be extended to a unique element $\widehat{x} = (x+D, g)$ in $\widehat{\mathrm{Div}}(\mathcal{X})_{\mathbb{Q}}$ satisfying the following conditions:

- the bundle $\mathcal{O}(\widehat{x}) \otimes \bar{\mathcal{L}}^{-1}$ is flat;
- for any finite place v of F, the component D_v of D on the special fiber of $\mathcal{X}$ over v satisfies
$$D_v \cdot c_1(\mathcal{L}) = 0;$$
- for any infinite place v,
$$\int_{X_v(\mathbb{C})} g c_1(\bar{\mathcal{L}}) = 0.$$

We define now *Green's function* $g_v(x, y)$ on $(X(F) \times X(F) - \text{diagonal})$ by

$$g_v(x, y) = (\widehat{x} \cdot \widehat{y})_v / \log q_v, \tag{9.6}$$

where $\log q_v = 1$ or 2 if v is real or complex. It is easy to see that $g_v(x, y)$ is symmetric, does not depend on the model $\mathcal{X}$ of X, and is stable under base change. Thus we have a well-defined Green's function on $X(\bar{F})$ for each place v of F.

10. Decomposition of Heights

We now want to apply the general theory of the previous section to intersections of CM-points to Shimura curves $X = X(N,K)$ over a totally real field F as defined in Section 6. Recall that X has the form

$$X = G(F)_+\backslash\mathcal{H} \times G(\mathbb{A}_f)/U(N,K) \cup \{\text{cusps}\} \tag{10.1}$$

which is a smooth and projective curve over F but may not be connected.

To define Green's function we need to extend the Hodge class ξ in $\mathrm{Pic}(X)_{\mathbb{Q}}$ to a class in $\widehat{\mathrm{Pic}}(X) \otimes \mathbb{Q}$. Notice that $\xi \in \mathrm{Pic}(X)_{\mathbb{Q}}$ is *Eisenstein* under the action of Hecke operators:

$$\mathrm{T}_a\xi = \sigma_1(a) \cdot \xi, \qquad \sigma_1(a) := \deg \mathrm{T}_a = \sum_{b|a} \mathrm{N}(b), \tag{10.2}$$

for any integral idele a prime to the level of X.

It is an interesting question to construct a class $\widehat{\xi}$ to extend ξ such that the above equation holds for $\widehat{\xi}$. But in [17], Corollary 4.3.3, we have constructed an extension $\widehat{\xi}$ of ξ such that

$$\mathrm{T}_a\widehat{\xi} = \sigma_1(a)\widehat{\xi} + \phi(a) \tag{10.3}$$

where $\phi(a) \in \widehat{\mathrm{Pic}}(F)$ is a σ_1-*derivation*, i.e., for any coprime a', a''

$$\phi(a'a'') = \sigma(a')\phi(a'') + \sigma(a'')\phi(a').$$

We have the following general definition.

Definition 10.1. Let $\mathbb{N}_F$ denote the semigroup of nonzero ideals of $\mathcal{O}_F$. For each $a \in \mathbb{N}_F$, let $|a|$ denote the *inverse norm* of a:

$$|a|^{-1} = \#\mathcal{O}_F/a.$$

For a fixed ideal M, let $\mathbb{N}_F(M)$ denote the sub-semigroup of ideals prime to M.

A function f on $\mathbb{N}_F(M)$ is called *quasi-multiplicative* if

$$f(a_1a_2) = f(a_1) \cdot f(a_2)$$

for all coprime $a_1, a_2 \in \mathbb{N}_F(M)$. For a quasi-multiplicative function f, let $\mathcal{D}(f)$ denote the set of all f-*derivations*, that is the set of all linear combinations

$$g = cf + h,$$

where c is a constant and h satisfies

$$h(a_1a_2) = h(a_1)f(a_2) + h(a_2)f(a_1)$$

for all $a_1, a_2 \in \mathbb{N}_F(M)$ with $(a_1, a_2) = 1$.

For a representation Π, the Fourier coefficient $\widehat{\Pi}(a)$ is defined to be

$$\widehat{\Pi}(a) := W_{\Pi,f}\begin{pmatrix} a\delta^{-1} & 0 \\ 0 & 1 \end{pmatrix},$$

where $W_{\Pi,f}$ is the product of Whittaker newvectors at finite places. In other words, $\widehat{\Pi}(a)$ is defined such that the finite part of L-series has expansion

$$L_f(s,\Pi) = \sum \widehat{\Pi}(a)|a|^{s-1/2}.$$

Then $\widehat{\Pi}(a)$ is quasi-multiplicative.

We can now define Green's functions g_v on divisors on $X(\bar{F})$ which are disjoint at the generic fiber for each place v of F. Let's try to decompose the heights of our Heegner points. The linear functional

$$a \longrightarrow |a|\langle x, \mathrm{T}_a x\rangle$$

is now the Fourier coefficient of a cuspform Ψ of weight 2:

$$\widehat{\Psi}(a) = |a|\langle x, \mathrm{T}_a x\rangle. \tag{10.4}$$

In the following we want to express this height in terms of intersections modulo some Eisenstein series and theta series.

Let $\widehat{P}_\chi$ be the arithmetic closure of P_χ with respect to $\widehat{\xi}$. Then the Hodge index formula (9.5) gives

$$\begin{aligned} |a|\langle x, \mathrm{T}_a x\rangle &= -|a|\big(\widehat{P}_\chi - \deg(P_\chi)\widehat{\xi},\, \mathrm{T}_a\widehat{P}_\chi - \deg(\mathrm{T}_a\widehat{P}_\chi)\widehat{\xi}\big) \\ &= -|a|(\widehat{P}_\chi, \mathrm{T}_a\widehat{P}_\chi) + \widehat{E}(a), \end{aligned} \tag{10.5}$$

where $\widehat{E}$ is a certain derivation of Eisenstein series.

The divisor P_χ and $\mathrm{T}_a P_\chi$ have some common components. We want to compute its contribution in the intersections. Let $r_\chi(a)$ denote the Fourier coefficients of the theta series associated to χ:

$$r_\chi(a) = \sum_{b|a} \chi(b). \tag{10.6}$$

Then the divisor

$$\mathrm{T}_a^0 P_\chi := \mathrm{T}_a P_\chi - r_\chi(a) P_\chi \tag{10.7}$$

is disjoint with P_χ.

It follows that $\widehat{\Psi}(a)$ is essentially given by a sum of local intersections

$$-\frac{1}{[L:F]} \sum_v \sum_{\iota\in \mathrm{Gal}(H_c/F)} g_v(\mathrm{T}_a^0 P_\chi^\iota, P_\chi^\iota)|a| \log q_v$$

modulo some derivations of Eisenstein series, and theta series of weight 1. We can further simplify this sum by using the fact that the Galois action of $\mathrm{Gal}(K^{\mathrm{ab}}/F)$ is given by the composition of the class field theory map

$$\nu : \mathrm{Gal}(K^{\mathrm{ab}}/F) \longrightarrow N_T(F)\backslash N_T(\mathbb{A}_f),$$

and the left multiplication of the group $N_T(\mathbb{A}_f)$. Finally we obtain:

$$\widehat{\Psi}(a) = -|a|\sum_v g_v(P_\chi, \mathrm{T}_a^0 P_\chi)\log q_v \pmod{\mathcal{D}(\sigma_1)+\mathcal{D}(r_\chi)}. \tag{10.8}$$

Assume that D is squarefree, and that $X(N,K)$ has a regular canonical integral model over $\mathcal{O}_K$, which is the case when $\mathrm{ord}_v(N) = 1$ if v is not split in K or $v \mid 2$. We can use the theory of Gross on canonical or quasi-canonical lifting to compute $g_v(P_\chi, \mathrm{T}_a^0 P_\chi)$ and to prove that the functional

$$\widehat{\Phi} - \frac{2^{g+1}}{\sqrt{\mathrm{N}(D)}}\widehat{\Psi}$$

vanishes modulo derivations of Eisenstein series or theta series. It then follows that this functional is actually zero by the following lemma:

LEMMA 10.2. *Let $f_1, \dots, f_r$ be quasi-multiplicative functions on $\mathbb{N}_F(ND)$, and assume they are all distinct. Then the sum*

$$\mathcal{D}(f_1) + \mathcal{D}(f_2) + \cdots + \mathcal{D}(f_r)$$

is a direct sum.

Thus we can prove the Gross–Zagier formula in the case D is squarefree and $X(N,K)$ has regular canonical model over $\mathcal{O}_K$. This is the main result in our *Annals* paper [17].

11. Construction of the Kernels

From this section to the end, we want to explain how to prove the Gross–Zagier formula of Sections 6 and 7 for the general case. We will start with a kernel construction. As explained earlier, there is no good construction of kernels in level N. The best we can do is to construct some *nice kernel* in level ND in the sense that the Fourier coefficients are *symmetric* and easy to compute, and that the projection of this form in $\Pi(\phi)$ is *recognizable.*

Recall from (8.5) that we have an integral expression of the Rankin–Selberg convolution:

$$L(s,\chi,\phi) = \frac{|\delta|^{s-1/2}}{|U_0(ND)|}\int \phi(g)\theta(g)E(s,g)\,dg \tag{11.1}$$

To obtain a more symmetric kernel we have to apply Atkin–Lehner operators to $\theta(g)E(s,g)$. Let S be the set of finite places ramified in K. For each such set T of S, let h_T be an element in $\mathrm{GL}_2(\mathbb{A})$ which has component 1 outside T and has component

$$\begin{pmatrix} 0 & 1 \\ -t_v & 0 \end{pmatrix},$$

where t_v has the same order as $c(\omega_v)$ and such that $\omega_v(t_v) = 1$. Now one can show that

$$L(s, \chi, \phi) = \frac{\gamma_T(s)}{|U_0(ND)|} \int \phi(g)\theta(gh_T^{-1}\varepsilon)E(s, gh_T^{-1})\, dg \tag{11.2}$$

where γ_T is a certain exponential function of s. Finally we define the kernel function

$$\Theta(s, g) = 2^{-|S|} \sum_{T \subset S} \gamma_T(s)\theta(gh_T^{-1})E(s, gh_T^{-1}).$$

By our construction,

$$L(s, \chi, \phi) = (\phi, \bar{\Theta}(s, -))_{U_0(ND)}.$$

Now the functional equation of $L(s, \chi, \phi)$ follows from the following equation of the kernel function which can be proved by a careful analysis of Atkin–Lehner operators:

$$\Theta(s, g) = \varepsilon(s, \chi, \phi)\Theta(1 - s, g). \tag{11.3}$$

Assume that ϕ is cuspidal, then we may define the projection of $\bar{\Theta}$ in $\Pi(\phi)$ as a form $\varphi \in \Pi(\phi)$ such that

$$\int f\Theta\, dg = \int f\bar{\varphi}\, dg, \qquad \text{for all } f \in \Pi.$$

Since the kernel $\Theta(s, g)$ constructed above has level ND, its projection onto Π will have level DN and thus is a linear combination of the forms

$$\phi_a := \rho\begin{pmatrix} a^{-1} & 0 \\ 0 & 1 \end{pmatrix}\phi, \qquad (a \mid D). \tag{11.4}$$

PROPOSITION 11.1. *The projection of $\bar{\Theta}(s, g)$ on Π is given by*

$$\frac{L(s, \chi, \phi)}{(\phi_s^\sharp, \phi_s^\sharp)_{U_0(ND)}} \cdot \phi_s^\sharp,$$

where $\phi_s^\sharp$ is the unique nonzero form in the space of $\Pi(\phi)$ of level ND satisfying the identities

$$(\phi_s^\sharp, \phi_a) = \nu^*(a)_s(\phi_s^\sharp, \phi_s^\sharp) \qquad (a \mid D),$$

where

$$\nu^*(a)_s = \prod_{v|S} \frac{|a|_v^{s-1/2} + |a|_v^{1/2-s}}{2} \begin{cases} \nu(a), & \text{if } a|c(\omega), \\ 0, & \text{otherwise.} \end{cases}$$

Write $\phi^\sharp = \phi_{1/2}^\sharp$ and call it the *quasi-newform* with respect to χ.

The function $\Theta(s, g)$ has a Fourier expansion

$$\Theta(s, g) = C(s, g) + \sum_{\alpha \in F^\times} W\left(s, \begin{pmatrix} \alpha & 0 \\ 0 & 1 \end{pmatrix} g\right). \tag{11.5}$$

Since $\Theta(s,g)$ is a linear combination of the form

$$\Theta(s,g)=\sum_i \theta_i(g)E_i(g)$$

with $\theta_i\in\Pi(\chi)$ and $E_i(g)\in\Pi(|\cdot|^{s-1/2},|\cdot|^{1/2-s}\omega)$, the constant and Whittaker function of $\Theta(s,g)$ can be expressed precisely in terms of Fourier expansions of θ_i and $E_i(g)$.

More precisely, let

$$\theta_i(g)=\sum_{\xi\in F} W_{\theta_i}(\xi,g), \qquad E_i(g)=\sum_{\xi\in F} W_{E_i}(\xi,g), \tag{11.6}$$

be Fourier expansions of θ_i and E_i respectively into characters $\begin{pmatrix}1 & x\\ 0 & 1\end{pmatrix}\mapsto\psi(\xi x)$ on $N(\mathbb{A})$. Then

$$C(s,g)=\sum_{\xi\in F} C(s,\xi,g), \tag{11.7}$$

$$W(s,g)=\sum_{\xi\in F} W(s,\xi,g), \tag{11.8}$$

where

$$C(s,\xi,g)=\sum_i W_{\theta_i}(-\xi,g)\,W_{E_i}(\xi,g), \tag{11.9}$$

$$W(s,\xi,g)=\sum_i W_{\theta_i}(1-\xi,g)\,W_{E_i}(\xi,g). \tag{11.10}$$

The behavior of the degenerate term $C(s,\xi,g)$ can be understood very well. The computation shows that the complex conjugation of $\Theta(s,g)$ is finite at each cusp unless χ is a form $\nu\circ\mathrm{N}_{K/F}$ in which case, we need to remove two Eisenstein series in the space

$$E_1\in\Pi(\|\cdot\|^s,\|\cdot\|^{-s})\otimes\nu, \qquad E_2\in\Pi(\|\cdot\|^{1-s},\|\cdot\|^{s-1})\otimes\nu\omega.$$

We let $\Phi(s,g)$ denote $\bar\Theta(\frac12,g)$ if χ is not of form $\nu\circ\mathrm{N}_{K/F}$, or

$$\bar\Theta(s,g)-E_1-E_2$$

if it is. Then $\Phi(s,g)$ is a form with following growth:

$$\Phi\left(s,\begin{pmatrix}a & 0\\ 0 & 1\end{pmatrix}\right)=c_1(g)|a|^{s-1/2}+c_2(g)|a|^{1/2-s}+O_g(e^{-\varepsilon|a|})$$

where $c_1(g)$, $c_2(g)$, and O_g term are all smooth functions of g and s. It follows that the value or all derivatives of $\Phi(s,g)$ at $s=\frac12$ are L^2-forms.

With our very definition of $\Theta(s,g)$ in the last section, we are able to decompose the nondegenerate term:

$$W(s,\xi,g)=\bigotimes W_v(s,\xi_v,g_v). \tag{11.11}$$

An explicit computation gives the local functional equation

$$W_v(s,\xi_v,g_v) = \omega_v(1-\xi_v^{-1})(-1)^{\#\Sigma\cap\{v\}} W_v(1-s,\xi_v,g). \tag{11.12}$$

If Σ is even, we can compute the Fourier coefficients of $\Theta(\frac{1}{2},g)$ for

$$g = \begin{pmatrix} a\delta^{-1} & 0 \\ 0 & 1 \end{pmatrix} \tag{11.13}$$

very explicitly. The computation of the nondegenerate term $W(s,\xi,g)$ is reduced to local terms $W_v(s,\xi,g)$. By the functional equation, we need only consider those ξ such that

$$1-\xi^{-1} \in N(K_v^\times) \iff v \notin \Sigma.$$

The form Φ is holomorphic of weight 2 at infinite places where Π is of weight 2.

If Σ is odd, $\Theta(\frac{1}{2},g)$ vanishes by the functional equation. We want to compute its derivative $\Theta'(\frac{1}{2},g)$ at $s=\frac{1}{2}$. Let's now describe the central derivative for $W(s,\xi,g)$ for g of the form (11.13). Recall that $W(s,\xi,g)$ is a product of $W_v(s,\xi,g)$, and that $W_v(s,\xi,g)$ satisfies the functional equation (11.12). It follows that

$$W'(\tfrac{1}{2},g) = \sum_v W'(\tfrac{1}{2},g)_v, \tag{11.14}$$

where v runs through the places which are not split in K with

$$W'(\tfrac{1}{2},g)_v = \sum_\xi W^v(\tfrac{1}{2},\xi,g^v)\cdot W_v'(\tfrac{1}{2},\xi,g). \tag{11.15}$$

Here W^v is the product of W_ℓ over places $\ell \neq v$, W_v' is the derivative for the variable s, and $\xi \in F-\{0,1\}$ satisfies

$$1-\xi^{-1} \in N(K_w^\times) \iff w \notin {}_v\Sigma, \tag{11.16}$$

with ${}_v\Sigma$ given by

$${}_v\Sigma = \begin{cases} \Sigma\cup\{v\}, & \text{if } v\notin\Sigma, \\ \Sigma-\{v\}, & \text{if } v\in\Sigma. \end{cases}$$

All these terms can be computed explicitly. We need to find the holomorphic projection of $\bar\Theta'(\frac{1}{2},g)$. That is a holomorphic cusp form Φ of weight 2 such that $\bar\Theta'(\frac{1}{2},g)-\Phi$ is perpendicular to any holomorphic form.

PROPOSITION 11.2. *With respect to the standard Whittaker function for holomorphic weight 2 forms, the a-th Fourier coefficients $\widehat\Phi(a)$ of the holomorphic projection Φ of $\bar\Theta'(\frac{1}{2},g)$ is a sum*

$$\widehat\Phi(a) = A(a) + B(a) + \sum_v \widehat\Phi_v(a)$$

where

$$A \in \mathcal{D}(\widehat\Pi(\chi)\otimes\alpha^{1/2}),$$
$$B \in \mathcal{D}(\widehat\Pi(\alpha^{1/2}\nu,\alpha^{-1/2}\nu)) + \mathcal{D}(\widehat\Pi(\alpha^{1/2}\nu\omega,\alpha^{-1/2}\nu\omega)),$$

and the sum is over places of F which are not split in K, with $\widehat{\Phi}_v(a)$ given by the following formulas:

(i) *If v is a finite place, then $\widehat{\Phi}_v(a)$ is a sum over $\xi \in F$ with $0 < \xi < 1$ of the following terms:*

$$(2i)^g|(1-\xi)\xi|_\infty^{1/2} \cdot \bar{W}_f^v\left(\tfrac{1}{2}, \xi, \begin{pmatrix} a\delta_f^{-1} & 0 \\ 0 & 1 \end{pmatrix}\right) \cdot \bar{W}_v'\left(\tfrac{1}{2}, \xi, \begin{pmatrix} a\delta_f^{-1} & 0 \\ 0 & 1 \end{pmatrix}\right).$$

(ii) *If v is an infinite place, then $\widehat{\Phi}_v(a)$ is the constant term at $s = 0$ of a sum over $\xi \in F$ such that $0 < \xi_w < 1$ for all infinite places $w \neq v$ and $\xi_v < 0$ of the following terms:*

$$(2i)^g|\xi(1-\xi)|_\infty^{1/2} \cdot \bar{W}_f\left(\tfrac{1}{2}, \xi, \begin{pmatrix} a\delta_f^{-1} & 0 \\ 0 & 1 \end{pmatrix}\right) \cdot \int_1^\infty \frac{-dx}{x(1+|\xi|_v x)^{1+s}}.$$

12. Geometric Pairing

The key to prove the Gross–Zagier formula is to compare the Fourier coefficients of the kernel functions and the local heights of CM-points. These local heights are naturally grouped by definite quaternion algebras which are the endomorphism rings of the supersingular points in the reductions of modular or Shimura curves. Furthermore, the intersection of two CM-points at supersingular points is given by a *multiplicity function* depending only the *relative position* of these two CM-points. In this section, we would like to abstractly define this kind of pairing with respect to an arbitrary multiplicity function. We will describe the relative position of two CM-points by a certain parameter ξ which will relate the same parameter in the last section by a *local Gross–Zagier formula.*

Let G be an inner form of PGL_2 over F. This means that $G = B^\times/F^\times$ with B a quaternion algebra over F. Let K be a totally imaginary quadratic extension of F which is embedded into B. Let T denote the subgroup of G given by $K^\times/F^\times$. Then the set

$$C := T(F)\backslash G(\mathbb{A}_f) \tag{12.1}$$

is called the *set of CM-points.* This set admits a natural action by $T(\mathbb{A}_f)$ (resp. $G(\mathbb{A}_f)$) by left (resp. right) multiplication.

As in Sections 6 and 7, there is a map

$$\iota : C \longrightarrow M := G(F)_+\backslash\mathcal{H}^n \times G(\mathbb{A}_f) \tag{12.2}$$

from C to the Shimura variety defined by G sending the class of $g \in G(\mathbb{A}_f)$ to the class of (h_0, g), where $h_0 \in \mathcal{H}^n$ is fixed by T. This map is an embedding if G is not totally definite.

The set of CM-points has a topology induced from $G(\mathbb{A}_f)$ and has a unique $G(\mathbb{A}_f)$-invariant measure dx induced from the one on $G(\mathbb{A}_f)$. The space

$$\mathcal{S}(C) = \mathcal{S}(T(F)\backslash G(\mathbb{A}_f))$$

of locally constant functions with compact support is called the space of *CM-cycles* which admits a natural action by $T(\mathbb{A}_f) \times G(\mathbb{A}_f)$. There is a natural pairing between functions f on Shimura variety M and CM-cycles α by

$$(f, \alpha) = \int_C \bar{\alpha}(x) f(\iota x)\, dx. \tag{12.3}$$

Thus CM-cycles may serve as distributions or functionals on the space of functions on M. Of course this pairing is invariant under the action by $G(\mathbb{A}_f)$.

Since $T(F)\backslash T(\mathbb{A}_f)$ is compact, one has a natural decomposition

$$\mathcal{S}(C) = \oplus_\chi \mathcal{S}(\chi, C)$$

where the sum is over the characters of $T(F)\backslash T(\mathbb{A}_f)$. There is also a local decomposition for each character χ:

$$\mathcal{S}(\chi, C) = \bigotimes_v \mathcal{S}(\chi_v, G(F_v)). \tag{12.4}$$

In the following we will define a class of pairings on CM-cycles which are *geometric* since it appears naturally in the local intersection pairing of CM-points on Shimura curves . To do this, let's write CM-points in a slightly different way,

$$C = G(F)\backslash(G(F)/T(F)) \times G(\mathbb{A}_f), \tag{12.5}$$

then the topology and measure of C is still induced by those of $G(\mathbb{A}_f)$ and the *discrete* ones of $G(F)/T(F)$.

Let m be a *real valued* function on $G(F)$ which is $T(F)$-invariant and such that $m(\gamma) = m(\gamma^{-1})$. Then m can be extended to $G(F)/T(F) \times G(\mathbb{A}_f)$ such that

$$m(\gamma, g_f) = \begin{cases} m(\gamma), & \text{if } g_f = 1, \\ 0, & \text{otherwise.} \end{cases} \tag{12.6}$$

We now have a kernel function

$$k(x, y) = \sum_{\gamma \in G(F)} m(x^{-1}\gamma y) \tag{12.7}$$

on $C \times C$. Then we can define a pairing on $\mathcal{S}(C)$ by

$$\langle \alpha, \beta \rangle = \int_{C^2} \alpha(x) k(x, y) \bar{\beta}(y)\, dx\, dy := \lim_{U \to 1} \int_{C^2} \alpha(x) k_U(x, y) \bar{\beta}(y)\, dx\, dy, \tag{12.8}$$

where U runs through the open subgroup of $G(\mathbb{A}_f)$ and

$$k_U(x, y) = \mathrm{vol}(U)^{-2} \int_{U^2} k(xu, yv)\, du\, dv.$$

This pairing is called a *geometric pairing with multiplicity function m*. For two function α and β in $\mathcal{S}(T(F)\backslash G(\mathbb{A}_f))$, one has

$$\langle \alpha, \beta \rangle = \sum_{\gamma \in T(F)\backslash G(F)/T(F)} m(\gamma) \langle \alpha, \beta \rangle_\gamma \tag{12.9}$$

where

$$\langle \alpha, \beta \rangle_\gamma = \int_{T_\gamma(F)\backslash G(\mathbb{A}_f)} \alpha(\gamma y)\bar{\beta}(y)\, dy, \tag{12.10}$$

and where

$$T_\gamma := \gamma^{-1}T\gamma \cap T = \begin{cases} T & \text{if } \gamma \in N_T, \\ 1 & \text{otherwise,} \end{cases} \tag{12.11}$$

and where N_T is the normalizer of T in G. The integral $\langle \alpha, \beta \rangle_\gamma$ is called the *linking number* of α and β at γ.

Since both α and β are invariant under left-translation by $T(F)$, the linking number at γ depends only on the class of γ in $T(F)\backslash G(F)/T(F)$. Let's define a parameterization of this set by by writing $B = K + K\varepsilon$ where $\varepsilon \in B$ is an element such that $\varepsilon^2 \in F^\times$ and $\varepsilon x = \bar{x}\varepsilon$. Then the function

$$\xi(a + b\varepsilon) = \frac{\mathrm{N}(b\varepsilon)}{\mathrm{N}(a + b\varepsilon)}$$

defines an embedding

$$\xi : T(F)\backslash G(F)/T(F) \longrightarrow F. \tag{12.12}$$

Write

$$\langle \alpha, \beta \rangle_\gamma = \langle \alpha, \beta \rangle_\xi. \tag{12.13}$$

Notice that $\xi(\gamma) = 0$ (resp. 1) if and only if $\xi \in T$ (resp. $\xi \in N_T - T$). The image of $G(F) - N_T$ is the set of $\xi \in F$ such that $\xi \neq 0, 1$ and where for any place v of F,

$$1 - \xi^{-1} \in \begin{cases} \mathrm{N}(K_v^\times), & \text{if } B_v \text{ is split,} \\ F_v^\times - \mathrm{N}(K_v^\times) & \text{if } B_v \text{ is not split.} \end{cases} \tag{12.14}$$

We may write $m(\xi)$ for $m(\gamma)$ when $\xi(\gamma) = \xi$, and extend $m(\xi)$ to all F by setting $m(\xi) = 0$ if ξ is not in the image of the map in (12.12). Then

$$\langle \alpha, \beta \rangle = \sum_{\xi \in F} m(\xi) \langle \alpha, \beta \rangle_\xi. \tag{12.15}$$

Let χ be a character of $T(F)\backslash T(\mathbb{A}_f)$. The linking number is easy to compute if $\xi = 0$ or 1. The difficult problem is to compute $\langle \alpha, \beta \rangle_\xi$ when $\xi \neq 0, 1$.

If both α and β are decomposable,

$$\alpha = \bigotimes \alpha_v, \qquad \beta = \bigotimes \beta_v,$$

then we have a decomposition of linking numbers into local *linking numbers* when $\xi \neq 0, 1$:

$$\langle \alpha, \beta \rangle_\xi = \prod \langle \alpha_v, \beta_v \rangle_\xi, \tag{12.16}$$

where

$$\langle \alpha_v, \beta_v \rangle_\xi = \int_{G(F_v)} \alpha_v(\gamma y)\bar{\beta}_v(y)\, dy. \tag{12.17}$$

Notice that when $\gamma \notin N_T$, these local linking numbers depend on the choice of γ in its class in $T(F)\backslash G(F)/T(F)$ while their product does not. This problem can be solved by taking γ to be a *trace-free element* in its class which is unique up to conjugation by $T(F)$.

Notation. For a compact open compact subgroup U of $G(\mathbb{A}_f)$ (or $G(F_v)$) and two CM-cycles α and β, we write $\langle \alpha, \beta \rangle_U$ and $\langle \alpha, \beta \rangle_{\xi,U}$ for

$$\langle \alpha, \beta \rangle_U = |U|^{-1} \langle \alpha, \beta \rangle, \qquad \langle \alpha, \beta \rangle_{\xi,U} = |U|^{-1} \langle \alpha, \beta \rangle_\xi.$$

Similarly, for a CM-cycle α and a function f on M, we write

$$(f, \alpha)_U = |U|^{-1}(f, \alpha).$$

13. Local Gross–Zagier Formula

In this section, we would like to compute the linking numbers for some special CM-cycles and then compare with Fourier coefficients of the kernel functions. The construction of CM-cycles is actually quite simple and is given as follows.

We will fix one order A of B such that, for each finite place v,

$$A_v = \mathcal{O}_{K,v} + \mathcal{O}_{K,v}\lambda_v c(\chi_v), \tag{13.1}$$

where $\lambda_v \in B_v^\times$ is such that $\lambda_v x = \bar{x}\lambda_v$ for all $x \in K$, and $\mathrm{ord}(\det \lambda_v) = \mathrm{ord}_v(N)$.

Let Δ be a subgroup of $G(\mathbb{A}_f)$ generated by images of $\widehat{A}^\times$ and $K_v^\times$ for v ramified in K:

$$\Delta = \prod_{v \nmid c(\omega_v)} A_v^\times F_v^\times / F_v^\times \cdot \prod_{v | c(\omega_v)} A_v^\times K_v^\times / F_v^\times. \tag{13.2}$$

The character can be naturally extended to a character of Δ. The CM-cycle we need is defined by the function

$$\eta = \prod \eta_v, \tag{13.3}$$

with η_v supported on $T(F_v) \cdot \Delta_v$ and such that

$$\eta_v(tu) = \chi_v(t)\chi_v(u), \qquad t \in T(F_v), u \in \Delta_v. \tag{13.4}$$

Take $a \in \mathbb{A}_f^\times$ integral and prime to ND. We would like to compute the pairing $\langle \mathrm{T}_a \eta, \eta \rangle$. The Hecke operator here is defined as

$$\mathrm{T}_a \eta = \prod_{v|a} \mathrm{T}_{a_v} \eta_v, \qquad \mathrm{T}_{a_v} \eta_v(x) = \int_{H(a_v)} \eta_v(xg)\, dg, \tag{13.5}$$

where

$$H(a_v) := \{g \in M_2(\mathcal{O}_v) : |\det g| = |a_v|\}, \tag{13.6}$$

and dg is a measure such that $\mathrm{GL}_2(\mathcal{O}_v)$ has volume 1. Then we have the decomposition

$$\langle \mathrm{T}_a\eta,\eta\rangle_\Delta = \mathrm{vol}(T(F)\backslash T(\mathbb{A}_f)\Delta)\left(m(0)\mathrm{T}_a\eta(e)+m(1)\mathrm{T}_a\eta(\varepsilon)\delta_{\chi^2=1}\right) + \sum_{\xi\neq 0,1} m(\xi)\prod_v \langle \mathrm{T}_a\eta_v,\eta_v\rangle_{\xi,\Delta_v}, \quad (13.7)$$

where $\varepsilon \in N_T(F) - T(F)$.

Let v be a fixed finite place of F. We want to compute all terms in the right-hand side involving η_v. Notice that we have extended the definition to all $\xi \in F - \{0,1\}$ by insisting that $\langle \mathrm{T}_a\eta_v,\eta_v\rangle_{\xi,\Delta_v} = 0$ when ξ is not in the image of (12.12).

The computation of degenerate terms is easy. The nondegenerate term is given by the following local Gross–Zagier formula:

Proposition 13.1. *Let* $g = \begin{pmatrix} a\delta_v^{-1} & 0 \\ 0 & 1\end{pmatrix}$. *Then*

$$\bar{W}_v(\tfrac{1}{2},\xi,g) = |c(\omega_v)|^{1/2}\cdot \varepsilon(\omega_v,\psi_v)\chi_v(u)\cdot |(1-\xi)\xi|_v^{1/2}|a|\cdot \langle \mathrm{T}_a\eta_v,\eta_v\rangle_{\xi,\Delta_v},$$

where u is any trace-free element in $K^\times$.

Corollary 13.2. *Let $\langle\cdot,\cdot\rangle$ be the geometric pairing on the CM-cycle with multiplicity function m on F such that $m(\xi)=0$ if ξ is not in the image of (12.12). Assume that $\delta_v = 1$ for $v \mid \infty$. Then there are constants c_1, c_2 such that for an integral idele a prime to ND,*

$$|c(\omega)|^{1/2}|a|\langle \mathrm{T}_a\eta,\eta\rangle_\Delta = (c_1 m(0)+c_1 m(1))|a|^{1/2}W_f(g) + i^{[F:\mathbb{Q}]}\sum_{\xi\in F-\{0,1\}} |\xi(1-\xi)|_\infty^{1/2}\bar{W}_f(\tfrac{1}{2},\xi,g)m(\xi),$$

where $g = \begin{pmatrix} a\delta_f^{-1} & 0 \\ 0 & 1\end{pmatrix}$.

Remarks. If the kernel Θ had level N as in the original approach in Section 8, the CM-cycle we should consider is P_χ as in Section 6 corresponding to the function ζ supported in $T(\mathbb{A}_f)i_cU(N,K)$ such that $\zeta(ti_cu)=\chi(t)$. The computation of linking numbers for this divisor seems very difficult!

Our local formula is the key to the proof of the Gross–Zagier formula. But the formula is only proved under the condition that $c(\chi)$, $c(\omega)$, and N are coprime to each other. One may still expect that this local formula is still true in the general case considered by Waldspurger but with more than one term on the right-hand side. The main problem is to construct elements in

$$\mathcal{S}(\chi_v, G(F_v)) \quad \text{and} \quad \mathcal{W}(\Pi(\chi_v),\psi_v)\otimes\mathcal{W}(\Pi(|\cdot|^{s-1/2},|\cdot|^{1/2-s}\omega).$$

More precisely, we need to find an element

$$W = \sum W_{i1}\otimes W_{2i} \in \mathcal{W}(\Pi(\chi_v),\psi_v)\otimes\mathcal{W}(\Pi(|\cdot|^{s-1/2},|\cdot|^{1/2-s}\omega)$$

satisfying the following properties:

- Let Φ_i be the element in $\mathcal{S}(F_v^2)$ such that $W_{2i} = W_{\Phi_i}$. Then for any representation Π_v of $\mathrm{GL}_2(F_v)$ with a newform $W_v \in \mathcal{W}(\Pi, \psi_v)$,

$$L(s, \chi_v, \phi_v) = \sum \Psi(s, W_v, W_{1i}, \Phi_i)$$

  with notation in [16, § 2.5].
- Let's define

$$W(s, \xi, g) = \sum_i W_{1i}\left(\begin{pmatrix} 1-\xi & 0 \\ 0 & 1 \end{pmatrix} g\right) W_{2i}\left(\begin{pmatrix} \xi & 0 \\ 0 & 1 \end{pmatrix} g\right).$$

  Then $W(s, \xi, g)$ satisfies the following functional equation

$$W(s, \xi, g) = \omega_v(1 - \xi^{-1})\varepsilon_v(\Pi_v \otimes \chi_v, s) W(1 - s, \xi, g).$$

The next step is to find elements $q_j \in \mathcal{S}(\chi_v, G(F_v))$ such that the above local Gross–Zagier formula is true with $\langle \mathrm{T}_a \eta_v, \eta_v \rangle_\xi$ replaced by

$$\sum_j \langle \mathrm{T}_a q_j, q_j \rangle_\xi.$$

We may even assume that $a = 1$ in time. Thus, what really varies is the parameter ξ.

14. Gross–Zagier Formula in Level ND

The study of kernel functions, geometric pairing, and local Gross–Zagier formula in the last three sections suggests that it may be easier to prove a Gross–Zagier formula in level ND instead of level N directly. This is the main result in our *Asian Journal* paper [16].

Let's start with the case where Σ is odd. Thus, we are in the situation of Section 6. Let U be any open compact subgroup of Δ over which χ is trivial. Let X_U be corresponding Shimura curve or its compactification over F.

Recall that the CM-points corresponding to K on $X_U(\mathbb{C})$ form a set

$$C_U := G(F)_+ \backslash G(F)_+ \cdot h_0 \times G(\mathbb{A}_f)/U = T(F) \backslash G(\mathbb{A}_f)/U,$$

where $h_0 \in \mathcal{H}^+$ is the unique fixed point of the torus points $K^\times/F^\times$. Let η_U be a divisor on X_U with complex coefficient defined by

$$\eta_U = \sum_{x \in C_U} \eta(x)[x].$$

The Heegner class we want now is the class difference

$$y := [\eta_U - \deg(\eta_U)\xi] \in \mathrm{Jac}(X_U)(H_c) \otimes \mathbb{C}.$$

Notice that this class has character χ_Δ under the action by Δ on $\mathrm{Jac}(H_c)$. Let y_ϕ denote the ϕ-typical component of y. The main theorem in our *Asian Journal* paper [16] is this:

THEOREM 14.1. *Let $\phi^\sharp$ be the quasi-newform as in Section* 11. *Then*

$$\widehat{\phi^\sharp}(1)L'(1,\chi,\phi) = 2^{g+1}d_{K/F}^{-1/2}\cdot\|\phi^\sharp\|^2_{U_0(ND)}\cdot\|y_\phi\|^2_\Delta,$$

where

- *$d_{K/F}$ is the relative discriminant of K over F;*
- *$\|\phi^\sharp\|^2_{U_0(ND)}$ is the L^2-norm with respect to the Haar measure dg normalized (as in the Introduction) so that* $\mathrm{vol}(U_0(ND)) = 1$;
- *$\|y_\phi\|_\Delta$ is the Néron–Tate height of y_ϕ on X_U times $[\Delta : U]^{-1}$, which is independent of the choice of U;*
- *$\widehat{\phi^\sharp}(1)$ is the first Fourier coefficient of $\phi^\sharp$ as defined in* (8.1).

We now move to the situation in Section 7 where ϕ has possible nonholomorphic components, but we assume that the sign of the functional equation of $L(s,\chi,\phi)$ is $+1$, or equivalently, Σ is even. We have the variety X_U which is defined in the same way as in the odd case. Then we have a unique line of cuspidal functions ϕ_χ on X_U with the following properties:

- ϕ_χ has character χ_Δ under the action of Δ;
- for each finite place v not dividing $N\cdot D$, ϕ_χ is the eigenform for Hecke operators T_v with the same eigenvalues as ϕ.

We call ϕ_χ *a toric newform* associated to ϕ. See [16, § 2.3] for more details.

The CM-points on X_U, associated to the embedding $K \longrightarrow B$, form the infinite set

$$C_U := G(F)_+\backslash G(F)_+h_0\times G(\mathbb{A}_f)/U \simeq H\backslash G(\mathbb{A}_f)/U,$$

where h_0 is a point in $\mathcal{H}^n$ fixed by T and $H\subset G$ is the stabilizer of z in G. Notice that H is either isomorphic to T if $n\neq 0$ or $H = G$ if $n = 0$. In any case there is a finite map

$$\iota : C_U = T(F)\backslash G(\mathbb{A}_f)/U \longrightarrow M_U.$$

The Gross–Zagier formula for central value in level ND is the following:

THEOREM 14.2. *Let ϕ_χ be a toric newform such that $\|\phi_\chi\|_\Delta = 1$. Then*

$$\widehat{\phi^\sharp}(1)L(1,\chi,\phi) = 2^{g+n}d_{K/F}^{-1/2}\cdot\|\phi^\sharp\|^2_{U_0(ND)}\cdot|(\phi,\eta)_\Delta|^2,$$

where $\widehat{\phi^\sharp}(1)$ is the first Fourier coefficient by the same formula as (8.1) *with respect the standard Whittaker function defined in* (4.7) *and* (4.8), *and where*

$$(\phi_\chi,\eta)_\Delta := [\Delta : U]^{-1}\sum_{x\in C_U}\bar\eta(x)\phi_\chi(\iota(x)).$$

15. Green's Functions of Heegner Points

In this section we explain the proof of the central derivative formula for level ND stated in the last section. Just as explained in Section 8, the question is reduced to a comparison of the Fourier coefficients of the kernel and heights of CM-points. We need to show that, up to a constant and modulo some *negligible forms*, the newform Ψ with Fourier coefficient

$$\widehat{\Psi}(a) := |a|\langle \eta, \mathrm{T}_a \eta\rangle \tag{15.1}$$

is equal to the holomorphic cusp form Φ defined in Section 11 which represents the derivative of Rankin L-function $L'(\frac{1}{2}, \chi, \phi)$. Thus we need to show that the functional $\widehat{\Psi}$ on $\mathbb{N}_F(ND)$ is equal to the Fourier coefficient $\widehat{\Phi}(a)$.

As in Section 9, we would like to decompose the height pairing to Green's functions. It is more convenient work on the tower of Shimura curves than a single one. Let's first try to extend the theory of heights to the projective limit X_∞ of X_U. Let $\widehat{\mathrm{Pic}}(X_\infty)$ denote the direct limit of $\widehat{\mathrm{Pic}}(X_U)$ with respect to the pull-back maps. Then the intersection pairing can be extended to $\widehat{\mathrm{Pic}}(X_\infty)$ if we multiply the pairings on $\mathrm{Pic}(X_U)$ by the scale $\mathrm{vol}(U)$. Of course, this pairing depends on the choice of measure dg on $G(\mathbb{A})$ as in Introduction. For some fixed open compact subgroup U of $G(\mathbb{A}_f)$, we write $\langle z_1, z_2\rangle_U$ for the measure

$$\langle z_1, z_2\rangle_U = |U|^{-1}\langle z_1, z_2\rangle = [U : U']^{-1}\langle z_1, z_2\rangle_{X_{U'}}.$$

where z_1, z_2 are certain elements in $\widehat{\mathrm{Pic}}(X_\infty)$ realized on $X_{U'}$ for some $U' \subset U$. So defined pairing will depends only on the choice of U and gives the exact pairing on X_U.

Similarly, we can modify the local intersection pairing and extend the height pairing to $\mathrm{Jac}(X_\infty) = \mathrm{Pic}^0(X_\infty)$, which is the direct limit of $\mathrm{Pic}^0(X_U)$ where $\mathrm{Pic}^0(X_U)$ is the subgroup of $\mathrm{Pic}(X_U)$ of classes whose degrees are 0 on each connected component.

We can now define Green's functions g_v on divisors on $X_\infty(\bar{F})$ which are disjoint at the generic fiber for each place v of F by multiplying the Green's functions on X_U by $\mathrm{vol}(U)$. Notice that for two CM-divisors A and B on X_U with disjoint support represented by two functions α and β on $T(F)\backslash G(\mathbb{A}_f)$, the Green's function at a place v depends only on α and β. Thus, we may simply denote it as

$$g_v(A, B)_{U_v} = g_v(\alpha, \beta)_{U_v}.$$

Recall that η is a divisor on X_∞ defined by (13.3). As in Section 10, with P_χ replaced by η_U, we have

$$\widehat{\Psi} = \sum_v \widehat{\Psi}_v \mod \mathcal{D}(\sigma_1) + \mathcal{D}(r_\chi), \tag{15.2}$$

where v runs through the set of places of F, and

$$\widehat{\Psi}_v(a) := -|a| g_v(\eta, \mathrm{T}_a^0 \eta)_{\Delta_v} \log q_v. \tag{15.3}$$

Thus, it suffices to compare these local terms for each place v of F. We need only consider v which is not split in K, since $\widehat{\Phi}_v = 0$ and $\widehat{\Psi}_v$ is a finite sum of derivations of Eisenstein series when v is split in K.

Our main tool is the local Gross–Zagier formula in Section 13 for quaternion algebra ${}_vB$ with the ramification set

$$ {}_v\Sigma = \begin{cases} \Sigma \cup \{v\}, & \text{if } v \notin \Sigma, \\ \Sigma - \{v\}, & \text{if } v \in \Sigma. \end{cases} \tag{15.4} $$

Let ${}_vG$ denote the algebraic group ${}_vB^\times/F^\times$.

LEMMA 15.1. *For v an infinite place,*

$$ \widehat{\Phi}_v(a) = 2^{g+1}|c(\omega)|^{1/2}\widehat{\Psi}_v(a). $$

The idea of proof is to use the local Gross–Zagier formula to write both sides as the constant terms at $s = 0$ of two geometric pairings of divisors $\mathrm{T}_a\eta$ and η with two multiplicity functions:

$$ m_s^v(\xi) = \int_1^\infty \frac{dx}{x(1+|\xi|_v x)^{1+s}}, \qquad 2Q_s(\xi) = \int_1^\infty \frac{(1-x)^s\,dx}{x^{1+s}(1+|\xi|_v x)^{1+s}}. $$

It follows that the difference of two sides will be the constant term of a geometric pairing on $T(F)\backslash {}_vG(\mathbb{A}_f)$ with multiplicity function

$$ m_s^v - 2Q_s $$

which has no singularity and converges to 0 as $s \longrightarrow 0$. Notice that the Legendre function Q_s appears here because an explicit construction of Green's function at archimedean place.

We now consider unramified cases.

LEMMA 15.2. *Let v be a finite place prime to ND. Then there is a constant c such that*

$$ \widehat{\Phi}_v(a) - 2^{g+1}|c(\omega)|^{1/2}\widehat{\Psi}_v(a) = c\log|a|_v \cdot |a|^{1/2}\widehat{\Pi}(\chi)(a). $$

The proof is similar to the archimedean case. Write $a = \pi_v^n a'$ ($\pi_v \nmid a'$). Since the Shimura curve and CM-points all have good reduction, using Gross' theory of canonical lifting, we can show that $\widehat{\Psi}_v(a)$ is the geometric pairing of $\mathrm{T}_{a'}\eta$ and η on on $T(F)\backslash {}_vG(\mathbb{A}_f)$ with multiplicity function

$$ m_n(\xi) = \begin{cases} \frac{1}{2}\mathrm{ord}_v(\xi\pi_v^{1+n}), & \text{if } \xi \neq 0 \text{ and } \mathrm{ord}_v(\xi\pi_v^n) \text{ is odd}, \\ n/2, & \text{if } \xi \neq 0 \text{ and } n \text{ is even}, \\ 0, & \text{otherwise}. \end{cases} $$

On other hand, by using the local Gross–Zagier formula, we may also write Φ, up to a multiple of $|a|^{1/2}\widehat{\Pi}(\chi)\log|a|_v$, as a geometric pairing $\langle \mathrm{T}_{a'}\eta, \eta\rangle$ with multiplicity

$$ -2m_n(\xi)\log q_v. $$

It remains to treat the case where v is place dividing ND. In this case we will not be able to prove the identity as in the archimedean case, or in the unramified case, since there is no explicit regular model of Shimura curves we can use. But we can classify these contributions:

LEMMA 15.3. *For v a finite place dividing ND, we have*

$$\widehat{\Phi}_v(a) - 2^{g+1}|c(\omega)|^{1/2}\widehat{\Psi}_v(a) = c\,|a|^{1/2}\widehat{\Pi}(\chi)(a) +_v \widehat{f},$$

where c is a constant, and $_v\widehat{f}$ is a form on $_vG(F)\backslash_v G(\mathbb{A}_f)$. Moreover, the function $_vf$ has character χ under the right translation by $K_v^\times$.

Using the local Gross–Zagier formula, we still can show that $\widehat{\Phi}_v$ is equal the geometric local pairing

$$2^g|c(\omega))|^{1/2}|a|\langle\eta, \mathrm{T}_a\eta\rangle$$

for a multiplicity function $m(g)$ on $_vG(F)$ with singularity

$$\log|\xi|_v.$$

On other hand, it is not difficult to show that Green's function

$$\widehat{\Psi}_v(a) = -g_v(\eta, \mathrm{T}_a^0\eta)\log q_v$$

is also a geometric pairing for a multiplicity function with singularity

$$\frac{1}{2}\log|\xi|_v.$$

(This is equivalent to saying that $\xi^{1/2}$ is a local parameter in the v-adic space of CM-points). Thus the difference

$$\widehat{\Phi}_v(a) - 2^{g+1}|c(\omega)|^{1/2}\widehat{\Psi}_v(a)$$

is a geometric pairing without *singularity*. In other words, it is given by

$$\int_{[T(F)\backslash_v G(\mathbb{A}_f)]^2} \eta(x)k(x,y)\mathrm{T}_a\eta(y)\,dx\,dy,$$

for $k(x,y)$ a locally constant function of $(_vG(F)\backslash_v G(\mathbb{A}_f))^2$ which has a decomposition

$$k(x,y) = \sum_i c_i(x)f_i(y)$$

into eigenfunctions f_j for Hecke operators on $_vG(F)\backslash_v G(\mathbb{A}_f)$. It follows that the difference of two sides in the lemma is given by

$$\sum_i \lambda_i(a)\int_{T(F)\backslash_v G(\mathbb{A}_f)} \eta(x)c_i(x)\,dx \cdot \int_{T(F)\backslash_v G(\mathbb{A}_f)} f_i(y)\bar{\eta}(y)\,dy,$$

where $\lambda_i(a)$ is the eigenvalue of T_a for f_i. Thus, we may take

$$_vf = \sum_i \int_{T(F)\backslash_v G(\mathbb{A}_f)} \eta(x)c_i(x)\,dx \cdot \int_{T(F)\backslash_v G(\mathbb{A}_f)} f_i(y)\bar{\eta}(y)\,dy.$$

In summary, at this stage we have shown that the quasi-newform

$$\Phi - 2^{g+1}|c(\omega)|^{1/2}\Psi$$

has Fourier coefficients which are a sum of the following terms:

- derivations A of Eisenstein series,
- derivations B of theta series $\Pi(\chi) \otimes \alpha^{1/2}$,
- functions ${}_vf$ appearing in ${}_vG(F)\backslash{}_vG(\mathbb{A}_f)$ with character χ under the right translation of $K_v^\times$, where v are places dividing DN.

By the linear independence of Fourier coefficients of derivations of forms in Lemma 10.3, we conclude that $A = B = 0$.

Let Π be the representation generated by the form ϕ. All the projections of ${}_vf$'s in Π must vanish, by local results of Waldspurger and Gross–Prasad. See [16, § 2.3].

In summary we have shown that $\Phi - 2^{g+1}|c(\omega)|^{1/2}\Psi$ is an old form. By Proposition 11.1, the projection of this difference on $\Pi(\phi)$ is

$$\frac{L'(\frac{1}{2},\chi,\phi)}{(\phi^\sharp,\phi^\sharp)_{U_0(ND)}}\cdot\phi^\sharp - 2^{g+1}|c(\omega)|^{1/2}\cdot\langle y_\phi, y_\phi\rangle_\Delta\cdot\phi.$$

This is again an old form and has vanishing first Fourier coefficient. Theorem 14.1 follows by taking the first Fourier coefficient.

16. Spectral Decomposition

In this section we want to explain the proof for the central value formula, Theorem 14.2. The idea is copied from the odd case. Thus, we need to define a *height pairing of CM-cycles.* Since there is no natural arithmetic and geometric setting for heights corresponding to nonholomorphic forms, we would like to use the local Gross–Zagier formula to *suggest* a definition of height. Indeed, by corollary 13.2, modulo some Einstein series of type $\Pi(\|\cdot\|^{1/2},\|\cdot\|^{-1/2})\otimes\eta$ with η quadratic, the kernel $\Phi(g) := \Phi(\frac{1}{2},g)$ has a Whittaker function satisfying

$$W_\Phi\left(g_\infty\cdot\begin{pmatrix} a\delta^{-1} & 0\\ 0 & 1\end{pmatrix}\right) = |c(\omega)|^{1/2}|a|\langle \mathrm{T}_a\eta,\eta\rangle_\Delta(g_\infty), \tag{16.1}$$

where $g_\infty \in \mathrm{GL}_2(F_\infty)$ is viewed as a parameter, and a is a finite integral idele which prime to ND, and the pairing $\langle\,\cdot,\cdot\,\rangle_\Delta$ is defined by the multiplicity function

$$m(\xi,g_\infty) := \prod_{v|\infty} m_v(\xi,g_v), \tag{16.2}$$

with each $m_v(\xi, g)$ the Whittaker function of weight k_v whose value at $\begin{pmatrix} a & 0 \\ 0 & 1 \end{pmatrix}$ is

$$m_v\left(\xi, \begin{pmatrix} a & 0 \\ 0 & 1 \end{pmatrix}\right) = \begin{cases} 4|a|e^{-2\pi a}, & \text{if } 1 \geq \xi \geq 0,\ a > 0,\ k_v = 2, \\ 4|a|e^{2\pi a(2\xi - 1)}, & \text{if } a\xi \leq \min(0, a),\ k_v = 0, \\ 0, & \text{otherwise.} \end{cases} \tag{16.3}$$

This suggests a definition of the height pairing for CM-cycles by the above multiplicity function. This height pairing is no long valued in numbers but in Whittaker functions. In the case where all $k_v = 2$, then this Whittaker function is twice the standard one. Thus, we can get a pairing with values in $\mathbb{C}$. We would like to have a good understanding of decomposition of this height pairing according to eigenforms on X_U. By (12.8), we need only decompose the kernel k_U defined in (12.7). This decomposition is actually very simple: As Whittaker functions on $\mathrm{GL}_2(F_\infty)$,

$$k_U(x, y)(g_\infty) = 2^{[F:\mathbb{Q}]+n} \sum_{\phi_i} W_i(g_\infty) \cdot \phi_i(x)\bar{\phi}_i(y) + 2^{[F:\mathbb{Q}]+n} \int_{\mathfrak{M}} W_{\mathfrak{m}}(g_\infty) E_{\mathfrak{m}}(x) \bar{E}_{\mathfrak{m}}(y)\, d\mathfrak{m}, \tag{16.4}$$

where n is the number of places where $k_v = 0$, and the sum is over all cuspidal eigenforms ϕ_i of Laplacian and Hecke operators on $G(F)\backslash G(\mathbb{A})/U$ such that $\|\phi_i\|_\Delta = 1$, and W_i are standard Whittaker function for ϕ_i. Here the integration is nontrivial only when $n = g$ then $\mathfrak{M}$ is a measured space parameterizing an orthogonal basis of Eisenstein series of norm 1. (See Section 18 for more details.) Thus for a cuspidal eigenform ϕ,

$$\frac{1}{|\Delta|} \int_{G(F)\backslash G(\mathbb{A})} k(x, y)\phi(y)\, dy = 2^{[F:\mathbb{Q}]+n} W_\phi(g_\infty)\phi(x).$$

It follows that for any two CM-cycles α and β on X_U, the height pairing has a decomposition

$$\langle \alpha, \beta \rangle = 2^{[F:\mathbb{Q}]+n} \sum_{\phi_i} W_i(g_\infty) \cdot (\phi_i, \bar{\alpha})_\Delta (\bar{\phi}_i, \beta)_\Delta + 2^{[F:\mathbb{Q}]+n} \int_{\mathfrak{M}} W_{\mathfrak{m}}(g_\infty)(E_{\mathfrak{m}}, \bar{\alpha})_\Delta (\bar{E}_{\mathfrak{m}}, \beta)_\Delta\, d\mathfrak{m}. \tag{16.5}$$

This leads us to define the following form of $\mathrm{PGL}_2(\mathbb{A})$ of weight k_v at v:

$$H(\alpha, \beta) = 2^{[F:\mathbb{Q}]+n} \sum_{\phi_i} \phi_i^{\mathrm{new}}(g_\infty) \cdot (\phi_i, \bar{\alpha})_\Delta (\bar{\phi}_i, \beta)_\Delta + 2^{[F:\mathbb{Q}]+n} \int_{\mathfrak{M}} E_{\mathfrak{m}}^{\mathrm{new}}(g_\infty)(E_{\mathfrak{m}}, \bar{\alpha})_\Delta (\bar{E}_{\mathfrak{m}}, \beta)_\Delta\, d\mathfrak{m}, \tag{16.6}$$

where ϕ_i^{new} (resp. E_λ^{new}) is the *newform* of weight $(2, \dots, 2, 0, \dots, 0)$ in the representation Π_i of $\mathrm{PGL}_2(\mathbb{A})$ corresponding to the representation Π_i' of $G(\mathbb{A})$ generated by ϕ_i (resp. $E_{\mathfrak{m}}$) via Jacquet–Langlands theory. With α replaced by

$\mathrm{T}_a\alpha$ in (16.5), one obtains the usual relation between height pairing and Fourier coefficient:

$$|a|\langle \mathrm{T}_a\alpha, \beta\rangle_\Delta(g_\infty) = 2^{[F:\mathbb{Q}]+n} W_{H(\alpha,\beta)}\left(g_\infty \cdot \begin{pmatrix} a\delta^{-1} & 0 \\ 0 & 1 \end{pmatrix}\right). \qquad (16.7)$$

Let Ψ denote the form $2^{[F:\mathbb{Q}]+n}|c(\omega)|^{1/2}H(\eta,\eta)$ having a decomposition

$$\Psi = 2^{[F:\mathbb{Q}]+n}|c(\omega)|^{1/2}\sum_i \phi_i^{\mathrm{new}}\, |(\phi_i,\eta)_\Delta|^2 + 2^{[F:\mathbb{Q}]+n}|c(\omega)|^{1/2}\int_{\mathfrak{M}} E_{\mathfrak{m}}^{\mathrm{new}}\, |(E_{\mathfrak{m}},\eta)|^2\, d\mathfrak{m}. \qquad (16.8)$$

Since η has a character χ under the action by Δ, we may require that ϕ_i (resp. $E_{\mathfrak{m},\chi}$) has character χ under the action by Δ. For a given ϕ^{new} of level N, then ϕ_i with $\phi_i^{\mathrm{new}} = \phi^{\mathrm{new}}$ must be the toric newform as in Section 14.

Equations (16.1) and (16.7) shows that, modulo certain Eisenstein series in the space $\Pi(\|\cdot\|^{1/2}, \|\cdot\|^{-1/2})\otimes\eta$ with $\eta^2 = 1$, the form $\Phi - \Psi$ has vanishing Fourier coefficient at g such that

$$g_f = \begin{pmatrix} \delta^{-1}a & 0 \\ 0 & 1 \end{pmatrix}$$

with integral a prime to ND. Thus $\Phi - \Psi$ is an old form.

Let ϕ be the newform as in Theorem 14.2. By Proposition 11.1 and formula (16.6), the projection of $\Phi - \Psi$ in $\Pi(\phi)$ is given by

$$\frac{L(\frac{1}{2},\chi,\phi)}{\|\phi^\sharp\|^2_{U_0(ND)}}\phi^\sharp - 2^{[F:\mathbb{Q}]+n}|c(\omega)|^{1/2}\cdot|(\eta,\phi_\chi)_\Delta|^2\phi.$$

Thus, we have proven the Gross–Zagier formula, Theorem 14.2, by computing the first Fourier coefficient of the above form.

17. Lowering Levels

Now it is remains to deduce GZF(N) (the Gross–Zagier formulas for level N in Sections 6–7) from GZF(ND) (formulas in Section 14). Our plan is as follows:

1. Show GZF(N) up to a certain universal function of local parameters.
2. Prove GZF(ND) for Eisenstein series for level ND thus get GZF(N) for Eisenstein series with the same universal function.
3. Prove GZF(N) for Eisenstein series directly by evaluating the periods thus get the triviality of the universal function.

In this section we are doing the first step.

PROPOSITION 17.1. *For each $v \mid D$, there is a rational function $Q_v(t) \in \mathbb{C}(t)$ depending only on χ_v which takes 1 at $t = 0$ and is regular for*

$$|t| < |\pi_v|^{1/2} + |\pi_v|^{-1/2},$$

such that both Gross–Zagier formulas in Sections 6–7 are true after multiplying the left-hand side by

$$C(\chi) \prod_{\mathrm{ord}_v(D)>0} Q_v(\lambda_v),$$

where $C(\chi)$ *is a constant depends only* χ, *and* λ_v *the parameter appeared in the L-function*:

$$L_v(s,\phi) = \frac{1}{1-\lambda_v|\pi_v|^s + |\pi_v|^{2s}}.$$

The idea of proof is to show that, in the comparison of GZF(N) and GZF(ND), all four quantities

$$\widehat{\phi^\sharp}(1), \qquad \frac{\|\phi^\sharp\|^2_{U_0(ND)}}{\|\phi\|^2_{U_0(N)}}, \qquad \frac{|(\eta,\phi_\chi)_\Delta|^2}{|i_\chi(\widetilde{\phi})|^2}, \qquad \frac{\|x_\phi\|^2_\Delta}{\|y_\phi\|^2_{U_0(N)}},$$

are universal functions described in the Proposition, and that the last two quantities have the same functions. Here the last two fractions are considered as ratios since the denominators may be 0.

Let's try to localize the definition of quasi-newform in Section 11. For each finite place v, let Π_v be the local component of $\Pi(\phi)$ at v. Then Π_v is a unitary representation as $\Pi = \bigotimes \Pi_v$ is. Fix a Hermitian form for the Whittaker model $\mathcal{W}(\Pi_v,\psi_v)$ such that the norm of the new vector is 1 for almost all v. The product of this norm induces a norm on Π which is proportional to the L^2-norm on Π. Now we can define the quasi-newform

$$W_v^\sharp \in \mathcal{W}(\Pi_v,\psi_v) \tag{17.1}$$

to be a certain form of level D_v. Recall that the space of forms of level D_v has a basis consisting of forms

$$W_{vi}(g) = W_v\left(g\begin{pmatrix}\pi_v^{-i} & 0\\ 0 & 1\end{pmatrix}\right), \qquad 0\le i\le \mathrm{ord}_v(D_v), \tag{17.2}$$

where $W_{v0}=W_v$ is the newform. Then $W_v^\sharp$ is the unique nonzero form of level D_v satisfying the equations

$$(W_v^\sharp, W_{vi} - \nu_v^i W_v^\sharp) = 0, \qquad 0\le i\le \mathrm{ord}_v(D_v), \tag{17.3}$$

where $\nu_v = 0$ if v is not ramified in K; otherwise $\nu_v = \chi_v(\pi_{K,v})$. It is not difficult to show that if we write

$$W_v^\sharp = \sum c_{vi} W_{v,i},$$

then c_{vi} is rational function of quantities

$$\alpha_{vi} := (W_{vi},W_v)/(W_v,W_v).$$

The quantities α_{vi} do not depend on the choice of pairing $(\cdot,\cdot)$ on Whittaker models. On the other hand, it is easy to show that $\phi^\sharp$ has the Whittaker function as product of $W_v^\sharp$.

It follows that both $\widehat{\phi^\sharp}(1)$ and

$$\frac{\|\phi^\sharp\|^2_{U_0(ND)}}{\|\phi\|^2_{U_0(N)}} = \frac{\|\phi^\sharp\|^2_{U_0(N)}}{\|\phi\|^2_{U_0(N)}} \cdot [U_0(1) : U_0(D)]$$

are the products of rational functions at v of quantities α_{vi}. It remains to show that the α_{vi} are rational functions of λ_v. Let $U_v = \mathrm{GL}_2(\mathcal{O}_v)$. Then

$$\begin{aligned}\alpha_{vi} &= (W_v, W_v)^{-1}\mathrm{vol}(U_v)^{-1}\int_{U_v} (\rho(u)W_{vi}, \rho(u)W_v)\, du\\ &= (\rho(t_{vi})W_v, W_v)/(W_v, W_v),\end{aligned}$$

where t_{vi} is the Hecke operator corresponding the constant function $\mathrm{vol}(H_{vi})^{-1}$ on

$$H_{vi} = U_v \begin{pmatrix} \pi^{-i} & 0 \\ 0 & 1 \end{pmatrix} U_v.$$

It is well known that W_v is an eigenform under t_{vi} with eigenvalue a rational function of λ_v. This shows that the α_{vi} are rational functions of λ_v. Since we have used only the unitary property of the local representation Π_v for $v \mid D$, the so obtained rational functions as in theorem for these quantities are regular for any λ_v as long as Π_v is unitary. In other words, these functions are regular at λ_v satisfying

$$|\lambda_v| < |\pi_v|^{1/2} + |\pi_v|^{-1/2}.$$

It remains to compare the last two quantities in the odd and even cases, respectively. Obviously the ratio of normalizations of measures is given by

$$|U(N, K)|/|\Delta|,$$

which equals a product of constants at places dividing D. Thus we may take the same measure in the comparison. Let's define a function ζ on CM-points $T(F)\backslash G(\mathbb{A}_f)$ supported on $T(\mathbb{A})i_cU(N, K)$ such that

$$\zeta(ti_cu) = \chi(t), \qquad t \in T(\mathbb{A}_f), u \in U(N, K).$$

Then the CM-points in $\mathrm{GZF}(N)$ are defined by ζ, and those in $\mathrm{GZF}(ND)$ by η. The key to proving our result is to compare these CM-cycles.

Recall that for a finite place v and a compactly supported, locally constant function h on $G(F_v)$, one defines the Hecke operator $\rho(h)$ on CM-cycles by

$$\rho(h) = \int_{G(F_v)} h(g)\rho(g)\, dg.$$

Let U_1 and U_2 be the compact subgroups of $G(\mathbb{A}_f)$ defined by $U_i = \prod U_{iv}$ and

$$\begin{aligned}U_{1v} &= (\mathcal{O}_{c_v} + c_v\mathcal{O}_{K,v}\lambda_v)^\times,\\ U_{2v} &= U(N, K)_v^\times.\end{aligned}$$

LEMMA 17.2. *For each finite place v, let h_v denote the constant function* $\mathrm{vol}(U_{1v})^{-1}$ *on $G(F_v)$ supported on $U_{2,v}i_{c,v}^{-1}$. Then*

$$\zeta_v = \rho(h_v)\eta_v.$$

Before we prove this lemma, let us see how to use this lemma to finish the proof of our Proposition. First assume we are in the even case. Then we have

$$(\widetilde{\phi}, P_\chi) = (\widetilde{\phi}, \zeta)_{U(N,K)}.$$

Here the product is taken as pairings between CM-cycles and functions. Let Ψ_ζ be the form $H(\zeta, \zeta)$ defined in (16.6). Then Ψ_ζ has a decomposition

$$\Psi_\zeta = \sum_i \phi_i^{\mathrm{new}} |(\zeta, \phi_i)|^2 + \text{Eisenstein series}, \tag{17.4}$$

where ϕ_i is an orthonormal basis of eigenforms on $X(N, K)$. For $\phi_i^{\mathrm{new}} = \phi$ with level N, ϕ_i must be the test form $\widetilde{\phi}$ as in Section 7.

Now the equation (16.7) implies that the Hecke operator the adjoint of $\rho(h)$ is $\rho(h^\vee)$ with $h^\vee(g) = \bar{h}(g^{-1})$. It follows that,

$$H(\zeta, \zeta) = H(\rho(h)\eta, \rho(h)\eta) = H(\rho(h^\vee * h)\eta, \eta).$$

Since η has character χ under the action by Δ, we may replace $h^\vee * h$ by a function h_0 which has character (χ, χ^{-1}) by actions of Δ from both sides and invariant under conjugation. Now

$$\begin{aligned} \Psi_\zeta &= \sum \phi_i^{\mathrm{new}} (\rho(h_0)\eta, \phi_i)\overline{(\eta, \phi_i)} + \cdots \\ &= \sum \phi_i(\eta, \rho(h_0)\phi_i)\overline{(\eta, \phi_i)} + \cdots. \end{aligned} \tag{17.5}$$

Since η and $\rho(h_0)\eta$ both have character χ under the action by Δ, we may replacing ϕ_i by functions $\phi_{i,\chi}$ which has character χ under χ. For $\phi_i^{\mathrm{new}} = \phi^{\mathrm{new}}$ with level N, $\phi_{i,\chi}$ must be the toric newform ϕ_χ as in Section 14. Of course,

$$\rho(h_0)\phi_\chi = \prod_{v|D} P_v(\lambda_v)\phi_\chi \tag{17.6}$$

as ϕ_χ with P_v some polynomial functions. From (17.4)–(17.6), we obtain

$$|(\zeta, \widetilde{\phi})|^2 = \prod_{v|D} P_v(\lambda_v) \cdot |(\eta, \phi_\chi)|^2.$$

In the odd case, the proof is same but simpler with $H(\zeta, \zeta)$ defined as a holomorphic cusp form of weight 2 with Fourier coefficients given by the height pairings of $\langle \mathrm{T}_a x, x\rangle$ for two CM-divisors after minus some multiple of Hodge class. Then we end up with the expression

$$\Psi_\zeta = \sum_i \phi_i \|x_{\phi_i}\|^2,$$

where ϕ_i are newforms of level dividing N. The same reasoning as above shows that

$$\Psi_\zeta = H(\rho(h_0)y, y) = \sum \phi_i \cdot \langle y_{\phi_i}, \rho(h_0)y_{\phi_i}\rangle.$$

Thus we have

$$\|x_\phi\|^2 = \langle y_\phi, \rho(h_0)y_\phi\rangle = \prod_{v|D} P_v(\lambda_v)\|y_\phi\|^2.$$

It remains to prove the lemma. By definition,

$$\rho(h_v)\eta_v(g) = \mathrm{vol}(U_{1v})^{-1} \int_{U_{2v}} \eta_v(gui_{c,v}^{-1})\, du.$$

If $\rho(h_v)\eta_v(g) \neq 0$, then

$$gU_{1,v}i_{c,v}^{-1} \in T(F_v)U_{1,v}$$

or equivalently,

$$g \in T(F_v)U_{1,v}i_{c,v}U_{2,v}.$$

By (iii) in the following lemma, we have $g \in T(F_v)i_{c,v}U_{2,v}$. Write $g = ti_{c,v}u_g$. It follows that

$$\rho(h_v)\eta_v(g) = \chi(t)\mathrm{vol}(U_{1v})^{-1} \int_{U_{2v}} \eta_v(i_{c,v}ui_{c,v}^{-1})\, du.$$

By (iii) in the following lemma again, the integral is the same as

$$\int_{U_{1v}} \eta_v(u)\, du.$$

Thus we have

$$\rho(h_v)\eta_v = \zeta_v.$$

LEMMA 17.3. *For v split in B, there is an isomorphism*

$$\mu : M_2(F_v) \longrightarrow \mathrm{End}_{F_v}(K_v)$$

with compatible embedding of K_v and such that

$$\mu(M_2(\mathcal{O}_v)) = \mathrm{End}_{\mathcal{O}_v}(\mathcal{O}_{K,v}).$$

Moreover, let $i_{\pi^n} \in \mathrm{GL}_2(F_v)$ be such that

$$\mu(i_{\pi^n})(\mathcal{O}_{K,v}) = \mathcal{O}_{\pi^n} = \mathcal{O}_v + \pi_v^n\mathcal{O}_{K,v}.$$

Then

(i) $$\mathrm{GL}_2(F_v) = \coprod_{n\geq 0} K_v^\times i_{\pi^n}\mathrm{GL}_2(\mathcal{O}_v),$$

(ii) $$i_{\pi^n}\mathrm{GL}_2(\mathcal{O}_v)i_{\pi_v^n}^{-1} \cap K_v^\times = \mathcal{O}_{\pi^n}^\times,$$

(iii) $$i_{\pi^n}U_{2,v}i_{\pi^n}^{-1} \cap K_v^\times U_{1,v} = U_{1,v}.$$

PROOF. Indeed for any given embedding $K_v \longrightarrow M_2(F_v)$ such that $\mathcal{O}_{K,v}$ maps to $M_2(\mathcal{O}_v)$, then F_v^2 becomes a K_v-module of rank 1 such that $\mathcal{O}_v^2$ is stable under $\mathcal{O}_{K,v}$. Then we find an isomorphism $\mathcal{O}_v^2 \simeq \mathcal{O}_{K,v}$ as $\mathcal{O}_{K,v}$ module. This induces the required isomorphism $\mu : B_v \longrightarrow \mathrm{End}_{F_v}(K_v)$.

Now, for any $g \in \mathrm{GL}_2(F_v)$, let $t \in g(\mathcal{O}_{K,v})$ be the elements with minimal order. Then $\mu(t^{-1}g)(\mathcal{O}_{K,v})$ will be an order of $\mathcal{O}_{K,v}$, say $\mathcal{O}_{\pi^n}$. Thus

$$\mu(t^{-1}g)(\mathcal{O}_{K,v}) = \mathcal{O}_{\pi^n} = \mu(i_{\pi^n})(\mathcal{O}_{K,v}).$$

It follows that

$$g \in t i_{\pi^n} \mathrm{GL}_2(\mathcal{O}_v).$$

The first equality follows.

For the second equality, let $t \in K_v$. Then

$$i_{\pi^n}^{-1} t i_{\pi_n} \in \mathrm{GL}_2(\mathcal{O}_v)$$

if and only if

$$\mu(i_{\pi^n}^{-1} t i_{\pi^n})\mathcal{O}_K = \mathcal{O}_K,$$

or equivalently

$$\mu(t i_{\pi^n})\mathcal{O}_K = \mu(i_{\pi^n})\mathcal{O}_K, \qquad t\mathcal{O}_{\pi^n} = \mathcal{O}_{\pi^n}.$$

This is equivalent to the fact that $t \in \mathcal{O}_{\pi^n}^\times$.

It remains to show the last equality. First we want to show

$$i_{c,v}^{-1} U_{1,v} i_{c,v} \subset U_{2,v}.$$

To see this, we need to show that

$$\mu(i_{c,v}^{-1} u i_{c,v})\mathcal{O}_{K,v} = \mathcal{O}_{K,v}$$

for each $u \in U_{1,v}$. This is equivalent to

$$\mu(u i_{c,v})\mathcal{O}_{K,v} = i_{c,v}\mathcal{O}_{K,v},$$

or equivalently,

$$\mu(u)\mathcal{O}_{c,v} = \mathcal{O}_{c,v}.$$

Thus is clear from the fact that $u = t(1 + cM_2(\mathcal{O}_v))$ for some $t \in \mathcal{O}_c^\times$.

Now the last equality follows easily: since

$$i_{\pi_v^n} U_{2,v} i_{\pi_v^n}^{-1} \cap K_v^\times = \mathcal{O}_{\pi_v^n}$$

and

$$i_{\pi_v^n} U_{2,v} i_{\pi_v^n}^{-1} \supset U_{1,v},$$

it follows that

$$i_{\pi_v^n} U_{2,v} i_{\pi_v^n}^{-1} \cap K_v^\times U_{1,v} = U_{1,v}. \qquad \square$$

18. Continuous Spectrum

In this section, we would like to extend GZF(ND) (the Gross–Zagier formula in level ND, Theorem 14.2) to Eisenstein series in the continuous spectrum. Recall that the space of L^2-forms on $\mathrm{PGL}_2(F)\backslash\mathrm{PGL}_2(\mathbb{A})$ is a direct sum of cusp forms, characters, and Eisenstein series corresponding to characters (μ, μ^{-1}). We say two characters μ_1, μ_2 are connected if $\mu_1 \cdot \mu_2^{\pm}$ is trivial on the subgroup $\mathbb{A}^1$ of norm 1. Thus each connected component is a homogeneous space of $\mathbb{R}$ or $\mathbb{R}/\pm 1$. See [3] for more details.

We now fix a component containing a character (μ, μ^{-1}). Without loss of generality, we assume that μ^2 is not of form $|\cdot|^t$ for some $t \neq 0$. Then the space $\mathrm{Eis}(\mu)$ of L^2-form corresponding to this component consists of the forms

$$E(g) = \int_{-\infty}^{\infty} E_t(g)\, dt \tag{18.1}$$

where $E_t(g)$ is the Eisenstein series corresponding to characters $(\mu|\cdot|^{it}, \mu^{-1}|\cdot|^{-it})$. For the uniqueness of this integration we assume that $E_t(g) = 0$ if $t < 0$ and $\mu^2 = 1$. Now the two elements $E_1(g)$ and $E_2(g)$ has inner product given by

$$(E_1, E_2) = \int_{-\infty}^{\infty} (E_{1t}, E_{2t})_t dt, \tag{18.2}$$

where $(\cdot, \cdot)_t$ is some Hermitian form on the space

$$\Pi_t := \Pi(\mu| \cdot |^{it}, \mu^{-1}| \cdot |^{-it}).$$

This Hermitian norm is unique up to constant multiple as the representation is irreducible. The precise definition of this norm is not important to us.

Now we want to compute the Rankin–Selberg convolution of $E \in \mathrm{Eis}(\mu)$ with θ as in Section 8. Assume that χ is not of form $\nu \cdot \mathrm{N}_{K/F}$. Then θ is a cusp form and the kernel function Θ is of L^2-form as its constant term has exponential decay near cusp. Thus it makes sense to compute $(E, \bar{\Theta})$.

For ϕ a function on $\mathbb{R}$ (or $\mathbb{R}_+$ when $\mu^2 = 1$), let's write E_ϕ for element in $\mathrm{Eis}(\mu)$ with form $E_t(g) = \phi(t) E_t^{\mathrm{new}}(g)$ with $E_t^{\mathrm{new}}(g)$ a newform in $\Pi(\mu| \cdot |^s, \mu^{-1}| \cdot |^{-s})$ and $\phi(s) \in \mathbb{C}$, then we still have

$$\begin{aligned}(E_\phi, \bar{\Theta}_s) &= \int_0^{\infty} L(s, \Pi_t \otimes \chi)\phi(t)\, dt \\ &= \int_0^{\infty} L(s + it, \mu \otimes \chi) L(s + it, \mu^{-1} \otimes \chi)\phi(t)\, dt.\end{aligned}$$

If $s = \frac{1}{2}$, we obtain

$$(E_\phi, \bar{\Theta}_{1/2}) = \int \left|L(\tfrac{1}{2} + it, \mu \otimes \chi)\right|^2 \phi(t)\, dt. \tag{18.3}$$

The form Θ has level D. For any a dividing D, let's define

$$E_{\phi,a} = \rho\begin{pmatrix} a^{-1} & 0 \\ 0 & 1 \end{pmatrix} E_\phi.$$

Then the space of Eisenstein series is generated by $E_{\phi,a}$. We can define so called *quasi-newforms* by the formula

$$E_\phi^\sharp = \int_0^\infty E_t^\sharp \phi(t)\, dt. \tag{18.4}$$

One can show that the projection of $\Phi := \bar{\Theta}_{1/2}$ on the continuous spectrum corresponding to μ is $E_\phi^\sharp$ with ϕ given by

$$\phi(t) = \frac{\left|L(\frac{1}{2} + it, \mu \otimes \chi)\right|^2}{\|E_t^\sharp\|^2}. \tag{18.5}$$

We now study the geometric pairing in Section 16. Formula (16.8) shows that the continuous contribution for representation Π_t in the form Ψ is given by

$$2^{2g}|c(\omega)|^2 E_\psi,$$

where

$$\psi(t) = (E_{t,\chi}, \eta). \tag{18.6}$$

Here $E_{t,\chi}$ is a form toric form of norm 1 with respect to Δ.

Again $E_\phi^\sharp - 2^{2g}|c(\omega)|^{1/2}E_\psi$ will be an old form. Its first Fourier coefficient vanishes. Thus the Gross–Zagier formula can be extended to Eisenstein series:

PROPOSITION 18.1. *Assume that χ is not of form $\nu \circ \mathrm{N}_{K/F}$ with ν a character of $F^\times\backslash\mathbb{A}^\times$. Then*

$$\widehat{E_t^\sharp}(1)|L(\tfrac{1}{2} + it, \chi)|^2 = 2^{2g}|c(\omega)|^{1/2} \cdot \|E_t^\sharp\|^2|(E_{\chi,t}, \eta)|^2.$$

Also the proof of Proposition 17.1 is purely local, and so can be extended to Eisenstein series:

PROPOSITION 18.2. *Assume that χ is not of form $\nu \circ \mathrm{N}_{K/F}$ with ν a character of $F^\times\backslash\mathbb{A}^\times$. Let*

$$\lambda_v(t) = \mu_v(\pi_v)|\pi_v|^{it} + \mu_v(\pi_v)^{-1}|\pi_v|^{-it}, \quad \textit{and} \quad E_t^* := \|E_t\| \cdot \widetilde{E}_t,$$

then

$$c(\chi) \prod_{\mathrm{ord}_v(D)>0} Q_v(\lambda_v(t)) = \frac{2^{2g}}{\sqrt{\mathrm{N}(D)}}\left|\frac{(E_t^*, P_\chi)}{L(\frac{1}{2} + it, \chi)}\right|^2.$$

Notice that when μ is unramified $\widehat{R}$ is conjugate to $M_2(\widehat{\mathcal{O}}_F)$. The form E_t^* is obtained from E_t by $\rho(j)$ for a certain $j \in G(\mathbb{A})$ satisfying (19.1) and (19.2) below. Thus the formula does not involve the definition of hermitian forms on Π_t.

19. Periods of Eisenstein Series

In this section we want to compute the periods of Eisenstein series appearing in the GZF(N) (up to a universal function), Proposition 18.2. Our result shows that all the universal functions are trivial, and thus end up the proof of GZF(N).

First let's describe the main result. Let μ be a unramified quasi-character of $F^\times\backslash\mathbb{A}^\times$. Let R be a maximal order of $M_2(F)$ containing $\mathcal{O}_K$. Let E be the newform in $\Pi(\mu,\mu^{-1})$. Let $j\in G(\mathbb{A})$ such that

$$j_\infty \mathrm{SO}_2(\mathbb{R}) j_\infty^{-1} = T(\mathbb{R}) \tag{19.1}$$

and

$$j_f \mathrm{GL}_2(\widehat{\mathcal{O}}) j_f^{-1} = R. \tag{19.2}$$

Then the form $E^*(g) := E(gj)$ is invariant under $T(\mathbb{R})\cdot\widehat{R}^\times$. Let $\lambda\in K$ be a nonzero trace-free element. Then one can show that $\mathrm{ord}_v(\lambda/D)$ for all finite place v is always even. We thus assume that $4\lambda/D$ has a square root at a finite place and that $D_v=-1$ when $v\mid\infty$.

PROPOSITION 19.1. *Assume that* $\chi\neq\mu_K:=\mu\circ\mathrm{N}_{K/F}$. *Then*

$$(E^*,P_\chi) = 2^{-g}\mu(\delta^{-1}\sqrt{4\lambda/D})|4\lambda/D|^{1/4}L(\tfrac12,\bar\chi\cdot\mu_K).$$

Before we go to the proof of this result, let's see how to use Proposition 19.1 to complete the proof of GZF(N). Combined Propositions 19.1 and 18.2 with $\mu(x)=|x|^{it}$ ($t\in\mathbb{R}$), we obtain

$$C(\chi)\prod_{\mathrm{ord}_v(D)>0} Q_v(\lambda_v(t)) = 1 \qquad \text{for all } t\in\mathbb{R}.$$

Notice that each $\lambda_v(t)$ is a rational function of p^{ti} where p is prime number divisible by v. Since functions p^{ti} for different primes p are rationally independent, we obtain that for each prime p

$$\prod_{\mathrm{ord}_v(D_p)>0} Q_v(\lambda_v(t)) = \mathrm{const}$$

where $D_p=\prod_{v|p}D_v$.

It is not difficult to show that for each χ_v we can find a finite character χ' of $\mathbb{A}_K^\times/K^\times\mathbb{A}^\times$ such that the following conditions are verified:

- $c(\chi')$ is prime to N, $c(\omega)$;
- χ' is unramified at all $w\mid p, w\neq v$;
- χ' is not of form $\nu\circ\mathrm{N}_{K/F}$.

If we apply the above result to χ' then we found that Q_v is constant thus is 1. Thus we have shown:

PROPOSITION 19.2. *All polynomials* $Q_v(t)$ *and* $C(\chi)$ *are constant* 1.

Now GZF(N) follows from Proposition 17.1.

We now start the proof of Proposition 19.1 from the integral

$$(E^*, P_\chi) = \int_{T(F)\backslash T(\mathbb{A}_f)} \chi^{-1}(x) E^*(x_\infty x i_c)\, dx$$

where $x_\infty \in \mathcal{H}^g$ is fixed by $T(\mathbb{R})$ and $i_c \in G(\mathbb{A}_f)$ is an element such that

$$i_c R i_c^{-1} \cap K = \mathcal{O}_c.$$

Here we pick up a measure dt on $T(\mathbb{A})$ with local decomposition $dt = \bigotimes_v dt_v$ such that $T(\mathbb{R})$ and $T(\mathcal{O}_{c,v})$ all have volume 1. Write $h = i_c j$. It follows that

$$(E^*, P_\chi) = \int_{T(F)\backslash T(\mathbb{A})} \bar{\chi}(x) E(xh)\, dx.$$

Since E is obtained by analytic continuation from the newform in the Eisenstein series in $\Pi(\mu|\cdot|^s, \mu^{-1}|\cdot|^s)$ with $\operatorname{Re} s \gg 0$, we thus need only compute the periods for quasi-character μ with big exponent. In this case,

$$E(g) = \sum_{\gamma \in P(F)\backslash G(F)} f(\gamma g)$$

with

$$f(g) = \mu^{-1}(\delta) f_\Phi,$$

where $\Phi = \bigotimes \Phi_v \in \mathcal{S}(\mathbb{A}^2)$ is the standard element: Φ_v is the characteristic function of $\mathcal{O}_v^2$ if $v \nmid \infty$, and $\Phi_v(x,y) = e^{-\pi(x^2+y^2)}$ if $v \mid \infty$. It is not difficult to show that the embedding $T \longrightarrow G$ defines an bijective map

$$T(F) \simeq P(F)\backslash G(F).$$

Thus

$$(E^*, P_\chi) = \int_{T(\mathbb{A})} \chi^{-1}(x) f(xh)\, dx.$$

This is of course the product of local integrals

$$i_{\chi_v}(f_v) = \int_{T(F_v)} \chi_v(x^{-1}) f_v(xh_v)\, dx.$$

Recall that f_v is defined as follows:

$$f_v(g) = \mu(\delta_v^{-1} \cdot \det g) |\det g|^{1/2} \int_{F_v^\times} \Phi[(0,t)g] \mu^2(t)|t| d^\times t.$$

It follows that

$$\begin{aligned} i_{\chi_v}(f_v) &= \int_{T(F_v)} \chi_v(x^{-1}) \mu(\delta_v^{-1} \det xh_v) |\det xh_v|^{1/2} \int_{F_v^\times} \Phi[(0,t)xh_v] \mu^2(t)|t| d^\times t\, dx \\ &= \mu(\delta_v^{-1} \det h_v) |\det h_v|^{1/2} \int_{K_v^\times} \bar{\chi}_v \mu_K(x) |x|_K^{1/2} \Phi_{K_v}(x)\, dx \\ &= \mu(\delta_v^{-1} \det h_v) |\det h_v|^{1/2} Z(\tfrac{1}{2}, \bar{\chi} \cdot \mu_K, \Phi_{K_v}), \end{aligned}$$

where for $x \in K_v^\times$,

$$\Phi_{K_v}(x) = \Phi_v[(0,1)xh_v].$$

Thus the period computation is reduced to the computation of local zeta functions.

Let v be a finite place. The map $x \longrightarrow (0,1)x$ defines an isomorphism between K and F^2 with compatible actions by K. Thus we have two lattices $\mathcal{O}_F^2$ and $\mathcal{O}_c$ in K. The element h_f as a class in

$$\widehat{K}^\times \backslash \mathrm{GL}_2(\mathbb{A}_f)/\mathrm{GL}_2(\widehat{\mathcal{O}}_F)$$

is determined by the property that

$$h_f M_2(\widehat{\mathcal{O}}_F)h_f^{-1} \cap K = \mathcal{O}_c.$$

We may take h_f such that

$$(0,1)\mathcal{O}_c h_f = \mathcal{O}_F^2.$$

It follows that $\Phi_{K,v}$ is the characteristic function of $\widehat{\mathcal{O}}_c$. Now the zeta function is easy to compute:

$$Z(\tfrac{1}{2}, \bar{\chi}\cdot\mu_K, \Phi_{K,c}) = \int_{\mathcal{O}_{c_v}} \chi\cdot\mu_K(x)|x|_K^{1/2}d^\times x.$$

We get the standard L-function if $c = 0$.

We assume now that $c > 0$ thus K_v/F_v is an unramified extension. First we assume that K_v is a field. We decompose the set $\mathcal{O}_c$ into the disjoint union of $\mathcal{O}_{c,n}$ of subset of elements of order n. Then

$$Z(\tfrac{1}{2}, \bar{\chi}\cdot\mu_K, \Phi_{K,c}) = \sum_{n\geq 0}\mu(\pi_v)^{2n}|\pi_v|^n \int_{\mathcal{O}_{c,n}} \chi(x)d^\times x.$$

Write $\mathcal{O}_{K,v} = \mathcal{O}_v + \mathcal{O}_v\lambda$ then

$$\mathcal{O}_c = \mathcal{O}_v + \pi_v^c\mathcal{O}_v\lambda.$$

If $n \geq c$, then $\mathcal{O}_{c,n} = \pi^n\mathcal{O}_K^\times$. The integral vanishes as χ has conductor π^c. If $n < c$ then

$$\mathcal{O}_{c,n} = |\pi_v|^n\mathcal{O}_v^\times(1 + \pi_v^{c-n}\mathcal{O}_K).$$

The integration on $\mathcal{O}_{n,c}$ vanishes unless $n = 0$ as χ has conductor π^c. Thus the total contribution is

$$\mathrm{vol}(\mathcal{O}_{c_v}^\times) = 1.$$

We assume now that K_v/F_v is split. Then $K_v = F_v^2$ and $\mathcal{O}_c$ consists of integral elements (a,b) such that $a \equiv b \pmod{\pi_v^c}$. Write $\chi = (\nu, \nu^{-1})$ then ν has conductor π_v^c. It follows that

$$Z(\tfrac{1}{2}, \bar{\chi}\cdot\mu_K, \Phi_{K,c}) = \int_{(a,b)\in\mathcal{O}_c} \nu(a/b)\mu(ab)|ab|^{1/2}d^\times a d^\times b.$$

For a fixed $b \in \mathcal{O}_v$, the condition in a is as follows:

$$\begin{cases} a \in \pi^c \mathcal{O}_F, & \text{if } b \in \pi^c \mathcal{O}_v, \\ a \in b(1+\pi^{c-n}\mathcal{O}_v), & \text{if } b \in \pi^n \mathcal{O}_v^\times \text{ with } n < c. \end{cases}$$

Since ν has conductor c, the only case gives nontrivial contribution is when $b \in \mathcal{O}_v^\times$ and $a \in b(1+\pi^c\mathcal{O}_v)$. The contribution is given by

$$\mathrm{vol}(\mathcal{O}_{c_v}^\times) = 1.$$

To compute $\det h_v$, we write $K = F + F\sqrt{\lambda}$ and make the following embedding $K \longrightarrow M_2(F)$:

$$a + b\lambda \mapsto \begin{pmatrix} a & b\lambda \\ b & a \end{pmatrix}. \tag{19.3}$$

Then $\mathcal{O}_F^2$ corresponding to the lattice

$$\mathcal{O}_F + \mathcal{O}_F\sqrt{\lambda}.$$

Thus h_f satisfies

$$(0,1)(\mathcal{O}_v + \mathcal{O}_v\sqrt{\lambda}) = (0,1)\mathcal{O}_{c_v} h_f.$$

It follows that

$$\mathrm{disc}(\mathcal{O}_v + \mathcal{O}_v\sqrt{\lambda}) = \mathrm{disc}(\mathcal{O}_{c_v}) \det h_v^2.$$

Thus

$$\det h_v = \sqrt{4\lambda/D_v}$$

for a suitable D_v in its class modulo $\mathcal{O}_v^\times$ such that $4\lambda/D_v$ does have a square root in $F_v^\times$. In summary we have shown that

$$i_v(f_v) = L(\tfrac{1}{2}, \bar{\chi} \otimes \mu)\mu(\delta_v\sqrt{4\lambda/D_v})|4\lambda/D_v|^{1/4}. \tag{19.4}$$

It remains to compute the periods at archimedean places v. For equation (19.1), we may take

$$h_v = \begin{pmatrix} |\lambda_v|^{1/2} & 0 \\ 0 & 1 \end{pmatrix}.$$

Then it is easy to see that

$$\Phi_{K,v}(x) = e^{-\pi|x|^2}.$$

Assume that $\mu(x) = |x|^t$, $\chi = 1$, and notice that the measure on $K_v^\times = \mathbb{C}^\times$ is induced from the standard $d^\times x$ from $\mathbb{R}^\times$ and one from $\mathbb{C}^\times/\mathbb{R}^\times$ with volume one. Thus the measure has the form $dr\, d\theta/\pi r$ for polar coordinates $re^{i\theta}$. It follows that

$$\begin{aligned} Z(\tfrac{1}{2}, \bar{\chi}\cdot\mu_K, \Phi_{K_v}) &= \int_{\mathbb{C}^\times} e^{-\pi r^2} r^{2t+1} \frac{dr\, d\theta}{\pi r} = \pi^{-1/2-t}\Gamma(t+\tfrac{1}{2}) \\ &= \mu(2)2^{-1/2}L(\tfrac{1}{2}, \bar{\chi}\cdot\mu_K). \end{aligned}$$

The period at v is then given by

$$i_v(f_v) = 2^{-1}\mu_v(|4\lambda|^{1/2})|4\lambda_v|^{1/4}L(\tfrac{1}{2}, \bar{\chi}\mu_K). \tag{19.5}$$

Setting $D_v = -1$ for archimedean places, we obtain the same formula as (19.4). The proof of the proposition is completed.

Acknowledgments

I thank N. Vatsal and H. Xue for pointing out many inaccuracies in our previous paper [16] (especially the absence of the first Fourier coefficient of the quasi-newform in the main formulas); to B. Gross for his belief in the existence of a formula in level N and for his many very useful suggestions in preparation of this article; to D. Goldfeld and H. Jacquet for their constant support and encouragement.

References

[1] M. Bertolini and H. Darmon, "Heegner points on Mumford–Tate curves", *Invent. Math.* **126** (1996), 413–456.

[2] H. Carayol, "Sur la mauvaise réduction des courbes de Shimura", *Compositio Math.* **59** (1986), 151–230.

[3] S. Gelbart and H. Jacquet, "Forms of GL(2) from the analytic point of view", pp. 213–251 in *Automorphic forms, representations and L-functions* (Corvallis, OR, 1977), Proc. Sympos. Pure Math. **33**, Part 1, edited by A. Borel and W. Casselman, Providence, Amer. Math. Soc., 1979.

[4] D. Goldfeld and S. Zhang, "The holomorphic kernel of the Rankin–Selberg convolution", *Asian J. Math.* **3** (1999), 729–747.

[5] B. Gross, *Heegner points and representation theory*, pp. 37–65 in *Heegner Points and Rankin L-Series*, edited by Henri Darmon and Shou-Wu Zhang, *Math. Sci. Res. Inst. Publications* **49**, Cambridge U. Press, New York, 2004.

[6] B. H. Gross and D. Zagier, "Heegner points and derivatives of L-series", *Invent. Math.*, **84** (1986), 225–320.

[7] H. Jacquet, "Automorphic forms on GL_2", part II, Lecture Notes in Math. **289**, Berlin, Springer, 1972.

[8] H. Jacquet and R. Langlands, "Automorphic forms on GL(2)," Lecture Notes in Math. **114**, Berlin, Springer, 1970.

[9] J. Tate, "On the conjectures of Birch and Swinnerton-Dyer and a geometric analog", pp. 415–440 (exposé 306) in *Séminaire Bourbaki*, 1965/66, Benjamin, 1967; reprint, Paris, Société Mathématique de France, 1995.

[10] V. Vatsal, "Special values for Rankin L-functions", pp. 165–190 in *Heegner Points and Rankin L-Series*, edited by Henri Darmon and Shou-Wu Zhang, *Math. Sci. Res. Inst. Publications* **49**, Cambridge U. Press, New York, 2004.

[11] V. Vatsal, "Uniform distribution of Heegner points", *Invent. Math.* **148**:1 (2002), 1–46.

[12] J. L. Waldspurger, "Sur les values de certaines fonctions L automorphes en leur centre de symmétre", *Compositio Math.* **54** (1985), 173–242.

[13] J. L. Waldspurger, "Correspondence de Shimura et quaternions", *Forum Math.* **3** (1991), 219–307.

[14] Y. Tian, "Euler systems on Shimura curves", Thesis at Columbia University (2003).

[15] H. Xue, "Central values for twisted Rankin L-functions", Thesis at Columbia University (2002).

[16] S. Zhang, "Gross–Zagier formula for GL_2", *Asian J. Math.* **5** (2001), 183–290.

[17] S. Zhang, "Heights of Heegner points on Shimura curves", *Annals of mathematics* (2) **153**:1 (2000), 27–147.

[18] S. Zhang, "Elliptic curves, L-functions, and CM-points", pp. 179–219 in *Current Developments in Mathematics* (2001), Cambridge (MA), International Press, 2002.

Shou-Wu Zhang
Department of Mathematics
Columbia University
New York, NY 10027
United States
szhang@math.columbia.edu

Heegner Points and Rankin L-Series
MSRI Publications
Volume **49**, 2004

Special Cycles and Derivatives of Eisenstein Series

STEPHEN S. KUDLA

A man hears what he wants to hear and disregards the rest.
– Paul Simon and Art Garfunkel, *The Boxer*

This article is an expanded version of a lecture given at the conference on Special Values of Rankin L-Series at MSRI in December of 2001. I have tried to retain some of the tone of an informal lecture. In particular, I have attempted to outline, in very broad terms, a program involving relations among

(i) algebraic cycles,
(ii) Eisenstein series and their derivatives, and
(iii) special values of Rankin–Selberg L-functions and their derivatives,

ignoring many important details and serious technical problems in the process. I apologize at the outset for the very speculative nature of the picture given here. I hope that, in spite of many imprecisions, the sketch will provide a context for a variety of particular cases where precise results have been obtained. Recent results on one of these, part of an ongoing joint project with Michael Rapoport and Tonghai Yang on which much of the conjectural picture is based, are described in Yang's article [79]. A less speculative discussion of some of this material can be found in [42; 44; 45].

I thank my collaborators B. Gross, M. Harris, J. Millson, S. Rallis, M. Rapoport and T. Yang for generously sharing their mathematical ideas and for their support over many years. I also thank R. Borcherds, J.-B. Bost, J. Cogdell, J. Funke, R. Howe, D. Kazhdan, K. Keating, J. Kramer, U. Kühn, J.-S. Li, J. Schwermer, and D. Zagier for helpful discussions, comments and suggestions. Finally, I thank Henri Darmon and Shou-Wu Zhang for organizing such an enjoyable and inspiring program and to MSRI for its ever excellent hospitality.

Partially supported by NSF grant DMS-9970506 and by a Max-Planck Research Prize from the Max-Planck Society and Alexander von Humboldt Stiftung.

I. An Attractive Family of Varieties

1. Shimura Varieties of Orthogonal Type

We begin with the following data:

$$\begin{aligned} &V,\ (\ ,\ ) = \text{inner product space over } \mathbb{Q},\\ &\text{sig}(V) = (n,2),\\ &\quad G = \text{GSpin}(V),\\ &\quad D = \{w \in V(\mathbb{C}) \mid (w,w) = 0,\ (w,\bar{w}) < 0\}/\mathbb{C}^\times \ \subset\ \mathbb{P}(V(\mathbb{C})),\\ &\quad n = \dim_{\mathbb{C}} D, \end{aligned} \tag{1.1}$$

This data determines a Shimura variety $M = \text{Sh}(G, D)$, with a canonical model over $\mathbb{Q}$, where, for $K \subset G(\mathbb{A}_f)$ a compact open subgroup,

$$M_K(\mathbb{C}) \simeq G(\mathbb{Q})\backslash\big(D \times G(\mathbb{A}_f)/K\big). \tag{1.2}$$

Note that $D = D^+ \cup D^-$ is a union of two copies of a bounded domain of type IV; see [68, p. 285]. They are interchanged by the complex conjugation $w \mapsto \bar{w}$. If we let $G(\mathbb{R})^+$ be the subgroup of $G(\mathbb{R})$ which preserves D^+ and write

$$G(\mathbb{A}) = \coprod_j G(\mathbb{Q})G(\mathbb{R})^+ g_j K, \tag{1.3}$$

then

$$M_K(\mathbb{C}) \simeq \coprod_j \Gamma_j\backslash D^+, \tag{1.4}$$

where $\Gamma_j = G(\mathbb{Q}) \cap G(\mathbb{R})^+ g_j K g_j^{-1}$. Thus, for general K, the quasi-projective variety M_K can have many components and the individual components are only rational over some cyclotomic extension. The action of the Galois group on the components is described, for example, in [17; 62].

M_K is quasi-projective of dimension n over $\mathbb{Q}$, and projective if and only if the rational quadratic space V is anisotropic. By Meyer's Theorem, this can only happen for $n \le 2$. In the range $3 \le n \le 5$, we can have $\text{witt}(V) = 1$, where witt(V) is the dimension of a maximal isotropic $\mathbb{Q}$-subspace of V. For $n \ge 6$, witt(V) $= 2$. A nice description of the Baily–Borel compactification of $\Gamma\backslash D^+$ and its toroidal desingularizations can be found in [59].

For small values of n, the M_K's include many classical varieties, for example:

$n = 1$, modular curves and Shimura curves [41];
$n = 2$, Hilbert–Blumenthal surfaces and quaternionic versions [51];
$n = 3$, Siegel 3-folds and quaternionic analogues [53; 73, 25];
$n \le 19$, moduli spaces of K3 surfaces [5].

Of course, such relations are discussed in many places; see for example, [18], and the Appendix below for $n = 1$.

More generally, one could consider quadratic spaces V over a totally real field $\boldsymbol{k}$ with $\mathrm{sig}(V_{\infty_i}) = (n, 2)$ for $\infty_i \in S_1$ and $\mathrm{sig}(V_{\infty_i}) = (n{+}2, 0)$ for $\infty_i \in S_2$, where $S_1 \cup S_2$ is a disjoint decomposition of the set of archimedean places of $\boldsymbol{k}$. If $S_2 \neq \varnothing$, then the varieties M_K are always projective. Such compact quotients are considered in [46; 47; 40]. For a discussion of automorphic forms in this situation from a classical point of view, see [70].

2. Algebraic Cycles

An attractive feature of this family of Shimura varieties is that they have many algebraic cycles; in fact, there are sub-Shimura varieties of the same type of all codimensions. These can be constructed as follows.

Let $\mathcal{L}_D$ be the homogeneous line bundle over D with

$$\mathcal{L}_D \setminus \{0\} = \{w \in V(\mathbb{C}) \mid (w, w) = 0,\ (w, \bar{w}) < 0\}, \tag{2.1}$$

so that $\mathcal{L}_D$ is the restriction to D of the bundle $\mathcal{O}(-1)$ on $\mathbb{P}(V(\mathbb{C}))$. We equip $\mathcal{L}_D$ with an hermitian metric $\|\ \|$ given by $\|w\|^2 = |(w, \bar{w})|$. The action of $G(\mathbb{R})$ on D lifts in a natural way to an action on $\mathcal{L}_D$, and hence, this bundle descends to a line bundle $\mathcal{L}$ on the Shimura variety M. For example, for a given compact open subgroup K, $\mathcal{L}_K \to M_K$, has a canonical model over $\mathbb{Q}$ (see [32; 62]), and

$$\mathcal{L}_K(\mathbb{C}) \simeq G(\mathbb{Q})\backslash\big(\mathcal{L}_D \times G(\mathbb{A}_f)/K\big). \tag{2.2}$$

Any rational vector $x \in V(\mathbb{Q})$ defines a section s_x over D of the dual bundle $\mathcal{L}_D^\vee$ by the formula

$$(s_x, w) = (x, w), \tag{2.3}$$

and, for $x \neq 0$, the (possibly empty) divisor[1] in D of this section is given by

$$\mathrm{div}(s_x) = \{w \in D \mid (x, w) = 0\}/\mathbb{C}^\times =: D_x \subset D. \tag{2.4}$$

Assuming that $Q(x) := \frac{1}{2}(x, x) > 0$ and setting

$$V_x = x^\perp \tag{2.5}$$

and

$$G_x = \mathrm{GSpin}(V_x) = \text{stabilizer of } x \text{ in } G,$$

there is a sub-Shimura variety

$$Z(x) : \mathrm{Sh}(G_x, D_x) \to \mathrm{Sh}(G, D) = M \tag{2.6}$$

giving a divisor $Z(x)_K$, rational over $\mathbb{Q}$, on M_K for each K.

If $Q(x) \leq 0$, and $x \neq 0$, then the section s_x is never zero on D so that $D_x = \varnothing$. If $x = 0$, then we formally set $D_x = D$, and take $Z(0) = M$.

---

[1]This is a rational quadratic divisor in Borcherds' terminology [3].

More generally, given an r-tuple of vectors $x \in V(\mathbb{Q})^r$ define V_x, G_x, D_x by the same formulas. If the matrix

$$Q(x) = \tfrac{1}{2}\big((x_i, x_j)\big) \tag{2.7}$$

is positive semidefinite of rank $r(x)$ (this being the dimension of the subspace of V spanned by the components of $x = (x_1, x_2, \dots, x_r) \in V(\mathbb{Q})^r$), then the restriction of $(\ ,\ )$ to V_x has signature $(n - r(x), 2)$, and there is a corresponding cycle $Z(x) : \mathrm{Sh}(G_x, D_x) \to \mathrm{Sh}(G, D) = M$, of codimension $r(x) = \mathrm{rk}(Q(x)) \le r$.

For $g \in G(\mathbb{A}_f)$, we can also make a "translated" cycle $Z(x, g)$, where, at level K,

$$\begin{aligned} Z(x, g; K) : G_x(\mathbb{Q})\backslash\big(D_x \times G_x(\mathbb{A}_f)/K_x^g\big) &\to G(\mathbb{Q})\backslash\big(D \times G(\mathbb{A}_f)/K\big) \simeq M_K(\mathbb{C}), \\ G_x(\mathbb{Q})(z, h)K_x^g \quad &\mapsto \quad G(\mathbb{Q})(z, hg)K, \end{aligned} \tag{2.8}$$

where we write $K_x^g = G_x(\mathbb{A}_f) \cap gKg^{-1}$ for short. This cycle is again rational over $\mathbb{Q}$.

Finally, we form certain weighted combinations of these cycles, essentially by summing over integral x's with a fixed matrix of inner products [40]. More precisely, suppose that a K-invariant Schwartz function[2] $\varphi \in S(V(\mathbb{A}_f)^r)^K$ on r copies of the finite adeles $V(\mathbb{A}_f)$ of V and $T \in \mathrm{Sym}_r(\mathbb{Q})_{\ge 0}$ are given. Let

$$\Omega_T = \{\ x \in V^r \mid Q(x) = T,\ r(x) = \operatorname{rank} T\ \}, \tag{2.9}$$

and, assuming that $\Omega_T(\mathbb{Q})$ is nonempty, write

$$\Omega_T(\mathbb{A}_f) \cap \mathrm{supp}(\varphi) = \coprod_j K g_j x \tag{2.10}$$

for $x \in \Omega_T(\mathbb{Q})$ and $g_j \in G(\mathbb{A}_f)$. Then there is a cycle $Z(T, \varphi; K)$ in M_K defined by

$$Z(T, \varphi; K) = \sum_j \varphi(g_j^{-1}x)\, Z(x, g_j; K) \tag{2.11}$$

of codimension $\mathrm{rank}(T) =: r(T)$, given by a weighted combination of the $Z(x)$'s for x with $Q(x) = T$.

These weighted cycles have nice properties [40]. For example, if $K' \subset K$ and $\mathrm{pr} : M_{K'} \to M_K$ is the corresponding covering map, then

$$\mathrm{pr}^* Z(T, \varphi; K) = Z(T, \varphi; K'). \tag{2.12}$$

Thus it is reasonable to drop K from the notation and write simply $Z(T, \varphi)$.

EXAMPLE. The classical Heegner divisors, traced down to $\mathbb{Q}$, arise in the case $n = 1$, $r = 1$. A detailed description is given in the Appendix.

---

[2] For example, for $r = 1$, φ might be the characteristic function of the closure in $V(\mathbb{A}_f)$ of a coset $\mu + L$ of a lattice $L \subset V$.

3. Modular Generating Functions

In this section, we discuss the generating functions that can be constructed from the cycles $Z(T,\varphi)$ by taking their classes either in cohomology or in Chow groups. The main goal is to prove that such generating functions are, in fact, modular forms. Of course, these constructions are modeled on the work of Hirzebruch and Zagier [36] on generating functions for the cohomology classes of curves on Hilbert–Blumenthal surfaces.

3.1. Classes in cohomology. The cycles defined above are very special cases of the locally symmetric cycles in Riemannian locally symmetric spaces studied some time ago in a long collaboration with John Millson [46; 47; 48]. The results described in this section are from that joint work. For $T \in \mathrm{Sym}_r(\mathbb{Q})_{\geq 0}$ and for a weight function φ, there are cohomology classes

$$[\,Z(T,\varphi)\,] \in H^{2r(T)}(M_K) \quad \text{and} \quad [\,Z(T,\varphi)\,] \cup [\mathcal{L}^\vee]^{r-r(T)} \in H^{2r}(M_K), \tag{3.1}$$

where $r(T)$ is the rank of T and $[\mathcal{L}^\vee] \in H^2(M_K)$ is the cohomology class of the dual $\mathcal{L}^\vee$ of the line bundle $\mathcal{L}$. Here we view our cycles as defining linear functionals on the space of compactly supported closed forms, and hence these classes lie in the absolute cohomology $H^\bullet(M_K)$ of $M_K(\mathbb{C})$ with complex coefficients.

In [48], we proved:

THEOREM 3.1. *For $\tau = u+iv \in \mathfrak{H}_r$, the Siegel space of genus r, the holomorphic function*

$$\phi_r(\tau,\varphi) = \sum_{T\in \mathrm{Sym}_r(\mathbb{Q})_{\geq 0}} [\,Z(T,\varphi)\,] \cup [\mathcal{L}^\vee]^{r-r(T)}\, q^T,$$

is a Siegel modular form of genus r and weight $n/2+1$ valued in $H^{2r}(M_K)$. Here $q^T = e(\mathrm{tr}(T\tau))$.

IDEA OF PROOF. The main step is to construct a theta function *valued in the closed (r,r)-forms on $M_K(\mathbb{C})$*. Let $A^{(r,r)}(D)$ be the space of smooth (r,r)-forms on D, and let $S(V(\mathbb{R})^r)$ be the Schwartz space of $V(\mathbb{R})$. The group $G(\mathbb{R})$ acts naturally on both of these spaces. For $\tau \in \mathfrak{H}_r$, there is a Schwartz form (see [46; 47; 48])

$$\varphi^r_\infty(\tau) \in \big[\,S(V(\mathbb{R})^r) \otimes A^{(r,r)}(D)\,\big]^{G(\mathbb{R})} \tag{3.2}$$

with the following properties:

(i) For all $x \in V(\mathbb{R})^r$, $d\varphi^r_\infty(\tau,x) = 0$, i.e., $\varphi^r_\infty(\tau,x)$ is a closed form on D.
(ii) For $g \in G(\mathbb{R})$ and $x \in V(\mathbb{R})^r$,

$$g^*\varphi^r_\infty(\tau,x) = \varphi^r_\infty(\tau,g^{-1}x). \tag{3.3}$$

Thus, for example, $\varphi_\infty(\tau,x) \in A^{(r,r)}(D)^{G(\mathbb{R})_x}$ is a closed $G(\mathbb{R})_x$-invariant form on D. Note that $\varphi^r_\infty(\tau)$ is *not* holomorphic in τ. For any $\varphi \in S(V(\mathbb{A}_f)^r)^K$, the

Siegel theta function

$$\theta_r(\tau,\varphi) := \sum_{x\in V(\mathbb{Q})^r} \varphi_\infty(\tau,x)\varphi(x) \in A^{(r,r)}(M_K) \tag{3.4}$$

is a closed (r,r)-form on $M_K(\mathbb{C})$ and, by the standard argument based on Poisson summation, is modular of weight $n/2+1$ for a subgroup $\Gamma' \subset \mathrm{Sp}_r(\mathbb{Z})$. Finally, the cohomology class

$$\phi_r(\tau,\varphi) = [\theta_r(\tau,\varphi)] \tag{3.5}$$

of the theta form (3.4) coincides with the *holomorphic* generating function of the Theorem and hence this generating function is also modular of weight $n/2+1$. □

The Schwartz forms satisfy the cup product identity:

$$\varphi_\infty^{r_1}(\tau_1)\wedge\varphi_\infty^{r_2}(\tau_2) = \varphi_\infty^{r_1+r_2}(\left(\begin{smallmatrix}\tau_1 & \\ & \tau_2\end{smallmatrix}\right)), \tag{3.6}$$

where the left side is an element of the space $S(V(\mathbb{R})^{r_1})\otimes S(V(\mathbb{R})^{r_2})\otimes A^{(r,r)}(D)$, with $r = r_1+r_2$, and $\tau_j \in \mathfrak{H}_{r_j}$. Hence, for weight functions $\varphi_j \in S(V(\mathbb{A}_f)^{r_j})$, one has the identity for the theta forms

$$\theta_{r_1}(\tau_1,\varphi_1)\wedge\theta_{r_2}(\tau_2,\varphi_2) = \theta_r(\left(\begin{smallmatrix}\tau_1 & \\ & \tau_2\end{smallmatrix}\right),\ \varphi_1\otimes\varphi_2). \tag{3.7}$$

Passing to cohomology, (3.7) yields the pleasant identity, [40]:

$$\phi_{r_1}(\tau_1,\varphi_1)\cup\phi_{r_2}(\tau_2,\varphi_2) = \phi_{r_1+r_2}(\left(\begin{smallmatrix}\tau_1 & \\ & \tau_2\end{smallmatrix}\right),\ \varphi_1\otimes\varphi_2), \tag{3.8}$$

for the cup product of the generating functions valued in $H^\bullet(M_K)$. Comparing coefficients, we obtain the following formula for the cup product of our classes. Suppose that $T_1 \in \mathrm{Sym}_{r_1}(\mathbb{Q})_{>0}$ and $T_2 \in \mathrm{Sym}_{r_2}(\mathbb{Q})_{>0}$. Then

$$[Z(T_1,\varphi_1)]\cup[Z(T_1,\varphi_2)] = \sum_{\substack{T\in\mathrm{Sym}_r(\mathbb{Q})_{\geq 0}\\ T=\left(\begin{smallmatrix}T_1 & * \\ * & T_2\end{smallmatrix}\right)}} [Z(T,\varphi_1\otimes\varphi_2)]\cup[\mathcal{L}^\vee]^{r-\mathrm{rk}(T)}. \tag{3.9}$$

3.2. Classes in Chow groups. We can also take classes of the cycles in the usual Chow groups[3]. For this, when V is anisotropic so that M_K is compact, we consider the classes

$$\{\,Z(T,\varphi)\,\} \in \mathrm{CH}^{r(T)}(M_K) \qquad \text{and} \qquad \{\,Z(T,\varphi)\,\}\cdot\{\mathcal{L}^\vee\}^{r-r(T)} \in \mathrm{CH}^r(M_K) \tag{3.10}$$

in the Chow groups of M_K, and corresponding generating functions

$$\phi_r^{\mathrm{CH}}(\tau,\varphi) = \sum_{T\in\mathrm{Sym}_r(\mathbb{Q})_{\geq 0}} \{\,Z(T,\varphi)\,\}\cdot\{\mathcal{L}^\vee\}^{r-r(T)}\,q^T \tag{3.11}$$

[3]We only work with rational coefficients.

valued in $\mathrm{CH}^r(M_K)_{\mathbb{C}}$. Here $\cdot$ denotes the product in the Chow ring $\mathrm{CH}^\bullet(M_K)$, and $\{\mathcal{L}^\vee\} \in \mathrm{Pic}(M_K) \simeq \mathrm{CH}^1(M_K)$ is the class of $\mathcal{L}^\vee$. Note that, for the cycle class map:

$$cl : \mathrm{CH}^r(M_K) \to H^{2r}(M_K), \tag{3.12}$$

we have

$$\phi_r^{\mathrm{CH}}(\tau, \varphi) \mapsto \phi_r(\tau, \varphi), \tag{3.13}$$

so that the generating function $\phi_r^{\mathrm{CH}}(\tau, \varphi)$ "lifts" the cohomology valued function $\phi_r(\tau, \varphi)$, which is modular by Theorem 3.1.

If V is isotropic, let $\widetilde{M}_K$ be a smooth toroidal compactification of M_K, [59]. Let $Y_K = \widetilde{M}_K \setminus M_K$ be the compactifying divisor, and let $\mathrm{CH}^1(\widetilde{M}_K, Y_K)$ be the quotient of $\mathrm{CH}^1(\widetilde{M}_K)$ by the subspace generated by the irreducible components of Y_K. We use the same notation for the classes $\{\, Z(T, \varphi) \,\}$ of our cycles in this group.

The following result is due to Borcherds [6; 7].

Theorem 3.2. *For $r = 1$ and for a K-invariant weight function $\varphi \in S(V(\mathbb{A}_f))^K$, the generating function*

$$\phi_1^{\mathrm{CH}}(\tau, \varphi) = \{\mathcal{L}^\vee\} + \sum_{t>0} \{\, Z(t, \varphi) \,\}\, q^t$$

is an elliptic modular form of weight $n/2 + 1$ valued in $\mathrm{CH}^1(\widetilde{M}_K, Y_K)$.

Proof. Since the result as stated is not quite in [6], we indicate the precise relation to Borcherds' formulation. For a lattice $L \subset V$ on which the quadratic form $Q(x) = \frac{1}{2}(x, x)$ is integer-valued, let $L^\vee = \{x \in V \mid (x, L) \subset \mathbb{Z}\}$ be the dual lattice. Let $S_L \subset S(V(\mathbb{A}_f))$ be the finite dimensional subspace spanned by the characteristic functions φ_λ of the closures in $V(\mathbb{A}_f)$ of the cosets $\lambda + L$ where $\lambda \in L^\vee$. Every $\varphi \in S(V(\mathbb{A}_f))$ lies in some S_L for sufficiently small L. There is a (finite Weil) representation ρ_L of a central extension Γ' of $\mathrm{SL}_2(\mathbb{Z})$ on S_L. Suppose that F is a holomorphic function on $\mathfrak{H}$, valued in S_L, which is modular of weight $1 - n/2$, i.e., for all $\gamma' \in \Gamma'$

$$F(\gamma'(\tau)) = j(\gamma', \tau)^{2-n} \rho_L(\gamma') F(\tau), \tag{3.14}$$

where $j(\gamma', \tau)$ with $j(\gamma', \tau)^2 = (c\tau + d)$ is the automorphy factor attached to γ', and $\left(\begin{smallmatrix} a & b \\ c & d \end{smallmatrix}\right)$ is the projection of γ' to $\mathrm{SL}_2(\mathbb{Z})$. The function F is allowed to have a pole of finite order at ∞, i.e., F has a Fourier expansion of the form

$$F(\tau) = \sum_{\lambda \in L^\vee/L} \sum_{m \in \mathbb{Q}} c_\lambda(m)\, q^m\, \varphi_\lambda, \tag{3.15}$$

where only finitely many coefficients $c_\lambda(m)$ for $m < 0$ can be nonzero. Note that, by the transformation law, $c_\lambda(m)$ can only be nonzero when $m \equiv -Q(\lambda) \mod \mathbb{Z}$. For any such F where, in addition, all $c_\lambda(-m)$ for $m \geq 0$ are in $\mathbb{Z}$, Borcherds [6; 7; 3] constructs a meromorphic function $\Psi(F)$ on D with the properties:

(i) There is an integer N such that, for any F, $\Psi(F)^N$ is a meromorphic automorphic form of weight $k = N\,c_0(0)/2$, i.e., a meromorphic section of $\mathcal{L}^{\otimes k}$.

(ii)
$$\operatorname{div}(\Psi(F)^2) = \sum_{\lambda\in L^\vee/L}\sum_{m>0} c_\lambda(-m)\, Z(m,\varphi_\lambda). \tag{3.16}$$

In [6], Borcherds defines a rational vector space $\mathrm{CHeeg}(M_K)$ with generators $y_{m,\lambda}$, for $\lambda \in L^\vee/L$ and $m>0$ with $m \equiv Q(\lambda) \mod \mathbb{Z}$, and $y_{0,0}$ and relations

$$c_0(0)y_{0,0} + \sum_{\lambda}\sum_{m>0} c_\lambda(-m)\, y_{m,\lambda}, \tag{3.17}$$

as F runs over the quasi-modular forms of weight $1-n/2$, as above. Under the assumption that a certain space of vector valued forms has a basis with rational Fourier coefficients, Borcherds proved that the space $\mathrm{CHeeg}(M_K)$ is finite dimensional and that the generating function

$$\phi_1^B(\tau, L) = y_{0,0} + \sum_{\lambda}\sum_{m>0} y_{m,\lambda}\, q^m\, \varphi_\lambda^\vee, \tag{3.18}$$

valued in $\mathrm{CHeeg}(M_K)\otimes S_L^\vee$, is a modular form of weight $n/2+1$ for Γ'. Here $S_L^\vee$ is the dual space of S_L, with the dual representation $\rho_L^\vee$ of Γ'. William McGraw [61] recently proved that the necessary basis exists.

To finish the proof of our statement, we choose a nonzero (meromorphic) section Ψ_0 of $\mathcal{L}$ and define a map

$$\begin{aligned} \mathrm{CHeeg}(M_K) &\to \mathrm{CH}(\widetilde{M}_K, Y_K), \\ y_{m,\lambda} &\mapsto Z(m,\varphi_\lambda), \\ y_{0,0} &\mapsto -\operatorname{div}(\Psi_0). \end{aligned} \tag{3.19}$$

This is well defined, since a relation (3.17) is mapped to

$$-c_0(0)\operatorname{div}(\Psi_0) + \operatorname{div}(\Psi(F)^2) = N^{-1}\operatorname{div}(\Psi(F)^{2N}\,\Psi_0^{-2k}) \equiv 0, \tag{3.20}$$

since $\Psi(F)^{2N}\,\Psi_0^{-2k}$ is a meromorphic *function* on M_K. (Of course, one needs to check that it extends to a meromorphic function on $\widetilde{M}_K$.) Since the generating function $\phi_1^{\mathrm{CM}}(\tau,\varphi)$ is a finite linear combination of components of Borcherds' generating function, it is modular for some suitable subgroup of Γ', as claimed. Note that $\{\mathcal{L}^\vee\} = \{-\operatorname{div}(\Psi_0)\}$. □

PROBLEM 1. *Is $\phi_r^{\mathrm{CH}}(\tau,\varphi)$ a Siegel modular form for $r>1$?*

PROBLEM 2. *Does the cup product formula like (3.8) still hold?*

PROBLEM 3. *When V is isotropic, define classes $\{\widetilde{\mathcal{L}}\} \in \mathrm{Pic}(\widetilde{M}_K)$,*

$$\{\widetilde{Z}(T,\varphi)\} \in \mathrm{CH}^{r(T)}(\widetilde{M}_K) \quad \text{and} \quad \{\widetilde{Z}(T,\varphi)\}\cdot\{\mathcal{L}^\vee\}^{r-r(T)} \in \mathrm{CH}^r(\widetilde{M}_K)$$

so that the resulting generating function $\phi_r^{\mathrm{CH}}(\tau,\varphi)$ is modular.

Additional information about the map

$$\mathrm{CHeeg}(M_K)/\mathbb{Q}\, y_{0,0} \to \mathrm{CH}(\widetilde{M}_K, Y_K)/\mathbb{Q}\{\mathcal{L}^\vee\}, \tag{3.21}$$

e.g., concerning injectivity, was obtained by Bruinier [12; 13].

4. Connections with Values of Eisenstein Series

To obtain classical scalar valued modular forms, one can apply linear functionals to the modular generating functions valued in cohomology. For a moment, we again assume that we are in the case of compact quotient. Then, using the class

$$[\mathcal{L}^\vee] \in H^2(M_K), \tag{4.1}$$

and the composition

$$H^{2r}(M_K) \times H^{2(n-r)}(M_K) \to H^{2n}(M_K) \xrightarrow{\deg} \mathbb{C}, \tag{4.2}$$

of the cup product and the degree map, we have:

$$\deg(\,\phi_r(\tau,\varphi) \cup [\mathcal{L}^\vee]^{n-r}\,) = \int_{M_K} \theta_r(\tau,\varphi) \wedge \Omega^{n-r} := I_r(\tau,\varphi), \tag{4.3}$$

where Ω is the Chern form of the line bundle $\mathcal{L}^\vee$ for its natural metric.

Now, the Siegel–Weil formula [76; 50] relates the integral $I_r(\tau,\varphi)$ of a theta function determined by a Schwartz function $\varphi \in S(V(\mathbb{A}_f)^r)^K$ to a special value of a Siegel Eisenstein series $E_r(\tau,s,\varphi)$, also associated to φ. The parameter s in this Eisenstein series is normalized as in Langlands, so that there is a functional equation with respect to $s \mapsto -s$, and the halfplane of absolute convergence is $\mathrm{Re}(s) > (r+1)/2$. Note that, to apply the Siegel–Weil formula, we must first relate the integral of the theta *form* occurring in (4.3) to the adèlic integral of the theta *function* occurring in the Siegel–Weil theory; see [43, Section 4]. Hence, we obtain:

THE VOLUME FORMULA. *In the case of compact quotient* (see [40]), *we have*

$$\deg(\,\phi_r(\tau) \cup [\mathcal{L}^\vee]^{n-r}\,) \overset{(1)}{=} I_r(\tau,\varphi) \overset{(2)}{=} \mathrm{vol}(M,\Omega^n) \cdot E_r(\tau,s_0,\varphi), \tag{4.4}$$

where

$$s_0 = \frac{n+1-r}{2}. \tag{4.5}$$

In fact, this formula should hold in much greater generality, i.e., when V is isotropic. First of all, the theta integral is termwise convergent whenever Weil's condition $r < n+1-\mathrm{witt}(V)$ holds, and so the identity (2) in (4.4) is then valid. The result of [49] can be applied and the argument given in [43] for the case $r = 1$ carries over to prove the following.

THEOREM 4.1. *When $r < n+1-\mathrm{witt}(V)$, there is an identity*

$$\sum_{T\geq 0} \mathrm{vol}(Z(T,\varphi),\Omega^{n-r(T)})\, q^T = \mathrm{vol}(M,\Omega^n)\cdot E_r(\tau,s_0,\varphi).$$

It remains to give a cohomological interpretation of the left side of this identity in the noncompact case.

Some sort of regularization of the theta integral, say by the method of [50], is needed to obtain an extension of (2) to the range $r \geq n+1-\mathrm{witt}(V)$, i.e., to the cases $r = n-1$ and n when $\mathrm{witt}(V) = 2$ or the case $r = n$, if $\mathrm{witt}(V) = 1$. For example, in the case of modular curves, where $n = r = 1$, it was shown by Funke [20] that the theta integral coincides with Zagier's nonholomorphic Eisenstein series of weight $\frac{3}{2}$; see [80]. In this case, there are definitely (non-holomorphic!) correction terms that do not have an evident cohomological meaning, although they are consistent with a suitable arithmetic Chow group formulation; see Yang's article [79]. Recent work of Funke and Millson [21] considered the pairing of the theta form with closed forms not of compact support in the case of arithmetic quotients of hyperbolic n-space.

EXAMPLES. 1. If $n = 1$ and V is anistropic, so that $M = M_K$ is a Shimura curve over $\mathbb{Q}$, then

$$\mathrm{vol}(M_K)\cdot E_1(\tau,\tfrac{1}{2},\varphi) = \deg(\phi_1(\tau,\varphi)) = \mathrm{vol}(M_K,\Omega) + \sum_{t>0} \deg(Z(t,\varphi))\, q^t \quad (4.6)$$

is a special value at $s = \frac{1}{2}$ of an Eisenstein series of weight $\frac{3}{2}$, and the $Z(t,\varphi)$'s are Heegner type 0-cycles on M_K (see the Appendix). This identity is described in more detail in [55].

2. If $n = 2$ and V has $\mathrm{witt}(V) = 1$ or is anisotropic, so that M_K is a Hilbert–Blumenthal surface for some real quadratic field or a compact analogue, then

$$\mathrm{vol}(M_K)\cdot E_1(\tau,1,\varphi) = \deg(\phi_1(\tau,\varphi)\cup\Omega) = \mathrm{vol}(M_K) + \sum_{t>0} \mathrm{vol}(Z(t,\varphi),\Omega)\, q^t \quad (4.7)$$

is the special value at $s = 1$ of an Eisenstein series of weight 2, and the $Z(t,\varphi)$'s are Hirzebruch–Zagier type curves [73] on M_K.

3. If $n = 2$ and V is anisotropic, then

$$\begin{aligned}\mathrm{vol}(M_K)\cdot E_2(\tau,\tfrac{1}{2},\varphi) &= \deg(\phi_2(\tau,\varphi))\\ &= \mathrm{vol}(M_K) + \sum_{\substack{T\in \mathrm{Sym}_2(\mathbb{Q})_{\geq 0}\\ r(T)=1}} \mathrm{vol}(Z(T,\varphi),\Omega)\, q^T + \sum_{T>0} \deg(Z(T,\varphi))\, q^T \quad (4.8)\end{aligned}$$

is the special value at $s = \frac{1}{2}$ of an Eisenstein series of weight 2 and genus 2, and, for $T > 0$, the $Z(T,\varphi)$'s are 0-cycles. Gross and Keating [28] observed such a phenomenon in the split case as well.

4. If $n = 3$, and for V with $\mathrm{witt}(V) = 2$, M_K is a Siegel modular 3-fold. Then, for $r = 1$, the $Z(t,\varphi)$'s are combinations of Humbert surfaces, and the

identity of Theorem 4.1 asserts that their volumes are the Fourier coefficients of an Eisenstein series of weight $\frac{5}{2}$; see [74; 43].

II. Speculations on the Arithmetic Theory

The main idea is that many of the phenomena described above have an analogue in arithmetic geometry, where the varieties M are replaced by integral models $\mathcal{M}$ over Spec $(\mathbb{Z})$, the cycles $Z(T, \varphi)$ are replaced by arithmetic cycles on $\mathcal{M}$, and the classes of these cycles are taken in arithmetic Chow groups $\widehat{CH}^r(\mathcal{M})$; see [24; 72]. One could then define a function $\widehat{\phi}_r$ valued in $\widehat{CH}^r(\mathcal{M})$, lifting the modular generating function ϕ_r valued in cohomology. The main goal would be to prove the modularity of $\widehat{\phi}_r$ and to find analogues of the identities discussed above, where the values of the Eisenstein series occurring in Section 4 are replaced by their derivatives, i.e., by the second terms in their Laurent expansions.

At this point, I am going to give an idealized picture which ignores many serious technical problems involving: (i) the existence of good integral models; (ii) bad reduction and the possible bad behavior of cycles at such places; (iii) non-compactness, boundary contributions; (iv) extensions of the Gillet–Soulé theory [24] of arithmetic Chow groups $\widehat{CH}^r(\mathcal{M})$ to allow singular metrics (see [9; 57; 15]); and (iv) suitable definitions of Green functions, etc.

Nevertheless, the idealized picture can serve as a guide and, with sufficient effort, one can obtain rigorous results in various particular cases; see [41; 51; 52; 53; 54; 55]. In all of these cases, we only consider a good maximal compact subgroup K and a specific weight function φ determined by a nice lattice, so, in the discussion to follow, we will suppress both K and φ from the notation.

5. Integral Models and Cycles

Suppose that we have:

$$\begin{aligned}
\mathcal{M} &= \text{ a regular model of } M \text{ over Spec}\,(\mathbb{Z}),\\
\widehat{CH}^{\bullet}(\mathcal{M}) &= \text{its (extended) arithmetic Chow groups},\\
\hat{\omega} &= \text{ extension of the metrized line bundle } \mathcal{L}^{\vee} \text{ to } \mathcal{M},\\
\hat{\omega} &\in \widehat{\text{Pic}}(\mathcal{M}) \simeq \widehat{CH}^1(\mathcal{M})\\
\mathcal{Z}(T) &= \text{ an extension of } Z(T) \text{ on } M \text{ to a cycle on } \mathcal{M}, \text{ so that}
\end{aligned} \tag{5.1}$$

$$\begin{array}{ccccc}
 & \mathcal{Z}(T) & \rightarrow & \mathcal{M} & \\
 & \uparrow & & \uparrow & \\
Z(T) = & \mathcal{Z}(T)_{\mathbb{Q}} & \rightarrow & \mathcal{M}_{\mathbb{Q}} & = M \qquad \text{(generic fibers).}
\end{array}$$

Finally, to obtain classes in the Gillet–Soulé arithmetic Chow groups $\widehat{CH}^r(\mathcal{M})$ from the $\mathcal{Z}(T)$'s, we need Green forms; see [24; 11; 8]. Based on the constructions for $r = 1$ ([41]) and for $r = 2$, $n = 1$ ([42]), we suppose that these have the form:

$$\begin{aligned} \tau &= u + iv \in \mathfrak{H}_r, \\ \Xi(T, v) &= \text{ Green form for } Z(T), \text{ depending on } v, \\ \widehat{\mathcal{Z}}(T, v) &= (\mathcal{Z}(T), \Xi(T, v)) \in \widehat{CH}^{r(T)}(\mathcal{M}). \end{aligned} \tag{5.2}$$

In all cases done so far [54; 41; 52; 51; 53] we have $0 \le n \le 3$, M is of PEL type and the model $\mathcal{M}$ is obtained by extending the moduli problem over $\mathbb{Q}$ to a moduli problem over Spec $(\mathbb{Z})$ or, at least, Spec $(\mathbb{Z}[N^{-1}])$ for a suitable N. The cycles $\mathcal{Z}(T)$ are defined by imposing additional endomorphisms satisfying various compatibilities, the special endomorphisms. See [42; 44] for further discussion.

With such a definition, it can happen that $\mathcal{Z}(T)$ is non-empty, even when $\mathcal{Z}(T)_{\mathbb{Q}} = Z(T)$ is empty. For example, purely vertical divisors can occur in the fibers of bad reduction of the arithmetic surfaces attached to Shimura curves, [52]. In addition, there can be cases where $\mathcal{Z}(T)$ is empty, but $\Xi(T, v)$ is a nonzero smooth form on $M(\mathbb{C})$, so that there are classes

$$\widehat{\mathcal{Z}}(T, v) = (0, \Xi(T, v)) \in \widehat{CH}^r(\mathcal{M}) \tag{5.3}$$

"purely vertical at infinity", even for T not positive semi-definite; see [41, 42].

Finally, we define the *arithmetic theta function*

$$\widehat{\phi}_r(\tau) = \sum_{T \in \mathrm{Sym}_r(\mathbb{Q})} \widehat{\mathcal{Z}}(T, v) \cdot \hat{\omega}^{r - r(T)} \, q^T \in \widehat{CH}^r(\mathcal{M}), \tag{5.4}$$

where $\cdot$ denotes the product in the arithmetic Chow ring $\widehat{CH}^{\bullet}(\mathcal{M})$. Note that this function is not holomorphic in τ, since the Green forms depend on v. Under the restriction maps

$$\mathrm{res} : \widehat{CH}^r(\mathcal{M}) \to \mathrm{CH}^r(\mathcal{M}_{\mathbb{Q}}) \to H^{2r}(M), \tag{5.5}$$

we have

$$\widehat{\phi}_r(\tau) \mapsto \phi_r^{\mathrm{CH}}(\tau) \mapsto \phi_r(\tau), \tag{5.6}$$

so that $\widehat{\phi}_r$ lifts ϕ_r^{CH} and ϕ_r to the arithmetic Chow group.

PROBLEM 4. *Can the definitions be made so that $\widehat{\phi}_r(\tau)$ is a Siegel modular form of weight $n/2 + 1$ valued in $\widehat{CH}^r(\mathcal{M})$, lifting ϕ_r and ϕ_r^{CH}?*

At present, this seems out of reach, especially for $1 < r < n + 1$.

PROBLEM 5. *Is there an intersection product formula for the arithmetic Chow ring:*

$$\widehat{\phi}_{r_1}(\tau_1) \cdot \widehat{\phi}_{r_2}(\tau_2) \overset{??}{=} \widehat{\phi}_{r_1 + r_2}\begin{pmatrix} \tau_1 & \\ & \tau_2 \end{pmatrix} \tag{5.7}$$

lifting the cup product relation (3.8) *in cohomology?*

6. Connections with Derivatives of Eisenstein Series

As in the standard Gillet–Soulé theory, suppose[4] that there is an arithmetic degree map

$$\widehat{\deg} : \widehat{CH}^{n+1}(\mathcal{M}) \to \mathbb{C}, \tag{6.1}$$

and a height pairing

$$\langle\ ,\ \rangle : \widehat{CH}^{r}(\mathcal{M}) \times \widehat{CH}^{n+1-r}(\mathcal{M}) \to \mathbb{C}, \qquad \langle \widehat{\mathcal{Z}}_1, \widehat{\mathcal{Z}}_2\rangle = \widehat{\deg}(\widehat{\mathcal{Z}}_1 \cdot \widehat{\mathcal{Z}}_2). \tag{6.2}$$

These can be used to produce "numerical" generating functions from the $\widehat{\phi}_r$'s.

Let

$$\mathcal{E}_r(\tau, s) = C(s)\, E_r(\tau, s, \varphi_0) \tag{6.3}$$

be the Siegel–Eisenstein series of weight $n/2 + 1$ and genus r associated to φ_0, our standard weight function, with suitably normalizing factor $C(s)$ (see [55] for an example of this normalization). The choice of $C(s)$ becomes important in the cases in which the leading term in nonzero. Then the following *arithmetic volume formula* is an analogue of the volume formula of Theorem 4.1 above:

PROBLEM 6. *For a suitable definition of* $\mathcal{E}_r(\tau, s)$, *show that*

$$\mathcal{E}_r'(\tau, s_0) \overset{??}{=} \langle\, \widehat{\phi}_r(\tau), \hat{\omega}^{n+1-r}\,\rangle = \sum_T \widehat{\deg}(\widehat{\mathcal{Z}}(T, v) \cdot \hat{\omega}^{n+1-r(T)}\,)\, q^T. \tag{6.4}$$

where $s_0 = (n{+}1{-}r)/2$ *is the critical value of* s *occurring in the Siegel–Weil formula. Here* r *lies in the range* $1 \le r \le n + 1$.

REMARKS. (i) Identity (6.4) can be proved without knowing that $\widehat{\phi}_r$ is modular, and one can obtain partial results by identifying corresponding Fourier coefficients on the two sides.

(ii) One can view the quantities $\widehat{\deg}(\, \widehat{\mathcal{Z}}(T, v) \cdot \hat{\omega}^{n+1-r(T)}\,)$ as arithmetic volumes or heights [11].

(iii) Assuming that $C(s_0) = \mathrm{vol}(M)$, the leading term

$$\mathcal{E}_r(\tau, s_0) = \mathrm{vol}(M)\, E_r(\tau, s_0) \tag{6.5}$$

of the normalized Eisenstein series at $s = s_0$ is just the generating function for geometric volumes, via Theorem 4.1.

(iv) In the case $r = n + 1$, so that $\widehat{\phi}_r(\tau) \in \widehat{CH}^{n+1}(\mathcal{M})$, the image of $\widehat{\phi}_{n+1}$ in cohomology or in the usual Chow ring of $\mathcal{M}_{\mathbb{Q}}$ is identically zero, since this group vanishes. On the other hand, the Eisenstein series $E_{n+1}(\tau, s)$ is incoherent in the sense of [41], [42], the Siegel–Weil point is $s_0 = 0$, and $E_{n+1}(\tau, 0)$ is also identically zero. Thus the geometric volume identity is trivially valid. The arithmetic volume formula would then be

$$\widehat{\deg}(\widehat{\phi}_{n+1}(\tau)) \overset{??}{=} \mathcal{E}_{n+1}'(\tau, 0). \tag{6.6}$$

[4] Recall that, in the noncompact cases, we have to use some extended theory of arithmetic Chow groups [15], which allows the singularities of the natural metric on $\hat{\omega}$.

Example 1: Moduli of CM elliptic curves [54]. Here $n = 0$, $r = 1$, V is a negative definite quadratic form given by the negative of the norm form of an imaginary quadratic field $\boldsymbol{k}$, and $\mathcal{M}$ is the moduli stack of elliptic curves with CM by $O_{\boldsymbol{k}}$, the ring of integers of $\boldsymbol{k}$. For $t \in \mathbb{Z}_{>0}$, the cycle $\mathcal{Z}(t)$ is either empty or is a 0-cycle supported in a fiber $\mathcal{M}_p$ for a prime p determined by t. The identity

$$\widehat{\deg}(\widehat{\phi}_1(\tau)) = \mathcal{E}_1'(\tau, 0) \tag{6.7}$$

for the central derivative of an incoherent Eisenstein series of weight 1 is proved in [54], in the case in which $\boldsymbol{k}$ has prime discriminant. The computation of the arithmetic degrees is based on the result of Gross, [26], which is also the key to the geometric calculations in [31].

Remark. In the initial work on the arithmetic situation [41] and in the subsequent joint papers with Rapoport [53; 51], the main idea was to view the central derivative of the incoherent Eisenstein series, restricted to the diagonal, as giving the height pairing of cycles in complementary degrees; see formula (6.10) below. At the Durham conference in 1996, Gross insisted that it would be interesting to consider the "simplest case", $n = 0$. Following his suggestion, we obtained the results of [54] and came to see that the central derivative should *itself* have a nice geometric interpretation, as a generating function for the arithmetic degrees of 0-cycles on $\mathcal{M}$, without restriction to the diagonal. This was a crucial step in the development of the picture discussed here.

Example 2: Curves on arithmetic surfaces. Here $n = 1$, $r = 1$, V is the space of trace zero elements of an indefinite division quaternion algebra over $\mathbb{Q}$, and $\mathcal{M}$ is the arithmetic surface associated to a Shimura curve. For $t \in \mathbb{Z}_{>0}$, the cycle $\mathcal{Z}(t)$ is a divisor on $\mathcal{M}$ and can have vertical components. The identity

$$\langle\, \widehat{\phi}_1(\tau), \hat{\boldsymbol{\omega}}\,\rangle = \mathcal{E}_1'(\tau, \tfrac{1}{2}) \tag{6.8}$$

is proved in [55]. Here $\mathcal{E}_1(\tau, s)$ is a normalized Eisenstein series of weight $\frac{3}{2}$. An unknown constant occurs in the definition of the class $\widehat{\mathcal{Z}}(0, v)$ in the constant term of the generating function. This constant arises because we do not have, at present, an explicit formula for the quantity $\langle \hat{\boldsymbol{\omega}}, \hat{\boldsymbol{\omega}} \rangle$ for the arithmetic surface attached to a Shimura curve. In the analogous example for modular curves, discussed in Yang's talk [79], the quantity $\langle \hat{\boldsymbol{\omega}}, \hat{\boldsymbol{\omega}} \rangle$ is known, thanks to the work of Ulf Kühn [57] and Jean-Benoit Bost [9; 10], independently. The computation of such arithmetic invariants via an arithmetic Lefschetz formula is discussed in [60]. The identity (6.8) is the first arithmetic case in which the critical point s_0 for $\mathcal{E}(\tau, s)$ is not zero and and the leading term $\mathcal{E}(\tau, \frac{1}{2})$ does not vanish. It is also the first case in which a truely global quantity, the pairing $\langle\, \widehat{\mathcal{Z}}(t, v), \hat{\boldsymbol{\omega}}\,\rangle$ for a horizontal cycle $\mathcal{Z}(t)$, must be computed; it is determined as the Faltings height of a CM elliptic curve [55].

Example 3: 0-cycles on arithmetic surfaces. In the case $n = 1$, $r = 2$, $\widehat{\phi}_2(\tau)$ is a generating function for 0-cycles on the arithmetic surface $\mathcal{M}$. The combination of [41], joint work with Rapoport [52], and current joint work with Rapoport and Yang [56] comes very close to proving the identity

$$\widehat{\deg}(\widehat{\phi}_2(\tau)) \overset{??}{=} \mathcal{E}_2'(\tau, 0), \tag{6.9}$$

again up to an ambiguity in the constant term of the generating function due to the lack of a formula for $\langle\hat{\omega}, \hat{\omega}\rangle$. For $T > 0$ and p-regular, as defined in [42], the cycle $\mathcal{Z}(T)$ is a 0-cycle concentrated in a single fiber $\mathcal{M}_p$ for a prime p determined by T. In this case, the computation of $\widehat{\deg}(\mathcal{Z}(T))$ amounts to a counting problem and a problem in the deformation theory of p-divisible groups. The latter is a special case of a deformation problem solved by Gross and Keating [28]. On the analytic side, the computation of the corresponding term in the central derivative of the Eisenstein series amounts to the *same* counting problem and the computation of the central derivative of a certain Whittaker function on $\mathrm{Sp}_2(\mathbb{Q}_p)$. This later computation depends on the explicit formulas due to Kitaoka [37] for the representation densities of T by unimodular quadratic forms of rank $4 + 2j$; see [42, Section 5] for a more detailed discussion.

Example 4: Siegel modular varieties. Here $n = 3$. (See [53].) The Shimura variety M attached to a rational quadratic space V of signature $(3, 2)$ is, in general, a "twisted" version of a Siegel 3-fold. The canonical model M over $\mathbb{Q}$ can be obtained as a moduli space of polarized abelian varieties of dimension 16 with an action of a maximal order O_C in the Clifford algebra of V. A model $\mathcal{M}$ over $\mathrm{Spec}\,\mathbb{Z}[N^{-1}]$, for a suitable N can likewise be defined as a moduli space [53]. The possible generating functions and their connections with Eisenstein series are as follows [44]:

r			
1	$\mathcal{Z}(t)_{\mathbb{Q}} =$ Humbert surface	$\hat{\phi}_1(\tau) = \hat{\omega} + ? + \sum_{t\neq 0} \widehat{\mathcal{Z}}(t, v)\, q^t$	$\langle\hat{\phi}_1(\tau), \hat{\omega}^3\rangle \overset{?}{=} \mathcal{E}_1'(\tau, \frac{3}{2})$
2	$\mathcal{Z}(T)_{\mathbb{Q}} =$ curve	$\hat{\phi}_2(\tau) = \hat{\omega}^2 + ? + \sum_{T\neq 0} \widehat{\mathcal{Z}}(T, v)\, q^T$	$\langle\hat{\phi}_2(\tau), \hat{\omega}^2\rangle \overset{?}{=} \mathcal{E}_2'(\tau, 1)$
3	$\mathcal{Z}(T)_{\mathbb{Q}} =$ 0-cycle	$\hat{\phi}_3(\tau) = \hat{\omega}^3 + ? + \sum_{T\neq 0} \widehat{\mathcal{Z}}(T, v)\, q^T$	$\langle\hat{\phi}_2(\tau), \hat{\omega}\rangle \overset{?}{=} \mathcal{E}_3'(\tau, \frac{1}{2})$
4	$\mathcal{Z}(T)_{\mathbb{Q}} = \varnothing$	$\hat{\phi}_4(\tau) = \hat{\omega}^4 + ? + \sum_{T\neq 0} \widehat{\mathcal{Z}}(T, v)\, q^T$	$\widehat{\deg}\,\hat{\phi}_4(\tau) \overset{?}{=} \mathcal{E}_4'(\tau, 0)$

The Siegel–Eisenstein series $\mathcal{E}_r(\tau, s)$ and, conjecturally, the generating functions $\hat{\phi}_r(\tau)$ have weight $\frac{5}{2}$ and genus r, and the last column in the chart gives the "arithmetic volume formula" of Problem 6 in each case. Some evidence for the last of these identities was obtained in joint work with M. Rapoport [53].

In the case of a prime p of good reduction a model of M over $\mathrm{Spec}\,(\mathbb{Z}_p)$ is defined in [53], and cycles are defined by imposing special endomorphisms. For $r = 4$, the main results of [53] give a criterion for $\mathcal{Z}(T)$ to be a 0–cycle in a fiber

$\mathcal{M}_p$ and show that, when this is the case, then $\widehat{\deg}((\mathcal{Z}(T),0))\,q^T = \mathcal{E}'_{4,T}(\tau,0)$. The calculation of the left hand side is again based on the result of Gross and Keating, [28]. For $r=1$, the results of [43, in particular Sections 5 and 6], are consistent with the identity in the first row, which involve arithmetic volumes of divisors.

Example 5: Divisors. For any n, when $r=1$, the arithmetic volume formula predicts that the second term $\mathcal{E}'_1(\tau,n/2)$ in the Laurent expansion of an elliptic modular Eisenstein series of weight $n/2+1$ at the point $s_0=n/2$ has Fourier coefficients involving the arithmetic volumes $\langle\widehat{\mathcal{Z}}(t,v),\hat{\omega}^n\rangle$ of divisors on the integral model $\mathcal{M}$ of M. The first term $\mathcal{E}_1(\tau,n/2)$ in the Laurent expansion at this point has Fourier coefficients involving the usual volumes of the corresponding geometric cycles. For example, for a suitable choice of V, $\mathcal{E}(\tau,n/2)$ is a familiar classical Eisenstein series, e.g., $E_2(\tau)$ (non-holomorphic), $E_4(\tau)$, $E_6(\tau)$, etc., for $\dim(V)$ even, and Cohen's Eisenstein series $E_{n/2+1}(\tau)$, [16], for $\dim(V)$ odd. This means that the second term in the Laurent expansion of such classical Eisenstein series should contain information from arithmetic geometry! Again, related results are obtained in [43].

An Important Construction. We conclude this section with an important identity which relates the generating function for height pairings with that for arithmetic degrees. Suppose that $n=2r-1$ is odd. Then the various conjectural identities above, in particular (5.7) and (6.6), lead to the formula:

$$\begin{aligned}\langle\widehat{\phi}_r(\tau_1),\widehat{\phi}_r(\tau_2)\rangle &= \widehat{\deg}\big(\widehat{\phi}_r(\tau_1)\cdot\widehat{\phi}_r(\tau_2)\big)\\ &= \widehat{\deg}\,\widehat{\phi}_{2r}\big(\big(\begin{smallmatrix}\tau_1 & \\ & \tau_2\end{smallmatrix}\big)\big) && \text{(via (5.7))}\\ &= \mathcal{E}'_{2r}\big(\big(\begin{smallmatrix}\tau_1 & \\ & \tau_2\end{smallmatrix}\big),0\big) && \text{(via (6.6))}\end{aligned}\tag{6.10}$$

relating the height pairing of the series $\widehat{\phi}_r(\tau)\in\widehat{CH}^r(\mathcal{M})$ in the middle degree with the restriction of the central derivative of the Siegel–Eisenstein series $\mathcal{E}_{2r}(\tau,s)$ of genus $2r$ and weight $r+\frac{1}{2}$. This weight is always half-integral. These series are the "incoherent" Eisenstein series discussed in [41] and [42]. The conjectural identity (6.10) will be used in an essential way in the next section.

III. Derivatives of L-Series

In this part, we explain how the modularity of the arithmetic theta functions and the conjectural relations between their inner products and derivatives of Siegel Eisenstein series might be connected with higher dimensional Gross–Zagier type formulas expressing central derivatives of certain L-functions in terms of height pairings of special cycles. These formulas should be analogues of those connecting *values* of certain L-functions to inner products of theta lifts, such as the Rallis inner product formula [67; 58; 50, Section 8].

7. Arithmetic Theta Lifts

Suppose that $f \in S^{(r)}_{n/2+1}$ is a holomorphic Siegel cusp form of weight $n/2+1$ and genus r for some subgroup $\Gamma' \subset \mathrm{Sp}_r(\mathbb{Z})$. Then, assuming the existence of the generating function $\widehat{\phi}_r$ valued in $\widehat{CH}^r(\mathcal{M})$ and that this function is also modular for Γ', we can define an *arithmetic theta lift*:

$$\begin{aligned}\hat{\theta}_r(f) &:= \langle\, f, \widehat{\phi}_r \,\rangle_{\mathrm{Pet}} \\ &= \int_{\Gamma'\backslash\mathfrak{H}_r} f(\tau)\overline{\widehat{\phi}_r(\tau)}\,\det(v)^{n/2+1}\,d\mu(\tau) \in \widehat{CH}^r(\mathcal{M}),\end{aligned}$$

where $\langle\ ,\ \rangle_{\mathrm{Pet}}$ is the Petersson inner product. Thus, we get a map

$$S^{(r)}_{n/2+1} \to \widehat{CH}^r(\mathcal{M}), \qquad f \mapsto \hat{\theta}_r(f).$$

This map is an arithmetic analogue of a correspondence like the Shimura lift [69] from forms of weight $\frac{3}{2}$ to forms of weight 2, which can be defined by integration against a classical theta function, [63]. For example, if f is a Hecke eigenform, then $\hat{\theta}_r(f)$ will also be an Hecke eigenclass.

8. Connections with Derivatives of L-Functions

Restricting to the case $n = 2r-1$, where the target is the arithmetic Chow group $\widehat{CH}^r(\mathcal{M})$ in the middle dimension, we can compute the height pairing of the classes $\hat{\theta}_r(f)$ using identity (6.10) above:

$$\begin{aligned}\langle\, \hat{\theta}_r(f_1), \hat{\theta}_r(f_2)\,\rangle &= \langle\ \langle f_1, \widehat{\phi}_r\rangle_{\mathrm{Pet}}, \langle f_2, \widehat{\phi}_r\rangle_{\mathrm{Pet}}\ \rangle \\ &= \big\langle\, f_1 \otimes \bar{f}_2, \langle \widehat{\phi}_r(\tau_1), \widehat{\phi}_r(\tau_2)\rangle\,\big\rangle_{\mathrm{Pet}} \\ &= \big\langle\, f_1 \otimes \bar{f}_2, \mathcal{E}'_{2r}\big(\big(\begin{smallmatrix}\tau_1 & \\ & \tau_2\end{smallmatrix}\big), 0\big)\,\big\rangle_{\mathrm{Pet}} \qquad \text{by (6.10)} \\ &= \frac{\partial}{\partial s}\big\langle\, f_1 \otimes \bar{f}_2, \mathcal{E}_{2r}\big(\big(\begin{smallmatrix}\tau_1 & \\ & \tau_2\end{smallmatrix}\big), s\big)\big\rangle_{\mathrm{Pet}}\Big|_{s=0}.\end{aligned} \tag{8.1}$$

Here we use the hermitian extension of the height pairing (6.2) to $\widehat{CH}^1(\mathcal{M})_{\mathbb{C}}$ taken to be conjugate linear in the second argument. Aficionados of Rankin–Selberg integrals will now recognize in the last line of (8.1) the *doubling integral* of Rallis and Piatetski-Shapiro [64], and, in classical language, of Böcherer [1] and Garrett [22]:

$$\big\langle\, f_1 \otimes \bar{f}_2, \mathcal{E}_{2r}\big(\big(\begin{smallmatrix}\tau_1 & \\ & \tau_2\end{smallmatrix}\big), s\big)\,\big\rangle_{\mathrm{Pet}} = \langle f_1, f_2\rangle_{\mathrm{Pet}}\, L(s+\tfrac{1}{2}, \pi)\, B(s), \tag{8.2}$$

where

$$
\begin{aligned}
S^{(r)}_{r+1/2} \ni f \longleftrightarrow F = {} & \text{ automorphic form for } H(\mathbb{A}), \text{ for } H = \mathrm{SO}(r{+}1, r);\\
& \text{under the analogue of the Shimura–Waldspurger correspondence}\\
& \text{between forms of weight } r + \tfrac{1}{2} \text{ on } Mp_r \text{ and forms on } \mathrm{SO}(r{+}1, r);\\
\pi = {} & \text{the irreducible automorphic cuspidal representation attached to } F;\\
L(s, \pi) = {} & \text{the degree } 2r \text{ Langlands } L\text{-function attached to } \pi\\
& \text{and the standard representation of the } L\text{-group } H^\vee = Sp_r(\mathbb{C})\\
B(s) = {} & \text{contribution of bad local zeta integrals.}
\end{aligned}
$$

Considerations of local theta dichotomy [35; 39] control the local root numbers of $L(s, \pi)$ so that the global root number is -1 and $L(\frac{1}{2}, \pi) = 0$. Combining (8.1) and (8.2), we obtain the *arithmetic inner product formula*:

$$\big\langle \hat{\theta}_r(f), \hat{\theta}_r(f) \big\rangle = \langle f, f \rangle_{\mathrm{Pet}} \, L'\big(\tfrac{1}{2}, \pi\big) \, B\big(\tfrac{1}{2}\big). \tag{8.3}$$

Of course, this is only conjectural! For a general discussion of what one expects of such central critical values, see [27].

Example 1: The Gross–Kohnen–Zagier formula. In the case $n = r = 1$, we have

$$
\begin{aligned}
M &= \text{Shimura curve},\\
\mathcal{M} &= \text{integral model},\\
f &= \text{weight } \tfrac{3}{2},\\
F &= \text{corresponding form of weight 2 (assumed a normalized newform)},\\
\pi &= \text{associated automorphic representation of } PGL_2(\mathbb{A}),\\
\hat{\theta}_1(f) &\in \widehat{CH}^1(\mathcal{M}),\\
L(s, \pi) &= L\big(s{+}\tfrac{1}{2}, F\big),\\
&= \text{the standard Hecke } L\text{-function (with } s \mapsto 1{-}s \text{ functional eq.)}
\end{aligned}
\tag{8.4}
$$

— in other words, the Langlands L-function normalization. In this case, identity (8.3) becomes

$$\langle \, \hat{\theta}_1(f), \hat{\theta}_1(f) \, \rangle = \|f\|^2 \, L'(1, F) \, B(\tfrac{1}{2}) \tag{8.5}$$

This is essentially the Gross–Kohnen–Zagier formula [29; Theorem C].

Example 2: Curves on Siegel 3-folds. The next example is $n = 3$ and $r = 2$.

Then:

$$\begin{aligned} M &= \text{Siegel 3-fold}, \\ \mathcal{M} &= \text{arithmetic 4-fold}, \\ f &= \text{a Siegel cusp form of weight } \tfrac{5}{2} \text{ and genus 2}, \\ \pi &= \text{corresponding automorphic representation of } O(3,2), \\ L(s,\pi) &= \text{the degree 4 } L\text{-function of } \pi. \end{aligned} \tag{8.6}$$

The cycles $\mathcal{Z}(T)$ in the generating function $\widehat{\phi}_2(\tau)$ are now Shimura curves on the generic fiber $M = \mathcal{M}_{\mathbb{Q}}$, extended to arithmetic surfaces in the arithmetic 4-fold $\mathcal{M}$, and

$$\hat{\theta}_2(f) \in \widehat{CH}^2(\mathcal{M}). \tag{8.7}$$

Then identity (8.3) says that the the central derivative of the degree 4 L-function $L(s,\pi)$ is expressible in terms of the height pairing $\langle \hat{\theta}_2(f), \hat{\theta}_2(f)\rangle$ of the class $\hat{\theta}_2(f)$ i.e., made out of the f-eigencomponents of "curves on a Siegel 3-fold". Of course, the proof of such a formula by the method outlined here requires that we prove the relevant versions of (5.7), (6.6) and (6.10), and, above all, the modularity of the codimension 2 generating function $\widehat{\phi}_2(\tau)$. Needless to say, this remains very speculative!

Example 3: The central derivative of the triple product L-function. This case involves a slight variant of the previous pattern. If we take V of signature $(2,2)$, we have

$$M = \begin{cases} M_1 \times M_1, \quad M_1 = \text{modular curve or Shimura curve}, \\ \text{Hilbert–Blumenthal surface} \\ \text{compact Hilbert–Blumenthal type surface} \end{cases} \tag{8.8}$$

where, in the two cases in the first line, the discriminant of V is a square and $\text{witt}(V) = 2$ or 0, respectively, while, in the second two cases, $\boldsymbol{k} = \mathbb{Q}(\sqrt{\text{discr}\, V})$ is a real quadratic field and $\text{witt}(V) = 1$ or 0 respectively. Then, $\mathcal{M}$ is an arithmetic 3-fold, and, conjecturally, the generating function

$$\widehat{\phi}_1(\tau) \in \widehat{CH}^1(\mathcal{M}) \tag{8.9}$$

is a modular form of weight 2 for a subgroup $\Gamma' \subset \mathrm{SL}_2(\mathbb{Z})$. Note that, on the generic fiber, the cycles $\mathcal{Z}(t)_{\mathbb{Q}}$ are the Hirzebruch–Zagier curves [51]. For a cusp form $f \in S_2(\Gamma')$, we obtain a class

$$\hat{\theta}_1(f) \in \widehat{CH}^1(\mathcal{M}). \tag{8.10}$$

Consider the trilinear form on $\widehat{CH}^1(\mathcal{M})$ defined by

$$\langle \hat{z}_1, \hat{z}_2, \hat{z}_3 \rangle := \widehat{\deg}(\hat{z}_1 \cdot \hat{z}_2 \cdot \hat{z}_3). \tag{8.11}$$

Then, for a triple of cusp forms of weight 2,

$$\begin{aligned}\langle\hat\theta(f_1),\hat\theta(f_2),\hat\theta(f_3)\rangle &= \langle f_1f_2f_3, \langle\widehat{\phi}_1(\tau_1),\widehat{\phi}_1(\tau_2),\widehat{\phi}_1(\tau_3)\rangle\rangle_{\mathrm{Pet}}\\ &= \Big\langle f_1f_2f_3, \widehat{\deg}\Big(\widehat{\phi}_3\Big(\begin{pmatrix}\tau_1&&\\&\tau_2&\\&&\tau_3\end{pmatrix}\Big)\Big)\Big\rangle_{\mathrm{Pet}}\\ &= \Big\langle f_1f_2f_3, \mathcal{E}_3'\Big(\begin{pmatrix}\tau_1&&\\&\tau_2&\\&&\tau_3\end{pmatrix},0\Big)\Big\rangle_{\mathrm{Pet}}\\ &= \frac{\partial}{\partial s}\Big\langle f_1f_2f_3, \mathcal{E}_3\Big(\begin{pmatrix}\tau_1&&\\&\tau_2&\\&&\tau_3\end{pmatrix},s\Big)\Big\rangle_{\mathrm{Pet}}\Big|_{s=0}.\end{aligned} \tag{8.12}$$

If we assume that the f_i's are newforms with associated cuspidal automorphic representations π_i, $i=1,2,3$, then the integral in the last line is (apart from the fact that one must actually work with the similitude group GSp_3) the integral representation of the following triple product L-function (see [23, 65, 30, 2]):

$$\Big\langle f_1f_2f_3, \mathcal{E}_3\Big(\begin{pmatrix}\tau_1&&\\&\tau_2&\\&&\tau_3\end{pmatrix},s\Big)\Big\rangle_{\mathrm{Pet}} = B(s)\,L\big(s+\tfrac12,\pi_1\otimes\pi_2\otimes\pi_3\big). \tag{8.13}$$

The results of D. Prasad on dichotomy for local trilinear forms [66] control the local root numbers and the "target" space V.

Here, in addition to the modularity of the generating function $\widehat{\phi}_1(\tau)$, we have used the conjectural identities

$$\widehat{\phi}_1(\tau_1)\cdot\widehat{\phi}_1(\tau_2)\cdot\widehat{\phi}_1(\tau_3) \overset{??}{=} \widehat{\phi}_3\Big(\begin{pmatrix}\tau_1&&\\&\tau_2&\\&&\tau_3\end{pmatrix}\Big), \tag{8.14}$$

analogous to (5.7), and

$$\widehat{\deg}\,\widehat{\phi}_3(\tau) \overset{??}{=} \mathcal{E}_3'(\tau,0), \tag{8.15}$$

analogous to (6.6). The equality of certain coefficients on the two sides of (8.15) follows from the result of Gross and Keating [28] and the formulas of Kitaoka [38]. See also [51].

In fact, one of the starting points of my long crusade to establish connections between heights and Fourier coefficients of central derivatives of Siegel–Eisenstein series was an old joint project with Gross and Zagier, of which [30] was a preliminary "exercise". The other was my collaboration with Michael Harris on Jacquet's conjecture about the central value of the triple product L-function [33; 34], based in turn on a long collaboration with Steve Rallis on the Siegel–Weil formula. And, of course, the geometric picture which serves as an essential guide comes from joint work with John Millson. I would like to thank them all, together with my current collaborators Michael Rapoport and Tonghai Yang, for their generosity with their ideas, advice, encouragement, support and patience.

Appendix: Shimura Curves

In this appendix, we illustrate some of our basic constructions in the case of modular and Shimura curves. In particular, this allows us to make a direct connection with classical Heegner points, one of the main themes of the conference.

In the case of a rational quadratic space V of signature $(1,2)$, the varieties of part I are the classical Shimura curves. Let B be an indefinite quaternion algebra over $\mathbb{Q}$, and let

$$V = \{x \in B \mid \operatorname{tr}(x) = 0\}, \qquad Q(x) = \nu(x) = -x^2. \tag{A.1}$$

The associated bilinear form is $(x, y) = \operatorname{tr}(xy^\iota)$, where $x \mapsto x^\iota$ is the main involution on B. The action of $B^\times$ on V by conjugation induces an isomorphism

$$B^\times \xrightarrow{\sim} G = \operatorname{GSpin}(V). \tag{A.2}$$

We fix an isomorphism

$$B_\mathbb{R} = B \otimes_\mathbb{Q} \mathbb{R} \xrightarrow{\sim} M_2(\mathbb{R}), \tag{A.3}$$

and obtain an identification

$$\mathbb{P}^1(\mathbb{C}) \setminus \mathbb{P}^1(\mathbb{R}) \xrightarrow{\sim} D, \qquad z \mapsto w(z) = \begin{pmatrix} z & -z^2 \\ 1 & -z \end{pmatrix} \quad \operatorname{mod} \mathbb{C}^\times. \tag{A.4}$$

Let S be the set of the primes p for which $B_p = B \otimes_\mathbb{Q} \mathbb{Q}_p$ is a division algebra and let $D(B) = \prod_{p \in S} p$. For a fixed maximal order O_B of B, there is an isomorphism

$$\begin{aligned} B(\mathbb{A}_f) &\xrightarrow{\sim} \Big(\prod_{p \in S} B_p\Big) \times M_2(\mathbb{A}_f^S), \\ O_B \otimes_\mathbb{Z} \widehat{\mathbb{Z}} &\xrightarrow{\sim} \Big(\prod_{p \in S} O_{B,p}\Big) \times M_2(\widehat{\mathbb{Z}}^S). \end{aligned} \tag{A.5}$$

For an integer N prime to $D(B)$, let R be the Eichler order of discriminant $ND(B)$ with

$$R \otimes_\mathbb{Z} \widehat{\mathbb{Z}} \xrightarrow{\sim} \Big(\prod_{p \in S} O_{B,p}\Big) \times \big\{x \in M_2(\widehat{\mathbb{Z}}^S) \mid c \equiv 0 \,(\operatorname{mod} N)\big\}. \tag{A.6}$$

Then, for the compact open subgroup $K = (R \otimes_\mathbb{Z} \widehat{\mathbb{Z}})^\times \subset G(\mathbb{A}_f)$, the quotient

$$X_0^B(N) := M_K(\mathbb{C}) \simeq G(\mathbb{Q}) \backslash \big(D \times G(\mathbb{A}_f)/K\big) \simeq \Gamma \backslash D^+, \tag{A.7}$$

where $\Gamma = G(\mathbb{Q})^+ \cap K = R^\times$ is the analogue for B of the modular curve $X_0(N)$. Of course, when $B = M_2(\mathbb{Q})$, we need to add the cusps. The 0-cycles $Z(t, \varphi; K)$ are weighted combinations of CM–points. These can be described as follows. If we identify $V(\mathbb{Q})$ with a subset of $B_\mathbb{R} = M_2(\mathbb{R})$, then, for

$$\begin{aligned} x &= \begin{pmatrix} b & 2c \\ -2a & -b \end{pmatrix} \in V(\mathbb{Q}) \subset M_2(\mathbb{R}), \qquad Q(x) = -(b^2 - 4ac), \\ D_x &= \big\{z \in \mathbb{P}^1(\mathbb{C}) \setminus \mathbb{P}^1(\mathbb{R}) \;\big|\; (x, w(z)) = -2(az^2 + bz + c) = 0\big\}. \end{aligned} \tag{A.8}$$

For general B, the coordinates a, b and c of x need not lie in $\mathbb{Q}$. For $d > 0$, let

$$\Omega_d = \big\{x \in V \mid Q(x) = d\big\} \tag{A.9}$$

and note that, if $x_0 \in \Omega_d(\mathbb{Q})$, then

$$\Omega_d(\mathbb{A}_f) = G(\mathbb{A}_f) \cdot x_0 = K \cdot \Omega_d(\mathbb{Q}). \tag{A.10}$$

By [40, Lemma 2.2(iii)], if $x \in V(\mathbb{Q})$, $g \in G(\mathbb{A}_f)$ and $\gamma \in G(\mathbb{Q})$, then, for the cycle defined by (2.8) above,

$$Z(\gamma x, \gamma g; K) = Z(x, g; K). \tag{A.11}$$

Thus, for any $\varphi \in S(V(\mathbb{A}_f))^K$, the weighted 0–cycle $Z(d, \varphi; K)$ on $X_0^B(N)$ is given by

$$Z(d, \varphi; K) = \sum_r \varphi(x_r)\, Z(x_r, 1; K), \tag{A.12}$$

with the notation of (2.11), where

$$\operatorname{supp}(\varphi) \cap \Omega_d(\mathbb{A}_f) = \coprod_r K \cdot x_r, \qquad x_r \in \Omega_d(\mathbb{Q}). \tag{A.13}$$

For example, if $L^\vee$ is the dual lattice of $L := R \cap V(\mathbb{Q})$, there is a Schwartz function

$$\varphi_\mu = \operatorname{char}(\mu + \widehat{L}) \in S(V(\mathbb{A}_f)) \tag{A.14}$$

for each coset $\mu + L$, for $\mu \in L^\vee$. Here $\widehat{L} = L \otimes_{\mathbb{Z}} \widehat{\mathbb{Z}}$. The group Γ acts on $L^\vee/L$, and each Γ-orbit $\mathcal{O}$ defines a K-invariant weight function

$$\varphi_{\mathcal{O}} = \sum_{\mu \in \mathcal{O}} \varphi_\mu \quad \in S(V(\mathbb{A}_f))^K. \tag{A.15}$$

Then,

$$Z(d, \varphi_{\mathcal{O}}; K) = \sum_{\substack{x \in L^\vee \cap \Omega_d(\mathbb{Q}) \\ x + L \in \mathcal{O} \\ \bmod \Gamma}} \operatorname{pr}(D_x^+), \tag{A.16}$$

where $D_x^+ = D_x \cap D^+$ and $\operatorname{pr} : D^+ \to \Gamma\backslash D^+ = X_0^B(N)$ is the projection. Here each point $\operatorname{pr}(D_x^+)$ is to be counted with multiplicity e_x^{-1}, where $2e_x$ is the order of the stablizer of x in Γ.

The Heegner cycles studied by Gross, Kohnen, and Zagier [29] can be recovered from this formalism in the case $B = M_2(\mathbb{Q})$. Of course, we take the standard identification $B_{\mathbb{R}} = M_2(\mathbb{R})$ and the maximal order $O_B = M_2(\mathbb{Z})$. For $x \in V$, we let

$$y = \tfrac{1}{2} J^{-1} x = \frac{1}{2}\begin{pmatrix} & -1 \\ 1 & \end{pmatrix}\begin{pmatrix} b & 2c \\ -2a & -b \end{pmatrix} = \frac{1}{2}\begin{pmatrix} 2a & b \\ b & 2c \end{pmatrix} = \begin{pmatrix} a & b/2 \\ b/2 & c \end{pmatrix}. \tag{A.17}$$

This is the matrix for the quadratic form denoted by $[a, b, c]$ in [29, p. 504]. Moreover, if $g \in SL_2(\mathbb{Z})$, then the action of g on $[a, b, c]$ is given by $y \mapsto {}^t g y g$, and this amounts to

$$x \mapsto g^{-1} x g \tag{A.18}$$

on the original x. Let

$$L = \left\{ x = \begin{pmatrix} b & c \\ -a & -b \end{pmatrix} \in M_2(\mathbb{Z}) \mid a \equiv b \equiv 0 \bmod 2N \text{ and } c \equiv 0 \bmod 2 \right\}, \tag{A.19}$$

and, for a coset $r \in \mathbb{Z}/2N\mathbb{Z}$, let $\varphi_{N,r} \in S(V(\mathbb{A}_f))$ be the characteristic function of the set

$$\begin{pmatrix} r & \\ & -r \end{pmatrix} + \hat{L} \quad \subset V(\mathbb{A}_f). \tag{A.20}$$

Note that the function $\varphi_{N,r}$ is K-invariant.

The set $\operatorname{supp}(\varphi) \cap \Omega_d(\mathbb{Q})$ of x's that contribute to $Z(d, \varphi; K)$ equals

$$\left\{ x = \begin{pmatrix} b & 2c \\ -2a & b \end{pmatrix} \mid b^2 - 4ac = -d, \ a \equiv 0 \bmod N, \ b \equiv r \bmod 2N \right\}. \tag{A.21}$$

This set is mapped bijectively to the set

$$\mathcal{Q}_{N,r,d} = \left\{ y = \begin{pmatrix} a & b/2 \\ b/2 & c \end{pmatrix} \;\middle|\; b^2 - 4ac = -d, \ a \equiv 0 \bmod N \text{ and } b \equiv r \bmod 2N \right\} \tag{A.22}$$

under the map $x \mapsto y$ described above. Therefore $Z(d, \varphi; K)$ is precisely the image in $\Gamma_0(N)\backslash D^+$ of the set of roots z in D^+, identified with the upper half plane, of the quadratic equations $az^2 + bz + c = 0$, with $[a, b, c] \in \mathcal{Q}_{N,r,d}$. Note that, if

$$\mathcal{Q}^+_{N,r,d} = \{ y \in \mathcal{Q}_{N,r,d} \mid a > 0 \}, \tag{A.23}$$

then

$$\mathcal{Q}_{N,r,d} \simeq \mathcal{Q}^+_{N,r,d} \cup \mathcal{Q}^+_{N,-r,d}. \tag{A.24}$$

The set of roots, counted with multiplicity, for $[a, b, c] \in \mathcal{Q}^+_{N,r,d}$ is denoted by $\mathcal{P}_{-d,r}$ in [29, p. 542], and

$$\mathcal{P}^*_{-d,r} = \mathcal{P}_{-d,r} \cup \mathcal{P}_{-d,-r}, \tag{A.25}$$

where points are counted with the sum of their multiplicities in the two sets. We conclude:

Proposition A.1. *Fix N and $r \bmod 2N$, and let $\varphi_{N,r}$ be as above. Then, for $K = K_0(N)$, as above,*

$$Z(d, \varphi_{N,r}; K) = \begin{cases} \mathcal{P}_{-d,r} + \mathcal{P}_{-d,-r} & \text{if } -d \equiv r^2 \bmod 4N \\ 0 & \text{otherwise,} \end{cases}$$

as 0-cycles on $X_0(N)$.

References

[1] S. Böcherer, "Über die Funktionalgleichung automorpher L-Funktionen zur Siegelschen Modulgruppe", *J. Reine Angew. Math.* **362** (1985), 146–168.

[2] S. Böcherer and R. Schulze-Pillot, "On the central critical value of the triple product L-function", pp. 1–46 in *Number theory* (Paris, 1993–1994), London Math. Soc. Lecture Note Ser. **235**, Cambridge, Cambridge Univ. Press, 1996.

[3] R. Borcherds, "Automorphic forms on $O_{s+2,2}(\mathbf{R})$ and infinite products", *Invent. math.* **120** (1995), 161–213.

[4] R. Borcherds, "Automorphic forms with singularities on Grassmannians", *Invent. math.* **132** (1998), 491–562.

[5] R. Borcherds, "Families of K3 surfaces", *J. Alg. Geom.* **7** (1998), 183–193.

[6] R. Borcherds, "The Gross–Kohnen–Zagier theorem in higher dimensions", *Duke Math. J.* **97** (1999), 219–233.

[7] R. Borcherds, "Correction to: "The Gross–Kohnen–Zagier theorem in higher dimensions"", *Duke Math. J.* **105** (2000), 183–184.

[8] J.-B. Bost, "Théorie de l'intersection et théorème de Riemann–Roch arithmétiques", pp. 43–88 (exp. 731) in Sém. Bourbaki, Astérisque **201–203**, 1991.

[9] J.-B. Bost, "Potential theory and Lefschetz theorems for arithmetic surfaces", *Ann. Sci. École Norm. Sup.* **32** (1999), 241–312.

[10] J.-B. Bost, Lecture at Univ. of Maryland, November 11, 1998.

[11] J.-B. Bost, H. Gillet and C. Soulé, "Heights of projective varieties and positive Green forms", *J. Amer. Math. Soc.* **7** (1994), 903–1027.

[12] J. H. Bruinier, "Borcherds products and Chern classes of Hirzebruch–Zagier divisors", *Invent. Math.* **138** (1999), 51–83.

[13] J. H. Bruinier, *Borcherds products on $O(2,1)$ and Chern classes of Heegner divisors*, Lecture Notes in Math. **1780**, New York, Springer, 2002.

[14] J. H. Bruinier, J. Burgos, and U. Kühn, "Borcherds products and arithmetic intersection theory on Hilbert modular surfaces", preprint, 2003.

[15] J. Burgos, J. Kramer and U. Kühn, "Cohomological arithmetic Chow rings", preprint, 2003.

[16] H. Cohen, "Sums involving the values at negative integers of L-functions of quadratic characters", *Math. Ann.* **217** (1975), 271–285.

[17] P. Deligne, "Travaux de Shimura", exp. 389 in *Sém. Bourbaki* 1970/71, Lecture Notes in Math. **244**, Berlin, Springer, 1971

[18] E. Freitag and C. F. Hermann, "Some modular varieties of low dimension", *Advances in Math,* **152** (2000), 203–287.

[19] J. Funke, "Rational quadratic divisors and automorphic forms", Thesis, Univ. of Maryland, 1999.

[20] J. Funke, "Heegner Divisors and non-holomorphic modular forms", *Compositio Math.* **133** (2002), 289–321.

[21] J. Funke and J. Millson, "Cycles in non-compact manifolds of hyperbolic type and Fourier coefficients of Siegel modular forms", *Manuscripta Math.* **107** (2002), 409–444.

[22] P. Garrett, "Pullbacks of Eisenstein series; applications", in *Automorphic forms of several variables* (Tanaguchi Symposium, Katata, 1983), Boston, Birkhäuser, 1984.

[23] P. Garrett, "Decomposition of Eisenstein series: Rankin triple products", *Ann. Math.* **125** (1987), 209–235.

[24] H. Gillet and C. Soulé, "Arithmetic intersection theory", *Publ. Math. IHES* **72** (1990), 93–174.

[25] V. Gritsenko and V. Nikulin, "Siegel automorphic corrections of some Lorentzian Kac–Moody Lie algebras", *Amer. J. Math.* **119** (1997), 181–224.

[26] B. H. Gross, "On canonical and quasi-canonical lifting", *Invent. math.* **84** (1986), 321–326.

[27] B. H. Gross, "L-functions at the central critical point", pp. 527–535 in *Motives*, edited by Uwe Jannsen et al., *Proc. Symp. Pure Math* **55**, Amer. Math. Soc., Providence, RI, 1994.

[28] B. H. Gross and K. Keating, "On the intersection of modular correspondences", *Invent. Math.* **112** (1993), 225–245.

[29] B. H. Gross, W. Kohnen and D. Zagier, "Heegner points and derivatives of L-functions, II", *Math. Annalen* **278** (1987), 497–562.

[30] B. H. Gross and S. Kudla, "Heights and the central critical values of triple product L-functions", *Compositio Math.* **81** (1992), 143–209.

[31] B. H. Gross and D. Zagier, "Heegner points and the derivatives of L-series", *Inventiones math.* **84** (1986), 225–320.

[32] M. Harris, "Arithmetic vector bundles and automorphic forms on Shimura varieties I", *Invent. Math.* **82** (1985), 151–189.

[33] M. Harris and S. Kudla, "The central critical value of a triple product L-function", *Ann. Math.* **133** (1991), 605–672.

[34] M. Harris and S. Kudla, "On a conjecture of Jacquet", in *Contributions to automorphic forms, geometry and number theory,* edited by H. Hida, D. Ramakrishnan and F. Shahidi, Johns Hopkins Univ. Press, Baltimore, 2004; preprint, 2001, math.NT/0111238.

[35] M. Harris, S. Kudla, and W. Sweet, "Theta dichotomy for unitary groups", *Jour. of the AMS* **9** (1996), 941–1004.

[36] F. Hirzebruch and D. Zagier, "Intersection numbers of curves on Hilbert modular surfaces and modular forms", of Nebentypus *Invent. Math.* **36** (1976), 57–113.

[37] Y. Kitaoka, "A note on local densities of quadratic forms", *Nagoya Math. J.* **92** (1983), 145–152.

[38] Y. Kitaoka, "Fourier coefficients of Eisenstein series of degree 3", *Proc. of Japan Acad.* **60** (1984), 259–261.

[39] S. Kudla, "On the theta correspondence", unpublished notes from lectures at the European School of Group Theory, Sept. 1–15, 1996.

[40] S. Kudla, "Algebraic cycles on Shimura varieties of orthogonal type", *Duke Math. J.* **86** (1997), 39–78.

[41] S. Kudla, "Central derivatives of Eisenstein series and height pairings", *Ann. Math.* **146** (1997), 545–646.

[42] S. Kudla, "Derivatives of Eisenstein series and generating functions for arithmetic cycles", pp. 341–368 (exp. 876) in *Sém. Bourbaki*, Astérisque **276**, Paris, Soc. math. de France, 2002.

[43] S. Kudla, "Integrals of Borcherds forms", *Compositio Math* **137** (2003), 293–349.

[44] S. Kudla, "Eisenstein series and arithmetic geometry", pp. 173–183 in *Proc. International Congress of Mathematicians* (Beijing, 2002), vol. II, Higher Ed. Press, Beijing, 2002.

[45] S. Kudla, "Modular forms and arithmetic geometry", pp. 135–180 in *Current Developments in Math.* 2002, International Press, Boston, 2003.

[46] S. Kudla and J. Millson, "The theta correspondence and harmonic forms I", *Math. Annalen* **274** (1986), 353–378.

[47] S. Kudla and J. Millson, "The theta correspondence and harmonic forms II", *Math. Annalen* **277** (1987), 267–314.

[48] S. Kudla and J. Millson, "Intersection numbers of cycles on locally symmetric spaces and Fourier", coefficients of holomorphic modular forms in several complex variables *Publ. Math. IHES* **71** (1990), 121–172.

[49] S. Kudla and J. Millson, "Tubes, cohomology with growth conditions and an application to the theta", correspondence *Canad. J. Math.* **40** (1988), 1–37.

[50] S. Kudla and S. Rallis, "A regularized Siegel–Weil formula: the first term identity", *Ann. Math.* **140** (1994), 1–80.

[51] S. Kudla and M. Rapoport, "Arithmetic Hirzebruch–Zagier cycles", *J. reine angew. Math.* **515** (1999), 155–244.

[52] S. Kudla and M. Rapoport, "Height pairings on Shimura curves and p-adic uniformization", *Invent. math.* **142** (2000), 153–223.

[53] S. Kudla and M. Rapoport, "Cycles on Siegel threefolds and derivatives of Eisenstein series", *Ann. Scient. Éc. Norm. Sup.* **33** (2000), 695–756.

[54] S. Kudla, M. Rapoport and T. Yang, "On the derivative of an Eisenstein series of weight 1", *Int. Math. Res. Notices* **1999**:7 (1999), 347–385.

[55] S. Kudla, M. Rapoport and T. Yang, "Derivatives of Eisenstein series and Faltings heights", to appear in *Compositio Math.*

[56] S. Kudla, M. Rapoport and T. Yang, in preparation.

[57] U. Kühn, "Generalized arithmetic intersection numbers", *J. reine angew. Math.* **534** (2001), 209–236.

[58] J.-S. Li, "Non-vanishing theorems for the cohomology of certain", arithmetic quotients *J. reine angew. Math.* **428** (1992), 177–217.

[59] E. Looijenga, "Compactifications defined by arrangements II: locally symmetric varieties of type IV", preprint, 2002, math.AG/0201218.

[60] V. Maillot and D. Roessler, "Conjectures sur les dérivées logarithmiques des fonctions L d'Artin aux entiers négatifs", preprint, 2001.

[61] W. J. McGraw, "On the rationality of vector-valued modular forms", *Math. Annalen* **326** (2003), 105–122.

[62] J. Milne , "Canonical models of (mixed) Shimura varieties and automorphic vector bundles", pp. 283–414 *Automorphic forms, Shimura varieties and L-functions, Perspect. Math.* **10**, Boston, Academic Press, 1990.

[63] S. Niwa, "Modular forms of half integral weight and the integral of certain theta-functions", *Nagoya Math. J.* **56** (1975), 147–161.

[64] I. I. Piatetski-Shapiro and S. Rallis, "L-functions for classical groups", pp. 1–52 in *Explicit constructions of automorphic L-functions*, Lecture Notes in Math. **1254**, New York, Springer, 1987.

[65] I. I. Piatetski-Shapiro and S. Rallis, "Rankin triple L-functions", *Compositio Math.* **64** (1987), 31–115.

[66] D. Prasad, "Trilinear forms for representations of $GL(2)$ and local", ε-factors *Compositio Math.* **75** (1990), 1–46.

[67] S. Rallis , "Injectivity properties of liftings associated to Weil representations", *Compositio Math.* **52** (1984), 139–169.

[68] I. Satake, *Algebraic structures of symmetric domains*, Publ. Math. Soc. of Japan **14**, Princeton, Univ. Press, 1980.

[69] G. Shimura, "On modular forms of half integral weight", *Ann. Math.* **97** (1973), 440–481.

[70] G. Shimura, "The arithmetic of certain zeta functions and automorphic forms on orthogonal groups", *Ann. Math.* **111** (1980), 313–375.

[71] G. Shimura, "Confluent hypergeometric functions on tube domains", *Math. Ann.* **260** (1982), 269–302.

[72] C. Soulé, D. Abramovich, J.-F. Burnol, and J.Kramer, *Lectures on Arakelov Geometry*, Cambridge Stud. Adv. Math. **33**, Cambridge, U. Press., 1992.

[73] G. van der Geer, "On the geometry of a Siegel modular threefold", *Math. Ann.* **260** (1982), 317–350.

[74] G. van der Geer, *Hilbert modular surfaces*, New York, Springer, 1988.

[75] J.-L. Waldspurger, "Correspondance de Shimura", *J. Math. Pures Appl.* **59** (1980), 1–132.

[76] A. Weil, "Sur la formule de Siegel dans la théorie des groupes classiques", *Acta Math.* **113** (1965), 1–87.

[77] T. Yang, "An explicit formula for local densities of quadratic forms", *J. Number Theory* **72** (1998), 309–356.

[78] T. Yang, "The second term of an Eisenstein series", to appear in *Proceedings of ICCM.*

[79] T. Yang, "Faltings heights and the derivative of Zagier's Eisenstein series", pp. 271–284 in *Heegner Points and Rankin L-Series*, edited by Henri Darmon and Shou-Wu Zhang, *Math. Sci. Res. Inst. Publications* **49**, Cambridge U. Press, New York, 2004.

[80] D. Zagier, "Nombres de classes et formes modulaires de poids 3/2", *C. R. Acad. Sc. Paris* **281** (1975), 883–886.

[81] D. Zagier, "Modular points, modular curves, modular surfaces and modular forms", pp. 225–248 in *Arbeitstagung Bonn*, 1984, Lecture Notes in Math. **1111**, Berlin, Springer, 1985.

Stephen S. Kudla
Mathematics Department
University of Maryland
College Park, MD 20742
United States
ssk@math.umd.edu

Heegner Points and Rankin L-Series
MSRI Publications
Volume **49**, 2004

Faltings Heights and the Derivative of Zagier's Eisenstein Series

TONGHAI YANG

D. Zagier discovered in 1975 [HZ] the following famous modular form of weight $\frac{3}{2}$ for $\Gamma_0(4)$:

$$E_{\mathrm{Zagier}}(\tau) = -\frac{1}{12} + \frac{1}{8\pi\sqrt{v}} + \sum_{m=1}^{\infty} H_0(m) q^m + \sum_{n>0} 2g(n,v) q^{-n^2}. \tag{0.1}$$

Here $\tau = u + iv$ is in the upper half plane, $q = e^{2\pi i \tau}$, $H_0(m)$ is the Hurwitz class number of binary quadratic forms of discriminant $-m$, and

$$g(n,v) = \frac{1}{16\pi\sqrt{v}} \int_1^{\infty} e^{-4\pi n^2 v r} r^{-3/2} dr. \tag{0.2}$$

This function can be obtained, via analytic continuation, as a special value of an Eisenstein series $\mathcal{E}(\tau, s)$ at $s = \frac{1}{2}$. In this note, we will give an arithmetic interpretation to Zagier's Eisenstein series and its derivative at $s = \frac{1}{2}$, using Arakelov theory.

Let $\mathcal{M}$ be the Deligne–Rapoport compactification of the moduli stack over $\mathbb{Z}$ of elliptic curves [DR]. In Section 3 we will define a generating function of arithmetic Chow cycles of codimension 1 in $\mathcal{M}$ with real coefficients, in the sense of Bost and Kühn [Bos1,Kun]:

$$\hat{\phi}(\tau) = \sum_{m \in \mathbb{Z}} \hat{\mathcal{Z}}(m, v) q^m, \tag{0.3}$$

such that

$$2 \deg \hat{\phi}(\tau) = \mathcal{E}(\tau, \tfrac{1}{2}) = E_{\mathrm{Zagier}}(\tau), \tag{0.4}$$

and

$$4\langle \hat{\phi}, \hat{\omega} \rangle = \mathcal{E}'(\tau, \tfrac{1}{2}). \tag{0.5}$$

Here $\hat{\omega}$ is a normalized metrized Hodge bundle on $\mathcal{M}$, to be defined in Section 3, and $\langle\, , \rangle$ is the Gillet–Soule intersection pairing ([GS]; see also section 2). Bost's arithmetic Chow cycles with *real* coefficients are crucial here since, for example,

Partially supported by NSF grant DMS-0070476.

the negative Fourier coefficients of Zagier's Eisenstein series are clearly not rational numbers. This note is a slight variation of joint work with Stephen Kudla and Michael Rapoport [KRY2]. It confirms a conjecture of Kudla [Ku2].

1. The Chowla–Selberg Formula

Part of the formulas (0.4) and (0.5) can be viewed as a generalization of the Chowla–Selberg formula which we now describe.

For a positive integer $m \neq 0$, let $K_m = \mathbb{Q}(\sqrt{-m})$ and let $\mathcal{O}_m$ be the order in K_m of discriminant $-m$. Notice that $\mathcal{O}_m$ exists if and only if $m \equiv 0, -1 \bmod 4$. We can and will write $m = dn^2$ such that $-d$ is the fundamental discriminant of K_m and $n \geq 1$ is an integer.

Let $Z(m)$ be the set of isomorphic classes of elliptic curves E over $\mathbb{C}$ such that there is an embedding $\mathcal{O}_m \hookrightarrow \operatorname{End}(E)$. When $\mathcal{O}_m$ does not exist, we take $Z(m)$ to be empty. Let

$$\deg Z(m) = \sum_{E \in Z(m)} \frac{1}{\#\operatorname{Aut} E} \tag{1.1}$$

and

$$h_{\mathrm{Fal}}(Z(m)) = \sum_{E \in Z(m)} \frac{1}{\#\operatorname{Aut} E} h_{\mathrm{Fal}}(E), \tag{1.2}$$

where $h_{\mathrm{Fal}}(E)$ is the (renormalized) Faltings height, which measures, in some sense, the complexity of the elliptic curve E. It is defined as follows. Let L be a number field over which E is defined and has good reduction everywhere, and let ω be the Néron differential on E over $\mathcal{O}_L$. Then

$$h_{\mathrm{Fal}}(E) = -\frac{1}{2[L:\mathbb{Q}]} \sum_{\sigma : L \hookrightarrow \mathbb{C}} \log \left| \frac{i}{2\pi} e^{-C} \int_{E^\sigma(\mathbb{C})} \omega^\sigma \wedge \bar{\omega}^\sigma \right|, \tag{1.3}$$

where $C = \frac{1}{2}(\gamma + \log 4\pi)$ is a normalizing factor we have the liberty to add. Here γ is Euler's constant.

On the other hand, one can define a modified Dirichlet L-series for every integer $m \neq 0$

$$L(s, \chi_m) = L(s, \chi_d) \prod_{p \mid n} b_p(n, s) \tag{1.4}$$

where χ_d is the quadratic character associated to the quadratic field $K_m = K_d$, and $L(s, \chi_d)$ is the usual Dirichlet L-series of χ_d, and

$$b_p(n, s) = \frac{1 - \chi_d(p)X + \chi_d(p)p^k X^{1+2k} - (pX^2)^{1+k}}{1 - pX^2} \tag{1.5}$$

with $X = p^{-s}$ and $k = \operatorname{ord}_p n$. This L-series occurs in the Fourier coefficients of an Eisenstein series as we will see in Section 3. The complete L-series

$$\Lambda(s, \chi_m) = |m|^{s/2} \pi^{-(s+a)/2} \Gamma\Big(\frac{s+a}{2}\Big) L(s, \chi_m) \tag{1.6}$$

has an analytic continuation and the functional equation

$$\Lambda(s, \chi_m) = \Lambda(1 - s, \chi_m). \tag{1.7}$$

Here $a = (1 + \operatorname{sign} m)/2$. Now we can state part of our formula as

THEOREM 1.1. *With the notation as above, one has for $m \geq 1$:*

(1) $2 \deg Z(m) = \Lambda(0, \chi_m) = H_0(m)$.
(2) $4 h_{\mathrm{Fal}}(Z(m)) = -\Lambda'(0, \chi_m)$.

Only orders of the form $\mathcal{O}_{4m}$ are considered in [KRY2].

SKETCH OF PROOF. Part (1) is basically the analytic class number formula and is well-known. Part (2) is basically [KRY1, Corollary 10.12]. We give an outline for this special case for the reader's convenience.

Step 0: When $-m = -d$ is the fundamental discriminant of K_m, the formula is just the Chowla–Selberg formula (see [Gro1; Gro2; Col] for this interpretation and geometric proof): For an elliptic curve E with CM by the ring $\mathcal{O}_d$ of integers, one has

$$2 h_{\mathrm{Fal}}(E) = -\frac{\Lambda'(0, \chi_d)}{\Lambda(0, \chi_d)}.$$

In particular, the height $h_{\mathrm{Fal}}(E)$ is independent of the choice of the elliptic curves. Combining this with (1), one proves (2) for this case.

Step 1: In the general case of non-fundamental discriminants, the Chowla–Selberg formula was considered by Nakkajima and Taguchi [NT]. An elliptic curve E with CM by $K = K_d$ is of type c if $\operatorname{End} E \cong \mathcal{O}_{dc^2}$. Let E be an elliptic curve of type c. Choose a CM elliptic curve E_0 of type 1 together with an isogeny $u_L : E_0 \longrightarrow E$. Let L be a number field over which E, E_0 and u_L are defined and have good reduction everywhere. Extend u_L to an isogeny u on their Néron models over $\mathcal{O}_L$ with kernel N

$$0 \longrightarrow N \longrightarrow \mathcal{E}_0 \longrightarrow \mathcal{E} \longrightarrow 0.$$

Then Raynaud's isogeny theorem [Ra, p. 205] asserts that

$$h_{\mathrm{Fal}}(E) = h_{\mathrm{Fal}}(E_0) + \tfrac{1}{2} \log(\deg u_L) - \frac{1}{[L:\mathbb{Q}]} \log |\varepsilon^*(\Omega_{N/\mathcal{O}_L})|.$$

The term $\log |\varepsilon^*(\Omega_{N/\mathcal{O}_L})|$ can be computed locally and was done in [KRY1, Section 10]. In this special case, we can choose E_0 so that u_L is of degree c by the theory of complex multiplication, and thus only [KRY1, Propositions 10.1 and 10.3] are needed. Indeed, if $E_c = \mathbb{C}/c^{-1}\mathcal{O}_{c^2 d}$, which can actually be defined over H_c, the ring class field of $\mathcal{O}_{c^2 d}$, then $E_1 = \mathbb{C}/\mathcal{O}_d$, and the desired map $E_c \longrightarrow E_1$ is induced by the identity map on $\mathbb{C}$. In general, every elliptic curve E of type c is a Galois conjugate E_c^σ of E_c by some $\sigma \in \operatorname{Gal}(H_c/K)$, and thus the desired map is $E_c^\sigma \longrightarrow E_1^\sigma$. The end result is the following formula [KRY1, Theorem 10.7]:

$$2h_{\mathrm{Fal}}(E) = 2h_{\mathrm{Fal}}(E_0) + \log c - \sum_p \frac{(1-p^{-\operatorname{ord}_p c})(1-\chi_d(p))}{(1-p^{-1})(p-\chi_d(p))}\log p.$$

In particular, the Faltings height of an elliptic curve E of type c depends only on c. In [KRY1], abelian surfaces are considered and an extra isogeny has to be studied.

Step 2: Now combining terms together with Step 0 produces

$$2h_{\mathrm{Fal}}(Z(m)) = 2\deg Z(m)\Bigg(-\frac{\Lambda'(0,\chi_d)}{\Lambda(0,\chi_d)} + \frac{\sum_{c|n} c\prod_{l|c}(1-\chi_d(l)l^{-1})\sum_{p|c}\eta_p(\operatorname{ord}_p c)\log p}{\prod_{p|n} b_p(n,0)}\Bigg).$$

Here

$$\eta_p(r) = r - \frac{(1-p^{-r})(1-\chi_d(p))}{(1-p^{-1})(p-\chi_d(p))}.$$

On the other hand,

$$-\frac{\Lambda'(0,\chi_m)}{\Lambda(0,\chi_m)} = -\frac{\Lambda'(0,\chi_d)}{\Lambda(0,\chi_d)} + \sum_{p|n}\left(\log|n|_p - \frac{b_p'(n,0)}{b_p(n,0)}\right).$$

Now what is needed is to verify an algebraic identity, which was done in [KRY1, Lemma 10.9] using induction on the number of prime factors of n. □

2. Bost's L_1^2-Arithmetic Divisors and Intersection Theory

As a background for section 3, we briefly review Bost's L_1^2-arithmetic divisors with real coefficients and the corresponding intersection theory for the convenience of the readers. We refer to [Bos1] for detail. A similar theory was also developed by Kühn [Kun].

Let $\mathcal{M}$ be an arithmetic surface over $\mathbb{Z}$, and let $M = \mathcal{M}(\mathbb{C})$ be the corresponding Riemann surface. A generalized function ϕ on M is (locally) L_1^2 if both ϕ and $\partial\phi$ are (locally) L^2, i.e., square integrable. For example, $\log\log|z|^{-1}$ is locally L_1^2 near $z=0$, but $\log|z|$ is not. Similarly, $\log \operatorname{Im} z$ is locally L_1^2 at $z = i\infty$. It is also known that if $\phi \in L_1^2$ then e^ϕ is L^p for every $p < \infty$. A current α on M is called (locally) L_{-1}^2 if it can be written (locally) as

$$\alpha = \partial\beta$$

for some (locally) L^2 1-form β.

Bost's arithmetic Chow groups. Let $\mathcal{Z}^1(\mathcal{M})$ be the free abelian group of Weil divisors of $\mathcal{M}$, and let $\mathcal{Z}^1_{\mathbb{R}}(\mathcal{M}) = \mathcal{Z}^1(\mathcal{M}) \otimes \mathbb{R}$. An L^2_1-arithmetic divisor on $\mathcal{M}$ is a pair $\hat{D} = (D, g)$ with $D \in \mathcal{Z}^1(\mathcal{M})$ and g is a L^2_1-Green function for D in the following sense: There is a usual C^∞-Green function g_0 for D and a L^2_1-generalized function $\phi \in L^2_1(M)$ such that

$$g = g_0 + \phi. \tag{2.1}$$

Equivalently, for any local holomorphic coordinate z on an open neighborhood of M, one has

$$g = \phi + \sum_{P \in U} n_P \log |z - z(P)|^{-2} \tag{2.2}$$

for some $\phi \in L^2_1(U)$ if $D = \sum n_P P$. In this case,

$$\omega_1(\hat{D}) = dd^c g + \delta_D$$

is a "L^2_{-1}"-current on M of degree 2 [Bos1, p. 255]. Let $\widehat{\mathcal{Z}}^1(\mathcal{M})$ be the abelian group of L^2_1-arithmetic divisors in $\mathcal{M}$. Notice that (2.2) makes sense even if $n_P \in \mathbb{R}$. In such a case, we call (D, g) a L^2_1-arithmetic divisor with real coefficients, and denote the abelian group of all L^2_1-arithmetic divisors with real coefficients by $\widehat{\mathcal{Z}}^1_{\mathbb{R}}(\mathcal{M})$. For a rational function $f \in \mathbb{Q}(\mathcal{M})$,

$$\widehat{\mathrm{div}}(f) = (\mathrm{div}(f), -\log |f|^2)$$

is certainly a L^2_1-arithmetic divisor—a principal arithmetic divisor. Let $\widehat{\mathrm{CH}}^1(\mathcal{M})$ and $\widehat{\mathrm{CH}}^1_{\mathbb{R}}(\mathcal{M})$ be the quotient groups of $\widehat{\mathcal{Z}}^1(\mathcal{M})$ and $\widehat{\mathcal{Z}}^1_{\mathbb{R}}(\mathcal{M})$ respectively by the principal arithmetic divisors. There is a natural map

$$\widehat{\mathrm{CH}}^1(\mathcal{M}) \longrightarrow \widehat{\mathrm{CH}}^1_{\mathbb{R}}(\mathcal{M})$$

whose kernel is determined by [Bos1, Theorem 5.5], and the intersection pairing on $\widehat{\mathrm{CH}}^1(\mathcal{M})$ (to be reviewed below) factors through $\widehat{\mathrm{CH}}^1_{\mathbb{R}}(\mathcal{M})$.

The Arakelov–Gillet–Soule–Bost pairing. If $\hat{D}_1 = (D_1, g_1)$ and $\hat{D}_2 = (D_2, g_2)$ are L^2_1-arithmetic divisors on $\mathcal{M}$ such that $|D_1|$ and $|D_2|$ do not meet in the generic fiber $\mathcal{M}_{\mathbb{Q}}$, their intersection is defined as (see [Bos1, (5.8)])

$$\langle \hat{D}_1, \hat{D}_2 \rangle = \widehat{\deg}\, D_1.D_2 + \frac{1}{2} \int_M g_1 * g_2 \tag{2.3}$$

where the star product integral $\int_M g_1 * g_2$ is defined as follows [Bos1, (5.1) and (5.4)]. If both g_1 and g_2 are C^∞-Green functions, one has as usual

$$g_1 * g_2 = g_1 \omega_2 + g_2 \delta_{D_1}, \tag{2.4}$$

where

$$\omega_i = dd^c g_i + \delta_{D_i}$$

is a C^∞-form of degree 2 on M. If $\tilde{g}_i = g_i + \phi_i$ such that g_i are C^∞ and $\phi_i \in L_1^2(M)$, then

$$\int_M \tilde{g}_1 * \tilde{g}_2 = \int_M g_1 * g_2 + \int_M \phi_1 \omega_2 + \int_M \phi_2 \omega_1 + \frac{1}{2\pi i} \int_M \partial \phi_1 \wedge \bar{\partial} \phi_2. \tag{2.5}$$

It is checked in [Bos1, Section 5.1]) that this is well-defined, and $\langle \hat{D}, \widehat{\operatorname{div}}(f) \rangle = 0$. So (2.3) and (2.4) give an intersection pairing on $\widehat{\mathrm{CH}}^1(\mathcal{M})$ and $\widehat{\mathrm{CH}}^1_{\mathbb{R}}(\mathcal{M})$ [Bos1, Theorem 5.5].

L_1^2-metrized line bundles and Bost's arithmetic Picard group. A L_1^2-metrized line bundle is a pair $\bar{\mathcal{L}} = (\mathcal{L}, \|\cdot\|)$ where $\mathcal{L}$ is as usual a line bundle on $\mathcal{M}$ and $\|\cdot\|$ is a L_1^2-metric on $L = \mathcal{L} \otimes \mathbb{C}$ (invariant under complex conjugation) in the following sense: There is a C^∞ metric $\|\cdot\|_0$ on L together with an $\phi \in L_1^2(L)$ such that

$$\|\cdot\|^2 = \|\cdot\|_0^2 e^{-\phi}. \tag{2.6}$$

The natural map $\widehat{\operatorname{Pic}}(\mathcal{M}) \longrightarrow \widehat{\mathrm{CH}}^1(\mathcal{M})$ given by

$$\bar{\mathcal{L}} \mapsto \hat{c}_1(\bar{\mathcal{L}}) = (\operatorname{div}(s), -\log \|s\|^2) \tag{2.7}$$

extends to the L_1^2-case, where s is a meromorphic section of $\mathcal{L}$. Let $\bar{\mathcal{L}}_0 = (\mathcal{L}, \|\cdot\|_0)$, then

$$\hat{c}_1(\bar{\mathcal{L}}) = \hat{c}_1(\bar{\mathcal{L}}_0) + (0, \phi \circ s), \tag{2.8}$$

and the corresponding first Chern form is

$$c_1(\bar{\mathcal{L}}) = c_1(\bar{\mathcal{L}}_0) + dd^c(\phi \circ s). \tag{2.9}$$

Via the map (2.7), we have then an intersection theory

$$\widehat{\operatorname{Pic}}(\mathcal{M}) \times \widehat{\mathrm{CH}}^1_{\mathbb{R}}(\mathcal{M}) \longrightarrow \mathbb{C}.$$

It can be computed directly as follows:

$$\langle \bar{\mathcal{L}}, (D, g) \rangle = h_{\bar{\mathcal{L}}}(D) + \frac{1}{2} \int_{\mathcal{M}(\mathbb{C})} g c_1(\bar{\mathcal{L}}), \tag{2.10}$$

where the height function $h_{\bar{\mathcal{L}}}(D)$ is defined as in [KRY1, pages 15-16]. The change from arithmetic surfaces to stacks is the same as in [KRY1, Section 4]

3. The Main Result

Let $\mathcal{M}_0$ be the moduli stack over $\mathbb{Z}$ of elliptic curves considered by Deligne and Rapoport [DR]. Over $\mathbb{C}$, it is the same as the orbifold

$$\mathcal{M}_0(\mathbb{C}) = [\mathrm{SL}_2(\mathbb{Z}) \backslash \mathbb{H}],$$

where each elliptic curve $E \in \mathcal{M}_0(\mathbb{C})$ is counted with multiplicity $1/(\# \operatorname{Aut} E)$, i.e., with the order of its stabilizer in $\mathrm{SL}_2(\mathbb{Z})$. Let $\mathcal{M}$ be the Deligne-Rapoport compactification of $\mathcal{M}_0$, so $\mathcal{M}(\mathbb{C})$ is $\mathcal{M}_0(\mathbb{C})$ plus the cusp ∞.

For an integer $m \geq 1$ (with $m \equiv -1, 0 \bmod 4$), let $\mathcal{Z}(m)$ be the moduli stack over $\mathbb{Z}$ of elliptic curves E such that there is $\mathcal{O}_m \hookrightarrow \mathrm{End}(E)$. Then $\mathcal{Z}(m)$ is a divisor on $\mathcal{M}$, and one has $\mathcal{Z}(m)(\mathbb{C}) = [Z(m)]$ with each elliptic curve counted $1/(\#\,\mathrm{Aut}\,E)$ times. This explains the weight in defining $\deg Z(m)$ and $h_{\mathrm{Fal}}(Z(m))$. We can view $\mathcal{Z}(m)$ as giving a class in $\mathrm{CH}^1(\mathcal{M})$.

For every integer m and a positive real number $v > 0$, Kudla constructed in [Ku1] a function $\Xi(m, v)$ on $\mathcal{M}_0(\mathbb{C})$. We will review this construction in Section 4 and sketch a proof of the following proposition.

PROPOSITION 3.1. (1) *When $m > 0$, $\Xi(m, v)$ is also smooth at the cusp ∞ and is a Green's function for $\mathcal{Z}(m)$ on $\mathcal{M}(\mathbb{C})$. That is, there is a C^∞ $(1,1)$-form $\omega(m, v)$ on M such that*

$$dd^c g + \delta_{Z(m)} = [\omega(m, v)]$$

as currents.

(2) *When $-m > 0$ is not a square, $\Xi(m, v)$ is smooth everywhere, including at the unique cusp ∞.*

(3) *When $-m = n^2 > 0$ is square, $\Xi(m, v)$ is smooth in the upper half plane but is singular at the cusp ∞. As a current, it satisfies the Green's equation*

$$\partial\bar{\partial}\Xi(m, v) + g(n, v)\delta_\infty = [\omega(m, v)] \tag{3.1}$$

for a L^2_{-1}-form $\omega(m, v)$ on M of degree 2. Here $g(n, v)$ is given in (0.2).

(4) *When $m = 0$, $\Xi(0, v)$ is smooth in the upper half plane but is singular at the cusp ∞. As a current, it satisfies the Green's equation*

$$\partial\bar{\partial}\Xi(0, v) + \frac{1}{16\pi\sqrt{v}}\,\delta_\infty = [\omega(0, v)]$$

for a L^2_{-1}-form $\omega(0, v)$ on M of degree 2.

Because of this proposition, we can define the arithmetic Chow cycles with real coefficients $\hat{\mathcal{Z}}(m, v) \in \widehat{\mathrm{CH}}^1_{\mathbb{R}}(\mathcal{M})$ for $m \neq 0$ and $v > 0$ via

$$\hat{\mathcal{Z}}(m, v) = \begin{cases} (\mathcal{Z}(m), \Xi(m, v)) & \text{if } m > 0, \\ (0, \Xi(m, v)) & \text{if } -m > 0 \text{ is not a square}, \\ (g(n, v)\cdot\infty, \Xi(m, v)) & \text{if } -m = n^2 > 0. \end{cases} \tag{3.2}$$

To define $\hat{\mathcal{Z}}(0, v)$, we need the metrized Hodge bundle $\hat{\omega}$ on $\mathcal{M}$. Let $\mathcal{E}$ be the universal elliptic curve over $\mathcal{M}$ with zero section ε. Then the Hodge line bundle is $\omega = \varepsilon^*\Omega_{\mathcal{E}/\mathcal{M}}$. Notice that ω^2 is the relative differential bundle $\Omega_{\mathcal{M}/\mathbb{Z}}$, which is associated to modular forms of weight 2. So the Hodge bundle is associated to modular forms of weight one. The metric on $\omega_{\mathbb{C}}$ is defined as follows. For a section α of $\omega_{\mathbb{C}}$, α_z at $z \in \mathcal{M}(\mathbb{C})$ corresponds to a holomorphic 1-form on the associated elliptic curve $\mathcal{E}_z$, we define

$$\|\alpha_z\|^2 = \left|\frac{i}{2\pi}e^{-C}\int_{\mathcal{E}_z(\mathbb{C})} \alpha_z \wedge \bar{\alpha_z}\right| \tag{3.3}$$

where C is the constant as in (1.3). We remark that $\hat{\omega} = (\omega, \| \ \|) \in \widehat{\mathrm{Pic}}(\mathcal{M})$ has singularity at the cusp ∞. See [Bos2] and [Kun] for detailed discussion on this issue. In these papers, Bost and Kühn independently computed the self-intersection number of $\hat{\omega}$, and their result is

$$\langle \hat{\omega}, \hat{\omega} \rangle = \tfrac{1}{2}\zeta(-1) + \zeta'(-1) + \tfrac{1}{12}C, \tag{3.4}$$

where ζ is the usual Riemann zeta function and C is the constant in (1.3). In view of the map (2.7), we may view $\hat{\omega} \in \widehat{\mathrm{CH}}^1_{\mathbb{R}}(\mathcal{M})$. Its first Chern form $c_1(\hat{\omega})$ is

$$\frac{1}{4\pi}\frac{dx \wedge dy}{y^2}.$$

Finally, we define

$$\hat{\mathcal{Z}}(0, v) = \left(\frac{1}{16\pi\sqrt{v}}\infty, \Xi(0, v)\right) - \hat{\omega} - (0, \log v) \in \widehat{\mathrm{CH}}^1_{\mathbb{R}}(\mathcal{M}) \tag{3.5}$$

and the generating function of arithmetic cycles

$$\hat{\phi}(\tau) = \sum_{m \in \mathbb{Z}} \hat{\mathcal{Z}}(m, v) q^m. \tag{3.6}$$

According to [Ku2], $\hat{\phi}(\tau)$ should be a modular form of weight 3/2 valued in the arithmetic Chow group. In particular, for any linear functional f on the arithmetic Chow group,

$$f(\hat{\phi}(\tau)) = \sum_{m \in \mathbb{Z}} f(\hat{\mathcal{Z}}(m, v)) q^m$$

should be a scalar modular form of weight 3/2. The main theorem below asserts that it is true when f is the degree map or the intersection map with $\hat{\omega}$. Here the degree map is given by

$$\deg \hat{D} = \int_{\mathcal{M}(\mathbb{C})} \omega_1(\hat{D}),$$

where $\omega_1(\hat{D}) = dd^c g + \delta_D$ if $\hat{D} = (D, g)$. To be more precise, we need to introduce the Eisenstein series. Following Hirzebruch and Zagier [HZ, pp. 91, 93] and using their notation, we let

$$E(\tau, s) = \sum_{\substack{m > 0 \\ (m, 2n) = 1}} \frac{\left(\frac{n}{m}\right)\left(\frac{-1}{m}\right)^{1/2}}{(m\tau + n)^{3/2}|m\tau + n|^{2s}} \tag{3.7}$$

and

$$\mathcal{F}(\tau, s) = -\frac{1}{96}\left((1 - i)E(s, \tau) - i\tau^{3/2}|\tau|^{-2s}E\left(s, -\frac{1}{4\tau}\right)\right). \tag{3.8}$$

They are non-holomorphic modular forms of weight $\frac{3}{2}$. We renormalize it as

$$\mathcal{E}(\tau, s) = \frac{(\frac{1}{2}+s)\Lambda(1+2s)}{\Lambda(2)}\left(\frac{v}{2}\right)^{(1/2)(s-1/2)}\mathcal{F}(\tau, \tfrac{1}{2}(s-\tfrac{1}{2})) \tag{3.9}$$

We refer to [KRY1; KRY2] for adelic construction of such Eisenstein series, which seems more natural.

THEOREM 3.2. *Let the notation be as above.*

(1) $\mathcal{E}(\tau, s) = \mathcal{E}(\tau, -s)$.
(2) $\mathcal{E}(\tau, \frac{1}{2}) = E_{\text{Zagier}}(\tau) = 2\deg(\hat{\phi}(\tau))$.
(3) $\mathcal{E}'(\tau, \frac{1}{2}) = 4\phi_{\text{height}}(\tau)$, *where*

$$\phi_{\text{height}}(\tau) = \langle \hat{\phi}(\tau), \hat{\omega} \rangle = \sum_m \langle \hat{\mathcal{Z}}(m, v), \hat{\omega} \rangle q^m.$$

In particular, when $m > 0$ the unfolding of the height pairing gives

$$\langle \hat{\mathcal{Z}}(m, v), \hat{\omega} \rangle = h_{\text{Fal}}(Z(m)) + \frac{1}{8\pi}\int_{\mathcal{M}(\mathbb{C})} \Xi(m, v)\frac{dx\,dy}{y^2}. \tag{3.10}$$

So just like the degree, the generating function of the Faltings' height $\sum_{m>0} h_{\text{Fal}}(Z(m))q^m$ is part of a modular form of weight 3/2. Moreover, we also give some arithmetic interpretation of the negative terms of Zagier's Eisenstein series as well as its 'derivative'.

We would like to point out three interesting features of Theorem 3.2. Firstly $s = \frac{1}{2}$ is *not* the symmetric center although it is a critical point. Secondly, both the value and the derivative of the Eisenstein series have interesting arithmetic meanings. In particular, the derivative here is *not* the leading term but the second term. Finally, since Zagier's Eisenstein series has real numbers $g(n, v)$ as its Fourier coefficients, any arithmetic interpretation of the Eisenstein series as degree generating function would be forced to consider arithmetic Chow cycles with *real* coefficients as we did here and in [KRY2]. It is a little amusing to us that our Greens' function $\Xi(m, v)$, originally defined for the Heegner cycle $Z(m)$ for $m > 0$, gives such a cycle and matches up perfectly with the Fourier coefficients when $-m = n^2 > 0$.

COROLLARY 3.3. *Let $\mathcal{E}(\tau, s)$ be as in Theorem 3.2. Then, for $m > 0$,*

(1) $2\deg Z(m) = \mathcal{E}_m(\tau, \frac{1}{2})q^{-m}$;
(2) $4h_{\text{Fal}}(Z(m)) = \lim_{v\mapsto\infty} \mathcal{E}'_m(\tau, \frac{1}{2})q^{-m}$.

PROOF. (1) follows directly from Theorem 3.2. (2) follows from Theorem 3.2, (3.7) and the fact that the integral in (3.7) goes to zero when v goes to infinity. □

Now we describe the Fourier expansion of the Eisenstein series. Let

$$\Psi(a, b, t) = \frac{1}{\Gamma(a)}\int_0^\infty e^{-tr}(1+r)^{b-a-1}r^{a-1}dr \tag{3.11}$$

be the classical second confluent hypergeometric function for $a > 0$ and $t > 0$, and let

$$\Psi_n(s,t) = \Psi\big(\tfrac{1}{2}(1+n+s),\, 1+s,\, t\big). \tag{3.12}$$

This function satisfies the functional equation

$$\Psi_n(-s,t) = z^s \Psi_n(s,t). \tag{3.13}$$

THEOREM 3.4. *The Eisenstein series $\mathcal{E}(\tau, s)$ in Theorem 3.2 has the following Fourier expansion*

$$\mathcal{E}(\tau,s) = \sum_{m \equiv 0,-1 \bmod 4} A_m(v,s) q^m,$$

where

(1) *for $m > 0$,*

$$A_m(v,s) = \Lambda(\tfrac{1}{2} - s, \chi_m)(4\pi m v)^{(1/2)(s-1/2)} \Psi_{-3/2}(s, 4\pi m v);$$

(2) *for $m < 0$,*

$$A_m(v,s) = \frac{(s^2 - \frac{1}{4})\Lambda(\frac{1}{2} - s, \chi_m)(4\pi|m|v)^{(1/2)(s-1/2)} \Psi_{\frac{3}{2}}(s, 4\pi|m|v)}{4\sqrt{\pi} e^{4\pi m v}};$$

(3) *the constant term is given by*

$$A_0(v,s) = -\frac{1}{2\pi}(G(s) + G(-s))$$

with

$$G(s) = (4v)^{(1/2)(1/2-s)}(s + \tfrac{1}{2})\Lambda(1+2s).$$

4. Construction of the Green's Function $\Xi(m,v)$

Although the construction is quite general [Ku1], we stick to the special case at hand.

Let

$$V = \{x = \begin{pmatrix} b & c \\ -a & -b \end{pmatrix} \in M_2(\mathbb{Q}) : \operatorname{tr} x = 0\}, \tag{4.1}$$

with the quadratic form

$$Q(x) = \det x = ac - b^2.$$

It has signature $(1,2)$. Let D be the set of negative 2-planes in $V(\mathbb{R})$. Then D is in bijection with the upper half plane $\mathbb{H}$, given by

$$z = g(i) \in \mathbb{H} \longleftrightarrow z = \{g z_1 g^{-1}, g z_2 g^{-1}\} \in D$$

for any $g \in \mathrm{GL}_2(\mathbb{R})$, where

$$z_1 = \begin{pmatrix} 1 & 0 \\ 0 & -1 \end{pmatrix}, \quad z_2 = \begin{pmatrix} 0 & 1 \\ 1 & 0 \end{pmatrix}.$$

Here $\{v_1, v_2\}$ denotes the subspace of $V(\mathbb{R})$ spanned by v_1 and v_2 in $V(\mathbb{R})$. We will use z to stand for both the complex number in $\mathbb{H}$ and the associated negative two-plane by abuse of notation. Given $z \in \mathbb{H} = D$, one has the orthogonal decomposition

$$V(\mathbb{R}) = z \oplus z^{\perp}, \quad x = (\mathrm{pr}_z\, x, \mathrm{pr}_{z^{\perp}}\, x).$$

For $x \in V(\mathbb{R})$ as in (4.1) and $z \in \mathbb{H}$, define

$$R(x, z) = -(\mathrm{pr}_z\, x, \mathrm{pr}_z\, x) = \frac{1}{2}(az\bar{z} + b(z + \bar{z}) + c - \det x). \tag{4.2}$$

Then $R(x, z) \geq 0$ and it equals zero if and only if $z \perp x$. Notice that when $x > 0$ (meaning $Q(x) > 0$), $D_x = x^{\perp}$ is a negative 2-plane and thus a point in D. Notice that $R(x, z) = 0$ if and only if $z = D_x$ in this case.

Instead of the maximal integral lattice $M_2(\mathbb{Z}) \cap V$ used in [KRY2], we choose the lattice

$$L = \{x = \begin{pmatrix} b & 2c \\ -2a & -b \end{pmatrix} : a, b, c \in \mathbb{Z}\} \subset V \tag{4.3}$$

(notice that $Q(x) = 4ac - b^2$), and for $m \neq 0$, let

$$L(m) = \{x \in L : \det x = m\}.$$

Then, for $m > 0$, one has the following identification

$$[L(m)/\operatorname{SL}_2(\mathbb{Z})] \leftrightarrow \mathcal{Z}(m)(\mathbb{C}), \quad x \mapsto D_x$$

where $x \in L(m)/\operatorname{SL}_2(\mathbb{Z})$ is counted with multiplicity $1/(\#\Gamma_x)$, and Γ_x is the stabilizer of x in $\operatorname{SL}_2(\mathbb{Z})$. For every integer m and real number $v > 0$, we define

$$\Xi(m, v)(z) = \frac{1}{2} \sum_{0 \neq x \in L(m)} \rho(x\sqrt{v}, z) \tag{4.4}$$

where

$$\rho(x, z) = \int_1^{\infty} e^{-2\pi R(x,z) r} \frac{dr}{r} = -\mathrm{Ei}(-2\pi R(x, z)). \tag{4.5}$$

Proposition 11.1 of [Ku1] asserts that as currents on D, one has

$$dd^c \rho(x, z) + \delta_{D_x} = [\phi_{KM}(x)] \tag{4.6}$$

where D_x is empty if $(x, x) \leq 0$ and

$$\phi_{KM}(x) = \left((x, z)^2 - \frac{1}{2\pi}\right) e^{-2\pi R(x,z)} \frac{i}{2} \frac{dz d\bar{z}}{(\mathrm{Im\, z})^2} \tag{4.7}$$

is a smooth $(1, 1)$-form in the upper half plane. Set

$$\omega(m, v) = \frac{1}{2} \sum_{0 \neq x \in L(m)} \phi_{KM}(x\sqrt{v}). \tag{4.8}$$

Note that, when $m = 0$, we sum over the set of nonzero null vectors $L(0) - \{0\}$. When $m > 0$, the series (4.8) is absolutely convergent and thus

$$dd^c\Xi(m, v) + \delta_{\mathcal{Z}(m)} = [\omega(m, v)]. \tag{4.9}$$

This proves Proposition 3.1(1). The same is true when $-m > 0$ is not a square. However, when $-m \geq 0$ is a square, the series (4.8) is convergent not termwise integrable. It turns out that $\Xi(m, v)$ has a logarithmic singularity at the cusp ∞ in this case and is in $L^1(\mathcal{M}(\mathbb{C}))$ according to Funke [Fu]. From this, one can check that $[\omega(m, v)]$ is a L^1 current and $(\mathcal{Z}(m, v), \Xi(m, v))$ is an arithmetic divisor with real coefficient. Furthermore, Funke proved [Fu, Propositions 4.7, 4.8] that

$$\int_{\mathcal{M}(\mathbb{C})} \omega(m, v) = g(n, v) \tag{4.10}$$

in this case. This proves Proposition 3.1.

5. The proof of Theorem 3.2

Now we sketch a proof of Theorem 3.2 and refer to [KRY2] for details. The functional equation follows from Theorem 3.4 and functional equations (1.7) and (3.13).

Checking the identity $E_{\text{Zagier}}(\tau) = \deg\hat{\phi}(\tau)$ amounts to verifying that

$$\deg \hat{\mathcal{Z}}(m, v) = \int_{\mathcal{M}(\mathbb{C})} \omega(m, v) = \begin{cases} \frac{1}{2}H_0(m) & \text{if } m > 0, \\ 0 & \text{if } -m > 0 \text{ is not a square}, \\ g(n, v) & \text{if } -m = n^2 > 0, \\ 1/(16\pi\sqrt{v}) & \text{if } m = 0. \end{cases}$$

When $m < 0$, this is basically (4.10). When $m > 0$,

$$\deg \hat{\mathcal{Z}}(m, v) = \deg \mathcal{Z}(m) = \tfrac{1}{2}H_0(m)$$

is just Theorem 1.1(1).

The identity $\mathcal{E}(\tau, \frac{1}{2}) = E_{\text{Zagier}}(\tau)$ follows from Theorem 3.4 directly. Indeed, when $m > 0$, the fact $\Psi_{-3/2}(\frac{1}{2}, t) = 0$ implies that

$$A_m(v, \tfrac{1}{2}) = \Lambda(0, \chi_{-4m})q^m = E_{\text{Zagier}, m}(\tau).$$

When $-m > 0$, one has [KRY1, page 84]

$$\Psi_{\frac{3}{2}}(\frac{1}{2}, 4\pi|m|v) = \frac{e^{4\pi|m|v}}{\sqrt{4\pi|m|v}} \int_1^\infty e^{-4\pi|m|vr} r^{-3/2} dr.$$

So $A_m(v, \frac{1}{2}) = 0$ unless $\Lambda(s, \chi_{-4m})$ has a pole at $s = 0$, i.e., $-m = n^2$ is a square. If $-m = n^2 > 0$, then

$$\Lambda(s, \chi_{-4m}) = \zeta(s)n^s \prod_{p|n} b_p(n, s)$$

and

$$\lim_{s \longrightarrow 0} s\Lambda(-s, \chi_m) = n.$$

Therefore

$$A_m(v, \tfrac{1}{2}) = \frac{1}{8\pi\sqrt{v}} \int_1^\infty e^{-4\pi|m|vr} r^{-3/2} dr = E_{\text{Zagier},m}(\tau) q^{-m}.$$

The case $m = 0$ is similar.

Part (3) of Theorem 3.2 is new and is proved in a term-by-term manner. The case $m > 0$ is basically Theorem 1.1 together with a routine calculation of the integral

$$\int_{\mathcal{M}(\mathbb{C})} \Xi(m, v) \frac{dx\, dy}{y^2}. \tag{5.1}$$

The case where $-m > 0$ is not a square is also a routine calculation of the integral (5.1), and is done in [KRY1, Section 12]. The case $-m = 0$ involves the self-intersection of the Hodge bundle $\hat{\omega}$, which was given by (3.4).

Finally, when $-m = n^2 > 0$ is a square, the proof goes as follows. Notice [Bos2; Kun] that

$$\hat{c}_1(\hat{\omega}) = \tfrac{1}{12}(\infty, -\log|\Delta|^2) + (0, C) \tag{5.2}$$

where C is the constant defined in (1.3). So

$$\widehat{\mathcal{Z}}(m, v) = 12g(n, v)\hat{c}_1(\hat{\omega}) + (0, \beta(n, v)) \tag{5.3}$$

with

$$\beta(n, v) = \Xi(-n^2, v) - 12g(n, v)C - g(n, v)\log|\Delta(\tau)|^2. \tag{5.4}$$

Notice that $\beta(n, v)$ is smooth at ∞. Therefore

$$\langle \widehat{\mathcal{Z}}(m, v), \hat{\omega} \rangle = 12g(n, v)\langle \hat{\omega}, \hat{\omega} \rangle + \frac{1}{8\pi} \int_{\mathcal{M}(\mathbb{C})} \beta(n, v) \frac{dx\, dy}{y^2}. \tag{5.5}$$

Since the self-intersection $\langle \hat{\omega}, \hat{\omega} \rangle$ is computed by Bost and Kühn, the key is thus to compute the integral in (5.5) and compare it with $A'(-n^2, \frac{1}{2})$. The computation turns out to be quite involving and amusing to us.

Acknowledgment

The author thanks S. Kudla and M. Rapoport for allowing him to report on joint work here and for stimulating discussion during the joint work. He thanks Kudla for constant encouragement and inspiration on this subject. Finally, the author thanks the organizers of this conference, H. Darmon and Shou-Wu Zhang and MSRI for this wonderful conference.

References

[Bor] R. Borcherds, "The Gross–Kohnen–Zagier theorem in higher dimensions", *Duke Math J.* **97** (1999), 219–233.

[Bos1] J.-B. Bost, "Potential theory and Lefschetz theorems for arithmetic surfaces", *Ann. Sci. École Norm. Sup.* **32** (1999), 241–312.

[Bos2] J.-B. Bost, unpublished.

[Col] P. Colmez, "Périodes des variétés abéliennes á multiplication complexe", *Ann. Math.* **138** (1993), 625–683.

[Fu] J. Funke, "Heegner Divisors and non-holomorphic modular forms", *Compositio Math.*, to appear.

[Gro1] B. Gross, *Arithmetic on elliptic curves with complex multiplication*, Lecture Notes in Mathematics **776**, Springer, Berlin, 1980.

[Gro2] B. Gross, "On the periods of Abelian integrals and a formula of Chowla and Selberg", *Invent. Math.* **45** (1978), 193–211.

[HZ] F. Hirzebruch and D. Zagier, "Intersection numbers of curves on Hilbert modular surfaces and modular forms of Nebentypus", *Invent. Math.* **36** (1976), 57–113.

[Ku1] S. Kudla, "Central derivatives of Eisenstein series and height pairings", *Ann. of Math.* **146** (1997), 545–646.

[Ku2] S. Kudla, "Special cycles and derivatives of Eisenstein series", pp. 241–269 in *Heegner Points and Rankin L-Series*, edited by Henri Darmon and Shou-Wu Zhang, *Math. Sci. Res. Inst. Publications* **49**, Cambridge U. Press, New York, 2004.

[KRY1] S. Kudla, M. Rapoport and T. Yang, "Derivatives of Eisenstein series and Faltings heights", preprint.

[KRY2] S. Kudla, M. Rapoport and T. Yang, in preparation.

[Kun] U. Kühn, "Generalized arithmetic intersection numbers", *J. reine angew. Math.* **534** (2001), 209–236.

[NT] Y. Nakkajima and Y. Taguchi, "A generalization of the Chowla–Selberg formula", *J. reine angew. Math.* **419** (1991), 119–124.

[Ra] M. Raynaud, "Hauteurs et isogénies", pp. 199–234 in *Séminaire sur les pinceaux arithmétiques: la conjecture de Mordell*, Astérisque **127**, Soc. math. de France, Paris, 1985.

Tonghai Yang
Department of Mathematics
University of Wisconsin
Madison, WI 53706
United States
thyang@math.wisc.edu

Heegner Points and Rankin L-Series
MSRI Publications
Volume **49**, 2004

Elliptic Curves and Analogies Between Number Fields and Function Fields

DOUGLAS ULMER

ABSTRACT. Well-known analogies between number fields and function fields have led to the transposition of many problems from one domain to the other. In this paper, we discuss traffic of this sort, in both directions, in the theory of elliptic curves. In the first part of the paper, we consider various works on Heegner points and Gross–Zagier formulas in the function field context; these works lead to a complete proof of the conjecture of Birch and Swinnerton-Dyer for elliptic curves of analytic rank at most 1 over function fields of characteristic > 3. In the second part of the paper, we review the fact that the rank conjecture for elliptic curves over function fields is now known to be true, and that the curves which prove this have asymptotically maximal rank for their conductors. The fact that these curves meet rank bounds suggests interesting problems on elliptic curves over number fields, cyclotomic fields, and function fields over number fields. These problems are discussed in the last four sections of the paper.

CONTENTS

1. Introduction 286
2. The Birch and Swinnerton-Dyer Conjecture over Function Fields 287
3. Function Field Analogues of the Gross–Zagier Theorem 289
4. Ranks over Function Fields 300
5. Rank Bounds 304
6. Ranks over Number Fields 306
7. Algebraic Rank Bounds 307
8. Arithmetic and Geometric Bounds I: Cyclotomic Fields 309
9. Arithmetic and Geometric Bounds II: Function Fields over Number Fields 310
Acknowledgements 312
References 312

This paper is based upon work supported by the National Science Foundation under Grant DMS-0070839.

1. Introduction

The purpose of this paper is to discuss some work on elliptic curves over function fields inspired by the Gross–Zagier theorem and to present new ideas about ranks of elliptic curves from the function field case which I hope will inspire work over number fields.

We begin in Section 2 by reviewing the current state of knowledge on the conjecture of Birch and Swinnerton-Dyer for elliptic curves over function fields. Then in Section 3 we discuss various works by Rück and Tipp, Pál, and Longhi on function field analogues of the Gross–Zagier formula and related work by Brown. We also explain how suitably general Gross–Zagier formulas together with my "geometric nonvanishing" results lead to a theorem of the form: the Birch and Swinnerton-Dyer conjecture for elliptic curves over function fields of curves over finite fields of characteristic > 3 holds for elliptic curves with analytic rank at most 1.

In Sections 4 and 5 we move beyond rank one and explain that the rank conjecture holds for elliptic curves over function fields: there are (nonisotrivial) elliptic curves with Mordell–Weil group of arbitrarily large rank. Moreover, these curves meet an asymptotic bound due to Brumer for the rank in terms of the conductor. So in the function field case, we know precisely the asymptotic growth of ranks of elliptic curves ordered by the size of their conductors. In fact, there are two bounds, one arithmetic, the other geometric, and both are sharp.

The rest of the paper is devoted to presenting some interesting problems suggested by the existence and sharpness of these two types of rank bounds. In Section 6 we state a conjecture which says roughly that Mestre's bound on the ranks of elliptic curves over $\mathbb{Q}$ and suitable generalizations of it over number fields are asymptotically sharp. Next, we note that the Mestre bound and even more so the Brumer bound can be "reformulated as" algebraic statements. For example, the Brumer bound can be interpreted as a statement about the eigenvalues of Frobenius on étale cohomology. It is therefore natural to ask for an algebraic proof; reformulating the bounds into statements that might admit an algebraic proof leads to interesting questions which are explained in Section 7.

Finally, in Sections 8 and 9 we discuss possible rank bounds over cyclotomic fields and over function fields over number fields. More precisely, we discuss pairs of ranks bounds, one "arithmetic" the other "geometric," for pairs of fields like $\mathbb{Q}^{p\text{-cyc}}/\mathbb{Q}$ or $\overline{\mathbb{Q}}(\mathcal{C})/\mathbb{Q}(\mathcal{C})$ where $\mathcal{C}$ is a curve over $\mathbb{Q}$. In both cases, one rank bound is known (arithmetic in the first case, geometric in the second) and the other bound has yet to be considered.

2. Review of the Birch and Swinnerton-Dyer Conjecture over Function Fields

We assume that the reader is familiar with elliptic curves over number fields, but perhaps not over function fields, and so in this preliminary section we set up some background and review the Birch and Swinnerton-Dyer conjecture. For many more details, examples, etc., we refer to [Ulmb].

Let $\mathcal{C}$ be a smooth, geometrically connected, projective curve over a finite field $\mathbb{F}_q$ and set $F = \mathbb{F}_q(\mathcal{C})$. Let E be an elliptic curve over F, i.e., a curve of genus one defined as usual by an affine Weierstrass equation

$$y^2 + a_1xy + a_3y = x^3 + a_2x^2 + a_4x + a_6$$

($a_i \in F$) with the point at infinity $[0,1,0]$ as origin; the discriminant Δ and j-invariant are given by the usual formulas (see, e.g., [Tat72]) and we of course assume that $\Delta \neq 0$. We say that E is *constant* if it is defined over $\mathbb{F}_q$, i.e., if it is possible to choose the Weierstrass model so that the $a_i \in \mathbb{F}_q$. Equivalently, E is constant if there exists an elliptic curve E_0 defined over $\mathbb{F}_q$ such that

$$E \cong E_0 \times_{\operatorname{Spec} \mathbb{F}_q} \operatorname{Spec} F.$$

In this case we say that E is based on E_0. We say that E is *isotrivial* if it becomes isomorphic to a constant curve after a finite extension of F; this is easily seen to be equivalent to the condition $j(E) \in \mathbb{F}_q$. Finally, we say that E is *nonisotrivial* if $j(E) \notin \mathbb{F}_q$.

Let $\mathfrak{n}$ be the conductor of E. This is an effective divisor on $\mathcal{C}$ which is divisible only by the places where E has bad reduction. More precisely, v divides $\mathfrak{n}$ to order 1 at places where E has multiplicative reduction and to order at least 2 at places where E has additive reduction and to order exactly 2 at these places if the characteristic of F is > 3. The reduction, exponent of conductor, and minimal model of E at places of F can be computed by Tate's algorithm [Tat72].

The Mordell–Weil theorem holds for E, namely $E(F)$ is a finitely generated abelian group. This can be proven in a manner entirely analogous to the proof over number fields, using Selmer groups and heights, or by more geometric methods; see [Nér52]. Also, both the rank conjecture (that for a fixed F, the rank of $E(F)$ can be arbitrarily large) and the torsion conjecture (that there is a bound on the order of the torsion subgroup of $E(F)$ depending only on the genus of F) are known to be true in this context. For the rank conjecture, see [Ulm02] and Section 4 below. The torsion conjecture was proven by Levin [Lev68], who showed that there is an explicit bound of the form $O(\sqrt{g}+1)$ for the order of the torsion subgroup of a nonisotrivial elliptic curve over F, where g is the genus of F. More recently, Thakur [Tha02] proved a variant bounding the order of torsion in terms of the *gonality* of $\mathcal{C}$, i.e., the smallest degree of a nonconstant map to $\mathbb{P}^1$.

The L-function of E is defined by the Euler product

$$L(E/F,s) = \prod_{v \nmid \mathfrak{n}} (1 - a_v q_v^{-s} + q_v^{1-2s})^{-1}$$

$$\times \prod_{v | \mathfrak{n}} \begin{cases} (1-q_v^{-s})^{-1} & \text{if } E \text{ has split multiplicative reduction at } v, \\ (1+q_v^{-s})^{-1} & \text{if } E \text{ has nonsplit multiplicative reduction at } v, \\ 1 & \text{if } E \text{ has additive reduction at } v. \end{cases}$$

Here q_v is the cardinality of the residue field $\mathbb{F}_v$ at v and the number of points on the reduced curve is $\#E(\mathbb{F}_v) = q_v + 1 - a_v$. The product converges absolutely in the half-plane $\operatorname{Re} s > \frac{3}{2}$, has a meromorphic continuation to the s plane, and satisfies a functional equation for $s \mapsto 2 - s$. If E is not constant, then $L(E/F,s)$ is a polynomial in s of degree $4g - 4 + \deg \mathfrak{n}$ and thus an entire function of s. (All this comes from Grothendieck's analysis of L-functions. See the last section of [Mil80] for more details.)

The Birch and Swinnerton-Dyer conjecture in this context asserts that

$$\operatorname{Rank} E(F) \stackrel{?}{=} \operatorname{ord}_{s=1} L(E/F,s)$$

and, setting $r = \operatorname{ord}_{s=1} L(E/F,s)$, that the leading coefficient is

$$\frac{1}{r!} L^{(r)}(E/F,1) \stackrel{?}{=} \frac{|\text{Ш}| R\tau}{|E(F)_{\mathrm{tor}}|^2}$$

where Ш is the Tate–Shafarevitch group, R is a regulator constructed from heights of a set of generators of $E(F)$, and τ is a certain Tamagawa number (an analogue of a period). We will not enter into the details of the definitions of these objects since they will play little role in what follows; see [Tat66b] for more details.

Much more is known about this conjecture in the function field case than in the number field case. Indeed, we have

$$\operatorname{Rank} E(F) \leq \operatorname{ord}_{s=1} L(E/F,s) \tag{2–1}$$

and the following assertions are equivalent:

1. Equality holds in (2–1).
2. The ℓ primary part of Ш is finite for any one prime ℓ ($\ell = p$ is allowed).
3. Ш is finite.

Moreover, if these equivalent conditions are satisfied, then the refined conjecture on the leading coefficient of the L-series is true. The "prime-to-p" part of this was proven by Artin and Tate [Tat66b]. More precisely, they showed that equality holds in (2–1) if and only if the ℓ primary part of Ш is finite for any one prime $\ell \neq p$ if and only if the ℓ primary part of Ш is finite for every $\ell \neq p$, and that if these conditions hold, the refined formula is correct up to a power of p. Milne proved the stronger statement above in [Mil75] for $p \neq 2$; due to later improvements in p-adic cohomology, his argument applies essentially verbatim to the case $p = 2$ as well.

These results were obtained by considering the elliptic surface $\mathcal{E} \to \mathcal{C}$ attached to E, which can be characterized as the unique smooth, proper surface over $\mathbb{F}_q$ admitting a flat and relatively minimal morphism to $\mathcal{C}$, with generic fiber E/F. Another key ingredient is Grothendieck's analysis of L-functions, which gives a cohomological interpretation of the ζ-function of $\mathcal{E}$ and the L-function of E.

Equality in (2–1), and therefore the full Birch and Swinnerton-Dyer conjecture, is known to hold in several cases (but certainly not the general case!): If it holds for E/K where K is a finite extension of F, then it holds for E/F (this is elementary); it holds for constant, and thus isotrivial, E (this follow from [Tat66a]); and it holds for several cases most easily described in terms of $\mathcal{E}$, namely if $\mathcal{E}$ is a rational surface (elementary), a $K3$ surface [ASD73], or if $\mathcal{E}$ is dominated by a product of curves (see [Tat94]). The rational and $K3$ cases are essentially those where the base field F is $\mathbb{F}_q(t)$ and the coefficients a_i in the defining Weierstrass equation of E have small degree in t.

3. Function Field Analogues of the Gross–Zagier Theorem

In this section we will give some background on modularity and Heegner points and then discuss various works on Gross–Zagier formulas in the function field context. Our treatment will be very sketchy, just giving the main lines of the arguments, but we will give precise references where the reader may find the complete story. Throughout, we fix a smooth, proper, geometrically connected curve $\mathcal{C}$ over a finite field $\mathbb{F}_q$ of characteristic p and we set $F = \mathbb{F}_q(\mathcal{C})$.

3.1. Two versions of modularity. Recall that for elliptic curves over $\mathbb{Q}$ there are two (not at all trivially!) equivalent statements expressing the property that an elliptic curve E of conductor N is modular:

1. There exists a modular form f (holomorphic of weight 2 and level $\Gamma_0(N)$) such that $L(E, \chi, s) = L(f, \chi, s)$ for all Dirichlet characters χ.
2. There exists a nonconstant morphism $X_0(N) \to E$ (defined over $\mathbb{Q}$).

(We note that over $\mathbb{Q}$, the equalities $L(E, \chi, s) = L(f, \chi, s)$ in (1) are implied by the *a priori* weaker statement that $L(E, s) = L(f, s)$. But this implication fails over higher degree number fields and over function fields and it is the stronger assertion that we need.)

In the next two subsections we will explain the analogues of these two statements in the function field context. In this case, the relationship between the two statements is a little more complicated than in the classical case. For example, the relevant automorphic forms are complex valued and thus are not functions or sections of line bundles on the analogue of $X_0(\mathfrak{n})$, which is a curve over F. Nevertheless, analogues of both modularity statements are theorems in the function field case.

3.2. Analytic modularity. We begin with (1). Let $\mathbb{A}_F$ be the adèle ring of F and $\mathcal{O}_F \subset \mathbb{A}_F$ the subring of everywhere integral adèles. Then for us, automorphic forms on GL_2 over F are functions on $\mathrm{GL}_2(\mathbb{A}_F)$ which are invariant under left translations by $\mathrm{GL}_2(F)$ and under right translations by a finite index subgroup K of $\mathrm{GL}_2(\mathcal{O}_F)$. In other words, they are functions on the double coset space

$$\mathrm{GL}_2(F)\backslash\mathrm{GL}_2(\mathbb{A}_F)/K. \tag{3–1}$$

These functions may take values in any field of characteristic zero; to fix ideas, we take them with values in $\overline{\mathbb{Q}}$ and we fix embeddings of $\overline{\mathbb{Q}}$ into $\mathbb{C}$ and into $\overline{\mathbb{Q}}_\ell$ for some $\ell \neq p$. The subgroup K is the analogue of the level in the classical setting and the most interesting case is when K is one of the analogues $\Gamma_0(\mathfrak{m})$ or $\Gamma_1(\mathfrak{m})$ of the Hecke congruence subgroups where $\mathfrak{m}$ is an effective divisor on $\mathcal{C}$. If $\psi : \mathbb{A}^\times/F^\times \to \overline{\mathbb{Q}}_\ell^{\times}$ is an idèle class character and f is an automorphic form, we say f has central character ψ if $f(zg) = \psi(z)f(g)$ for all $z \in Z(\mathrm{GL}_2(\mathbb{A}_F)) \cong \mathbb{A}_F^\times$ and all $g \in \mathrm{GL}_2(\mathbb{A}_F)$. The central character plays the role of weight: When k is a positive integer and $\psi(z) = |z|^{-k}$ (where $|\cdot|$ is the adèlic norm), f is analogous to a classical modular form of weight k. The basic reference for this point of view is [Wei71]; see Chapter III for definitions and first properties. For a more representation-theoretic point of view, see [JL70].

If we single out a place ∞ of F and assume that $K = \Gamma_0(\infty\mathfrak{n})$ where $\mathfrak{n}$ is prime to ∞, then there is an analogue of the description of classical modular forms as functions on the upper half plane. Namely, an automorphic form f may be viewed as a function (or a section of line bundle if the ∞ component of ψ is nontrivial) on a finite number of copies of the homogeneous space $\mathrm{PGL}_2(F_\infty)/\Gamma_0(\infty)$ which has the structure of an oriented tree. (Compare with $\mathrm{PGL}_2(\mathbb{R})/O_2(\mathbb{R}) \cong \mathbb{H}$.) The corresponding functions are invariant under certain finite index subgroups of $\mathrm{GL}_2(A) \subset \mathrm{GL}_2(F_\infty)$ where $A \subset F$ is the ring of functions regular outside ∞. The various copies of the tree are indexed by a generalized ideal class group of A. This point of view is most natural when $F = \mathbb{F}_q(t)$ and ∞ is the standard place $t = \infty$, in which case there is just one copy of the tree and this description is fairly canonical. In the general case, there are several copies of the tree and choices must be made to identify automorphic forms with functions on trees. Using this description (or suitable Hecke operators) one may define the notion of a form being "harmonic" or "special" at ∞. Namely, sum of the values over edges with a fixed terminus should be zero. This is an analogue of being holomorphic. See [DH87, Chap. 5], [GR96, Chap. 4], or [vdPR97, Chap. 2] for details.

Automorphic forms have Fourier expansions, with coefficients naturally indexed by effective divisors on $\mathcal{C}$. There are Hecke operators, also indexed by effective divisors on $\mathcal{C}$ and the usual connection between eigenvalues of Hecke operators and Fourier coefficients of eigenforms holds. There is a notion of cusp form and for a fixed K and ψ the space of cusp forms is finite dimensional. An automorphic form f gives rise to an L-function $L(f, s)$, which is a complex valued function of a complex variable s. If f is a cuspidal eigenform, this L-function has an Euler product, an analytic continuation to an entire function of s, and satisfies a functional equation. See [Wei71] for all of this except the finite dimensionality, which follows easily from reduction theory. See [Ser80, Chap. II] for the finite dimensionality when $F = \mathbb{F}_q(t)$ and [HLW80] for an explicit dimension formula in the general case.

The main theorem of [Wei71] is a "converse" theorem which says roughly that a Dirichlet series with a suitable analytic properties is the L-function of an automorphic form on GL_2. (The function field case is Theorem 3 of Chapter VII.) The most important requirement is that sufficiently many of the twists of the given Dirichlet series by finite order characters should satisfy functional equations. This result was also obtained by representation theoretic methods in [JL70]. Also, see [Li84] for an improved version, along the lines of [Wei71].

Now let E be an elliptic curve over F. By Grothendieck's analysis of L-functions, we know that the Dirichlet series $L(E, s)$ is meromorphic (entire if E is nonisotrivial) and its twists satisfy functional equations. In [Del73, 9.5–9.7], Deligne verified the hypotheses of Weil's converse theorem. The main point is to check that the functional equations given by Grothendieck's theory are the same as those required by Weil. The form f_E associated to E is characterized by the equalities $L(E, \chi, s) = L(f_E, \chi, s)$ for all finite order idèle class characters χ. It is an eigenform for the Hecke operators and is a cusp form if E is nonisotrivial. Its level is $\Gamma_0(\mathfrak{m})$ where $\mathfrak{m}$ is the conductor of E and it has central character $|\cdot|^{-2}$ (i.e., is analogous to a form of weight 2). If E has split multiplication at ∞, then f_E is special at ∞. The construction of f_E from E is the function field analogue of (1) above.

3.3. Geometric modularity. We now turn to Drinfeld modules and (2). There is a vast literature on Drinfeld modules and we will barely scratch the surface. The primary reference is [Dri74] and there are valuable surveys in [DH87] and [GvdPRG97].

Fix a place ∞ of F and define A to be the ring of elements of F regular away from ∞. Let F_∞ denote the completion of F at ∞ and C the completion of the algebraic closure of F_∞. The standard example is when $F = \mathbb{F}_q(t)$, ∞ is the standard place $t = \infty$, and $A = \mathbb{F}_q[t]$.

Let k be a ring of characteristic p equipped with a homomorphism $A \to k$. Let $k\{\tau\}$ be the ring of noncommutative polynomials in τ, with commutation relation $\tau a = a^p \tau$. There is a natural inclusion $\varepsilon : k \hookrightarrow k\{\tau\}$ with left inverse $D : k\{\tau\} \to k$ defined by $D(\sum_n a_n \tau^n) = a_0$. If R is any k-algebra, we may make the additive group of R into a module over $k\{\tau\}$ by defining $(\sum_n a_n \tau^n)(x) = \sum_n a_n x^{p^n}$.

A Drinfeld module over k (or elliptic module as Drinfeld called them) is a ring homomorphism $\phi : A \to k\{\tau\}$ whose image is not contained in k and such that $D \circ \phi : A \to k$ is the given homomorphism. The characteristic of ϕ is by definition the kernel of the homomorphism $A \to k$, which is a prime ideal of A. It is convenient to denote the image of $a \in A$ by ϕ_a rather than $\phi(a)$. If $A = \mathbb{F}_q[t]$ then ϕ is determined by ϕ_t, which can be any element of $k\{\tau\}$ of positive degree with constant term equal to the image of t under $A \to k$. For a general A and k equipped with $A \to k$ there may not exist any Drinfeld modules and if they do exist, they may not be easy to find. As above, a Drinfeld module ϕ turns any k-algebra into an A-module by the rule $a \cdot x = \phi_a(x)$.

It turns out that $a \mapsto \phi_a$ is always injective and there exists a positive integer r, the rank of ϕ, such that $p^{\deg_\tau(\phi_a)} = |a|_\infty^r = \#(A/a)^r$. If ϕ and ϕ' are Drinfeld modules over k, a homomorphism $u : \phi \to \phi'$ is by definition an element $u \in k\{\tau\}$ such that $u\phi_a = \phi'_a u$ for all $a \in A$ and an isogeny is a nonzero homomorphism. Isogenous Drinfeld modules have the same rank and characteristic. See [Dri74, § 2] or [DH87, Chap. 1].

We will only consider Drinfeld modules of rank 2. These objects are in many ways analogous to elliptic curves. For example, if k is an algebraically closed field and $\mathfrak{p} \subset A$ is a prime ideal, then we have an isomorphism of A-modules

$$\phi[\mathfrak{p}](k) := \{x \in k | \phi_a(x) = 0 \text{ for all } a \in \mathfrak{p}\} \cong (A/\mathfrak{p})^e$$

where $0 \le e \le 2$ and $e = 2$ if the characteristic of ϕ is relatively prime to $\mathfrak{p}$. A second analogy occurs with endomorphism rings: $\mathrm{End}(\phi)$, the ring of endomorphisms of ϕ, is isomorphic as A-module to either A, an A-order in an "imaginary" quadratic extension K of F, or an A-order in a quaternion algebra over F. Here "imaginary" means that the place ∞ of F does not split in K and an A-order in a division algebra D over F is an A-subalgebra R with $\mathrm{Frac}(R) = D$. The quaternion case can occur only if the characteristic of ϕ is nonzero, in which case the quaternion algebra is ramified precisely at ∞ and the characteristic of ϕ. A third analogy is the analytic description of Drinfeld modules over C: giving a Drinfeld module of rank 2 over C up to isomorphism is equivalent to giving a rank 2 A-lattice in C up to homothety by elements of $C^\times$. If ϕ corresponds to the lattice Λ, there is a commutative diagram

$$\begin{array}{ccccccccc}
0 & \longrightarrow & \Lambda & \longrightarrow & C & \xrightarrow{\exp_\Lambda} & C & \longrightarrow & 0 \\
 & & \downarrow a & & \downarrow a & & \downarrow \phi_a & & \\
0 & \longrightarrow & \Lambda & \longrightarrow & C & \xrightarrow{\exp_\Lambda} & C & \longrightarrow & 0
\end{array}$$

where $\exp_\Lambda : C \to C$ is the Drinfeld exponential associated to Λ. See [Dri74, §§ 2–3] or [DH87, Chaps. 1–2].

There is a natural generalization of all of the above to Drinfeld modules over schemes of characteristic p. Given an effective divisor $\mathfrak{n}$ on $\mathcal{C}$ relatively prime to ∞ (or equivalently, a nonzero ideal of A), there is a notion of "level $\mathfrak{n}$ structure" on a Drinfeld module. Using this notion, one may construct a moduli space $Y_0(\mathfrak{n})$ (a scheme if $\mathfrak{n}$ is nontrivial and a stack if $\mathfrak{n}$ is trivial) parameterizing Drinfeld modules of rank 2 with level $\mathfrak{n}$ structure, or equivalently, pairs of rank 2 Drinfeld modules connected by a "cyclic $\mathfrak{n}$-isogeny" $u : \phi \to \phi'$. (The notions of level $\mathfrak{n}$ structure and cyclic are somewhat subtle and a significant advance over the naive notions. Analogues of Drinfeld's notions were used in [KM85] to completely analyze the reduction of classical modular curves at primes dividing the level.) The curve $Y_0(\mathfrak{n})$ is smooth and affine over F and may be completed to a smooth, proper curve $X_0(\mathfrak{n})$. The added points ("cusps") can be interpreted in terms of certain degenerations of Drinfeld modules. The curve $X_0(\mathfrak{n})$ carries many of the structures familiar from the classical case, such as Hecke correspondences (indexed by effective divisors on $\mathcal{C}$) and Atkin–Lehner involutions. See [Dri74, § 5] and [DH87, Chap. 1, § 6]. The construction of the moduli space (or stack) is done very carefully in [Lau96, Chap. 1] and the interpretation of the cusps is given in [vdPT97].

The analytic description of Drinfeld modules over C yields an analytic description of the C points of $Y_0(\mathfrak{n})$. Namely, let Ω denote the Drinfeld upper half plane: $\Omega = \mathbb{P}^1(C) \setminus P^1(F_\infty)$. Then $Y_0(\mathfrak{n})(C)$ is isomorphic (as rigid analytic space) to a union of quotients of Ω by finite index subgroups of $\mathrm{GL}_2(A)$. The components of $Y_0(\mathfrak{n})(C)$ are indexed by a generalized ideal class group of A. More adelically, we have an isomorphism

$$Y_0(\mathfrak{n}) \cong \mathrm{GL}_2(F)\backslash\big(\mathrm{GL}_2(\mathbb{A}_F^f) \times \Omega\big)/\Gamma_0(\mathfrak{n})^f$$

where $\mathbb{A}_F^f$ denotes the "finite adèles" of F, namely the adèles with the component at ∞ removed, and similarly with $\Gamma_0(\mathfrak{n})^f$. See [Dri74, § 6] or [DH87, Chap. 3].

This description reveals a close connection between $Y_0(\mathfrak{n})$ and the description of automorphic forms as functions on trees (cf. (3–1)). Namely, there is a map between the Drinfeld upper half plane Ω and a geometric realization of the tree $\mathrm{PGL}_2(F_\infty)/\Gamma_0(\infty)$. Using this, Drinfeld was able to analyze the étale cohomology of $X_0(\mathfrak{n})$ as a module for $\mathrm{Gal}(\overline{F}/F)$ and the Hecke operators, in terms of automorphic forms of level $\Gamma_0(\mathfrak{n}\infty)$ which are special at ∞. (Drinfeld used an *ad hoc* definition of étale cohomology; for a more modern treatment, see [vdPR97].) This leads to one form of the Drinfeld reciprocity theorem: if f is an eigenform of level $\Gamma_0(\mathfrak{n}\infty)$ which is special at ∞, then there exists a factor A_f of the Jacobian $J_0(\mathfrak{n})$ of $X_0(\mathfrak{n})$, well-defined up to isogeny, such that

$$L(f, \chi, s) = L(A_f, \chi, s)$$

for all finite order idèle class characters χ of F. If the Hecke eigenvalues of f are rational integers, then A_f is an elliptic curve, and if f is a new, then E has conductor $\mathfrak{n}\infty$ and is split multiplicative at ∞. See [Dri74, § § 10–11], [DH87, Chaps. 4–5], and [GR96, Chap. 8].

So, starting with an elliptic curve E over F of level $\mathfrak{m} = \mathfrak{n}\infty$ which is split multiplicative at ∞, Deligne's theorem gives us an automorphic form f_E on GL_2 over F of level $\mathfrak{m}$ which is special at ∞ and which has integer Hecke eigenvalues.

From f_E, Drinfeld's construction gives us an isogeny class of elliptic curves A_{f_E} appearing in the Jacobian of $X_0(\mathfrak{n})$. Moreover, we have equalities of L-functions:

$$L(E,\chi,s) = L(f_E,\chi,s) = L(A_{f_E},\chi,s).$$

But Zarhin proved that the L-function of an abelian variety A over a function field (by which we mean the collection of all twists $L(A,\chi,s)$) determines its isogeny class. See [Zar74] for the case $p > 2$ and [MB85, XXI.2] for a different proof that works in all characteristics. This means that E is in the class A_{f_E} and therefore we have a nontrivial modular parameterization $X_0(\mathfrak{n}) \to E$.

In [GR96, Chap. 9] Gekeler and Reversat completed this picture by giving a beautiful analytic construction of $J_0(N)(C)$ and of the analytic parameterization $X_0(\mathfrak{n})(C) \to E(C)$. This is the analogue of the classical parameterization of an elliptic curve by modular functions. Recently, Papikian has studied the degrees of Drinfeld modular parameterizations and proved the analogue of the degree conjecture. See [Pap02] and forthcoming publications.

3.4. Heegner points and Brown's work. It was clear to the experts from the beginning that Heegner points, the Gross–Zagier formula, and Kolyvagin's work could all be extended to the function field case, using the Drinfeld modular parameterization, although nothing was done for several years. The first efforts in this direction were made by Brown in [Bro94].

Fix as usual F, ∞, A, and $\mathfrak{n}$, so we have the Drinfeld modular curve $X_0(\mathfrak{n})$. Let K/F be an imaginary quadratic extension and let B be an A-order in K. A Drinfeld–Heegner point with order B (or Heegner point for short) is by definition a point on $X_0(\mathfrak{n})$ corresponding to a pair $\phi \to \phi'$ connected by a cyclic $\mathfrak{n}$ isogeny such that $\mathrm{End}(\phi) = \mathrm{End}(\phi') = B$. These will exist if and only if there exists a proper ideal $\mathfrak{n}'$ of B (i.e., one such that $\{b \in K | b\mathfrak{n}' \subset \mathfrak{n}'\} = B$) with $B/\mathfrak{n}' \cong A/\mathfrak{n}$. The simplest situation is when every prime dividing $\mathfrak{n}$ splits in K and B is the maximal A-order in K, i.e., the integral closure of A in K. Assuming B has such an ideal $\mathfrak{n}'$, we may construct Heegner points using the analytic description of Drinfeld modules over C as follows. If $\mathfrak{a} \subset B$ is a nonzero proper ideal then the pair of Drinfeld modules ϕ and ϕ' corresponding to the lattices $\mathfrak{n}'\mathfrak{a}$ and $\mathfrak{a}$ in $K \hookrightarrow C$ satisfy $\mathrm{End}(\phi) = \mathrm{End}(\phi') = B$ and they are connected by a cyclic $\mathfrak{n}$-isogeny. The corresponding point turns out to depend only on B, $\mathfrak{n}'$, and the class of $\mathfrak{a}$ in $\mathrm{Pic}(B)$ and it is defined over the ring class field extension K_B/K corresponding to B by class field theory. The theory of complex multiplication of Drinfeld modules implies that $\mathrm{Gal}(K_B/K) \cong \mathrm{Pic}(B)$ acts on the Heegner points through its natural action on the class of $\mathfrak{a}$. Applying an Atkin–Lehner involution to a Heegner point is related to changing the choice of ideal $\mathfrak{n}'$ over $\mathfrak{n}$. All of this is discussed in [Bro94, § 2] in the context where $A = \mathbb{F}_q[t]$.

Taking the trace from K_B to K of a Heegner point and subtracting a suitable multiple of a cusp, we get a K-rational divisor of degree 0, and so a point $J_0(\mathfrak{n})(K)$. We write Q_K for the point so constructed when K is an imaginary quadratic extension of F in which every prime dividing $\mathfrak{n}$ splits and B is the maximal A-order in K. The point Q_K is well-defined, independently of the other choices ($\mathfrak{n}'$ and $\mathfrak{a}$), up to a torsion point of $J_0(\mathfrak{n})(K)$. If E is an elliptic curve over F of level $\mathfrak{n}\infty$ with split multiplicative reduction at ∞, then using the modular parameterization discussed above one obtains a point $P_K \in E(K)$, well-defined up to torsion.

Brown purports to prove, by methods analogous to those of Kolyvagin [Kol90], that if P_K is nontorsion, then the Tate–Shafarevitch group of E is finite and the rank of $E(K)$ is one. (He gives an explicit annihilator of the ℓ-primary part of Ш for infinitely many ℓ.) As we have seen, this implies that the Birch and Swinnerton-Dyer conjecture holds for E over K.

Unfortunately, Brown's paper is marred by a number of errors, some rather glaring. For example, the statement of the main theorem is not in fact what is proved and it is easily seen to be false if taken literally. Also, he makes the strange hypothesis that q, the number of elements in the finite ground field, is not a square. The source of this turns out to be a misunderstanding of quadratic reciprocity in the proof of his Corollary 3.4. In my opinion, although something like what Brown claims can be proved by the methods in his paper, a thorough revision is needed before his theorem can be said to have been proven.

There is another difficulty, namely that Brown's theorem does not give a very direct approach to the Birch and Swinnerton-Dyer conjecture. This is because it is rather difficult to compute the modular parameterization and thus the Heegner point, and so the hypotheses of Brown's theorem are hard to verify. (The difficulty comes from the fact that nonarchimedean integration seems to be of exponential complexity in the desired degree of accuracy, in contrast to archimedean integration which is polynomial time.) On the other hand it is quite easy to check whether the L-function of E vanishes to order 0 or 1, these being the only cases where one expects Heegner points to be of help. In fact the computation of the entire L-function of E is straightforward and (at least over the rational function field) can be made efficient using the existence of an automorphic form corresponding to E. See [TR92]. This situation is the opposite of that in the classical situation; cf. the remarks of Birch near the end of § 4 of his article in this volume [Bir04].

In light of this difficulty, a more direct and straightforward approach to the Birch and Swinnerton-Dyer conjecture for elliptic curves of rank ≤ 1 is called for. My interest in function field analogues of Gross–Zagier came about from an effort to understand Brown's paper and to find a better approach to BSD in this context.

3.5. Gross–Zagier formulas. Let us now state what the analogue of the Gross–Zagier formula [GZ86] should be in the function field context. Let E be an elliptic curve over F of conductor $\mathfrak{n}\infty$ and with split multiplicative reduction at ∞. Then for every imaginary quadratic extension K of F satisfying the Heegner hypotheses (namely that every prime dividing $\mathfrak{n}$ is split in K), we have a point $P_K \in E(K)$ defined using Heegner points on $X_0(\mathfrak{n})$ and the modular parameterization. The desired formula is then

$$L'(E/K,1) = a\langle P_K, P_K\rangle \tag{3–2}$$

where $\langle\, ,\rangle$ is the Néron–Tate canonical height on E and a is an explicit nonzero constant. Because equality of analytic and algebraic ranks implies the refined BSD conjecture, the exact value of a is not important for us.

The left hand side of this formula is also a special value of the L-function of an automorphic form (namely, the f such that $L(E,\chi,s) = L(f,\chi,s)$) and Equation (3–2) is a special case of a more general formula which applies to automorphic forms without the assumption that their Hecke eigenvalues are integers. Let S be the vector space of complex valued cuspidal automorphic forms on GL_2 over F which have level $\Gamma_0(\mathfrak{n}\infty)$, central character $|\cdot|^{-2}$, and which are special at ∞. (As discussed in Section 3.2, this is the analogue of $S_2(\Gamma_0(N))$.) Then we have a Petersson inner product

$$(\ ,\ ) : S \times S \to \mathbb{C}$$

which is positive definite Hermitian. For $f \in S$, let $L_K(f,s)$ be the L-function of the base change of f to a form on GL_2 over K. (This form can be shown to exist using a Rankin–Selberg integral representation and Weil's converse theorem.) Then the function $f \mapsto L'_K(f,1)$ is a linear map $S \to \mathbb{C}$ and so there exists a unique element $h_K \in S$ such that

$$(f, h_K) = L'_K(f,1)$$

for all $f \in S$.

For $h \in S$, let $c(h,\mathfrak{m})$ be the $\mathfrak{m}$-th Fourier coefficient of h. Then a formal Hecke algebra argument, as in the classical case, shows that the desired Gross–Zagier formula (3–2) (and its more general version mentioned above) follows from the following equalities between Fourier coefficients and heights on $J_0(\mathfrak{n})$:

$$c(h_K,\mathfrak{m}) = a\langle Q_K, T_{\mathfrak{m}} Q_K\rangle \tag{3–3}$$

for all effective divisors $\mathfrak{m}$ prime to $\mathfrak{n}\infty$. Here $T_{\mathfrak{m}}$ is the Hecke operator on $J_0(\mathfrak{n})$ indexed by $\mathfrak{m}$ and $\langle\, ,\rangle$ is the canonical height pairing on $J_0(\mathfrak{n})$.

From now on, by "Gross–Zagier formula" we will mean the sequence of equalities (3–3).

3.6. Rück–Tipp. Rück and Tipp were the first to write down a function field analogue of the Gross–Zagier formula [RT00]. They work over $F = \mathbb{F}_q(t)$ with q odd, and ∞ the standard place at infinity $t = \infty$ (so their ∞ has degree 1). They assume that $\mathfrak{n}$ is square free and that $K = F(\sqrt{D})$ where D is an *irreducible* polynomial in $\mathbb{F}_q[t]$. Under these hypotheses, they checked the equalities (3–3) for all $\mathfrak{m}$ prime to $\mathfrak{n}\infty$, which yields the formula (3–2). This gives some instances of the conjecture of Birch and Swinnerton-Dyer, under very restrictive hypotheses.

Their paper follows the method of Gross and Zagier [GZ86] quite closely (which is not to say that the analogies are always obvious or easy to implement!). They use the Rankin–Selberg method and a holomorphic projection operator to compute the Fourier coefficients of h_K. The height pairing is decomposed as a sum of local terms and, at finite places, the local pairing is given as an intersection number, which can be computed by counting isogenies between Drinfeld modules over a finite field. The local height pairing at ∞ is also an intersection number and one might hope to use a moduli interpretation of the points on the fiber at ∞ to calculate the local height. But to my knowledge, no one knows how to do this. Instead, Rück and Tipp compute the local height pairing using Green's functions on the Drinfeld upper half plane. This is a very analytic way of computing a rational number, but it matches well with the computations on the analytic side of the formula.

3.7. Pál and Longhi. Pál and Longhi worked (independently) on function field analogues of the Bertolini–Darmon [BD98] p-adic construction of Heegner points. Both work over a general function field F of odd characteristic. Let E be an elliptic curve over F with conductor $\mathfrak{n}\infty$ and which is split multiplicative at ∞. Let K be a quadratic extension in which ∞ is inert and which satisfies the Heegner hypotheses with respect to E. Also let H_n be the ring class field of K of conductor ∞^n and set $G = \varprojlim \operatorname{Gal}(H_n/K)$.

Pál [Pál00] used "Gross–Heegner" points, as in Bertolini–Darmon (following Gross [Gro87]), to construct an element $\mathcal{L}(E/K)$ in the completed group ring $\mathbb{Z}[\![G]\!]$ which interpolates suitably normalized special values $L(E/K, \chi, 1)$ for finite order characters χ of G. It turns out that $\mathcal{L}(E/K)$ lies in the augmentation ideal I of $\mathbb{Z}[\![G]\!]$ and so defines an element $\mathcal{L}'(E/K)$ in $I/I^2 \cong \mathcal{O}_{K_\infty}^\times/\mathcal{O}_{F_\infty}^\times \cong \mathcal{O}_{K_\infty,1}^\times$. (Here F_∞ and K_∞ are the completions at ∞ and $\mathcal{O}_{K_\infty,1}^\times$ denotes the 1-units in $\mathcal{O}_{K_\infty}$.) Since E is split multiplicative at ∞, we have a Tate parameterization $K_\infty^\times \to E(K_\infty)$ and Pál shows that the image of $\mathcal{L}'(E/K)$ in $E(K_\infty)$ is a global point. More precisely, if E is a "strong Weil curve," then Pál's point is $P_K - \overline{P}_K$ where P_K is the Heegner point discussed above and $\overline{P}_K$ is its "complex conjugate." It follows that if $\mathcal{L}'(E/K)$ is nonzero, then the Heegner point is of infinite order and so $\operatorname{Rank}_{\mathbb{Z}} E(K)$ is at least one. One interesting difference between Pál's work and [BD98] is that in the latter, there are 2 distinguished places, namely ∞, which is related to the classical modular parameterization, and p, which is related to the Tate parameterization. In Pál's work, the role of both of these primes is played by the prime ∞ of F. This means that his result is applicable in more situations than the naive analogy would predict — E need only have split multiplicative reduction at one place of F.

Longhi [Lon02] also gives an ∞-adic construction of a Heegner point. Whereas Pál follows [BD98], Longhi's point of view is closer to that of [BD01]. His ∞-adic L-element $\mathcal{L}(E/K)$ is constructed using ∞-adic integrals, following the approach of Schneider [Sch84] and a multiplicative version of Teitelbaum's Poisson formula [Tei91]. Unfortunately, there is as yet no connection between his ∞-adic $\mathcal{L}(E/K)$ and special values of L-functions.

Both of these works have the advantage of avoiding intricate height computations on Drinfeld modular curves, as in [GZ86]. (Pál's work uses heights of the much simpler variety considered in [Gro87].) On the other hand, they do not yet have any direct application to the conjecture of Birch and Swinnerton-Dyer, because presently we have no direct link between the ∞-adic L-derivative $\mathcal{L}'(E/K)$ and the classical L-derivative $L'(E/K, 1)$.

3.8. My work on BSD for rank 1. My interest in this area has been less in analogues of the Gross–Zagier formula or Kolyvagin's work over function fields *per se*, and more in their applications to the Birch and Swinnerton-Dyer conjecture itself. The problem with a raw Gross–Zagier formula is that it only gives the BSD conjecture with parasitic hypotheses. For example, to have a Drinfeld modular parameterization, and thus Heegner points, the elliptic curve must have split multiplicative reduction at some place and the existence of such a place presumably has nothing to do with the truth of the conjecture. Recently, I have proven a nonvanishing result which when combined with a suitable Gross–Zagier formula leads to a clean, general statement about Birch and Swinnerton-Dyer: "If E is an elliptic curve over a function field F of characteristic > 3 and $\operatorname{ord}_{s=1} L(E/F, s) \leq 1$, then the Birch and Swinnerton-Dyer conjecture holds for E." In the remainder of this section I will describe the nonvanishing result, and then give the statement and status of the Gross–Zagier formula I have in mind.

Thus, let E be an elliptic curve over F with $\operatorname{ord}_{s=1} L(E/F, s) \leq 1$; for purposes of BSD we may as well assume that $\operatorname{ord}_{s=1} L(E/F, s) = 1$ and that E is nonisotrivial. Because $j(E) \notin \mathbb{F}_q$, it has a pole at some place of F, i.e., E is potentially multiplicative there. Certainly we can find a finite extension F' of F such that E has a place of split multiplicative reduction and it will suffice to prove BSD for E over F'. But, to do this with Heegner points, we must be able to choose F' so that $\operatorname{ord}_{s=1} L(E/F', s)$, which is *a priori* ≥ 1, is equal to 1. This amounts to a nonvanishing statement for a (possibly nonabelian) twist of $L(E/F, s)$, namely $L(E/F', s)/L(E/F, s)$. Having done this, a similar issue comes up in the application of a Gross–Zagier formula, namely, we must find a quadratic extension K/F' satisfying the Heegner hypotheses such that $\operatorname{ord}_{s=1} L(E/K, s) = \operatorname{ord}_{s=1} L(E/F', s) = 1$. This amounts to a nonvanishing statement for quadratic twists of $L(E/F', s)$ by characters satisfying certain local conditions. This issue also comes up in the applications of the classical Gross–Zagier formula and is dealt with by automorphic methods. Recently, I have proven a very general nonvanishing theorem for motivic L-functions over function fields using algebro-geometric methods which when applied to elliptic curves yields the following result:

THEOREM 3.8.1. [Ulm03] *Let E be a nonconstant elliptic curve over a function field F of characteristic $p > 3$. Then there exists a finite separable extension F' of F and a quadratic extension K of F' such that the following conditions are satisfied*:

1. *E is semistable over F', i.e., its conductor is square-free.*
2. *E has split multiplicative reduction at some place of F' which we call ∞.*
3. *K/F' satisfies the Heegner hypotheses with respect to E and ∞. In other words, K/F' is split at every place $v \neq \infty$ dividing the conductor of E and it is not split at ∞.*
4. $\operatorname{ord}_{s=1} L(E/K, s)$ *is odd and at most* $\operatorname{ord}_{s=1} L(E/F, s) + 1$. *In particular, if* $\operatorname{ord}_{s=1} L(E/F, s) = 1$, *then* $\operatorname{ord}_{s=1} L(E/K, s) = \operatorname{ord}_{s=1} L(E/F', s) = 1$.

This result, plus a suitable Gross–Zagier formula, yields the desired theorem. Indeed, by point (2), E admits a Drinfeld modular parameterization over F' and by point (3) we will have a Heegner defined over K. Point (4) (plus GZ!) guarantees that the Heegner point will be nontorsion and so we have $\operatorname{Rank} E(K) \geq 1$. As we have seen, this implies BSD for E over K and thus also over F. Point (1) is included as it makes the needed GZ formula a little more tractable. Also, although it is not stated in the theorem, it is possible to specify whether the place ∞ of F' is inert or ramified in K and this too can be used to simplify the Gross–Zagier calculation.

Thus, the Gross–Zagier formula we need is in the following context: the base field F' is arbitrary but the level $\mathfrak{n}$ is square-free and we may assume that ∞ is inert (or ramified) in K. It would perhaps be unwise to write too much about a result which is not completely written and refereed, so I will just say a few words. The proof follows closely the strategy of Gross and Zagier, with a few simplifications due to Zhang [Zha01]. One computes the analytic side of (3–3) using the Rankin–Selberg method and a holomorphic projection and the height side is treated using intersection theory at the finite places and Green's functions at ∞. Because we work over an arbitrary function field, our proofs are necessarily adelic. Also, in the analytic part we emphasize the geometric view of automorphic forms, namely that they are functions on a moduli space of rank 2 vector bundles on $\mathcal{C}$. The full details will appear in [Ulma].

4. Ranks over Function Fields

We now move beyond rank 1 and consider the rank conjecture for elliptic curves over function fields. Recall from Section 2 the notions of constant, isotrivial, and nonisotrivial for elliptic curves over function fields. Our purpose in this section is to explain constructions of isotrivial and nonisotrivial elliptic curves over $\mathbb{F}_p(t)$ whose Mordell–Weil groups have arbitrarily large rank. These curves turn out to have asymptotically maximal rank, in a sense which we will explain in Section 5.

4.1. The Shafarevitch–Tate construction. First, note that if E is a constant elliptic curve over $F = \mathbb{F}_q(\mathcal{C})$ based on E_0, then $E(F) \cong \mathrm{Mor}_{\mathbb{F}_q}(\mathcal{C}, E_0)$ (morphisms defined over $\mathbb{F}_q$) and the torsion subgroup of $E(F)$ corresponds to constant morphisms. Since a morphism $\mathcal{C} \to E$ is determined up to translation by the induced map of Jacobians, we have $E(F)/tor \cong \mathrm{Hom}_{\mathbb{F}_q}(J(\mathcal{C}), E)$ where $J(\mathcal{C})$ denotes the Jacobian of $\mathcal{C}$.

The idea of Shafarevitch and Tate [TS67] was to take E_0 to be supersingular and to find a curve $\mathcal{C}$ over $\mathbb{F}_p$ which is hyperelliptic and such that $J(\mathcal{C})$ has a large number of isogeny factors equal to E_0. If E denotes the constant curve over $\mathbb{F}_p(t)$ based on E_0, then it is clear that $E(\mathbb{F}_p(t))$ has rank 0. On the other hand, over the quadratic extension $F = \mathbb{F}_p(\mathcal{C})$, $E(F)/tor \cong \mathrm{Hom}_{\mathbb{F}_q}(J(\mathcal{C}), E_0)$ has large rank. Thus if we let E' be the twist of E by the quadratic extension $F/\mathbb{F}_q(t)$, then $E'(\mathbb{F}_q(t))$ has large rank. Note that E' is visibly isotrivial.

To find such curves $\mathcal{C}$, Tate and Shafarevitch considered quotients of the Fermat curve of degree p^n+1 with n odd. The zeta functions of Fermat curves can be computed in terms of Gauss sums, and in the case of degree of the form p^n+1, the relevant Gauss sums are easy to make explicit. This allows one to show that the Jacobian is isogenous to a product of supersingular elliptic curves over $\overline{\mathbb{F}}_p$ and has a supersingular elliptic curve as isogeny factor to high multiplicity over $\mathbb{F}_p$.

We remark that the rank of the Shafarevitch–Tate curves goes up considerably under extension of the finite ground field: if the rank of $E'(\mathbb{F}_p(t))$ is r, then the rank of $E'(\overline{\mathbb{F}}_p(t))$ is of the order $2\log_p(r)r$.

It has been suggested by Rubin and Silverberg that one might be able to carry out a similar construction over $\mathbb{Q}(t)$, i.e., one might try to find hyperelliptic curves $\mathcal{C}$ defined over $\mathbb{Q}$ whose Jacobians have as isogeny factor a large number of copies of some elliptic curve. The obvious analogue of the construction above would then produce elliptic curves over $\mathbb{Q}(t)$ of large rank. In [RS01] they use this idea to find many elliptic curves of rank ≥ 3. Unfortunately, it is not at all evident that curves $\mathcal{C}$ such that $J(\mathcal{C})$ has an elliptic isogeny factor to high multiplicity exist, even over $\mathbb{C}$.

Back to the function field case: We note that isotrivial elliptic curves are very special and seem to have no analogue over $\mathbb{Q}$. Thus the relevance of the

Shafarevitch–Tate construction to the rank question over $\mathbb{Q}$ is not clear. In the next subsection we explain a construction of *nonisotrivial* elliptic curves over $\mathbb{F}_p(t)$ of arbitrarily large rank.

4.2. Nonisotrivial elliptic curves of large rank. In [Shi86], Shioda showed that one could often compute the Picard number of a surface which is dominated by a Fermat surface. He applied this to write down elliptic curves over $\overline{\mathbb{F}}_p(t)$ (with $p \equiv 3 \mod 4$) of arbitrarily large rank, using supersingular Fermat surfaces (i.e., those whose degrees divide p^n+1 for some n). I was able to use the idea of looking at quotients of Fermat surfaces and a different method of computing the rank to show the existence of elliptic curves over $\mathbb{F}_p(t)$ (any p) with arbitrarily large rank. Here is the precise statement:

THEOREM 4.2.1. [Ulm02] *Let p be a prime, n a positive integer, and d a divisor of p^n+1. Let q be a power of p and let E be the elliptic curve over $\mathbb{F}_q(t)$ defined by*

$$y^2 + xy = x^3 - t^d.$$

Then the j-invariant of E is not in $\mathbb{F}_q$, the conjecture of Birch and Swinnerton-Dyer holds for E, and the rank of $E(\mathbb{F}_q(t))$ is

$$\sum_{\substack{e|d \\ e\nmid 6}} \frac{\phi(e)}{o_e(q)} + \begin{cases} 0 & \text{if } 2\nmid d \text{ or } 4\nmid q-1 \\ 1 & \text{if } 2|d \text{ and } 4|q-1 \end{cases} + \begin{cases} 0 & \text{if } 3\nmid d \\ 1 & \text{if } 3|d \text{ and } 3\nmid q-1 \\ 2 & \text{if } 3|d \text{ and } 3|q-1. \end{cases}$$

Here $\phi(e)$ is the cardinality of $(\mathbb{Z}/e\mathbb{Z})^\times$ and $o_e(q)$ is the order of q in $(\mathbb{Z}/e\mathbb{Z})^\times$.

In particular, if we take $d = p^n+1$ and $q = p$, then the rank of E over $\mathbb{F}_p(t)$ is at least $(p^n-1)/2n$. On the other hand, if we take $d = p^n+1$ and q to be a power of p^{2n}, then the rank of E over $\mathbb{F}_q(t)$ is $d-1 = p^n$ if $6\nmid d$ and $d-3 = p^n-2$ if $6|d$. Note that the rank may increase significantly after extension of $\mathbb{F}_q$.

Here is a sketch of the proof: by old work of Artin and Tate [Tat66b], the conjecture of Birch and Swinnerton-Dyer of E is equivalent to the Tate conjecture for the elliptic surface $\mathcal{E} \to \mathbb{P}^1$ over $\mathbb{F}_q$ attached to E. (The relevant Tate conjecture is that $-\operatorname{ord}_{s=1} \zeta(\mathcal{E}, s) = \operatorname{Rank}_{\mathbb{Z}} NS(\mathcal{E})$ where $NS(\mathcal{E})$ denotes the Néron–Severi group of $\mathcal{E}$.) The equation of E was chosen so that $\mathcal{E}$ is dominated by the Fermat surface *of the same degree d*. (The fact that the equation of E has 4 monomials is essentially enough to guarantee that $\mathcal{E}$ is dominated by some Fermat surface; getting the degree right requires more.) Since the Tate conjecture is known for Fermat surfaces, this implies it also for $\mathcal{E}$ (and thus BSD for E). Next, a detailed analysis of the geometry of the rational map $F_d \dashrightarrow \mathcal{E}$ allows one to calculate the zeta function of $\mathcal{E}$ in terms of that of F_d, i.e., in terms of Gauss and Jacobi sums. Finally, because d is a divisor of p^n+1, the relevant Gauss sums are all supersingular (as in the Shafarevitch–Tate case) and can be made explicit. This gives the order of pole of $\zeta(\mathcal{E}, s)$ at $s = 1$ and thus the order of zero of $L(E/\mathbb{F}_q(t), s)$ at $s = 1$, and thus the rank.

We note that the proof does not explicitly construct any points, although it does suggest a method to do so. Namely, using the large automorphism group of the Fermat surface, one can write down curves which span $NS(F_d)$ and use these and the geometry of the map $F_d \dashrightarrow \mathcal{E}$ to get a spanning set for $NS(\mathcal{E})$ and thus a spanning set for $E(\mathbb{F}_q(t))$. It looks like an interesting problem to make this explicit, and to consider the heights of generators of $E(\mathbb{F}_q(t))$ and its Mordell–Weil lattice.

4.3. Another approach to high rank curves. The two main parts of the argument of Section 4.2 could be summarized as follows: (i) one can deduce the Tate conjecture for $\mathcal{E}$ and thus the BSD conjecture for E from the existence of a dominant rational map from the Fermat surface F_d to the elliptic surface $\mathcal{E}$ attached to E; and (ii) a detailed analysis of the geometry of the map $F_d \dashrightarrow \mathcal{E}$ allows one to compute the zeta function of $\mathcal{E}$ and thus the L-function of E, showing that it has a large order zero at $s = 1$.

Ideas of Darmon give an alternative approach to the second part of this argument (showing that the L-function has a large order zero at $s = 1$) and may lead (subject to further development of Gross–Zagier formulas in the function field case) to an alternative approach to the first part of the argument (the proof of BSD). Darmon's idea is quite general and leads to the construction of many elliptic curves over function fields of large rank (more precisely, provably of large analytic rank and conjecturally of large algebraic rank.) Here we will treat only the special case of the curve considered in Section 4.2 and we refer to his article in this volume [Dar04] for details of the general picture.

Let $q = p^n$ (p any prime), $d = q + 1$, and define $F = \mathbb{F}_q(u)$, $K = \mathbb{F}_{q^2}(u)$, and $H = \mathbb{F}_{q^2}(t)$ where $u = t^d$. Then H is Galois over F with dihedral Galois group. Indeed $\mathrm{Gal}(H/K)$ is cyclic of order d and because $q \equiv -1 \mod d$, the nontrivial element of $\mathrm{Gal}(K/F) \cong \mathrm{Gal}(\mathbb{F}_{q^2}/\mathbb{F}_q)$ acts on $\mathrm{Gal}(H/K)$ by inversion. Let E be the elliptic curve over F defined by the equation

$$y^2 + xy = x^3 - u.$$

Over H, this is the curve discussed in Section 4.2.

The L-function of E over H factors into a product of twisted L-functions over K:

$$L(E/H, s) = \prod_{\chi \in \hat{G}} L(E/K, \chi, s)$$

where the product is over the d characters of $G = \mathrm{Gal}(H/K)$. Because H/F is a dihedral extension and E is defined over F, we have the equality $L(E/K, \chi, s) = L(E/K, \chi^{-1}, s)$. Thus the functional equation

$$\begin{aligned} L(E/K, \chi, s) &= W(E/K, \chi,)q^{sd_{E,\chi}} L(E/K, \chi^{-1}, 2 - s) \\ &= W(E/K, \chi)q^{sd_{E,\chi}} L(E/K, \chi, 2 - s) \end{aligned}$$

(where $W(E/K,\chi)$ is the "root number" and $d_{E,\chi}$ is the degree of $L(E/K,\chi,s)$ as a polynomial in q^{-2s}) may force a zero of $L(E/K,\chi,s)$ at the critical point $s=1$. This is indeed what happens: A careful analysis shows that $W(E/K,\chi)$ is $+1$ if χ is trivial or of order exactly 6 and it is -1 in all other cases. Along the way, one also finds that $d_{E,\chi}$ is 0 if χ is trivial or of order exactly 6 and is 1 in all other cases. Thus $L(E/K,\chi,s)$ is equal to 1 if χ is trivial or of order exactly 6 and is equal to $(1-q^{-2s})$ and vanishes to order 1 at $s=1$ if χ is nontrivial and not of order exactly 6. We conclude that $\operatorname{ord}_{s=1} L(E/H,s)$ is $d-3$ if 6 divides d and $d-1$ if not.

Of course one also wants to compute the L-function of E over $H_0=\mathbb{F}_p(t)$. In this case, the L-function again factors into a product of twists, but the twists are by certain, generally nonabelian, representations of the Galois group of the Galois closure of $\mathbb{F}_p(t)$ over $\mathbb{F}_p(u)$. (The Galois closure is H and the Galois group is the semidirect product of $\operatorname{Gal}(H/K)$ with $\operatorname{Gal}(\mathbb{F}_{q^2}(u)/\mathbb{F}_p(u))$. See [Ulm03, § 3] for more on this type of situation.) We will not go into the details, but simply note that in order to compute the L-function $L(E/H_0,s)$ along the lines above, one needs to know the root numbers $W(E/\mathbb{F}_r(u),\chi)$ where $r=p^{o_e}$, o_e is the order of p in $\mathbb{Z}/e\mathbb{Z}$, and e is the order of the character χ. It turns out that each of the twisted L-functions has a simple zero at $s=1$.

This calculation of $L(E/H,s)$ and $L(E/\mathbb{F}_p(t),s)$ seems to be of roughly the same difficulty as the geometric one in [Ulm02] because the "careful analysis" of the root numbers $W(E/K,\chi)$ and $W(E/\mathbb{F}_r(u),\chi)$ is somewhat involved, especially if one wants to include the cases $p=2$ or 3. (I have only checked that the answer agrees with that in [Ulm02] when $p>3$.) Calculation of the root numbers requires knowing the local representations of decomposition groups on the Tate module at places of bad reduction and eventually boils down to analyzing some Gauss sums. The Shafarevitch–Tate lemma on supersingular Gauss sums (Lemma 8.3 of [Ulm02]) is a key ingredient.

Regarding the problem of verifying the BSD conjecture for E/H, note that K/F may be viewed as an "imaginary" quadratic extension, and H/K is the ring class extension of conductor $\mathfrak{n}=(0)(\infty)$. Because most of the twisted L-functions $L(E/K,\chi,s)$ vanish simply, we might expect to construct points in $(E(H)\otimes\mathbb{C})^\chi$ using Heegner points and show that they are nontrivial using a Gross–Zagier formula. But the relevant Gross–Zagier formula here would involve Shimura curve analogs of Drinfeld modular curves (since the extension K/F does not satisfy the usual Heegner hypotheses) and such a formula remains to be proven. Perhaps Darmon's construction will provide some motivation for the brave soul who decides to take on the Gross–Zagier formula in this context! On the other hand, Darmon's paper has examples of curves where Heegner points on standard Drinfeld modular curves should be enough to produce high rank elliptic curves over $\mathbb{F}_p(t)$.

5. Rank Bounds

We now return to a general function field $F = \mathbb{F}_q(\mathcal{C})$ and a general nonisotrivial elliptic curve E over F. Recall that the conductor $\mathfrak{n}$ of E is an effective divisor on $\mathcal{C}$ which is supported precisely at the places where E has bad reduction.

It is natural to ask how quickly the ranks of elliptic curves over F can grow in terms of their conductors. As discussed in Section 2, we have the inequality

$$\mathrm{Rank}_{\mathbb{Z}} E(F) \le \mathrm{ord}_{s=1} L(E/F, s).$$

Also, one knows that that $L(E/F, s)$ is a polynomial in q^{-s} of degree $4g-4+\deg\mathfrak{n}$ where g is the genus of $\mathcal{C}$. (This comes from Grothendieck's cohomological expression for the L-function and the Grothendieck–Ogg–Shafarevitch Euler characteristic formula.) Thus we have a bound

$$\mathrm{Rank}_{\mathbb{Z}} E(F) \le \mathrm{ord}_{s=1} L(E/F, s) \le 4g - 4 + \deg\mathfrak{n}. \tag{5–1}$$

This bound is geometric in the sense that it does not involve the size of the finite field $\mathbb{F}_q$; the same bound holds for $\mathrm{Rank}_{\mathbb{Z}} E(\overline{\mathbb{F}}_q(\mathcal{C}))$. On the other hand, as we have seen above, the rank can change significantly after extension of $\mathbb{F}_q$. It is thus natural to ask for a more arithmetic bound, i.e., one which is sensitive to q.

Such a bound was proven by Brumer [Bru92], using Weil's "explicit formula" technique, along the lines of Mestre's bound for the rank of an elliptic curve over $\mathbb{Q}$. Brumer's result is

$$\mathrm{Rank}_{\mathbb{Z}} E(F) \le \mathrm{ord}_{s=1} L(E/F, s) \le \frac{4g-4+\deg(\mathfrak{n})}{2\log_q \deg(\mathfrak{n})} + C\frac{\deg(\mathfrak{n})}{(\log_q \deg(\mathfrak{n}))^2} \tag{5–2}$$

Note that this bound is visibly sensitive to q and is an improvement on the geometric bound when $\deg\mathfrak{n}$ is large compared to q.

Here is a sketch of Brumer's proof: let $\Lambda(s) = q^{Ds/2}L(E/F, s)$ where $D = 4g-4+\deg\mathfrak{n}$. Then $\Lambda(s)$ is a Laurent polynomial in $q^{-s/2}$ and so is periodic in s with period $4\pi i/\ln q$; moreover, we have the functional equation $\Lambda(s) = \pm\Lambda(2-s)$. Our task is to estimate the order of vanishing at $s = 1$ of Λ or equivalently, the residue at $s = 1$ of the logarithmic derivative Λ'/Λ with respect to s. Let us consider the line integral

$$I = \oint \Phi d\log\Lambda = \oint \Phi\frac{\Lambda'}{\Lambda}\,ds$$

where Φ is a suitable test function to be chosen later and the contour of integration is

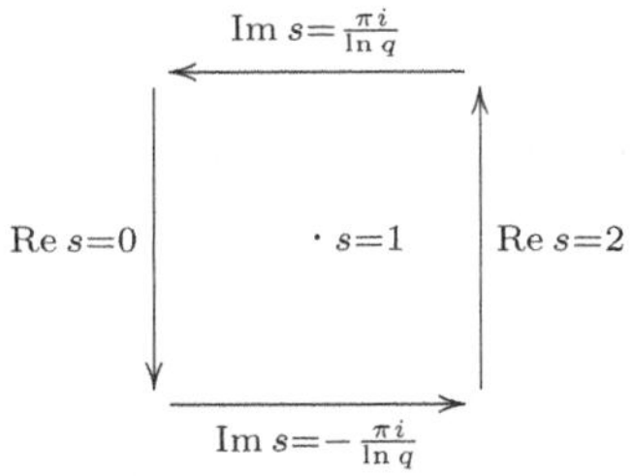

(Note that $d \log \Lambda$ is periodic with period $2\pi i/\ln q$. Also, we would have to shift the contour slightly if $L(E/F,s)$ has a zero at $1 \pm \pi i/\ln q$.) We assume that $\Phi(s)$ is nonnegative on the line $\operatorname{Re} s = 1$. By the Riemann hypothesis for $L(E/F,s)$, all the zeroes of $\Lambda(s)$ lie on this line and so

$$\Phi(1)\operatorname{ord}_{s=1} L(E/F,s) = \Phi(1)\operatorname{Res}_{s=1}\frac{\Lambda'}{\Lambda} \leq \sum_{\rho} \Phi(\rho) = I$$

where ρ runs over the zeroes of $L(E/F,s)$ inside the contour of integration counted with multiplicities. Now we assume in addition that Φ is a Laurent polynomial in q^{-s} (so periodic with period $2\pi i/\ln q$) and that it satisfies the functional equation $\Phi(s) = \Phi(2-s)$. Using the functional equation and periodicity of the integrand, the integral I is equal to

$$2\int_{2-\pi i/\ln q}^{2+\pi i/\ln q} \Phi\frac{\Lambda'}{\Lambda}\,ds. \tag{5–3}$$

Now the integration takes place entirely in the region of convergence of the Euler product defining $L(E/F,s)$ and so we can expand the integrand in a series and estimate the terms using the Riemann hypothesis for curves over finite fields. Finally, Brumer makes a clever choice of test function Φ which yields the desired estimate. (More precisely, he considers a sequence of test functions satisfying the hypotheses which when restricted to $\operatorname{Re} s = 1$ converge to the Dirac delta function at $s = 1$ — up to a change of variable, this is essentially the Fejér kernel — and then chooses Φ to be a suitable element of this sequence.)

Note the strongly analytic character of this proof. For example, it does not use the fact that there is massive cancellation in the series for $L(E/F,s)$ so that the L-function is really a polynomial in q^{-s} of degree D!

Let E_n be the curve of Theorem 4.2.1 with $d = p^n + 1$. Then it turns out that the degree of the conductor of E_n is $p^n + 2$ if $6|d$ and $p^n + 4$ if $6 \nmid d$. One sees immediately that the geometric bound is sharp when q is a power of p^{2n} and the main term of the arithmetic bound is met when $q = p$. Thus both the geometric and arithmetic bounds give excellent control on ranks.

The rest of this paper is devoted to considering various questions which arise naturally by analogy from the existence and sharpness of these two types of bounds, geometric and arithmetic.

6. Ranks over Number Fields

We now turn to analogous situations, starting with the case where the ground field F is either $\mathbb{Q}$ or a number field. Throughout, we assume that the L-series of elliptic curves have good analytic properties, namely analytic continuation, boundedness in vertical strips, and the standard functional equation. (This is of course now known for elliptic curves over $\mathbb{Q}$ by the work of Wiles and others, but is still open for a general number field F.) We also assume the conjecture of Birch and Swinnerton-Dyer so that "rank" can be taken to mean either analytic rank ($\operatorname{ord}_{s=1} L(E/F,s)$) or algebraic rank ($\operatorname{Rank}_{\mathbb{Z}} E(F)$); alternatively the reader may interpret each question or conjecture involving an unqualified "rank" to be two statements, one about analytic rank, the other about algebraic rank.

The Brumer bound discussed in the last section was modeled on work of Mestre [Mes86], who proved, along lines quite similar to those sketched above, a bound on analytic ranks of the following form:

$$\operatorname{ord}_{s=1} L(E/\mathbb{Q},s) = O\left(\frac{\log N}{\log\log N}\right) \tag{6–1}$$

where E is an elliptic curve over $\mathbb{Q}$ of conductor N. To see the analogy, note that the degree function on divisors is a kind of logarithm and so $\deg \mathfrak{n}$ is an analogue of $\log N$. To obtain this bound, Mestre assumes the Generalized Riemann Hypothesis for $L(E/\mathbb{Q},s)$ and he actually proves a more general statement about orders of vanishing for L-series of modular forms. Assuming good analytic properties and the generalized Riemann hypothesis, his argument extends readily to elliptic curves over number fields; in this case N should be replaced with the norm from F to $\mathbb{Q}$ of the conductor of E times the absolute value of the discriminant of F.

There is some evidence that the Mestre bound should be asymptotically sharp. First of all, it gives excellent results for small N. Secondly, in the function field case, the analogous bound is sharp. Moreover, the proof of the bound in the function field case does not use strongly any special features of that situation, such as the fact that the L-function is really a polynomial. Motivated by these facts, I make the following conjecture about the sharpness of the Mestre bound.

CONJECTURE 6.1. *Fix a number field F and for each positive integer N, define $r_F(N)$ by*

$$r_F(N) = \max\{\operatorname{Rank}_Z(E(F)) \mid E/F \text{ with } Norm_{F/\mathbb{Q}}(\mathfrak{n}_E) = N\}$$

where the maximum is taken over all elliptic curves E over F with conductor $\mathfrak{n}_E$ satisfying $Norm_{F/\mathbb{Q}}(\mathfrak{n}) = N$; if there are no such curves, we set $r_F(N) = 0$.

Then we have

$$\limsup_N \frac{r_F(N)}{\log N/\log\log N} > 0$$

By the generalization of the Mestre bound to number fields, the limit in the conjecture is finite.

If E is an elliptic curve over $\mathbb{Q}$, let $N_{\mathbb{Q}}(E)$ be its conductor and let $N_F(E)$ be the norm from F to $\mathbb{Q}$ of the conductor of E viewed as elliptic curve over F. Then there is a constant C depending only on F such that

$$1 \le \frac{N_{\mathbb{Q}}(E)^{[F:\mathbb{Q}]}}{N_F(E)} \le C$$

for all elliptic curves E over $\mathbb{Q}$. Indeed, if N is an integer, then $\mathrm{Norm}_{F/\mathbb{Q}}(N) = N^{[F:\mathbb{Q}]}$ and since the conductor of E over F is a divisor of $N_{\mathbb{Q}}(E)$ (viewed as an ideal of F), we have $1 \le N_{\mathbb{Q}}(E)^{[F:\mathbb{Q}]}/N_F(E)$. Since $N_{\mathbb{Q}}(F)$ divided by the conductor of E over F is divisible only by ramified primes and these occur with bounded exponents [BK94], there is a constant C such that

$$N_{\mathbb{Q}}(E)^{[F:\mathbb{Q}]}/N_F(E) \le C.$$

These inequalities show that a sequence of elliptic curves proving the conjecture over $\mathbb{Q}$ also proves the conjecture for a general number field F.

Finally, let us remark that there are experts who are skeptical about this conjecture. Certain probabilistic models predict that the denominator should be replaced by its square root, i.e., that the correct bound is

$$\mathrm{ord}_{s=1} L(E/\mathbb{Q}, s) \stackrel{?}{=} O\left(\left(\frac{\log N}{\log\log N}\right)^{1/2}\right).$$

On the other hand, certain random matrix models suggest that the Mestre bound is sharp. See the list of problems for the workshop on random matrices and L-functions at AIM, May 2001 (`http://aimath.org`) for more on this question.

7. Algebraic Rank Bounds

The Mestre and Brumer bounds are analytic in both statement and proof. It is interesting to ask whether they can be made more algebraic. For example, the Brumer bound is equivalent to a statement about the possible multiplicity of q as an eigenvalue of Frobenius on $H^1(\mathcal{C}, \mathcal{F})$ for a suitable sheaf $\mathcal{F}$, namely $R^1\pi_*\mathbb{Q}_\ell$ where $\pi : \mathcal{E} \to \mathcal{C}$ is the elliptic surface attached to E/F. It seems that statements like this might admit more algebraic proofs.

There is one situation where such algebraic proofs are available. Namely, consider an elliptic curve E over a number field or a function field F such that E has an F-rational 2-isogeny $\phi : E \to E'$. Then the Selmer group for multiplication by 2 sits in an exact sequence

$$\mathrm{Sel}(\phi) \to \mathrm{Sel}(2) \to \mathrm{Sel}(\check{\phi})$$

where $\check{\phi} : E' \to E$ is the dual isogeny. The orders of the groups $\mathrm{Sel}(\phi)$ and $\mathrm{Sel}(\check{\phi})$ can be crudely and easily estimated in terms of $\omega(N)$, the number of primes dividing the conductor N of E, and a constant depending only on F which involves the size of its class group and unit group. This yields a bound on the rank of the form

$$\mathrm{Rank}_{\mathbb{Z}} E(F) \leq C + 2\omega(N).$$

Note that this bound deserves to be called arithmetic because, for example, in the function field case $F = \mathbb{F}_q(\mathcal{C})$, $\omega(N)$ is sensitive to $\mathbb{F}_q$ since primes dividing N may split after extension of $\mathbb{F}_q$. Note also that it is compatible with the Mestre and Brumer bounds, since $\omega(N) = O(\log N/\log\log N)$ [HW79, p. 355] in the number field case and $\omega(N) = O(\deg N/\log\deg N)$ in the function field case.

It is tempting to guess that a similar bound (i.e., $\mathrm{Rank}\, E(F) = O(\omega(N))$) might be true in general, but there are several reason for skepticism. First of all, the estimation of the Selmer group above breaks down when there is no F-rational 2-isogeny. In this case, one usually passes to an extension field F' where such an isogeny exists. But then the "constant" C involves the units and class groups of F' and these vary with E since F' does. Given our current state of knowledge about the size of class groups, the bounds we obtain are not as good as the Mestre/Brumer bounds. This suggests that what is needed is a way to calculate or at least estimate the size of a Selmer group $\mathrm{Sel}(\ell)$ without passing to an extension where the multiplication by ℓ isogeny factors.

The second reason for skepticism is that such a bound would imply, for example, that there is a universal bound on the ranks of elliptic curves over $\mathbb{Q}$ of prime conductor. Although we have little information on the set of such curves (for example, it is not even known that this set is infinite), the experts seem to be skeptical about the existence of such a bound. One fact is that there is an elliptic curve over $\mathbb{Q}$ with prime conductor and rank 10 [Mes86], and so the constant in a bound of $O(\omega(N))$ would have to be at least 10, which does not seem very plausible. Also, in [BS96], Brumer and Silverman make a conjecture which contradicts an $O(\omega(N))$ bound—their conjecture implies that there should be elliptic curves with conductor divisible only by 2, 3, and one other prime and with arbitrarily large rank. There is no substantial evidence one way or the other for their conjecture, so some caution is necessary.

Lastly, wild ramification may have some role to play. Indeed, for $p = 2$ or 3 the curves of Section 4 have conductor which is divisible only by two primes ($t = 0$ and $t = \infty$) and yet their ranks are unbounded.

Despite all these reasons for skepticism about the existence of a bound of the form $\mathrm{Rank}_{\mathbb{Z}} E(F) \leq O(\omega(N))$, it is interesting to ask about the possibility of estimating ranks or Selmer groups directly, i.e., without reducing to isogenies of prime degree. It seems to me that there is some hope of doing this in the function field case, at least in the simplest context of a semistable elliptic curve over the rational function field. In this case, ideas from étale cohomology (e.g.,

[Mil80, pp. 211–214]) allow one to express a cohomology group closely related to the Selmer group as a product of local factors where the factors are indexed by the places of bad reduction of the elliptic curve.

Another approach over function fields is via p-descent. In this case, there is always a rational p-isogeny, namely Frobenius, but, in contrast to an ℓ-descent, the places of (good) supersingular reduction play a role more like places of bad reduction for ℓ-descents. This means that the output of a p-descent does not *a priori* give a bound for the rank purely in terms of the conductor and invariants of the ground field. More work will be required here to yield interesting results. See [Vol90] and [Ulm91] for foundational work on p-descents in characteristic p.

8. Arithmetic and Geometric Bounds I: Cyclotomic Fields

We now turn to some questions motivated by the existence of a *pair* of bounds, one geometric, one arithmetic. Let $\mathbb{Q}_n \subset \mathbb{Q}(\mu_{p^{n+1}})$ be the subfield with $\mathrm{Gal}(\mathbb{Q}_n/\mathbb{Q}) = \mathbb{Z}/p^n\mathbb{Z}$ and set $\mathbb{Q}^{p\text{-cyc}} = \bigcup_{n\geq 0}\mathbb{Q}_n$. This is the cyclotomic $\mathbb{Z}_p$-extension of $\mathbb{Q}$. As is well-known, the extension $\mathbb{Q}^{p\text{-cyc}}/\mathbb{Q}$ may be thought of as an analogue of the extension $\overline{\mathbb{F}}_q(\mathcal{C})/\mathbb{F}_q(\mathcal{C})$, and this analogy, noted by Weil [Wei79, p. 298], was developed by Iwasawa into a very fruitful branch of modern number theory. There has also been some traffic in the other direction, e.g., [MW83]. Let us consider the rank bounds of Section 5 in this light.

Mazur [Maz72], in analogy with Iwasawa's work, asked about the behavior of the Mordell–Weil and Tate–Shafarevitch groups of an elliptic curve (or abelian variety) defined over $\mathbb{Q}$ as one ascends the cyclotomic tower. For example, he conjectured that if E is an elliptic curve with good, ordinary reduction at p, then $E(\mathbb{Q}^{p\text{-cyc}})$ should be finitely generated. This turns out to be equivalent to the assertion that $\mathrm{Rank}_{\mathbb{Z}}\, E(\mathbb{Q}_n)$ is bounded as $n \to \infty$, i.e., it stabilizes at some finite n.

Today, by work of Rohrlich [Roh84], Kato [Kat00], Rubin [Rub98], and others, this is known to hold even without the assumption that E has ordinary reduction at p. (But we do continue to assume that E has good reduction, i.e., that p does not divide the conductor of E.)

Rohrlich proved the analytic version of this assertion, namely that the analytic rank $\mathrm{ord}_{s=1} L(E/\mathbb{Q}_n, s)$ is bounded as $n \to \infty$. (Rohrlich's paper is actually about the L-functions of modular forms, but by the work of Wiles and his school, it applies to elliptic curves.) Note that

$$L(E/\mathbb{Q}_n, s) = \prod_{\chi} L(E/\mathbb{Q}, \chi, s)$$

where χ ranges over characters of $\mathrm{Gal}(\mathbb{Q}_n/\mathbb{Q})$. So Rohrlich's theorem is that for any finite order character χ of $\mathrm{Gal}(\mathbb{Q}^{p\text{-cyc}}/\mathbb{Q})$ of sufficiently high conductor, $L(E/\mathbb{Q}, \chi, 1) \neq 0$. He proves this by considering the average of special values for conjugate characters $L(E/\mathbb{Q}, \chi^\sigma, 1)$ as σ varies over a suitable Galois group and showing that this average tends to 1 as the conductor of χ goes to infinity. Since $L(E/\mathbb{Q},\chi^\sigma,1)\neq 0$ if and only if $L(E/\mathbb{Q},\chi,1)\neq 0$, this implies $L(E/\mathbb{Q},\chi,1)\neq 0$.

Work of Rubin, Rubin–Wiles, and Coates–Wiles in the CM case and work of Kato in the non-CM case (see [Rub98, § 8.1] and the references there) allows us to translate this analytic result into an algebraic result. Namely, these authors show that $L(E/\mathbb{Q},\chi,1)\neq 0$ implies that $(E(\mathbb{Q}_n)\otimes\mathbb{C})^\chi=0$ where χ is a character of $\mathrm{Gal}(\mathbb{Q}_n/\mathbb{Q})$. This, together with Rohrlich's theorem implies that the rank of $E(\mathbb{Q}_n)$ stabilizes for large n.

Thus for an elliptic curve E over $\mathbb{Q}$ with good reduction at p, $\mathrm{Rank}_\mathbb{Z} E(\mathbb{Q}^{p\text{-cyc}})$ is finite and we may ask for a bound. Since there are only finitely many E of a given conductor, there is a bound purely in terms of p and N. The question then is what is the shape of this bound. For a fixed p, the results of Section 5 might lead one to guess that $\mathrm{Rank}_\mathbb{Z} E(\mathbb{Q}^{p\text{-cyc}})=O(\log N)$ (where the constant of course depends on p), but this is nothing more than a guess.

Rohrlich mentions briefly the issue of an effective bound for the smallest q such that $L(E/\mathbb{Q},\chi,1)\neq 0$ for all χ of conductor $p^n>q$. He obtains a bound of the form $q=CN^{170}$. Combined with the Mestre bound (6–1), this implies that $\mathrm{ord}_{s=1} L(E/\mathbb{Q}_n,s)$ is bounded for all n by a polynomial in N (which of course depends on p). This bound has recently been improved by Chinta [Chi02]. He points out that his Theorem 3 (or his Proposition 1 combined with Rohrlich's arguments) implies the following: If p is an odd prime where E has good reduction, then for every $\varepsilon>0$ there exist constants C_ε and e_ε such that

$$\mathrm{ord}_{s=1} L(E/\mathbb{Q}_n,s)\leq C_\varepsilon p^{e_\varepsilon} N^{1+\varepsilon}$$

for all n. The exponent e_ε may be taken to be linear in $1/\varepsilon$. This is of course a weaker bound than the guess $O(\log N)$; it might be interesting to try to establish the stronger bound on average. We remark that Chinta also shows the remarkable result that there exists an n_0 depending on E but *independent of* p such that $L(E/\mathbb{Q},\chi,1)\neq 0$ for all χ of conductor p^n, $n>n_0$.

9. Arithmetic and Geometric Bounds II: Function Fields over Number Fields

Let K be a number field and $\mathcal{C}$ a smooth, proper, geometrically connected curve over K. Let E be a nonisotrivial elliptic curve over $F=K(\mathcal{C})$ (i.e., $j(E)\notin K$). It is known that $E(F)$ is finitely generated [Nér52].

This finite generation, as well as a bound on the rank, can be obtained by considering the elliptic surface $\pi:\mathcal{E}\to\mathcal{C}$ attached to E/F. As in Section 2, $\mathcal{E}$ is the unique elliptic surface over $\mathcal{C}$ which is smooth and proper over K, with π flat, relatively minimal, and with generic fiber E/F. There is a close connection between the Mordell–Weil group $E(F)$ and the Néron–Severi group $NS(\mathcal{E})$. Using this, the cycle class map $NS(\mathcal{E})\to H^2(\mathcal{E}\times\overline{K},\mathbb{Q}_\ell)$ and an Euler characteristic formula, one obtains the same bound as in the positive characteristic case, namely:

$$\operatorname{Rank}_{\mathbb{Z}} E(F) \leq 4g - 4 + \deg \mathfrak{n} \tag{9–1}$$

where g is the genus of $\mathcal{C}$ and $\mathfrak{n}$ is the conductor of E.

This bound is geometric in that the number field K does not appear on the right hand side. In particular, the bound still holds if we replace K by $\overline{K}$:

$$\operatorname{Rank}_{\mathbb{Z}} E(\overline{K}(\mathcal{C})) \leq 4g - 4 + \deg \mathfrak{n}.$$

Using Hodge theory, this bound can be improved to $4g-4+\deg \mathfrak{n} - 2p_g$ where p_g is the geometric genus of $\mathcal{E}$, but this is again a geometric bound. It is reasonable to ask if there is a more arithmetic bound, improving (9–1).

There is some evidence that such a bound exists. Silverman [Sil00] considered the following situation: Let E be an elliptic curve over $F = K(t)$ and define $N^*(E)$ to be the degree of the part of the conductor of E which is prime to 0 and ∞. Alternatively, $N^*(E)$ is the sum of the number of points $t \in \overline{K}^{\times}$ where E has multiplicative reduction and twice the number of points $t \in \overline{K}^{\times}$ where E has additive reduction. Clearly $0 \leq \deg \mathfrak{n} - N^*(E) \leq 4$ and so the bound (9–1) gives $\operatorname{Rank}_{\mathbb{Z}} E(F) \leq N^*(E)$.

Now define E_n as the elliptic curve defined by the equation of E with t replaced by t^n. This is the base change of E by the the field homomorphism $K(t) \to K(t)$, $t \mapsto t^n$. Clearly $N^*(E_n) = nN^*(E)$ and so the geometric bound (9–1) gives $\operatorname{Rank}_{\mathbb{Z}} E_n(F) \leq nN^*(E)$.

Assuming the Tate conjecture (namely the equality $-\operatorname{ord}_{s=2} L(H^2(\mathcal{E}), s) = \operatorname{Rank}_{\mathbb{Z}} NS(\mathcal{E})$), Silverman proves by an analytic method that

$$\operatorname{Rank}_{\mathbb{Z}} E_n(F) \leq d_K(n) N^*(E)$$

where

$$d_K(n) = \sum_{d|n} \frac{\phi(d)}{[K(\mu_d) : K]}.$$

So, when $\mu_d \subset K$, $d_K(n) = n$ whereas if $K \cap \mathbb{Q}(\mu_n) = \mathbb{Q}$, then $d_K(n)$ is the number of divisors of n. Thus Silverman's theorem gives an arithmetic bound for ranks of a very special class of elliptic curves over function fields over number fields. I believe that there should be a much more general theorem in this direction.

There has been recent further work in this direction. Namely, Silverman [Sil03] has proven an interesting arithmetic bound on ranks of elliptic curves over unramified, abelian towers, assuming the Tate conjecture. In the special case where the base curve is itself elliptic and the tower is defined by the multiplication by n isogenies, he obtains a very strong bound, stronger than what is conjectured below. (See his Theorem 2.)

Silverman also formulates a beautiful and precise conjecture along the lines suggested above. Namely, he conjectures that there is an absolute constant C

such that for every nonisotrivial elliptic curve over $F = K(\mathcal{C})$ with conductor $\mathfrak{n}$,

$$\operatorname{Rank} E(F) \overset{?}{\leq} C \frac{4g - 4 + \deg \mathfrak{n}}{\log \deg \mathfrak{n}} \log |2\mathrm{Disc}(K/\mathbb{Q})|.$$

This conjecture is yet another instance of the fruitful interplay between function fields and number fields.

Acknowledgements

It is a pleasure to thank Guatam Chinta, Henri Darmon, Mihran Papikian, Joe Silverman, Dinesh Thakur, Adrian Vasiu and the referee for their comments, corrections, and references to the literature.

References

[ASD73] M. Artin and H. P. F. Swinnerton-Dyer, *The Shafarevich–Tate conjecture for pencils of elliptic curves on K3 surfaces*, Invent. Math. **20** (1973), 249–266. MR 54 #5240

[BD98] M. Bertolini and H. Darmon, *Heegner points, p-adic L-functions, and the Cerednik–Drinfeld uniformization*, Invent. Math. **131** (1998), 453–491. MR 99f:11080

[BD01] ———, *The p-adic L-functions of modular elliptic curves*, Mathematics unlimited—2001 and beyond, Springer, Berlin, 2001, pp. 109–170. MR 2002i:11061

[Bir04] B. Birch, *Heegner points: the beginnings*, Special values of Rankin L-series, *Math. Sci. Res. Inst. Publications* **49**, Cambridge U. Press, New York, 2004, pp. 1–10.

[BK94] A. Brumer and K. Kramer, *The conductor of an abelian variety*, Compositio Math. **92** (1994), 227–248. MR 95g:11055

[Bro94] M. L. Brown, *On a conjecture of Tate for elliptic surfaces over finite fields*, Proc. London Math. Soc. (3) **69** (1994), 489–514. MR 96b:11082

[Bru92] A. Brumer, *The average rank of elliptic curves. I*, Invent. Math. **109** (1992), 445–472. MR 93g:11057

[BS96] A. Brumer and J. H. Silverman, *The number of elliptic curves over* **Q** *with conductor N*, Manuscripta Math. **91** (1996), 95–102. MR 97e:11062

[Chi02] G. Chinta, *Analytic ranks of elliptic curves over cyclotomic fields*, J. Reine Angew. Math. **544** (2002), 13–24. MR 1 887 886

[Dar04] H. Darmon, *Heegner points and elliptic curves of large rank over function fields, Math. Sci. Res. Inst. Publications* **49**, Cambridge U. Press, New York, 2004, pp. 317–322.

[Del73] P. Deligne, *Les constantes des équations fonctionnelles des fonctions L*, Modular functions of one variable, II (Proc. Internat. Summer School, Univ. Antwerp, Antwerp, 1972), Springer, Berlin, 1973, pp. 501–597. Lecture Notes in Math., Vol. 349. MR 50 #2128

[DH87] P. Deligne and D. Husemoller, *Survey of Drinfel′d modules*, Current trends in arithmetical algebraic geometry (Arcata, Calif., 1985), Contemp. Math., vol. 67, Amer. Math. Soc., Providence, RI, 1987, pp. 25–91. MR 89f:11081

[Dri74] V. G. Drinfel′d, *Elliptic modules*, Mat. Sb. (N.S.) **94(136)** (1974), 594–627, 656. MR 52 #5580

[GR96] E.-U. Gekeler and M. Reversat, *Jacobians of Drinfeld modular curves*, J. Reine Angew. Math. **476** (1996), 27–93. MR 97f:11043

[Gro87] B. H. Gross, *Heights and the special values of L-series*, Number theory (Montreal, Que., 1985), CMS Conf. Proc., vol. 7, Amer. Math. Soc., Providence, RI, 1987, pp. 115–187. MR 89c:11082

[GvdPRG97] E.-U. Gekeler, M. van der Put, M. Reversat, and J. Van Geel (eds.), *Drinfeld modules, modular schemes and applications*, River Edge, NJ, World Scientific Publishing Co. Inc., 1997. MR 99b:11002

[GZ86] B. H. Gross and D. B. Zagier, *Heegner points and derivatives of L-series*, Invent. Math. **84** (1986), 225–320. MR 87j:11057

[HLW80] G. Harder, W. Li, and J. R. Weisinger, *Dimensions of spaces of cusp forms over function fields*, J. Reine Angew. Math. **319** (1980), 73–103. MR 81k:10042

[HW79] G. H. Hardy and E. M. Wright, *An introduction to the theory of numbers*, fifth ed., The Clarendon Press Oxford University Press, New York, 1979. MR 81i:10002

[JL70] H. Jacquet and R. P. Langlands, *Automorphic forms on* GL(2), Springer-Verlag, Berlin, 1970, Lecture Notes in Mathematics, Vol. 114. MR 53 #5481

[Kat00] K. Kato, *p-adic Hodge theory and values of zeta functions of modular forms*, preprint, 2000.

[KM85] Nicholas M. Katz and Barry Mazur, *Arithmetic moduli of elliptic curves*, Annals of Mathematics Studies, vol. 108, Princeton University Press, Princeton, NJ, 1985. MR 86i:11024

[Kol90] V. A. Kolyvagin, *Euler systems*, The Grothendieck Festschrift, Vol. II, Progr. Math., vol. 87, Birkhäuser Boston, Boston, MA, 1990, pp. 435–483. MR 92g:11109

[Lau96] G. Laumon, *Cohomology of Drinfeld modular varieties. Part I*, Cambridge Studies in Advanced Mathematics, vol. 41, Cambridge University Press, Cambridge, 1996, Geometry, counting of points and local harmonic analysis. MR 98c:11045a

[Lev68] M. Levin, *On the group of rational points on elliptic curves over function fields*, Amer. J. Math. **90** (1968), 456–462. MR 37 #6283

[Li84] W. Li, *A criterion on automorphic forms for* GL_1 *and* GL_2 *over global fields*, Seminar on number theory, Paris 1982–83 (Paris, 1982/1983), Progr. Math., vol. 51, Birkhäuser Boston, Boston, MA, 1984, pp. 161–172. MR 88a:11052

[Lon02] I. Longhi, *Non-Archimedean integration and elliptic curves over function fields*, J. Number Theory **94** (2002), 375–404. MR 1 916 280

[Maz72] B. Mazur, *Rational points of abelian varieties with values in towers of number fields*, Invent. Math. **18** (1972), 183–266. MR 56 #3020

[MB85] L. Moret-Bailly, *Pinceaux de variétés abéliennes*, Astérisque (1985), no. 129, 266. MR 87j:14069

[Mes86] J.-F. Mestre, *Formules explicites et minorations de conducteurs de variétés algébriques*, Compositio Math. **58** (1986), 209–232. MR 87j:11059

[Mil75] J. S. Milne, *On a conjecture of Artin and Tate*, Ann. of Math. (2) **102** (1975), 517–533. See also the addendum available at http://www.jmilne.org/math/. MR 54 #2659

[Mil80] ———, *Étale cohomology*, Princeton Mathematical Series, vol. 33, Princeton University Press, Princeton, N.J., 1980. MR 81j:14002

[MW83] B. Mazur and A. Wiles, *Analogies between function fields and number fields*, Amer. J. Math. **105** (1983), 507–521. MR 84g:12003

[Nér52] A. Néron, *Problèmes arithmétiques et géométriques rattachés à la notion de rang d'une courbe algébrique dans un corps*, Bull. Soc. Math. France **80** (1952), 101–166. MR 15,151a

[Pál00] A. Pál, *Drinfeld modular curves, Heegner points and interpolation of special values*, Ph.D. thesis, Columbia University, 2000.

[Pap02] Mihran Papikian, *On the degree of modular parametrizations over function fields*, J. Number Theory **97** (2002), no. 2, 317–349. MR 1 942 964

[Roh84] D. E. Rohrlich, *On L-functions of elliptic curves and cyclotomic towers*, Invent. Math. **75** (1984), 409–423. MR 86g:11038b

[RS01] K. Rubin and A. Silverberg, *Rank frequencies for quadratic twists of elliptic curves*, Experiment. Math. **10** (2001), 559–569. MR 2002k:11081

[RT00] H.-G. Rück and U. Tipp, *Heegner points and L-series of automorphic cusp forms of Drinfeld type*, Doc. Math. **5** (2000), 365–444 (electronic). MR 2001i:11057

[Rub98] K. Rubin, *Euler systems and modular elliptic curves*, Galois representations in arithmetic algebraic geometry (Durham, 1996), Cambridge Univ. Press, Cambridge, 1998, pp. 351–367. MR 2001a:11106

[Sch84] P. Schneider, *Rigid-analytic L-transforms*, Number theory, Noordwijkerhout 1983 (Noordwijkerhout, 1983), Lecture Notes in Math., vol. 1068, Springer, Berlin, 1984, pp. 216–230. MR 86b:11043

[Ser80] J.-P. Serre, *Trees*, Springer-Verlag, Berlin, 1980, Translated from the French by John Stillwell. MR 82c:20083

[Shi86] T. Shioda, *An explicit algorithm for computing the Picard number of certain algebraic surfaces*, Amer. J. Math. **108** (1986), 415–432. MR 87g:14033

[Sil00] J. H. Silverman, *A bound for the Mordell–Weil rank of an elliptic surface after a cyclic base extension*, J. Algebraic Geom. **9** (2000), 301–308. MR 2001a:11107

[Sil03] J. Silverman, *The rank of elliptic surfaces in unramified abelian towers*, preprint, 2003.

[Tat66a] J. T. Tate, *Endomorphisms of abelian varieties over finite fields*, Invent. Math. **2** (1966), 134–144. MR 34 #5829

[Tat66b] ———, *On the conjectures of Birch and Swinnerton-Dyer and a geometric analog*, Séminaire Bourbaki, Exp. No. 306, Vol. 9, Soc. Math. France, Paris, 1995, pp. 415–440. MR 1 610 977

[Tat72] ———, *Algorithm for determining the type of a singular fiber in an elliptic pencil*, Modular functions of one variable, IV (Proc. Internat. Summer School, Univ. Antwerp, Antwerp, 1972), Springer, Berlin, 1975, pp. 33–52. Lecture Notes in Math., Vol. 476. MR 52 #13850

[Tat94] ———, *Conjectures on algebraic cycles in l-adic cohomology*, Motives (Seattle, WA, 1991), Proc. Sympos. Pure Math., vol. 55, Amer. Math. Soc., Providence, RI, 1994, pp. 71–83. MR 95a:14010

[Tei91] J. T. Teitelbaum, *The Poisson kernel for Drinfel'd modular curves*, J. Amer. Math. Soc. **4** (1991), 491–511. MR 92c:11060

[Tha02] D. S. Thakur, *Elliptic curves in function field arithmetic*, Currents trends in number theory (Allahabad, 2000), Hindustan Book Agency, New Delhi, 2002, pp. 215–238. MR 1 925 657

[TR92] K.-S. Tan and D. Rockmore, *Computation of L-series for elliptic curves over function fields*, J. Reine Angew. Math. **424** (1992), 107–135. MR 93a:11057

[TS67] J. T. Tate and I. R. Shafarevic, *The rank of elliptic curves*, Dokl. Akad. Nauk SSSR **175** (1967), 770–773. MR 38 #5790

[Ulma] D. L. Ulmer, *Automorphic forms on* GL_2 *over function fields and Gross–Zagier theorems*, in preparation.

[Ulmb] ———, *Survey of elliptic curves over function fields*, in preparation.

[Ulm91] ———, *p-descent in characteristic p*, Duke Math. J. **62** (1991), 237–265. MR 92i:11068

[Ulm02] ———, *Elliptic curves with large rank over function fields*, Ann. of Math. (2) **155** (2002), 295–315. MR 1 888 802

[Ulm03] ———, *Geometric non-vanishing*, to appear in *Inventiones Math.*

[vdPR97] M. van der Put and M. Reversat, *Automorphic forms and Drinfeld's reciprocity law*, Drinfeld modules, modular schemes and applications (Alden-Biesen, 1996), World Sci. Publishing, River Edge, NJ, 1997, pp. 188–223. MR 99j:11059

[vdPT97] M. van der Put and J. Top, *Algebraic compactification and modular interpretation*, Drinfeld modules, modular schemes and applications (Alden-Biesen, 1996), World Sci. Publishing, River Edge, NJ, 1997, pp. 141–166. MR 99k:11084

[Vol90] J. F. Voloch, *Explicit p-descent for elliptic curves in characteristic p*, Compositio Math. **74** (1990), 247–258. MR 91f:11042

[Wei71] A. Weil, *Dirichlet series and automorphic forms*, Lecture Notes in Math., vol. 189, Springer-Verlag, New York, 1971.

[Wei79] ———, *Scientific works. Collected papers. Vol. I (1926–1951)*, Springer-Verlag, New York, 1979. MR 80k:01067a

[Zar74] Y. G. Zarhin, *A finiteness theorem for isogenies of abelian varieties over function fields of finite characteristic* (in Russian), Funkcional. Anal. i Priložen **8** (1974), 31–34. MR 50 #7162a

[Zha01] S.-W. Zhang, *Gross–Zagier formula for* GL_2, Asian J. Math. **5** (2001), 183–290. MR 1 868 935

Douglas Ulmer
Department of Mathematics
University of Arizona
Tucson, AZ 85721
United States
ulmer@math.arizona.edu

Heegner Points and Rankin L-Series
MSRI Publications
Volume **49**, 2004

Heegner Points and Elliptic Curves of Large Rank over Function Fields

HENRI DARMON

ABSTRACT. This note presents a connection between Ulmer's construction [Ulm02] of non-isotrivial elliptic curves over $\mathbb{F}_p(t)$ with arbitrarily large rank, and the theory of Heegner points (attached to parametrisations by Drinfeld modular curves, as sketched in Section 3 of Ulmer's article (see page ??). This ties in the topics in Section 4 of that article more closely to the main theme of this volume.

A review of the number field setting. Let K be a quadratic imaginary extension of $F = \mathbb{Q}$, and let $E_{/\mathbb{Q}}$ be an elliptic curve of conductor N. When all the prime divisors of N are split in K/F, the Heegner point construction (in the most classical form that is considered in [GZ], relying on the modular parametrisation $X_0(N) \longrightarrow E$) produces not only a canonical point on $E(K)$, but also a norm-coherent system of such points over all abelian extensions of K which are of "dihedral type". (An abelian extension H of K is said to be of *dihedral type* if it is Galois over $\mathbb{Q}$ and the generator of $\mathrm{Gal}(K/\mathbb{Q})$ acts by -1 on the abelian normal subgroup $\mathrm{Gal}(H/K)$.) The existence of this construction is consistent with the Birch and Swinnerton-Dyer conjecture, in the following sense: an analysis of the sign in the functional equation for $L(E/K,\chi,s) = L(E/K,\bar{\chi},s)$ shows that this sign is always equal to -1, for all complex characters χ of $G := \mathrm{Gal}(H/K)$. Hence

$$L(E/K,\chi,1) = 0 \quad \text{for all } \chi : G \longrightarrow \mathbb{C}^\times.$$

The product factorisation

$$L(E/H,s) = \prod_{\chi} L(E/K,\chi,s)$$

implies that

$$\mathrm{ord}_{s=1} L(E/H,s) \geq [H:K], \tag{1}$$

so that the Birch and Swinnerton-Dyer conjecture predicts that

$$\operatorname{rank}(E(H)) \stackrel{?}{\geq} [H:K]. \tag{2}$$

In fact, the G-equivariant refinement of the Birch and Swinnerton-Dyer conjecture leads one to expect that the rational vector space $E(H) \otimes \mathbb{Q}$ contains a copy of the regular representation of G.

It is expected in this situation that Heegner points account for the bulk of the growth of $E(H)$, as H varies over the abelian extensions of K of dihedral type. For example we have:

LEMMA 1. *If* $\operatorname{ord}_{s=1} L(E/H, s) \leq [H:K]$, *then the vector space* $E(H) \otimes \mathbb{Q}$ *has dimension* $[H:K]$ *and is generated by Heegner points.*

PROOF. For V any complex representation of G, let

$$V^{\chi} := \{v \in V \quad \text{such that } \sigma v = \chi(\sigma)v, \text{ for all } \sigma \in G\}.$$

Since equality is attained in (1), it follows that each $L(E/K, \chi, s)$ vanishes to order exactly one at $s = 1$. Zhang's extension of the Gross–Zagier formula to L-functions $L(E/K, s)$ twisted by (possibly ramified) characters of G [Zh01] shows that

$$\dim_{\mathbb{C}}(HP^{\chi}) = 1, \tag{3}$$

where HP denotes the subspace of $E(H) \otimes \mathbb{C}$ generated by Heegner points. Theorem 2.2 of [BD90], whose proof is based on Kolyvagin's method, then shows that

$$\dim_{\mathbb{C}}((E(H) \otimes \mathbb{C})^{\chi}) \leq 1. \tag{4}$$

The result follows directly from (3) and (4). □

0.1. The case $F = \mathbb{F}_q(u)$. As explained in Section 3 of [Ulm03], the Heegner point construction can be adapted to the case where $\mathbb{Q}$ is replaced by the rational function field $\mathbb{F}_q(u)$.

The basic idea of our construction is to start with an elliptic curve E_0 defined over $\mathbb{F}_p(u)$, and produce a Galois extension H of $\mathbb{F}_q(u)$ (for some power q of p) such that

(i) the Galois group of H over $\mathbb{F}_q(u)$ is isomorphic to a dihedral group of order $2d$;
(ii) H satisfies a suitable Heegner hypothesis relative to E_0 over $\mathbb{F}_q(u)$ so that the Birch and Swinnerton-Dyer conjecture implies an inequality like (2);
(iii) H is the function field of a curve of genus 0, say $H = F_q(t)$, so that E_0 yields a curve E over $\mathbb{F}_p(t)$ which acquires rank at least d over $\mathbb{F}_q(t)$.

A further argument is then made to show that the rank of E remains large over $\mathbb{F}_p(t)$, provided suitable choices of d and q have been made.

To illustrate the method, let p be an odd prime and let F_0 be the field $\mathbb{F}_p(u)$, with u an indeterminate. Let $K_0 = \mathbb{F}_p(v)$ be the quadratic extension of F_0

defined by $v + v^{-1} = u$. Choose an element $u_\infty \in \mathbb{P}_1(\mathbb{F}_p)$ such that the place $(u - u_\infty)$ is inert in K_0. (Such a u_∞ always exists when $p > 2$.) The chosen place u_∞ will play the role in our setting of the archimedean place of $\mathbb{Q}$ in the previous discussion. Note that K_0/F_0 becomes a quadratic "imaginary" extension with this choice of place at infinity, and that this continues to hold when $\mathbb{F}_p$ is replaced by $\mathbb{F}_q$ with $q = p^m$, provided that m is *odd*.

Let $E = E_u$ be an elliptic curve over F_0 having split multiplicative reduction at u_∞. Let $\mathcal{E}$ denote the Néron model of E over the subring $\mathcal{O} = \mathbb{F}_p[\frac{1}{u-u_\infty}]$ and let N denote its arithmetic conductor, viewed as a divisor of $\mathbb{P}_1 - \{u_\infty\}$. Suppose that

$$\text{all prime divisors of } N \text{ are split in } K_0/F_0, \tag{5}$$

which is the analogue of the classical Heegner hypothesis in our function field setting.

Finally, given any integer d, let o_d be the order of p in $(\mathbb{Z}/d\mathbb{Z})^\times$. Assume that

$$\text{the integer } o_d \text{ is odd.} \tag{6}$$

We then set $q = p^{o_d}$ and consider the extensions

$$F = \mathbb{F}_q(u); \quad K = \mathbb{F}_q(v); \quad H = \mathbb{F}_q(t), \text{ with } v = t^d.$$

Note that H/K is an abelian extension with Galois group $G = \mathrm{Gal}(H/K)$ isomorphic to $\mu_d(\mathbb{F}_q) \simeq \mathbb{Z}/d\mathbb{Z}$, and that this extension is of dihedral type, relative to the ground field F. Therefore the analysis of signs in functional equations that was carried out to conclude (1) carries over, mutatis mutandis, to prove the following.

PROPOSITION 2. *Assume the Birch and Swinnerton-Dyer conjecture over function fields. Then the rank of $E(H)$ is at least d. More precisely,*

$$\dim_{\mathbb{C}}\left((E(H) \otimes \mathbb{C})^\chi\right) \geq 1, \qquad \textit{for all } \chi : G \longrightarrow \mathbb{C}^\times.$$

One also wants to estimate the rank of E over the field $H_0 := \mathbb{F}_p(t)$. Let $\tilde{G} = \mathrm{Gal}(H/K_0)$; then $\tilde{G}$ is the semi-direct product $G \times \langle f \rangle$, where $\langle f \rangle \subset (\mathbb{Z}/d\mathbb{Z})^\times$ is the cyclic group of order o_d generated by the Frobenius element $f \in \mathrm{Gal}(H/H_0) = \mathrm{Gal}(\mathbb{F}_q/\mathbb{F}_p)$, which acts by conjugation on the abelian normal subgroup $G = \mu_d(\mathbb{F}_q)$ in the natural way. Since E is defined over K_0 (and even over F_0), the space $V := E(H) \otimes \mathbb{C}$ is a complex representation of $\tilde{G}$, and one may exploit basic facts about the irreducible representations of such a semi-direct product to obtain lower bounds for $E(H)^{f=1} = E(\mathbb{F}_p(t))$. More precisely, suppose that the character χ of G is one of the $\phi(d)$ *faithful* characters of G. Proposition 2 asserts that the space V^χ contains a non-zero vector v_χ. Note that V^χ is not preserved by the action of f, which sends V^χ to V^{χ^p}. Because of this, the vectors $v_\chi, fv_\chi, \ldots, f^{o_d-1}v_\chi$ are linearly independent since they belong to different eigenspaces for the action of G. Hence the vector

$$v_{[\chi]} = v_\chi + fv_\chi + \cdots + f^{o_d-1}v_\chi$$

is non-zero and belongs to $V^{f=1} = E(H_0) \otimes \mathbb{C}$. Futhermore the $v_{[\chi]}$ are linearly independent, as χ ranges over the f-orbits of faithful characters of G. Hence

$$\mathrm{rank}(E(\mathbb{F}_p(t)) \geq \phi(d)/o_d.$$

By taking into account the contributions coming from all the characters (and not just the faithful ones) one can obtain the following stronger estimate.

PROPOSITION 3. *Assume the Birch and Swinnerton-Dyer conjecture over function fields. Then*

$$\mathrm{rank}(E(\mathbb{F}_p(t)) \geq \sum_{e|d} \frac{\phi(e)}{o_e} \geq \frac{d}{o_d}. \tag{7}$$

PROOF. A complex character χ of G is said to be of level e if its image is contained in the group μ_e of eth roots of unity in $\mathbb{C}$ and in no smaller subgroup. Clearly the level e of χ is a divisor of d, the order o_e of p in $(\mathbb{Z}/e\mathbb{Z})^\times$ divides o_d, and there are exactly $\phi(e)$ distinct characters of G of level e. Note also that if χ is of level e, then f^{o_e} maps V^χ to itself. The same reasoning used to prove Proposition 2, but with d replaced by e, and q by p^{o_e}, shows that (under the Birch and Swinnerton-Dyer assumption)

$$V^\chi \quad \text{contains a non-zero vector fixed by } f^{o_e}.$$

If v_χ is such a vector, then just as before the vectors

$$v_{[\chi]} = v_\chi + f v_\chi + \cdots + f^{o_e - 1} v_\chi$$

form a linearly independent collection of $\phi(e)/o_e$ vectors in $E(\mathbb{F}_p(t)) \otimes \mathbb{C}$, as χ ranges over the f-orbits of characters of G of level e. Summing over all e dividing d proves the first inequality in (7). The second is obtained by noting that

$$\sum_{e|d} \frac{\phi(e)}{o_e} \geq \frac{1}{o_d} \sum_{e|d} \phi(e) = \frac{d}{o_d}. \qquad \square$$

REMARKS. 1. It is instructive to compare the bound (7) with the formula for the rank of Ulmer's elliptic curves which is given in Theorem 4.2.1 of [Ulm03].
2. Note that the expression which appears on the right of (7) can be made arbitrarily large by setting $d = p^n - 1$ with n odd, so that $o_d = n$.

Some examples. Elliptic curves satisfying the Heegner assumptions of the previous section are not hard to exhibit explicitly. For example, suppose for notational convenience that p is congruent to 3 modulo 4, and let $E[u]$ be a non-isotrivial elliptic curve over $\mathbb{F}_p(u)$ having good reduction everywhere except at $u = 0, 1$ and ∞, and having split multiplicative reduction at $u_\infty = 0$. There are a number of such curves, for example:

(i) An (appropriate twist of a) "universal" elliptic curve over the j-line in characteristic $p \neq 2, 3$, with $u = 1728/j$;
(ii) A "universal" curve over $X_0(2)$, or over $X_0(3)$;

(iii) The Legendre family $y^2 = x(x-1)(x-u)$ (corresponding to a universal family over the modular curve $X(2)$).
(iv) The curve $y^2 + xy = x^3 - u$ that is used in [Ulm03], in which the parameter space has no interpretation as a modular curve.

Choosing any parameter λ in $\mathbb{F}_p - \{0, \pm 1\}$, we see that the curve $E\left[\frac{u}{\lambda+\lambda^{-1}}\right]$ over $\mathbb{F}_p(u)$ satisfies all the desired properties, since two of the places $u = \infty$ and $\lambda + \lambda^{-1}$ dividing the conductor of E are split in K/F, while the third place $u = 0$, which lies below $v = \pm i$, is inert in K/F. (This is where the assumption $p \equiv 3 \pmod 4$ is used.) Hence Proposition 3 implies

COROLLARY 4. *Assume the Birch and Swinnerton-Dyer conjecture for function fields. Let $E[u]$ be any of the curves over $\mathbb{F}_p(u)$ listed above, and let λ be any element in $\mathbb{F}_p - \{0, \pm 1\}$. Then the curve*

$$E\left[\frac{t^d + t^{-d}}{\lambda + \lambda^{-1}}\right]$$

has rank at least d/o_d over $\mathbb{F}_p(t)$.

Dispensing with the Birch and Swinnerton-Dyer hypothesis. It may be possible, at least for some specific choices of $E[u]$ and of d, to remove the Birch and Swinnerton-Dyer assumption that appears in corollary 4, since the notion of Heegner points which motivated Proposition 2 also suggests a possible construction of a (hopefully, sufficiently large) collection of global points in $E(H)$. To produce explicit examples where the module *HP* generated by Heegner points in $E(H)$ has large rank, it may not be necessary to invoke the full strength of the theory described in Section 3 of [Ulm03] since quite often the mere knowledge that the Heegner point on $E(K)$ is of infinite order is sufficient to gain strong control over the Heegner points that appear in related towers. It appears worthwhile to produce explicit examples where Propositions 2 and 3 can be made unconditional thanks to the Heegner point construction.

REMARK. Crucial to the construction in this note is the fact that $\mathbb{P}_1$ has a large automorphism group, containing dihedral groups of arbitrarily large order. Needless to say, this fact breaks down when $\mathbb{F}_p(u)$ is replaced by $\mathbb{Q}$, which has no automorphisms. In this setting Heegner points are known to be a purely "rank one phenomenon", and are unlikely to yield any insight into the question of whether the rank of elliptic curves over $\mathbb{Q}$ is unbounded or not.

Remarks on Ulmer's construction. Let d be a divisor of $q+1$, where $q = p^n$. The curve

$$E_d : y^2 + xy = x^3 - t^d$$

studied in Theorem 4.2.1 of [Ulm03] is a pullback of the curve

$$E_0 : y^2 + xy = x^3 - u$$

by the covering $\mathbb{P}_1 \to \mathbb{P}_1$ given by $t \mapsto u := t^d$, a covering that becomes Galois (abelian) over $\mathbb{F}_{q^2}$. It is not hard on the other hand to see that the curve E_d does not arise as a pullback via any geometrically connected dihedral covering $\mathbb{P}_1 \to \mathbb{P}_1$. However, one may set

$$F = \mathbb{F}_q(u), \quad K = \mathbb{F}_{q^2}(u), \quad H = \mathbb{F}_{q^2}(t), \text{ with } u = t^d.$$

The congruence $q \equiv -1 \pmod{d}$ implies that $\mathrm{Gal}(H/F)$ is a dihedral group of order $2d$. Hence is becomes apparent a posteriori that the curves of [Ulm02] can be approached by a calculation of the signs in functional equations for the L-series of E_0 over K twisted by characters of $\mathrm{Gal}(H/K)$. (See the remarks in sec. 4.3 of [Ulm03] for further details on this calculation and its close connection with the original strategy followed in [Ulm02].)

It should be noted that the elliptic curves produced in our Corollary 4 have smaller rank-to-conductor ratios than the curves E_d in Theorem 4.2.1 of [Ulm03], which are essentially optimal in this respect.

References

[BD90] Massimo Bertolini and Henri Darmon, "Kolyvagin's descent and Mordell–Weil groups over ring class fields", *J. Reine Angew. Math.* **412** (1990), 63–74.

[GZ] Benedict Gross and Don Zagier, "Heegner points and derivatives of L-series", *Invent. Math.* **84**:2 (1986), 225–320.

[Ulm02] Douglas Ulmer, "Elliptic curves with large rank over function fields", *Ann. of Math.* (2) **155**:1 (2002), 295–315.

[Ulm03] Douglas Ulmer, "Elliptic curves and analogies between number fields and function fields", pp. 285–315 in *Heegner Points and Rankin L-Series*, edited by Henri Darmon and Shou-Wu Zhang, *Math. Sci. Res. Inst. Publications* **49**, Cambridge U. Press, New York, 2004.

[Zh01] Shou-Wu Zhang, "Gross–Zagier formula for GL_2", *Asian J. Math.* **5**:2 (2001), 183–290.

Henri Darmon
Mathematics and Statistics Department
McGill University
805 Sherbrooke W.
Montreal, Quebec
Canada H3A 2K6
darmon@math.mcgill.ca

Heegner Points and Rankin L-Series
MSRI Publications
Volume **49**, 2004

Periods and Points Attached to Quadratic Algebras

MASSIMO BERTOLINI, HENRI DARMON, AND PETER GREEN

Contents

Introduction 323
1. Classical Heegner Points 325
2. Heegner Points and p-adic Integration 330
3. Forms on $\mathcal{T}_p \times \mathcal{H}$ 341
4. Complex Periods and Heegner Points 344
5. p-adic Periods and Stark–Heegner Points 349
6. Heegner Points and Integration on $\mathcal{H}_p \times \mathcal{H}_q$ 354
7. Periods of Hilbert Modular Forms 361
References 365

Introduction

Let E be an elliptic curve over $\mathbb{Q}$ of conductor N. By the fundamental work initiated in [Wi] and brought to a successful conclusion in [BCDT], the curve E is equipped with a nonconstant complex uniformisation

$$\Phi : \mathcal{H}/\Gamma_0(N) \longrightarrow E(\mathbb{C}), \tag{0–1}$$

where $\mathcal{H}$ is the Poincaré upper half-plane and $\Gamma_0(N)$ is Hecke's congruence group of level N.

Fix a quadratic field K; when it is imaginary, the theory of complex multiplication combined with (0–1) yields the construction of a remarkable collection of points on E defined over certain ring class fields of K. These are the *Heegner points* recalled in Section 1 whose study is one of the themes of this proceedings volume.

When p is a prime that divides N exactly, and E a factor of the Jacobian of a Shimura curve attached to a quaternion algebra ramified at p, the uniformisation Φ of (0–1) admits a p-adic analogue based on theorems of Jacquet–Langlands and Cerednik–Drinfeld, in which $\mathcal{H}$ is replaced by Drinfeld's p-adic upper half plane $\mathcal{H}_p := \mathbb{C}_p - \mathbb{Q}_p$, and $\Gamma_0(N)$ by an appropriate $\{p\}$-arithmetic subgroup of a definite quaternion algebra over $\mathbb{Q}$. This somewhat less well-known theory is recalled in Section 2, where a p-adic variant of the complex Heegner point construction of Section 1 is described.

Retaining the assumption that $N = pM$ with $p \nmid M$, Section 3 introduces the $\{p\}$-arithmetic group $\Gamma \subset \mathrm{SL}_2(\mathbb{Q})$ defined by

$$\Gamma = \left\{ \begin{pmatrix} a & b \\ c & d \end{pmatrix} \in \mathrm{SL}_2(\mathbb{Z}[1/p]) \quad \text{such that } M \text{ divides } c \right\}.$$

This group acts both on $\mathcal{H}$ and on $\mathcal{H}_p$ by Möbius transformations, and the diagonal action of Γ on $\mathcal{H}_p \times \mathcal{H}$ is discrete. A *double integral* on $(\mathcal{H}_p \times \mathcal{H})/\Gamma$ is defined to be a $\mathbb{C}$ or $\mathbb{C}_p$-valued function of $(\tau_1, \tau_2, \tau_3, \tau_4) \in \mathcal{H}_p^2 \times \mathcal{H}^2$, denoted

$$\int_{\tau_1}^{\tau_2} \int_{\tau_3}^{\tau_4} \omega_f \in \mathbb{C} \text{ or } \mathbb{C}_p$$

satisfying the obvious additivity properties suggested by the notation, as well as being invariant under the diagonal action of Γ. The normalised eigenform f of weight 2 attached to E is shown in Section 3 to give rise in a natural way to a $\mathbb{C}$-valued double integral on $(\mathcal{H}_p \times \mathcal{H})/\Gamma$. This function can be viewed as encoding the complex periods of a "form ω of weight $(2,2)$" on $(\mathcal{H}_p \times \mathcal{H})/\Gamma$ attached to E.

Section 4 reformulates the results of Section 1 in terms of the $\mathbb{C}$-valued double integral attached to f. This amounts to revisiting the classical theory from a slightly different perspective, in which the first cohomology of $\Gamma_0(N)$ is replaced by the second cohomology of Γ. The Heegner points obtained in this construction are the same as those of Section 1. In particular, they are defined over the ring class fields of a quadratic field K in which the place ∞ is *inert* (i.e., K is *imaginary* quadratic) and the place p is split.

The main advantage of the (at first sight overly recundite) point of view developed in Sections 3 and 4 is that the places ∞ and p play more symmetrical roles in its formulation. Building on the techniques of Section 2 and using a theorem of Manin–Drinfeld already discussed for a different purpose in Section 4, a $\mathbb{C}_p$-valued double integral is attached to f in Section 5, leading to a theory in which the roles of p and ∞ are interchanged. The manipulations carried out in Section 4 make sense in this new context, at least formally, suggesting the possibility of using the $\mathbb{C}_p$-valued double integral to construct algebraic points on E defined over ring class fields of a quadratic field K in which p is inert and the prime ∞ is *split*. Because such a K is real quadratic, the resulting construction falls outside the scope of the theory of complex multiplication. For this reason the conjectures of Section 5 have so far eluded proof. But they are well-suited to numerical verifications, and can be used to calculate "Heegner points attached to real quadratic fields" [DG] in much the same way that Stark's conjecture can be used to approximate units in certain class fields by analytic means.

Section 6 transposes the theory of Section 5 to the product $\mathcal{H}_p \times \mathcal{H}_q$ of two nonarchimedean upper half planes. Here the situation is better understood, and the counterpart of the main conjectures of Section 5 can be proved entirely as a consequence the Cerednik–Drinfeld theory presented in Section 2. The relative ease in obtaining the results of Section 6 indicates that the mystery in Section 5 is intimately tied to the presence of an archimedean place (and only accessorily to that of a p-adic place) in the product of upper half planes on which the relevant integration theory is defined.

Section 7 reinforces this impression by modifying the setting of Section 5 in a different direction, replacing $\mathcal{H}_p \times \mathcal{H}$ by the product $\mathcal{H} \times \mathcal{H}$ of two archimedean upper half-planes, and the group Γ by a Hecke-type congruence subgroup of the Hilbert modular group $\mathrm{SL}_2(\mathcal{O}_F)$. Here F is a real quadratic field and $\mathcal{O}_F$ its ring of integers. The role of the double integrals on $(\mathcal{H}_p \times \mathcal{H})/\Gamma$ is played by the periods of a Hilbert modular form of weight $(2, 2)$ which is conjecturally attached to an elliptic curve over F. Such period integrals have already been defined and studied extensively in the classical literature. Yet the natural analogues of the conjectures of Section 5 in the Hilbert modular form setting have not been studied before and appear to transcend the classical theory of complex multiplication (as well as its generalisation to totally real fields proposed by Hilbert and his school, and rigorously developed by Shimura and Taniyama in the 50s and 60s). The article closes by proposing a conjecture which would allow the analytic construction of algebraic points on an elliptic curve E/F, points whose coordinates are defined over ring class fields of a quadratic extension K of F with two real and one complex place. A notable feature of this conjecture is that it provides a "modular" construction of algebraic points on an elliptic curve having everywhere good reduction over F, although such a curve is not expected in general to appear in the Jacobian of any modular or Shimura curve.

1. Classical Heegner Points

Let N be a positive integer as in the introduction, and let $X_0(N)$ be the modular curve classifying pairs $(\mathcal{E}, \mathcal{E}')$ of generalized elliptic curves equipped with a cyclic N-isogeny $\mathcal{E} \longrightarrow \mathcal{E}'$. Let $\mathcal{H}^* := \mathcal{H} \cup \mathbb{P}_1(\mathbb{Q})$ denote the extended upper half plane and let $\Gamma_0(N)$ be the Hecke congruence group of level N, acting on $\mathcal{H}^*$ by Möbius transformations. There is a well-known identification of Riemann surfaces

$$j : \mathcal{H}^*/\Gamma_0(N) \xrightarrow{\sim} X_0(N)(\mathbb{C}),$$

in which $\tau \in \mathcal{H}$ is mapped to the pair of elliptic curves $(\mathbb{C}/\langle\tau, 1\rangle, \mathbb{C}/\langle\tau, 1/N\rangle)$ related by the obvious cyclic N-isogeny.

Let $\mathcal{O}$ be an order in an imaginary quadratic subfield K of $\mathbb{C}$. Suppose N satisfies the *Heegner hypothesis* relative to $\mathcal{O}$, so that the ideal $N\mathcal{O}$ factors as a product $N\mathcal{O} = \mathcal{N}\bar{\mathcal{N}}$ of cyclic ideals of norm N. Let

$$\delta : \mathrm{Pic}(\mathcal{O}) \to X_0(N)(\mathbb{C})$$

be the natural map which to the class of a fractional ideal $\mathfrak{a}$ associates the pair $(\mathbb{C}/\mathfrak{a}, \mathbb{C}/\mathcal{N}^{-1}\mathfrak{a})$ of N-isogenous elliptic curves.

Write H for the ring class field attached to $\mathcal{O}$, so that the Artin reciprocity map of class field theory induces an identification

$$\mathrm{rec} : \mathrm{Pic}(\mathcal{O}) \xrightarrow{\sim} \mathrm{Gal}(H/K).$$

The maps δ and rec are related by the fundamental *Shimura reciprocity law:*

PROPOSITION 1.1. *For any* $\mathfrak{a} \in \mathrm{Pic}(\mathcal{O})$, *the point* $\delta(\mathfrak{a})$ *belongs to* $X_0(N)(H)$. *More precisely, for all* $\mathfrak{b} \in \mathrm{Pic}(\mathcal{O})$,

$$\delta(\mathfrak{b}^{-1}\mathfrak{a}) = \mathrm{rec}(\mathfrak{b}) \cdot \delta(\mathfrak{a}).$$

SKETCH OF PROOF. The set of Heegner points with endomorphism ring equal to $\mathcal{O}$ forms a $\mathrm{Pic}(\mathcal{O})$-affine space via the rule

$$\mathfrak{a} * (\mathcal{E}, \mathcal{E}') = (\mathrm{Hom}(\mathfrak{a}, \mathcal{E}), \mathrm{Hom}(\mathfrak{a}, \mathcal{E}')), \quad \mathfrak{a} \in \mathrm{Pic}(\mathcal{O}), \quad (\mathcal{E}, \mathcal{E}') \in X_0(N), \quad (1\text{–}1)$$

which on the level of complex tori is given by

$$\mathfrak{a} * (\mathbb{C}/\Lambda, \mathbb{C}/\mathcal{N}^{-1}\Lambda) = (\mathbb{C}/\mathfrak{a}^{-1}\Lambda, \mathbb{C}/\mathfrak{a}^{-1}\mathcal{N}^{-1}\Lambda).$$

Since there are finitely many Heegner points on $X_0(N)$ attached to $\mathcal{O}$ and since they are preserved under the action of $\mathrm{Aut}(\mathbb{C})$, they are algebraic. Furthermore, the action of $\mathrm{Gal}(\bar{K}/K)$ on these points commutes with the action of $\mathrm{Pic}(\mathcal{O})$ defined in (1–1), and hence is described by a homomorphism

$$\tilde{\delta} : \mathrm{Gal}(\bar{K}/K) \to \mathrm{Pic}(\mathcal{O}) \text{ satisfying } \tilde{\delta}(\sigma) * (\mathcal{E}, \mathcal{E}') = (\mathcal{E}^{\sigma}, \mathcal{E}'^{\sigma}),$$

for all Heegner points $(\mathcal{E}, \mathcal{E}')$ with $\mathrm{End}(\mathcal{E}) = \mathcal{O}$. This already shows that the Heegner points are defined over an abelian extension, namely the extension $\tilde{H} = \bar{K}^{\ker(\tilde{\delta})}$ cut out by $\tilde{\delta}$. Now let $\mathfrak{p}$ be a prime of $\mathcal{O}$ unramified in $\tilde{H}$, let $\mathfrak{P}$ be a prime of $\tilde{H}$ above $\mathfrak{p}$ and assume $\mathcal{E}$ has good reduction at $\mathfrak{P}$. A case by case verification (see [Se1]) shows that the elliptic curve obtained by reducing $\mathcal{E}$ (mod $\mathfrak{P}$) and raising its coefficients to the $\#(\mathcal{O}_K/\mathfrak{p})$-power is isomorphic to $\mathrm{Hom}(\mathfrak{p}, \mathcal{E})$ reduced (mod $\mathfrak{P}$). Hence,

$$\tilde{\delta}(\mathrm{Frob}_{\mathfrak{p}}) = [\mathfrak{p}] \in \mathrm{Pic}(\mathcal{O}),$$

so that, by the Artin reciprocity law, $\tilde{\delta}^{-1} = \mathrm{rec}$. Thus $\tilde{H} = H$, and the result follows. For more details see [Se1]. □

While CM points on $X_0(N)$ are in natural correspondence with elements of Picard groups of orders in imaginary quadratic fields, for more general types of modular curves such as the Shimura curves introduced in Section 2, it is only the $\mathrm{Pic}(\mathcal{O})$-affine action which generalizes. We therefore make a shift in notation, replacing $\mathrm{Pic}(\mathcal{O})$ with the $\mathrm{Pic}(\mathcal{O})$-set $\mathrm{Emb}(\mathcal{O}, M_0(N))$, defined below.

Here $M_0(N)$ denotes the order of matrices in $M_2(\mathbb{Z})$ which are upper triangular modulo N. At first this seems to only add opacity to the discourse, but in fact the notation allows greater flexibility which is useful in the sequel.

There is a convenient dictionary between classes of projective modules of rank 1 over $\mathcal{O}$ and *embeddings* Ψ of the quadratic imaginary field $K \subset \mathbb{C}$ into the central simple algebra $M_2(\mathbb{Q})$ satisfying some additional technical hypotheses:

1. The embedding Ψ is said to be *optimal* (relative to N and $\mathcal{O}$) if $\Psi(K) \cap \mathrm{M}_0(N) = \Psi(\mathcal{O})$.
2. An orientation of the Eichler order $\mathrm{M}_0(N)$ is a surjective ring homomorphism $\mathrm{M}_0(N) \to \mathbb{Z}/N\mathbb{Z}$. One could for instance choose the map
$$\begin{pmatrix} a & b \\ c & d \end{pmatrix} \mapsto a \pmod{N}.$$
Then Ψ is said to be *oriented* (relative to $\mathcal{N}$ and the chosen orientation on $M_0(N)$) if the following diagram commutes:
$$\begin{array}{ccc} \mathcal{O} & \xrightarrow{\Psi} & M_0(N) \\ \downarrow & & \downarrow \\ \mathcal{O}/\mathcal{N} & \xrightarrow{\sim} & \mathbb{Z}/N\mathbb{Z}. \end{array}$$
3. There is a unique fixed point $\tau_\Psi \in K$ for the action of the group $\Psi(K^\times)$ by Möbius transformations with the property that
$$\Psi(\lambda)\begin{pmatrix} \tau_\Psi \\ 1 \end{pmatrix} = \lambda \begin{pmatrix} \tau_\Psi \\ 1 \end{pmatrix}. \tag{1–2}$$
The embedding Ψ is said to be *oriented at* ∞ if τ_Ψ belongs to $\mathcal{H}$, under the fixed embedding of K into $\mathbb{C}$.

An embedding satisfying the three conditions above is called an *oriented optimal embedding* of $\mathcal{O}$ into $M_0(N)$. Note that this notion depends on the fixed choice of the ideal $\mathcal{N}$ and of the orientation on $M_0(N)$, as well as on the fact that K is given as a subfield of $\mathbb{C}$.

If R is any subring of a matrix or quaternion algebra, denote by $R_1^\times$ the group of elements in R of (reduced) norm 1. The reader may check that the group $\Gamma_0(N) = M_0(N)_1^\times$ acts naturally on the set of oriented optimal embeddings by conjugation. Let $\mathrm{Emb}(\mathcal{O}, M_0(N))$ denote the set of $\Gamma_0(N)$-conjugacy classes of oriented optimal embeddings of $\mathcal{O}$ into $M_0(N)$.

If Ψ is an oriented optimal embedding of $\mathcal{O}$ into $M_0(N)$, equation (1–2) shows that the lattice $\mathbb{Z}\tau_\Psi \oplus \mathbb{Z} \subset \mathbb{C}$ is preserved under multiplication by $\mathcal{O}$; the optimality of Ψ implies that it is in fact a projective $\mathcal{O}$-module. Let $\mathfrak{a}_\Psi$ denote its class in $\mathrm{Pic}(\mathcal{O})$. Replacing Ψ by an embedding that is conjugate to it under $\Gamma_0(N)$ does not change the class $\mathfrak{a}_\Psi$, and hence the assignment $\Psi \mapsto \mathfrak{a}_\Psi$ is a well defined function from $\mathrm{Emb}(\mathcal{O}, M_0(N))$ to $\mathrm{Pic}(\mathcal{O})$.

LEMMA 1.2. *The assignment* $\Psi \mapsto \mathfrak{a}_\Psi$ *is a bijection between* $\mathrm{Emb}(\mathcal{O}, M_0(N))$ *and* $\mathrm{Pic}(\mathcal{O})$.

PROOF. Suppose that $\mathfrak{a}$ belongs to $\mathrm{Pic}(\mathcal{O})$. The class $\mathfrak{a}$ has a representative of the form $\langle \tau_\mathfrak{a}, 1\rangle$, with $\tau_\mathfrak{a} \in \mathcal{H}$, such that

$$\mathcal{N}^{-1}\langle \tau_\mathfrak{a}, 1\rangle = \langle \tau_\mathfrak{a}, 1/N\rangle. \tag{1–3}$$

Furthermore, this $\tau_\mathfrak{a}$ is determined by $\mathfrak{a}$ up to the action of $\Gamma_0(N)$. We may define an optimal embedding $\Psi_\mathfrak{a} : K \hookrightarrow M_2(\mathbb{Q})$ by setting

$$\Psi_\mathfrak{a}(\tau_\mathfrak{a}) = \begin{pmatrix} \mathrm{Tr}(\tau_\mathfrak{a}) & -\mathrm{Nm}(\tau_\mathfrak{a}) \\ 1 & 0 \end{pmatrix},$$

whose class in $\mathrm{Emb}(\mathcal{O}, M_0(N))$ depends only on $\mathfrak{a}$ and not on the choice of ideal representative. The condition (1–3) forces the embedding $\Psi_\mathfrak{a}$ to be oriented. The reader will check that assignment $\mathfrak{a} \mapsto \Psi_\mathfrak{a}$ defines a map $\mathrm{Pic}(\mathcal{O}) \to \mathrm{Emb}(\mathcal{O}, M_0(N))$, which is inverse to the map $\Psi \mapsto \mathfrak{a}_\Psi$. This proves the lemma. □

Thanks to Lemma 1.2, the regular action of $\mathrm{Pic}(\mathcal{O})$ on itself by left multiplication translates into an action of this group on $\mathrm{Emb}(\mathcal{O}, M_0(N))$, which we denote

$$(\mathfrak{a}, \Psi) \mapsto \Psi^\mathfrak{a}.$$

This action can be described adelically as follows. Given $\mathfrak{a} \in \mathrm{Pic}(\mathcal{O})$, let a be the finite idele of K which corresponds to it by the rule $a(\mathcal{O} \otimes \hat{\mathbb{Z}}) = \mathfrak{a} \otimes \hat{\mathbb{Z}}$. One can assume that $\mathrm{Norm}_\mathbb{Q}^K(a) \in \mathbb{Q}^\times \subset \mathbb{A}_\mathbb{Q}^{f\,\times}$. By strong approximation, there is an element $\gamma \in \mathrm{GL}_2(\mathbb{Q})$ such that

$$\gamma = \Psi(a)u \quad \text{with } u \in (M_0(N) \otimes \hat{\mathbb{Z}})_1^\times. \tag{1–4}$$

(Here Ψ has been extended in the natural way to a map $\mathbb{A}_K^f \to M_2(\mathbb{A}_\mathbb{Q}^f)$, still denoted by Ψ.)

LEMMA 1.3. *For any* γ *satisfying* (1–4),

$$\Psi^a = \gamma^{-1}\Psi\gamma.$$

As in the introduction, let $E/\mathbb{Q}$ be an elliptic curve of conductor N, equipped with a modular parametrisation of minimal degree

$$\Phi : X_0(N) \to E$$

mapping the cusp ∞ to the identity of E. Let Λ_E be the lattice generated by the periods of a Néron differential ω_E on E, and let

$$\eta : \mathbb{C}/\Lambda_E \to E(\mathbb{C}).$$

be the Weierstrass uniformisation attached to Λ_E, or, equivalently, to the choice of differential ω_E. (See [Si, VI.5].) Up to a nonzero rational constant — the

Manin constant, which is expected to be ± 1, a fact which will be assumed from now on — the pull-back $(\Phi \circ j)^*\omega_E$ is equal to $2\pi i f(t)dt$ where

$$f = \sum_{n=1}^{\infty} a_n e^{2\pi i n \tau}$$

is the normalized weight 2 cusp form attached to E. Suppose $\tau \in \mathcal{H}$ is such that $\mathcal{N}^{-1}\langle \tau, 1\rangle$ equals $\langle \tau, 1/N\rangle$ up to homothety. Then the results of [Si, VI.5] reveal that

$$\Phi(\delta(\langle \tau, 1\rangle)) = \eta\left(\int_{i\infty}^{\tau} (\Phi \circ j)^*\omega_E\right). \tag{1–5}$$

Given an embedding Ψ of K into $M_2(\mathbb{Q})$, we may define a *period*

$$J_\Psi := \int_{i\infty}^{\tau_\Psi} 2\pi i f(t)dt = \sum_{n=1}^{\infty} \frac{a_n}{n} e^{2\pi i n \tau_\Psi}. \tag{1–6}$$

where $\tau_\Psi \in \mathcal{H}$ is the fixed point attached to Ψ as earlier. Restating equation (1–5), the class $\mathfrak{a}_\Psi \in \mathrm{Pic}(\mathcal{O})$ of $\langle \tau_\Psi, 1\rangle$ satisfies $\Phi(\delta(\mathfrak{a}_\Psi)) = \eta(J_\Psi)$. Since the map Φ is defined over $\mathbb{Q}$, the following is a direct consequence of proposition 1.1:

THEOREM 1.4. *For all* $\Psi \in \mathrm{Emb}(\mathcal{O}, M_0(N))$, *the point* $\eta(J_\Psi)$ *belongs to* $E(H)$, *and for all* $\mathfrak{b} \in \mathrm{Pic}(\mathcal{O})$,

$$\eta(J_{\Psi^{\mathfrak{b}}}) = \mathrm{rec}(\mathfrak{b})^{-1} \cdot \eta(J_\Psi).$$

This proposition asserts that the *analytically defined* invariants J_Ψ of equation (1–6) map under η to a system of algebraic points on the associated elliptic curve. It is exactly this phenomenon (stripped of the algebro-geometric underpinnings provided by Proposition 1.1) that we seek to generalize in this article.

Accordingly, the next sections introduce more general types of periods attached to the following data:

1. an embedding of a quadratic algebra K into a quaternion algebra B;
2. a cusp form on B attached to an elliptic curve E;
3. an appropriate integration theory on a product of (archimedean and nonarchimedean) upper half-planes.

The images of such periods under the appropriate (Weierstrass or Tate) uniformisation are expected to yield a system of algebraic points on E which behave just like the system of Heegner points defined in this section, insofar as

1. they satisfy a variant of the Shimura reciprocity law, and
2. their heights are related to derivatives of Rankin L-series attached to E and K in the spirit of the classical Gross–Zagier formula.

This expectation can be proved by means of the theory of complex multiplication in the cases considered in Section 2, 4 and 6; it is only conjectural in the cases of Sections 5 and 7.

2. Heegner Points and p-adic Integration

This section explains a construction of generalised Heegner points arising from certain Shimura curve parametrisations, in terms of p-adic period integrals. Frequently this construction can be performed on the modular elliptic curve E considered in Section 1.

Forms on $\mathcal{H}_p$. Suppose in this section that the conductor N of E admits an integer factorisation of the form

$$N = pN^+N^-, \tag{2–1}$$

where p is prime, the three factors are pairwise relatively prime, and N^- is a squarefree integer divisible by an odd number of primes. In particular, E has multiplicative reduction at p.

Write B for the (definite) quaternion algebra over $\mathbb{Q}$ of discriminant N^- and R for an Eichler $\mathbb{Z}[1/p]$-order of B of level N^+. The order R is unique up to conjugation by elements of $B^\times$. (For more on the arithmetic theory of quaternion algebras, see [Vi].)

As in the introduction, let $\mathcal{H}_p$ denote Drinfeld's p-adic upper half plane. It is equal to $\mathbb{P}_1(\mathbb{C}_p) - \mathbb{P}_1(\mathbb{Q}_p) = \mathbb{C}_p - \mathbb{Q}_p$, and carries a natural structure of a *rigid analytic space.* The *rigid analytic functions* on $\mathcal{H}_p$ play the role in this discussion of holomorphic functions on $\mathcal{H}$; since this notion is so important for the sequel we recall its precise definition which requires the introduction of certain attendant structures on $\mathcal{H}_p$.

Let $\mathcal{T}_p$ be the Bruhat–Tits tree of $\mathrm{PGL}_2(\mathbb{Q}_p)$. It is a homogeneous tree of degree $p+1$, whose set $\mathcal{V}(\mathcal{T}_p)$ of vertices corresponds bijectively to the set of homothety classes of $\mathbb{Z}_p$-lattices of rank 2 in $\mathbb{Q}_p^2$. Two vertices v and v' are said to be *adjacent* if they can be represented by lattices L and L' such that the finite quotient L/L' has order p. If v and v' are adjacent, the ordered pair $e = (v, v')$ is called an *oriented edge*, with origin $s(e) = v$ and target $t(e) = v'$. Write $\mathcal{E}(\mathcal{T}_p)$ for the set of oriented edges of $\mathcal{T}_p$. Given $e = (v, v')$, let $\bar{e} = (v', v)$ be the opposite edge. Note that $\mathrm{PGL}_2(\mathbb{Q}_p)$ acts naturally on the left on $\mathcal{T}_p$, as a group of isometries. Let v^o be the vertex corresponding to standard lattice $\mathbb{Z}_p^2$. Given $g \in \mathrm{PGL}_2(\mathbb{Q}_p)$, the assignment $g \mapsto gv^o$ induces an identification

$$\mathrm{PGL}_2(\mathbb{Q}_p)/\mathrm{PGL}_2(\mathbb{Z}_p) \xrightarrow{=} \mathcal{V}(\mathcal{T}_p).$$

Likewise, let e^o be the oriented edge such that $s(e^o) = v^o$ and the stabilizer of e^o in $\mathrm{PGL}_2(\mathbb{Q}_p)$ is equal to the group $\Gamma_0(p\mathbb{Z}_p)$ of matrices in $\mathrm{PGL}_2(\mathbb{Z}_p)$ which are upper triangular modulo p. The assignment $g \mapsto ge^o$ induces an identification

$$\mathrm{PGL}_2(\mathbb{Q}_p)/\Gamma_0(p\mathbb{Z}_p) \xrightarrow{=} \mathcal{E}(\mathcal{T}_p).$$

Let

$$\mathrm{red} : \mathbb{P}_1(\mathbb{C}_p) \to \mathbb{P}_1(\bar{\mathbb{F}}_p)$$

be the reduction map modulo the maximal ideal of the ring of integers of $\mathbb{C}_p$. Define a subset A_{v^o} of $\mathcal{H}_p$ by

$$A_{v^o} := \{z \in \mathbb{P}_1(\mathbb{C}_p) : \mathrm{red}(z) \notin \mathbb{P}_1(\mathbb{F}_p)\}.$$

More generally, given any vertex v, choose $g \in \mathrm{PGL}_2(\mathbb{Q}_p)$ such that $v = gv^o$, and let $A_v := gA_{v^o}$. (Note that A_v does not depend on the choice of g.) The sets A_v, obtained by excising $p+1$ disjoint open disks from $\mathbb{P}_1(\mathbb{C}_p)$, are examples of so-called *connected affinoid domains* in $\mathcal{H}_p$. Furthermore, define

$$W_{]e^o[} := \{z \in \mathbb{P}_1(\mathbb{C}_p) : 1 < |z|_p < p\} \subset \mathcal{H}_p.$$

For any $e \in \mathcal{E}(\mathcal{T}_p)$, written as $e = ge^o$ for $g \in \mathrm{PGL}_2(\mathbb{Q}_p)$, set $W_{]e[} := gW_{]e^o[}$. The set $W_{]e[}$ is called the *oriented wide open annulus* attached to e, the orientation corresponding to the choice of the disk

$$D_e := \{z : |g^{-1}z|_p \geq p\} \subset \mathbb{P}_1(\mathbb{C}_p)$$

in the complement of $W_{]e[}$. Finally, if $e = (v, v')$, define the *standard affinoid subset* attached to e to be

$$A_{[e]} := A_v \cup W_{]e[} \cup A_{v'}.$$

The sets $A_{[e]}$ give a covering of $\mathcal{H}_p$ by standard affinoid subdomains whose intersection relations are reflected in the incidence relations on the tree. This allows us to define a *reduction map*

$$r : \mathcal{H}_p \to \mathcal{V}(\mathcal{T}_p) \cup \mathcal{E}(\mathcal{T}_p), \tag{2–2}$$

by sending z to v (resp. to e) if z belongs to A_v (resp. to $W_{]e[}$).

DEFINITION 2.1. A function $f : \mathcal{H}_p \to \mathbb{C}_p$ is called *rigid analytic* if its restriction to each affinoid subset $A_{[e]}$ is a uniform limit (with respect to the sup norm) of a sequence of rational functions having poles outside $A_{[e]}$.

Fixing an embedding ι of B into $M_2(\mathbb{Q}_p)$ yields a natural action on $\mathcal{H}_p$ of the group

$$\Gamma := \iota(R_1^\times) \subset \mathrm{SL}_2(\mathbb{Q}_p),$$

where $R_1^\times$ denotes the group of norm 1 elements in $R^\times$. The rigid analytic structure of $\mathcal{H}_p$ yields a rigid analytic structure on the quotient

$$X_\Gamma := \mathcal{H}_p/\Gamma \tag{2–3}$$

which, by a p-adic variant of the Riemann existence theorem (or the GAGA principle) defines a curve over $\mathbb{C}_p$. This curve can be defined over $\mathbb{Q}_p$ and is equipped with a ring of Hecke correspondences T_n, arising from the fact that Γ is arithmetic, whose definition is given in [JoLi2].

Furthermore, the Cerednik–Drinfeld theorem as formulated in [JoLi1], states that X_Γ (just like the Riemann surface $\mathcal{H}/\Gamma_0(N)$ discussed in Section 1) admits a model over $\mathbb{Q}$. This model is described as the Shimura curve attached to an

Eichler $\mathbb{Z}$-order of level N^+ in the indefinite quaternion algebra of discriminant N^-p, and is isomorphic to X_Γ over the quadratic unramified extension of $\mathbb{Q}_p$. (For more background on Shimura curves, the reader may profitably consult the articles by Gross and Zhang in this volume. The article of [BoCa] gives a detailed account of Drinfeld's approach to the proof of the Cerednik–Drinfeld theorem.)

When combined with the Jacquet–Langlands correspondence (see for example Section 3.5 of [BD3]), this result shows that the classical modular form

$$f_E = \sum_{n\geq 1} a_n e^{2\pi i n\tau}$$

attached to E as in Section 1 corresponds to a weight 2 rigid analytic modular form on $\mathcal{H}_p$ with respect to Γ. More precisely, writing T_n for the n-th Hecke correspondence on X_Γ, one has the following result which follows by combining work of Cerednik–Drinfeld and Jacquet–Langlands:

THEOREM 2.2. *There is a rigid analytic function*

$$f : \mathcal{H}_p \longrightarrow \mathbb{C}_p,$$

well-defined up to multiplication by elements of $\mathbb{C}_p^\times$, *such that:*

1. $f(\gamma z) = (cz+d)^2 f(z)$ *for all* $\gamma = \begin{pmatrix} a & b \\ c & d \end{pmatrix} \in \Gamma$,
2. $T_n f = a_n f$ *for all* $n \geq 1$ *with* $(n, N) = 1$.

The work of P. Schneider and J. Teitelbaum yields a concrete description, well-suited for numerical calculations, of the form f in terms of its associated boundary measure.

Given a rigid analytic differential form $\omega = f(z)\,dz$ on $\mathcal{H}_p$, and an edge e of $\mathcal{T}_p$, let (f_j) be a sequence of rational functions on $\mathbb{C}_p$ converging uniformly to f on $A_{[e]}$ and having poles outside $A_{[e]}$. The limit

$$\lim_{j\to\infty} \sum_{t\in D_e} \mathrm{res}_t(f_j(z)\,dz)$$

exists in $\mathbb{C}_p$ and depends only on f and e, not on the approximating sequence f_j. It is called the *p-adic annular residue* of f at e and is denoted $\mathrm{res}_e(f(z)\,dz)$.

A function κ on $\mathcal{E}(\mathcal{T}_p)$ with values in $\mathbb{C}_p$ is called a *harmonic cocycle* if

1. $\sum_{s(e)=v} \kappa(e) = 0$ for all vertices v (where the sum is taken over all the edges originating from v),
2. $\kappa(\bar{e}) = -\kappa(e)$ for all oriented edges e.

LEMMA 2.3. *Let f be a rigid analytic modular form of weight* 2 *on* $\mathcal{H}_p/\Gamma$. *The function*

$$\kappa_f : \mathcal{E}(\mathcal{T}_p) \longrightarrow \mathbb{C}_p, \quad e \mapsto \mathrm{res}_e(f(z)\,dz),$$

is a $\mathbb{C}_p$*-valued harmonic cocycle. Moreover,* κ_f *is* Γ*-invariant, that is,* $\kappa_f(\gamma e) = \kappa_f(e)$.

PROOF. The first statement follows directly from the residue theorem for rational differentials in light of the definition of the residue of f at e. The Γ-invariance of κ_f follows from the statement that

$$\operatorname{res}_{\gamma e}(f(z)\,dz) = \operatorname{res}_e(f(\gamma^{-1}z)d(\gamma^{-1}z)),$$

which is valid for all $\gamma \in \mathrm{SL}_2(\mathbb{Q}_p)$. See [Sch] for a more complete discussion which also applies to modular forms of higher weight. □

Write $C_{\mathrm{har}}^{\Gamma}$, respectively, $C_{\mathrm{har}}(\mathbb{Q})^{\Gamma}$ for the space of Γ-invariant harmonic cocycles with values in $\mathbb{C}_p$, respectively, $\mathbb{Q}$. Note that the second space defines a $\mathbb{Q}$-structure on the first. Moreover, both spaces are equipped with an induced action of the Hecke algebra $\mathbb{T}_{\Gamma}$ attached to the curve X_{Γ}. It follows that the one-dimensional subspace of $C_{\mathrm{har}}^{\Gamma}$ on which $\mathbb{T}_{\Gamma}$ acts via the character defined by the form f is generated by an element of $C_{\mathrm{har}}(\mathbb{Q})^{\Gamma}$. Since $\mathcal{T}_p/\Gamma$ is a finite graph, an element $C_{\mathrm{har}}^{\Gamma}$ is completely specified by its values on a finite set of orbit representatives. From now on, normalize the form f by requiring that the set of values of κ_f be contained in $\mathbb{Z}$ but in no proper ideal of $\mathbb{Z}$. This condition determines f up to sign.

Each ordered edge e of $\mathcal{T}_p$ gives rise to a compact open subset U_e of $\mathbb{P}_1(\mathbb{Q}_p)$ by the rule:

$$U_e := D_e \cap \mathbb{P}_1(\mathbb{Q}_p).$$

The cocycle κ_f then gives rise to a Γ-invariant measure μ_f on $\mathbb{P}_1(\mathbb{Q}_p)$ satisfying the relation

$$\int_{U_e} d\mu_f(x) = \kappa_f(e).$$

THEOREM 2.4 (TEITELBAUM). *The form f can be recovered from its boundary measure μ_f by the rule*

$$f(z) = \int_{\mathbb{P}_1(\mathbb{Q}_p)} \frac{1}{z-t}\,d\mu_f(t).$$

SKETCH OF PROOF. In view of the injectivity of the residue map, it is sufficient to check that the integral expression

$$I(z) := \int_{\mathbb{P}_1(\mathbb{Q}_p)} \frac{1}{z-t}\,d\mu_f(t) \tag{2–4}$$

is a rigid analytic modular form of weight 2 on $\mathcal{H}_p/\Gamma$ which has the same residues as f. To see that $I(z)$ is rigid analytic, note that the integral on the right in (2–4) is expressed as a limit of finite Riemann sums, each of which are rational functions having poles in $\mathbb{P}_1(\mathbb{Q}_p)$, and that these rational functions converge uniformly to $I(z)$ on any affinoid $A_{[e]}$. The definition of the p-adic annular residue given above makes it clear that

$$\operatorname{res}_e(I(z)\,dz) = \int_{U_e} d\mu_f(t).$$

The Γ-invariance of $I(z)\,dz$ now follows from a direct calculation using the Γ-invariance of the measure μ_f combined with the fact that $\mu_f(\mathbb{P}_1(\mathbb{Q}_p)) = 0$. See [Te, pp. 402–403] for a detailed proof and a more general statement which also applies to modular forms of higher weight. □

Choose a branch

$$\log_p : \mathbb{C}_p^\times \longrightarrow \mathbb{C}_p$$

of the p-adic logarithm, and define the *Coleman p-adic line integral* associated to this choice by the rule

$$\int_{\tau_1}^{\tau_2} f(z)\,dz := \int_{\mathbb{P}_1(\mathbb{Q}_p)} \log_p\left(\frac{t-\tau_2}{t-\tau_1}\right) d\mu_f(t), \qquad \tau_1, \tau_2 \in \mathcal{H}_p. \tag{2–5}$$

Note that the integrand is a locally analytic $\mathbb{C}_p$-valued function on $\mathbb{P}_1(\mathbb{Q}_p)$, so that the integral converges in $\mathbb{C}_p$ (see [Te, p. 401]). Equation (2–5) can be justified by the following *formal* calculation relying on Theorem 2.4:

$$\begin{aligned}\int_{\tau_1}^{\tau_2} f(z)\,dz &= \int_{\tau_1}^{\tau_2}\int_{\mathbb{P}_1(\mathbb{Q}_p)} \left(\frac{1}{z-t}\right) d\mu_f(t)\,dz \\ &= \int_{\mathbb{P}_1(\mathbb{Q}_p)}\int_{\tau_1}^{\tau_2} \left(\frac{dz}{z-t}\right) d\mu_f(t) \\ &= \int_{\mathbb{P}_1(\mathbb{Q}_p)} \log_p\left(\frac{t-\tau_2}{t-\tau_1}\right) d\mu_f(t).\end{aligned}$$

Because the p-adic measure μ_f comes from a harmonic cocycle taking values in $\mathbb{Z}$ and not just $\mathbb{Z}_p$, it is even possible to define the following *multiplicative refinement* of the Coleman line integral by formally exponentiating the expression in (2–5)

$$⨰_{\tau_1}^{\tau_2} f(z)\,dz := ⨰_{\mathbb{P}_1(\mathbb{Q}_p)} \left(\frac{t-\tau_2}{t-\tau_1}\right) d\mu_f(t) \in \mathbb{C}_p^\times, \tag{2–6}$$

where

$$⨰_{\mathbb{P}_1(\mathbb{Q}_p)} g(t)\,d\mu_f(t) := \lim_{\mathcal{C}=\{U_\alpha\}} \prod_\alpha g(t_\alpha)^{\mu_f(U_\alpha)},$$

the limit is taken over increasingly fine covers $\mathcal{C} = \{U_\alpha\}$ of $\mathbb{P}_1(\mathbb{Q}_p)$ by disjoint compact open subsets, and $t_\alpha \in U_\alpha$ is any collection of sample points. This limit of "Riemann products" converges in $\mathbb{C}_p^\times$ provided that the $\mathbb{C}_p^\times$-valued function $g(t)$ is locally analytic.

The multiplicative Coleman integral has the virtue over its more classical additive counterpart that it does not rely on a choice of p-adic logarithm, and carries more information. In fact, the two integrals are related by the formula

$$\int_{\tau_1}^{\tau_2} f(z)\,dz = \log_p\left(⨰_{\tau_1}^{\tau_2} f(z)\,dz\right). \tag{2–7}$$

The *unramified upper half plane*, denoted $\mathcal{H}_p^{\mathrm{nr}}$, is the set of $\tau \in \mathcal{H}_p$ whose image under the reduction map r of equation (2–2) belongs to $\mathcal{V}(\mathcal{T}_p)$, i.e.,

$$\mathcal{H}_p^{\mathrm{nr}} = \bigcup_{v \in \mathcal{V}(\mathcal{T}_p)} A_v.$$

Given two vertices v_1 and v_2 of $\mathcal{T}_p$, let $\sum_{e:v_1\to v_2}$ denote the sum taken over all the ordered edges e in the path joining v_1 and v_2. We will have use for the following closed formula for the p-adic valuation of the multiplicative Coleman integral, which we content ourselves with stating in the special case where the p-adic endpoints of integration belong to $\mathcal{H}_p^{\mathrm{nr}}$.

LEMMA 2.5. *Let τ_1 and τ_2 be points in $\mathcal{H}_p^{\mathrm{nr}}$. Then*

$$\operatorname{ord}_p\left(⨰_{\tau_1}^{\tau_2} f(z)\,dz\right) = \sum_{e:r(\tau_1)\to r(\tau_2)} \kappa_f(e). \tag{2–8}$$

PROOF. Given $\tau_1, \tau_2 \in \mathcal{H}_p^{\mathrm{nr}}$, and $t_1, t_2 \in \mathbb{P}_1(\mathbb{Q}_p)$, let

$$\begin{pmatrix} t_1 : \tau_1 \\ t_2 : \tau_2 \end{pmatrix} = \frac{(t_2-\tau_2)(t_1-\tau_1)}{(t_2-\tau_1)(t_1-\tau_2)}$$

denote the usual cross-ratio. It is useful to have a formula for the p-adic valuation of this cross-ratio in terms of the combinatorics of $\mathcal{T}_p$. An ordered path on $\mathcal{T}_p$ is a sequence (e_j) (either finite or infinite) of ordered edges of $\mathcal{T}_p$ satisfying $t(e_j) = s(e_{j+1})$ for all j for which e_j is defined. The natural intersection pairing on $\mathcal{E}(\mathcal{T}_p)$ defined by

$$e_1 \cdot e_2 = \begin{cases} 1 & \text{if } e_1 = e_2, \\ -1 & \text{if } e_1 = \bar{e}_2, \\ 0 & \text{otherwise} \end{cases}$$

extends by $\mathbb{Z}$-linearity to the set of paths of $\mathcal{T}_p$, and $\gamma_1 \cdot \gamma_2$ is finite provided that the paths γ_1 and γ_2 do not share a half line in common. Let $[v_1 \to v_2]$ denote the finite path joining the vertices v_1 and v_2, and let $(t_1 \to t_2)$ denote the geodesic joining the points t_1 and t_2 of $\mathbb{P}_1(\mathbb{Q}_p)$ viewed as ends of $\mathcal{T}_p$. We claim that

$$\operatorname{ord}_p \begin{pmatrix} t_1 : \tau_1 \\ t_2 : \tau_2 \end{pmatrix} = (t_1 \to t_2) \cdot [r(\tau_1) \to r(\tau_2)]. \tag{2–9}$$

Since the action of $\mathrm{PSL}_2(\mathbb{Q}_p)$ preserves both the cross-ratio and the inner product on paths, and acts 3-point transitively on $\mathbb{P}_1(\mathbb{Q}_p)$, it is enough to verify this formula when $t_1 = 0$ and $t_2 = \infty$. Let v_j (with $j \in \mathbb{Z}$) denote the vertices on $(0 \to \infty)$, numbered consecutively in such a way that $v_0 = v^o$ and that the end $(v_0, v_1, v_2, \ldots,)$ corresponds to 0 while the end $(v_0, v_{-1}, v_{-2}, \ldots)$ corresponds to ∞. Once this is done, let $\mathcal{T}^j$ denote the largest connected subtree of $\mathcal{T}_p$ containing v_j and no edge on $(0 \to \infty)$. It follows directly from the definition of the reduction map, that $r(\tau)$ belongs to $\mathcal{T}^0$ if and only if $\operatorname{ord}_p(\tau) = 0$. Since the Möbius transformation $\tau \mapsto p\tau$ preserves the geodesic $(0 \to \infty)$, sending v_j to v_{j+1} for all $j \in \mathbb{Z}$, it follows that

$$r(\tau) \in \mathcal{T}^j \quad \text{if and only if} \quad \operatorname{ord}_p(\tau) = j.$$

Suppose that τ_1 and τ_2 map to vertices of $\mathcal{T}^{n_1}$ and $\mathcal{T}^{n_2}$ respectively under the reduction map. A direct calculation shows that

$$\operatorname{ord}_p \begin{pmatrix} 0: \ \tau_1 \\ \infty : \tau_2 \end{pmatrix} = \operatorname{ord}_p(\tau_1/\tau_2) = n_1 - n_2,$$

while clearly

$$(0 \to \infty) \cdot [r(\tau_1) \to r(\tau_2)] = n_1 - n_2.$$

Formula (2–9) follows from this.

By the additivity of both members of the equality (2–8) to be proven, it may now be assumed that τ_1 and τ_2 reduce to adjacent vertices, so that the path joining them consists of a single oriented edge $e := (r(\tau_1), r(\tau_2))$. Equation (2–6) yields

$$\operatorname{ord}_p \left(\times\!\!\!\!\!\!\int_{\tau_1}^{\tau_2} f(z)\, dz \right) := \int_{\mathbb{P}_1(\mathbb{Q}_p)} \operatorname{ord}_p \left(\frac{t - \tau_2}{t - \tau_1} \right) d\mu_f(t). \tag{2–10}$$

Let

$$\phi(t) := \operatorname{ord}_p \left(\frac{t - \tau_2}{t - \tau_1} \right)$$

be the expression appearing as the integrand in the right hand side of this equation. Setting $t_1 = t$ and $t_2 = \infty$ in equation (2–9) shows that

$$\phi(t) = \begin{cases} +1 & \text{if } e \text{ belongs to } (\infty \to t), \\ -1 & \text{if } \bar{e} \text{ belongs to } (\infty \to t), \\ \ \ 0 & \text{otherwise.} \end{cases} \tag{2–11}$$

It follows that $\phi(t)$ is locally constant on the disjoint covering $\{U_e, U_{\bar{e}}\}$ of $\mathbb{P}_1(\mathbb{Q}_p)$. It may be assumed, possibly at the cost of interchanging z_1 and z_2, that e (and not $\bar{e}$) belongs to the path $(\infty \to t)$ for some t. Then, one has $\phi(t) = 1$ on U_e and $\phi(t) = 0$ on $U_{\bar{e}}$. Hence, by (2–10),

$$\operatorname{ord}_p \times\!\!\!\!\!\!\int_{\tau_1}^{\tau_2} f(z)\, dz = \int_{U_e} d\mu_f(t) = \kappa_f(e),$$

as was to be shown. (For more on such formulae, see for example the arguments in the proof of Lemma 5.6 of [Man2].) □

The multiplicative Coleman integral yields a rigid analytic uniformisation of E by the curve X_Γ defined in equation (2–3). Extending by linearity, the formula (2–6) gives rise to a map from the degree zero divisors on $\mathcal{H}_p$ to $\mathbb{C}_p^\times$. This map descends to a map from $\operatorname{Pic}^0(X_\Gamma)$ to $\mathbb{C}_p^\times/q^{\mathbb{Z}}$,

$$(\tau_2) - (\tau_1) \mapsto \times\!\!\!\!\!\!\int_{(\tau_2)-(\tau_1)} f(z)\, dz := \times\!\!\!\!\!\!\int_{\tau_1}^{\tau_2} f(z)\, dz \tag{2–12}$$

where $\mathrm{Pic}^0(X_\Gamma)$ denotes the jacobian of X_Γ, and where the p-adic period q is an element of $p\mathbb{Z}_p$. The quotient $\mathbb{C}_p^\times/q^{\mathbb{Z}}$ defines an elliptic curve over $\mathbb{C}_p$ isogenous to E.

From now on, assume (possibly after replacing E by a curve which is $\mathbb{Q}$-isogenous to it) that the Tate p-adic uniformisation

$$\eta_p : \mathbb{C}_p^\times \longrightarrow E(\mathbb{C}_p) \tag{2–13}$$

induces an isomorphism from $\mathbb{C}_p^\times/q^{\mathbb{Z}}$ onto $E(\mathbb{C}_p)$. Choose a correspondence $\theta \in \mathbb{T}$ which maps $\mathrm{Div}(X_\Gamma)$ to $\mathrm{Div}^0(X_\Gamma)$ (such as $T_\ell - (\ell+1)$ for a prime ℓ not dividing N). Replace the multiplicative integral of (2–12) by the new expression

$$\times\!\!\!\!\!\!\int_{\tau_1}^{\tau_2} f(z)\,dz := \times\!\!\!\!\!\!\int_{\theta((\tau_2)-(\tau_1))} f(z)\,dz. \tag{2–14}$$

The redefined integral of (2–14) extends to the whole Picard variety $\mathrm{Pic}(X_\Gamma)$ by the rule

$$(\tau) \mapsto \times\!\!\!\!\!\!\int^{\tau} f(z)\,dz := \times\!\!\!\!\!\!\int_{\theta\tau} f(z)\,dz. \tag{2–15}$$

The map of (2–15) should be viewed as the definition of the concept of a semi-indefinite (multiplicative) integral in the current context. Note the relation in $\mathbb{C}_p^\times/q^{\mathbb{Z}}$:

$$\left(\times\!\!\!\!\!\!\int^{\tau_2} f(z)\,dz\right) \div \left(\times\!\!\!\!\!\!\int^{\tau_1} f(z)\,dz\right) = \times\!\!\!\!\!\!\int_{\tau_1}^{\tau_2} f(z)\,dz.$$

The semi-indefinite integral can also be described in terms of values of p-adic theta functions (so that the definite integral is a ratio of such values).

The next result makes explicit the relation between the additive Coleman integral corresponding to a choice of logarithm $\log_p$ and the harmonic cocycle κ_f. Define the $\mathcal{L}$-invariant attached to f by the rule

$$\mathcal{L}_f := \frac{\log_p(q)}{\mathrm{ord}_p(q)}.$$

PROPOSITION 2.6. *Let z be any point in $\mathcal{H}_p$ and let v be any vertex of $\mathcal{T}_p$. For all $\gamma \in \Gamma$, the equality*

$$\int_z^{\gamma z} f(\tau)d\tau = \mathcal{L}_f \cdot \sum_{e:v\to\gamma v} \kappa_f(e) \tag{2–16}$$

holds.

PROOF. Note that the left-hand side of equation (2–16) does not depend on the choice of z, as a consequence of the Γ-invariance of the differential $f(\tau)d\tau$. In particular, z can be taken in $\mathcal{H}_p^{\mathrm{nr}}$. Similarly, the right-hand side of equation (2–16) does not depend on the choice of v. The theory of p-adic uniformisation recalled above yields

$$\times\!\!\!\!\!\!\int_z^{\gamma z} f(\tau)d\tau = q^{n_\gamma} \tag{2–17}$$

for all γ, where n_γ is an integer. Taking $\log_p$ and ord_p of (2–17), and comparing the resulting equalities, gives

$$\int_z^{\gamma z} f(\tau)d\tau = \mathcal{L}_f \cdot \mathrm{ord}_p \times\!\!\!\!\!\!\int_z^{\gamma z} f(\tau)d\tau. \tag{2–18}$$

Proposition 2.6 follows by combining (2–18) with Lemma 2.5. □

The period integral. Let $K/\mathbb{Q}$ be a quadratic algebra contained in $\mathbb{C}_p$, and assume that it satisfies the following modified Heegner hypothesis:

1. K is an imaginary quadratic field;
2. all the primes dividing N^- are inert in K;
3. all the primes dividing N^+ are split in K;
4. p is inert in K.

Conditions 1 and 2 are precisely what is needed to ensure the existence of an embedding of K into B, the definite quaternion algebra of discriminant N^- introduced at the start of this section.

Let $\mathcal{O}$ be any $\mathbb{Z}[1/p]$-order of K of conductor prime to N. (So that $\mathcal{O}$ is a subring of K which is free of rank two over $\mathbb{Z}[1/p]$.) Choose an Eichler $\mathbb{Z}[1/p]$-order R of level N^+ in B. An *orientation* on the ring $T = \mathcal{O}$ or $T = R$ is a surjective ring homomorphism

$$\mathfrak{o} : T \longrightarrow (\mathbb{Z}/N^+\mathbb{Z}) \times \prod_{\ell | N^-} \mathbb{F}_{\ell^2}.$$

Note that such an orientation exists on $\mathcal{O}$ thanks to conditions 2 and 3 satisfied by K.

Fix such orientations on $\mathcal{O}$ and on R once and for all. Let $\Psi : K \longrightarrow B$ be an embedding of algebras. By condition 4 satisfied by K, the group $\iota\Psi(K^\times)$ acting on $\mathcal{T}_p$ has precisely one fixed vertex v_Ψ. An embedding $\Psi : K \longrightarrow B$ is said to be an *oriented optimal embedding* of $\mathcal{O}$ into R (relative to the fixed orientations on $\mathcal{O}$ and R) if:

1. Ψ embeds $\mathcal{O}$ optimally as a subring of R;
2. The map Ψ is compatible with the orientations on $\mathcal{O}$ and R in the obvious sense;
3. The map Ψ is *oriented at p*, in the sense that

$$v_\Psi \text{ belongs to } \mathrm{SL}_2(\mathbb{Q}_p)v^o.$$

The conditions imposed on K above ensure the existence of oriented optimal embeddings. It can be verified, just as in Section 1, that the group $\Gamma = \iota(R_1^\times)$ acts on them by conjugation; as in that section, let $\mathrm{Emb}(\mathcal{O}, R)$ denote the set of Γ-conjugacy classes of oriented optimal embeddings of $\mathcal{O}$ into R. The period integral considered here is attached to the class of an embedding Ψ in $\mathrm{Emb}(\mathcal{O}, R)$.

Write K_p for the local field $K \otimes \mathbb{Q}_p$. The torus $K_p^\times$ acting on $\mathcal{H}_p$ via the embedding $\iota\Psi$ has a unique fixed point τ_Ψ satisfying

$$\iota\Psi(\alpha)\begin{pmatrix}\tau_\Psi\\ 1\end{pmatrix} = \alpha\begin{pmatrix}\tau_\Psi\\ 1\end{pmatrix}, \quad \text{for all } \alpha \in K_p^\times. \tag{2–19}$$

This fixed point belongs to $\mathcal{H}_p \cap K_p$; let $\bar{\tau}_\Psi$ denote its conjugate under the action of $\mathrm{Gal}(K_p/\mathbb{Q}_p)$.

DEFINITION 2.7. The period integrals associated to the embedding Ψ are defined to be

$$I_\Psi := \times\!\!\!\!\!\!\int_{\bar{\tau}_\Psi}^{\tau_\Psi} f(z)\,dz \in \mathbb{C}_p^\times \quad \text{and} \quad J_\Psi := \times\!\!\!\!\!\!\int^{\tau_\Psi} f(z)\,dz \in \mathbb{C}_p^\times/q^{\mathbb{Z}}.$$

The next lemma follows directly from formula (2–6).

LEMMA 2.8. *The periods I_Ψ and J_Ψ belong to $K_p^\times$ and $K_p^\times/q^{\mathbb{Z}}$, respectively.*

Note that J_Ψ and I_Ψ are related by the equality

$$J_\Psi/\bar{J}_\Psi = I_\Psi \pmod{q^{\mathbb{Z}}}.$$

The following theorem states that the images under η_p of the period integrals J_Ψ and I_Ψ define global points on the elliptic curve E. Let $w = 1$ if $E/\mathbb{Q}_p$ has split multiplicative reduction, and $w = -1$ otherwise. Let H be the ring class field of K attached to the order $\mathcal{O}$. Note that p is totally split in the extension H/K. Fix an embedding of H into $K_p \subset \mathbb{C}_p$. This amounts to choosing a prime $\mathfrak{p}$ of H above p. In particular $E(H)$ becomes a subgroup of $E(\mathbb{C}_p)$ thanks to this choice. Write $\sigma_\mathfrak{p}$ for the Frobenius element of $\mathfrak{p}$ in $\mathrm{Gal}(H/\mathbb{Q})$.

THEOREM 2.9 ([BD2]). *For each optimal embedding Ψ of $\mathcal{O}$ into R, the point $\eta_p(J_\Psi)$ is a global point P_Ψ in $E(H)$. In particular $\eta_p(I_\Psi)$ is equal to the global point $P_\Psi - w\sigma_\mathfrak{p} P_\Psi$. Furthermore, for all $\mathfrak{a} \in \mathrm{Pic}(\mathcal{O})$,*

$$\eta_p(J_{\Psi^\mathfrak{a}}) = \mathrm{rec}(\mathfrak{a})^{-1} \cdot \eta_p(J_\Psi).$$

The proof of this theorem follows from Drinfeld's moduli interpretation of the p-adic upper half plane $\mathcal{H}_p$. This interpretation implies that the point on the Shimura curve model of X_Γ which corresponds to τ_Ψ is a Heegner point defined over H. See chapter 5 of [BD2] for details.

REMARKS. **1.** Note that the Shimura reciprocity law described above for the periods J_Ψ does not extend in the natural way to the periods I_Ψ, because the substitution $\tau_\Psi \mapsto \bar{\tau}_\Psi$ *does not commute* with the action of $\mathrm{Pic}(\mathcal{O})$ on the optimal embeddings Ψ. More precisely, one has

$$\bar{\Psi}^{\mathfrak{a}^{-1}} = \overline{\Psi^\mathfrak{a}},$$

so that by the Shimura reciprocity law above

$$\eta_p(I_{\Psi^\mathfrak{a}}) = \eta_p(J_{\Psi^\mathfrak{a}} \div J_{\overline{\Psi}^{\mathfrak{a}^{-1}}}).$$

Thus the expression of the right is not equal in general to $\mathrm{rec}(\mathfrak{a})^{-1}\eta_p(I_\Psi)$ unless $\mathfrak{a} = \mathfrak{a}^{-1}$ in $\mathrm{Pic}(\mathcal{O})$.

2 (The relation with p-adic L-functions). The periods I_Ψ are notable for their simple relation with special values of certain anticyclotomic p-adic L-functions. More precisely, let $\log_p$ denote the standard p-adic logarithm

$$\log_p : K_{p,1}^\times \to K_p,$$

where $K_{p,1}^\times$ is the compact group of norm 1 elements in $K_p^\times$. In [BDIS], building on an idea of Schneider, a partial anticyclotomic p-adic L-function is attached to f and the embedding Ψ by taking a p-adic Mellin transform of the measure attached to the form f:

$$L_p(f,\Psi,s) = \int_{\mathbb{P}_1(\mathbb{Q}_p)} \left(\frac{t-\tau_\Psi}{t-\bar{\tau}_\Psi}\right)^{s-1} d\mu_f(t). \tag{2–20}$$

Note that the expression $\frac{t-\tau_\Psi}{t-\bar{\tau}_\Psi}$ in the integrand belongs to $K_{p,1}^\times$, so that the p-adic exponentiation that occurs there can be defined by

$$\alpha^s := \exp(s\log_p(\alpha)),$$

a definition which is independent of a choice of logarithm.

Equation (2–20) defining the p-adic L-function can be justified by noting that the measure μ_f satisfies the following interpolation formula with respect to classical special values, which for the sake of illustration we give only in the case where $\mathrm{Pic}(\mathcal{O})$ has cardinality 1 so that there is a unique optimal embedding Ψ of $\mathcal{O}$ into R up to conjugation by Γ. If $\chi : K_{p,1}^\times \longrightarrow \mathbb{C}_p^\times$ is a finite order character, viewed as a (complex or p-adic) character of $\mathrm{Gal}(\bar{K}/K)$ in the usual way, then

$$\left| \int_{\mathbb{P}_1(\mathbb{Q}_p)} \chi\left(\frac{t-\tau_\Psi}{t-\bar{\tau}_\Psi}\right) d\mu_f(t) \right|^2 \doteq L(E/K,\chi,1),$$

where the symbol $\doteq$ denotes equality up to a simple fudge factor. See for example [Zh] where this formula is proved.

It is apparent from the definition of the p-adic L-function given in (2–20) that

$$L_p'(f,\Psi,1) = \log_p(I_\Psi). \tag{2–21}$$

Combined with Theorem 2.9, which expresses $\eta_p(I_\Psi)$ in terms of Heegner points, equation (2–21) can be viewed as a rigid analytic variant of the Gross–Zagier formula. Recall that Zhang's generalisation of the Gross–Zagier theorem [Zh] relates the Néron–Tate height of the points P_Ψ to the derivative of certain complex Rankin L-series attached to f and to K.

It is interesting to note that while the period I_Ψ admits a natural interpretation in terms of the first derivative of an (anticyclotomic) p-adic L-function, a similar interpretation is not known for the periods J_Ψ even though these periods are more natural insofar as they satisfy a cleanly stated Shimura reciprocity law.

A clue for understanding this phenomenon is provided by the computations performed in [BD1], which relate I_Ψ to a (partial) p-adic regulator expressing the leading term of the partial p-adic L-function $L_p(f, \Psi, s)$. Following the ideas of [MTT], this regulator is defined on the extended Mordell–Weil group of E, by pairing P_Ψ with the Tate period q.

3. Forms on $\mathcal{T}_p \times \mathcal{H}$

One of the questions which motivates this article is the apparent difficulty of extending the theory of complex multiplication (and hence, the Heegner point construction) to the setting where K is a real quadratic field. The difficulty (at least on a superficial level) can be traced to two *fundamentally different* causes in the settings that were covered in the previous two sections.

1. In Section 1, the difficulty comes about from the fact that the real quadratic subalgebras of $M_2(\mathbb{Q})$ do not have fixed points on $\mathcal{H}^*$, so that the period integral (1–6) cannot be extended to this setting in any obvious way.
2. In Section 2, the difficulty can be ascribed to the fact that the discrete subgroups of $\mathrm{SL}_2(\mathbb{Q}_p)$ occuring in the Cerednik–Drinfeld theory come from unit groups of $\mathbb{Z}[1/p]$-orders of *definite* quaternion algebras; but such quaternion algebras contain no real quadratic subalgebras.

If B is an *indefinite* quaternion algebra over $\mathbb{Q}$ for which the prime p is split, and R a $\mathbb{Z}[1/p]$-order in B, one can of course always consider the group

$$\Gamma := \iota(R_1^\times),$$

but this group does not act discretely on $\mathcal{H}_p$ or on $\mathcal{H}$. Rather, it acts discretely on the product $\mathcal{H}_p \times \mathcal{H}$.

It is from this point of view that a theory of mixed p-adic and archimedean modular forms becomes germane to the concerns of this article. Ihara [I] and Stark [St] studied this question in their papers on mixed place modular forms, and indeed the classical theory of complex multiplication plays an important role in the work of Ihara. The problem of defining a sensible notion of analytic function on $\mathcal{H}_p \times \mathcal{H}$ seems inextricably linked to that of finding a well-behaved tensor product of $\mathbb{C}_p$ with $\mathbb{C}$, which appears to be a difficult task. However, there is a good compromise, at least in the special case where B is the split quaternion algebra $M_2(\mathbb{Q})$.

Assume for this section that the conductor N of E is of the form pM with $(p, M) = 1$. To fix ideas, define Γ as in the introduction:

$$\Gamma := \left\{ \begin{pmatrix} a & b \\ c & d \end{pmatrix} \in \mathrm{SL}_2(\mathbb{Z}[1/p]) \quad \text{such that } M \text{ divides } c \right\}. \tag{3–1}$$

In other words, if $R = \mathrm{M}_0(M) \otimes \mathbb{Z}[1/p]$ denotes the standard Eichler $\mathbb{Z}[1/p]$-order of level M in $\mathrm{M}_2(\mathbb{Q})$, consisting of the matrices with entries in $\mathbb{Z}[1/p]$

which are upper triangular modulo M, then $\Gamma = R_1^\times$. The group $\Gamma \subset \mathrm{SL}_2(\mathbb{Q})$ acts by linear fractional transformations both on $\mathcal{H}_p$ and on $\mathcal{H}$, yielding a discrete discontinuous action on the product $\mathcal{H}_p \times \mathcal{H}$. Moreover, Γ acts transitively on the set of unordered edges of $\mathcal{T}_p$, and has exactly two orbits on $\mathcal{E}(\mathcal{T}_p)$, represented by e^o and $\bar{e}^o$ respectively. Recall that v^o is the base vertex of $\mathcal{T}_p$, characterized by the property that its stabiliser Γ_{v^o} in Γ is equal to $\Gamma_0(M)$, and that e^o is the edge characterized by the properties source$(e^o) = v^o$ and $\Gamma_{e^o} = \Gamma_0(N)$.

In Theorem 2.4 of Section 2 we saw that rigid analytic modular forms on $\mathcal{H}_p/\Gamma$ are naturally associated, and completely determined by, certain $\mathbb{C}_p$-valued functions $\mathcal{E}(\mathcal{T}_p)$ which are Γ-invariant and satisfy the harmonic cocycle property:

$$\sum_{s(e)=v} \kappa(e) = 0 \text{ for all } v \in \mathcal{V}(\mathcal{T}_p) \quad \text{and} \quad \kappa(\bar{e}) = -\kappa(e) \text{ for all } e \in \mathcal{E}(\mathcal{T}_p).$$

Thus one might surmise that the role of this associated harmonic cocycle, for an "invariant differential 2-form on $(\mathcal{H}_p \times \mathcal{H})/\Gamma$", should be played by a function κ on $\mathcal{E}(\mathcal{T}_p)$ *with values in a space of complex analytic functions on* $\mathcal{H}$ satisfying the obvious harmonicity and Γ-invariance properties.

This brings us to the key definition of this section.

DEFINITION 3.1. A cusp form of weight 2 on $(\mathcal{H}_p \times \mathcal{H})/\Gamma$ is a function

$$f : \mathcal{E}(\mathcal{T}_p) \times \mathcal{H} \to \mathbb{C}$$

satisfying the following properties.

1. (Analyticity) For every edge e of $\mathcal{T}_p$, $f_e(z) := f(e,z)$ is a holomorphic function such that $f_{\bar{e}} = -f_e$, and
$$\sum_{s(e)=v} f_e = 0, \quad \text{for every vertex } v \in \mathcal{V}(\mathcal{T}_p). \tag{3–2}$$
2. (Γ-invariance) For all $\gamma \in \Gamma$, $f(\gamma e, \gamma z) d(\gamma z) = f(e,z)\, dz$.
3. (Cuspidality) For every edge e, the function f_e is a complex cusp form of weight 2 on Γ_e, the stabiliser of e in Γ.

Let $S_2((\mathcal{H}_p \times \mathcal{H})/\Gamma)$ denote the space of cusp form of weight 2 on $(\mathcal{H}_p \times \mathcal{H})/\Gamma$. As usual, write $S_2(\mathcal{H}/\Gamma_0(N))$ for the space of weight 2 cusp forms on $\Gamma_0(N)$. Note that one has two natural degeneracy maps

$$S_2(\mathcal{H}/\Gamma_0(N)) \rightrightarrows S_2(\mathcal{H}/\Gamma_0(M)).$$

The forms killed by both degeneracy maps are called *new at p*. Denote the subspace of forms that are new at p by $S_2^{p\text{-new}}(\mathcal{H}/\Gamma_0(N))$. The proposition below shows that $S_2((\mathcal{H}_p \times \mathcal{H})/\Gamma)$ is a familiar object, identified directly with the p-new subspace of $S_2(\mathcal{H}/\Gamma_0(N))$.

PROPOSITION 3.2. *The map* $f \mapsto f_{e^o}$ *defines an isomorphism*

$$S_2((\mathcal{H}_p \times \mathcal{H})/\Gamma) \xrightarrow{\sim} S_2^{p\text{-new}}(\mathcal{H}/\Gamma_0(N)).$$

PROOF. Since the stabiliser Γ_{e^o} of e in Γ is equal to $\Gamma_0(N)$, it follows from part 3 of Definition 3.1 that f_{e^o} belongs to $S_2(\mathcal{H}/\Gamma_0(N))$, so that the assignment $f \mapsto f_{e^o}$ defines a linear map $S_2((\mathcal{H}_p \times \mathcal{H})/\Gamma) \to S_2(\mathcal{H}/\Gamma_0(N))$. To see that this assignment is injective, note that if f_{e^o} is 0, then so is $f_{\bar{e}^o} = -f_{e^o}$, and hence

$$f_e = 0 \text{ for all } e \in \mathcal{E}(\mathcal{T}_p) = \Gamma e^o \cup \Gamma \bar{e}^o.$$

Next, we claim that the image of $S_2((\mathcal{H}_p \times \mathcal{H})/\Gamma)$ is precisely the space

$$S_2^{p\text{-new}}(\mathcal{H}/\Gamma_0(N)).$$

Let f_0 be a modular form in this space, and define

$$f_e(z)\,dz = f_0(\gamma^{-1}z)d(\gamma^{-1}z) \text{ for } e = \gamma e^o \in \Gamma e^0.$$

Extend this definition to $e \in \Gamma \bar{e}^o$ by setting $f_e = -f_{\bar{e}}$. It is easy to see that the collection $\{f_e\}$ satisfies all the properties of Definition 3.1 with the possible exception of the harmonicity condition given in equation (3–2). To see that this latter condition is satisfied as well, note that by the definition of p-new-forms, f_0 satisfies

$$\sum_{\gamma \in \Gamma_0(N)/\Gamma_0(M)} f_0(\gamma^{-1}z)d(\gamma^{-1}z) = 0 \tag{3–3}$$

and

$$\sum_{\gamma \in \Gamma_0(N)/\Gamma_0(M)} f_0(\gamma^{-1}\alpha^{-1}z)d(\gamma^{-1}\alpha^{-1}z) = 0, \tag{3–4}$$

where $\alpha \in \mathrm{GL}_2^+(\mathbb{Z}[1/p])$ is an element of the normaliser of $\Gamma_0(N)$ which does not belong to $\Gamma_0(N)$. Note that $\alpha e^o = \overline{e^o}$, so that $v^1 := \alpha v^o$ is the target of e^o. By strong approximation (which in this setting amounts to the Chinese Remainder theorem), the natural embedding of $\mathrm{SL}_2(\mathbb{Z})$ into $\mathrm{SL}_2(\mathbb{Z}_p)$ identifies the coset space $\Gamma_0(N)/\Gamma_0(M)$ with $\mathrm{SL}_2(\mathbb{Z}_p)/\Gamma_0(p\mathbb{Z}_p)$, so that

$$\{\gamma e^o \, : \, \gamma \in \Gamma_0(N)/\Gamma_0(M)\}$$

is a complete list of edges with source v^o, while

$$\{\alpha\gamma e^o \, : \, \gamma \in \Gamma_0(N)/\Gamma_0(M)\}$$

is a complete list of edges with edges with source v^1. Hence equations (3–3) and (3–4) just amount to the statement that $\{f_e\}$ is harmonic at v^o and v^1:

$$\sum_{e:s(e)=v^o} f_e = 0; \qquad \sum_{e:s(e)=v^1} f_e = 0. \tag{3–5}$$

Harmonicity at all other vertices now follows from the Γ-equivariance built into the definition of f in terms of f_0, in light of the fact that $\mathcal{V}(\mathcal{T}_p) = \Gamma v^o \cup \Gamma v^1$. Hence f is a cusp form of weight 2 on $(\mathcal{H}_p \times \mathcal{H})/\Gamma$, satisfying $f_{e^o} = f_0$, and therefore the image of the assigment $f \mapsto f_{e^o}$ contains the forms that are new at p. That these are the only forms in the image follows from the equivalence of equation (3–5) with equations (3–3) and (3–4). □

Thanks to Proposition 3.2, the space $S_2((\mathcal{H}_p \times \mathcal{H})/\Gamma)$ inherits an action of the Hecke operators T_n (for $\gcd(n, N) = 1$) from the familiar action of Hecke operators on $S_2^{p\text{-new}}(\mathcal{H}/\Gamma_0(N))$. This Hecke action can be described directly as in formula (46) of [Da].

Given $\gamma \in \mathrm{PGL}_2(\mathbb{Q}_p)$, the quantity $|\gamma|_p := \mathrm{ord}_p(\det(\gamma))$ is well-defined modulo 2. Let w denote the *negative* of the eigenvalue of the Atkin–Lehner involution W_p acting on f_0. Thus, w is equal to 1 (resp. -1) if the abelian variety attached to f_0 has split (resp. nonsplit) multiplicative reduction over $\mathbb{Q}_p$. The following lemma describes an invariance property of f under the action of the larger group $\iota(R_+^\times) \supset \Gamma$, where $R_+^\times$ denotes the group of matrices in $R^\times$ with positive determinant.

LEMMA 3.3. *For all* $\gamma \in \iota(R_+^\times)$,

$$f_{\gamma e}(\gamma z) d(\gamma z) = w^{|\gamma|_p} f_e(z)\, dz.$$

The proof is explained in Lemma 1.5 of [Da].

4. Complex Periods and Heegner Points

Recall that if κ is a $\mathbb{Z}$-valued harmonic cocycle on $\mathcal{T}_p$, and f is the associated rigid analytic function on $\mathcal{H}_p$, then by Lemma 2.5:

$$\mathrm{ord}_p\left(\times\!\!\!\!\!\!\int_{\tau_1}^{\tau_2} f(z)\, dz \right) = \sum_{e:r(\tau_1)\to r(\tau_2)} \kappa(e)$$

for τ_1 and τ_2 in $\mathcal{H}_p^{\mathrm{nr}}$. In light of the definitions in the previous section, a natural notion of $\mathbb{C}$-valued line integral of an analytic function on $\mathcal{H}_p \times \mathcal{H}$ comes out of imagining an integration theory taking values in $\mathbb{C}_p^\times \otimes \mathbb{C}$ and applying to this hypothetical integral the map $\mathrm{ord}_p : \mathbb{C}_p^\times \otimes \mathbb{C} \longrightarrow \mathbb{C}$. Thus, given $f \in S_2((\mathcal{H}_p \times \mathcal{H})/\Gamma)$, define the additive integral of this section by the rule:

$$\int_{\tau_1}^{\tau_2}\int_{\tau_3}^{\tau_4} \omega_f = \sum_{e:r(\tau_1)\to r(\tau_2)} \int_{\tau_3}^{\tau_4} 2\pi i f_e(z)\, dz \tag{4–1}$$

for $\tau_1, \tau_2 \in \mathcal{H}_p^{\mathrm{nr}}$ and $\tau_3, \tau_4 \in \mathcal{H}^*$. It is immediately apparent that this double integral is additive in both sets of variables of integration, and satisfies the Γ-invariance property:

$$\int_{\gamma\tau_1}^{\gamma\tau_2}\int_{\gamma\tau_3}^{\gamma\tau_4} \omega_f = \int_{\tau_1}^{\tau_2}\int_{\tau_3}^{\tau_4} \omega_f \quad \text{for all } \gamma \in \Gamma. \tag{4–2}$$

Having now a suitable notion of weight 2 cusp form on $\mathcal{H}_p \times \mathcal{H}$ and a theory of integration of such forms, one may begin to carry out the program described at the end of Section 1 by defining a period attached to a quadratic subalgebra K of $M_2(\mathbb{Q})$.

Let Γ act trivially on $\mathbb{C}$. A direct computation shows that the function $\tilde{d}_{x,\tau} : \Gamma \times \Gamma \to \mathbb{C}$ defined by

$$\tilde{d}_{x,\tau}(\gamma_1, \gamma_2) = \int_{\gamma_1 x}^{\gamma_1\gamma_2 x} \int_{\tau}^{\gamma_1 \tau} \omega_f \tag{4–3}$$

is a 2-cocycle, for all $x \in \mathcal{H}_p$ and $\tau \in \mathcal{H}$. Furthermore, the image of $\tilde{d}_{x,\tau}$ in $H^2(\Gamma, \mathbb{C})$ is independent of the base points $x \in \mathcal{H}_p$ and $\tau \in \mathcal{H}$ that were chosen to define it.

Assume for simplicity that the cusp form

$$f_0 := f_{e^o} \in S_2^{p\text{-new}}(\Gamma_0(N))$$

is attached to an elliptic curve $E/\mathbb{Q}$ of conductor N, so that in particular f_0 has integer Fourier coefficients. Let $\Lambda \subset \mathbb{C}$ be the image of the relative homology $H_1(X_0(N), \text{cusps}; \mathbb{Z})$ under the integration pairing with f_0. The following theorem plays a crucial role in the definition of the period integral J_Ψ defined in this section, as well as in a key construction of Section 5.

THEOREM 4.1 (MANIN–DRINFELD). *The subgroup $\Lambda \subset \mathbb{C}$ is a lattice which is commensurable with the Néron lattice Λ_E attached to E.*

SKECH OF PROOF. The key to this proof is the fact that, for all primes ℓ not dividing N, the correspondences $\ell + 1 - T_\ell$ map the relative homology $H_1(X_0(N), \text{cusps}; \mathbb{Z})$ to $H_1(X_0(N), \mathbb{Z})$. This implies that, for any such ℓ, $(\ell + 1 - a_\ell)\Lambda$ is contained in Λ_E. The reader is refered to [Man1] or [Maz] where the proof is explained in more detail, □

The lattice Λ plays a key role in the integration theory considered in this section, thanks to the following lemma.

LEMMA 4.2. *Let f be the cusp form of weight 2 on $\mathcal{H}_p \times \mathcal{H}$ attached to E, and let Λ be the lattice associated to E as above.*

1. *The complex line integral $\int_x^y f_e(z)\,dz$ belongs to Λ, for all $x, y \in \mathbb{P}_1(\mathbb{Q})$ and all $e \in \mathcal{E}(\mathcal{T}_p)$.*
2. *For all $\tau_1, \tau_2 \in \mathcal{H}_p^{\text{nr}}$ and for all $x, y \in \mathbb{P}_1(\mathbb{Q})$,*

$$\int_{\tau_1}^{\tau_2} \int_x^y \omega_f \quad \textit{belongs to } \Lambda.$$

PROOF. To show the first part, note that if e is any edge of $\mathcal{T}_p$, there exists an element γ in Γ such that

$$\gamma^{-1} e^o = e \quad \text{or} \quad \gamma^{-1} e^o = \bar{e}.$$

In any case

$$\int_x^y f_e(z)\,dz = \pm \int_{\gamma x}^{\gamma y} f_0(z)\,dz,$$

and the result follows from the definition of Λ. The second part is an immediate consequence of the first, in light of the definition of the double integral attached to f. $\square$

Part 2 of Lemma 4.2 is used in this section in the proof of the following proposition:

PROPOSITION 4.3. *The natural image of* $\tilde{d}_{x,\tau}$ *in* $H^2(\Gamma, \mathbb{C}/\Lambda)$ *is zero.*

PROOF. The proof is obtained by showing that the 2-cocycle $\tilde{d}_{x,\tau}$ viewed modulo Λ is the image by the coboundary map of an explicit 1-cochain. Defining $\xi : \Gamma \to \mathbb{C}$ by

$$\xi(\gamma) = \int_x^{\gamma x}\int_\infty^\tau \omega_f,$$

and using the fact that Γ acts trivially on $\mathbb{C}$, one finds

$$\begin{aligned} d\xi(\gamma_1,\gamma_2) &= \int_x^{\gamma_1 x}\int_\infty^\tau \omega_f + \int_x^{\gamma_2 x}\int_\infty^\tau \omega_f - \int_x^{\gamma_1\gamma_2 x}\int_\infty^\tau \omega_f \\ &= \int_{\gamma_1\gamma_2 x}^{\gamma_1 x}\int_\infty^\tau \omega_f + \int_x^{\gamma_2 x}\int_\infty^\tau \omega_f \\ &= \int_{\gamma_1\gamma_2 x}^{\gamma_1 x}\int_\infty^\tau \omega_f + \int_{\gamma_1 x}^{\gamma_1\gamma_2 x}\int_{\gamma_1\infty}^{\gamma_1\tau} \omega_f \\ &= \int_{\gamma_1 x}^{\gamma_1\gamma_2 x}\int_\tau^{\gamma_1\tau} \omega_f + \int_{\gamma_1 x}^{\gamma_1\gamma_2 x}\int_{\gamma_1\infty}^{\infty} \omega_f. \end{aligned}$$

The second term of this last expression belongs to Λ by part 2 of Lemma 4.2, and the result follows. $\square$

The 1-cochain ξ is determined by x and τ up to 1-cocycles. To make ξ completely canonical one shows that the group of 1-cocyles has finite exponent, so that the ambiguity in the definition of ξ can be removed by replacing ξ by an appropriate integer multiple.

LEMMA 4.4. *The group of* 1*-cocycles on* Γ *with coefficients in* $\mathbb{C}/\Lambda$ *has finite exponent.*

PROOF. Since Γ acts trivially on $\mathbb{C}/\Lambda$, the group of 1-cocycles on Γ with coefficients in $\mathbb{C}/\Lambda$ is given by $\mathrm{Hom}(\Gamma, \mathbb{C}/\Lambda) = \mathrm{Hom}(\Gamma^{\mathrm{ab}}, \mathbb{C}/\Lambda)$. But the abelianisation Γ^{ab} of Γ is finite by the corollary to Theorem 3 in [Se2]. $\square$

Let $\mathcal{O}$ be as in Section 2 a $\mathbb{Z}[1/p]$-order in an imaginary quadratic field K. Let $\mathcal{O}_0 := \mathcal{O} \cap \mathcal{O}_K$ be the maximal $\mathbb{Z}$-order in $\mathcal{O}$, and assume as in Section 1 that the level N satisfies the Heegner hypothesis, choosing as earlier a factorisation $N\mathcal{O}_0 = \mathcal{N}\overline{\mathcal{N}}$ into a product of cyclic ideals of K. Set

$$R := \mathrm{M}_0(M) \otimes \mathbb{Z}[1/p],$$

where the $\mathbb{Z}[1/p]$-algebra R is viewed naturally as a subring of $M_2(\mathbb{Q})$. An oriented optimal embedding of $\mathcal{O}$ into R is defined exactly as in Section 1, except that N is replaced with M, $\mathcal{N}$ with $\mathcal{M} := \mathcal{N}\mathcal{O}$ and $M_0(N)$ with R, and the following additional orientation condition at p is imposed. One asks that the normalized fixed point $\tau = \tau_\Psi$ for the action of $\Psi(K_p^\times)$ on $\mathbb{P}_1(\mathbb{Q}_p)$ belongs to $\mathbb{Z}_p$, with respect to a fixed embedding of K into $\mathbb{Q}_p$. Define $\mathrm{Emb}(\mathcal{O}, R)$ as earlier, with $\Gamma = R_1^\times$ replacing $\Gamma_0(N)$. Let $\Psi \in \mathrm{Emb}(\mathcal{O}, R)$ be an oriented optimal embedding. Note that $\Psi(K_p^\times)$ leaves invariant the geodesic on $\mathcal{T}_p$ joining the two fixed points of $\Psi(K_p^\times)$ in $\mathbb{P}_1(\mathbb{Q}_p)$. Let Γ_τ denote the stabiliser of τ in Γ.

LEMMA 4.5. *The group* Γ_τ *has rank one.*

PROOF. Since $\Psi(K^\times)$ is the stabiliser of τ in $\mathrm{GL}_2(\mathbb{Q})$ and Ψ is an optimal embedding of $\mathcal{O}$ into R, it follows that $\Gamma_\tau = \Psi(\mathcal{O}_1^\times)$. The Dirichlet unit theorem shows that $\mathcal{O}_1^\times$ is of rank one, with generator given by an appropriate power of the fundamental p-unit of the imaginary quadratic field K. □

Choose a generator u of $\mathcal{O}_1^\times$ and set $\gamma_\tau = \Psi(u)$. The calculations

$$\int_x^{\gamma_\tau x}\int_\infty^\tau \omega_f - \int_y^{\gamma_\tau y}\int_\infty^\tau \omega_f = \int_{\gamma_\tau y}^{\gamma_\tau x}\int_\infty^\tau \omega_f - \int_y^x\int_\infty^\tau \omega_f$$
$$= \int_y^x\int_{\gamma_\tau^{-1}\infty}^\tau \omega_f - \int_y^x\int_\infty^\tau \omega_f = 0 \pmod{\Lambda},$$

and

$$\int_x^{\gamma\gamma_\tau\gamma^{-1}x}\int_\infty^{\gamma\tau} \omega_f = \int_{\gamma^{-1}x}^{\gamma_\tau\gamma^{-1}x}\int_{\gamma^{-1}\infty}^\tau \omega_f = \int_x^{\gamma_\tau x}\int_\infty^\tau \omega_f \pmod{\Lambda},$$

show that $\xi(\gamma_\tau)$ depends only on the Γ-conjugacy class of Ψ, and not on the choice of x, and later of ξ, that were made to define it. Define

$$J_\Psi := \xi(\gamma_\tau) = \sum_{e:v\to\gamma_\tau v}\int_\infty^\tau 2\pi i f_e(t)dt, \tag{4–4}$$

where v is any choice of vertex. Note that the rightmost term in (4–4) is independent of the choice of v.

Write $p\mathcal{O}_0 = \mathfrak{p}\bar{\mathfrak{p}}$ and let ϖ be a uniformizer for $\mathfrak{p}\otimes\mathbb{Z}_p = \varpi(\mathcal{O}_0\otimes\mathbb{Z}_p)$.

LEMMA 4.6. *Any class* $\Psi \in \mathrm{Emb}(\mathcal{O}, R)$ *has a representative* Ψ_0 *whose restriction to* $\mathcal{O}_0$ *is an oriented optimal embedding (in the sense of Section 1) of* $\mathcal{O}_0$ *into* $M_0(N)$ *characterized by the property that* v^o *and* $\Psi(\varpi)v^o$ *are adjacent vertices connected by* e^o.

PROOF. Let e be any edge of $\mathcal{T}_p$ that lies on the path joining the two fixed points of $\Psi(K_{\mathfrak{p}}^\times)$ acting on $\mathbb{P}_1(\mathbb{Q}_p)$. By interchanging the fixed points if necessary, there exists an element $\gamma \in \Gamma$ such that $\gamma e = e^o$. The embedding $\Psi_0 := \gamma\Psi\gamma^{-1}$ yields the desired optimal embedding of $\mathcal{O}_0$ into $M_0(N)$. The lattice $\Psi(\varpi)\cdot\mathbb{Z}_p^2$ is homothetic to

$$\mathbb{Z}_p\begin{pmatrix}\tau_{\Psi_0}\\ 1\end{pmatrix} + \mathbb{Z}_p\begin{pmatrix}p\\ 0\end{pmatrix},$$

which is in the class of the vertex adjacent v^o on the end to τ_{Ψ_0}, and so the result follows. □

For $\Psi_0 \in \mathrm{Emb}(\mathcal{O}_0, \mathrm{M}_0(N))$, let J_{0,Ψ_0} denote the period attached to it in Section 1 (where it was denoted J_{Ψ_0}). Recall the sign w introduced before the statement of Lemma 3.3, and let t be twice the order of $\mathfrak{p}^2$ in $\mathrm{Pic}(\mathcal{O}_0)$.

Proposition 4.7. *Suppose* $\Psi \in \mathrm{Emb}(\mathcal{O}, R)$ *and let* $\Psi_0 \in \mathrm{Emb}(\mathcal{O}_0, \mathrm{M}_0(N))$ *be as in Lemma* 4.6. *Then*

$$J_\Psi = \sum_{j=0}^{t-1} w^j J_{0,\Psi_0^{\mathfrak{p}^j}} \pmod{\Lambda}.$$

Proof. Let $v_0, v_1, \ldots, v_t$ denote the consecutive vertices on the path joining v_0 to $v_t := \gamma_\Psi v_0$, and let e_j (for $0 \le j \le t-1$) denote the edge joining v_j to v_{j+1}. Let R_{v_0,v_t} denote the subring of $M_2(\mathbb{Q}_p)$ consisting of the matrices which fix both v_0 and v_t. It is a local Eichler order of level t in $M_2(\mathbb{Z}_p)$, and it fixes pointwise all the vertices $v_0, \ldots, v_t$ as well as the edges $e_0, \ldots, e_{t-1}$. Note that

$$\Psi(\varpi)(v_j) = v_{j+1}, \qquad \text{for } j = 0, \ldots, t.$$

By strong approximation, there exists an element $\gamma_* \in R_+^\times$ whose image in $\mathrm{PGL}_2(\mathbb{Q}_p)/R_{v_0,v_t}^\times$ is equal to that of $\Psi(\varpi)$. In particular,

$$\gamma_*(e_j) = e_{j+1}, \qquad \text{for } j = 0, \ldots, t-1.$$

By definition,

$$J_\Psi = \sum_{j=0}^{t-1} \int_\infty^{\tau_{\Psi_0}} f_{e_j}(z)\, dz \pmod{\Lambda}. \tag{4–5}$$

But

$$\int_\infty^{\tau_{\Psi_0}} f_{e_j}(z)\, dz = \int_\infty^{\tau_{\Psi_0}} f_{\gamma_*^j e_0}(z)\, dz = w^j \int_\infty^{\gamma_*^{-j}\tau_{\Psi_0}} f_0(z)\, dz \pmod{\Lambda}, \tag{4–6}$$

where the last equality follows from Lemma 3.3. On the other hand, by the description of the action of $\mathrm{Pic}(\mathcal{O}_0)$ on $\mathrm{Emb}(\mathcal{O}_0, R_0)$ given in Lemma 1.3, we have

$$\int_\infty^{\gamma_*^{-j}\tau_{\Psi_0}} f_0(z)\, dz = J_{0,\Psi_0^{\mathfrak{p}^j}}. \tag{4–7}$$

The proposition follows by combining equations (4–5), (4–6) and (4–7). □

Let H_0 denote the ring class field of K attached to the $\mathbb{Z}$-order $\mathcal{O}_0$, and let $H := H_0^{\mathrm{Frob}_\mathfrak{p}^2}$ be the "p-narrow" class field. The reciprocity map of class field theory gives a canonical identification

$$\mathrm{rec} : \mathrm{Pic}^{p+}(\mathcal{O}) \longrightarrow \mathrm{Gal}(H/K)$$

where $\mathrm{Pic}^{p+}(\mathcal{O})$ is defined just as the usual class group $\mathrm{Pic}(\mathcal{O})$, except that equivalence is only up to scalars whose norm has even valuation at p. Letting

η be the Weierstrass uniformisation of E as before, one obtains the following analogue of Theorem 1.4:

THEOREM 4.8. *For all* Ψ *in* $\mathrm{Emb}(\mathcal{O}, R)$, *the point* $\eta(J_\Psi)$ *belongs to* $E(H)$. *More precisely, for all* $\mathfrak{b} \in \mathrm{Pic}^{p+}(\mathcal{O})$,

$$\eta(J_{\Psi^{\mathfrak{b}}}) = \mathrm{rec}(\mathfrak{b})^{-1} \cdot \eta(J_\Psi).$$

It is worth noting that Proposition 4.7 implies that

$$\eta(J_\Psi)^{\mathrm{Frob}_p} = w\, \eta(J_\Psi),$$

so that the global point $\eta(J_\Psi)$ belongs to the w-eigenspace for the involution Frob_p acting on $E(H)$.

5. p-adic Periods and Stark–Heegner Points

We now modify the discussion of the previous section by interchanging the roles of the places p and ∞.

The first stage in carrying out this program is to replace the $\mathbb{C}$-valued double integral of equation (4–1) by a $\mathbb{C}_p$-valued integral

$$\int_{\tau_1}^{\tau_2}\int_{\tau_3}^{\tau_4} \omega_f \overset{?}{\in} \mathbb{C}_p \qquad (\tau_1, \tau_2 \in \mathcal{H}_p, \quad \tau_3, \tau_4 \in \mathcal{H}^*). \tag{5–1}$$

A natural way of doing so is to start by using the complex endpoints of integration τ_3, τ_4 to convert the form-valued harmonic cocycle $e \mapsto f_e$ of Section 4 into a $\mathbb{C}$-valued harmonic cocycle $\tilde{\kappa}_f\{\tau_3 \to \tau_4\}$ on $\mathcal{T}_p$ by the rule:

$$\tilde{\kappa}_f\{\tau_3 \to \tau_4\}(e) := \int_{\tau_3}^{\tau_4} f_e(z)\, dz. \tag{5–2}$$

The resulting distribution, denoted $\tilde{\mu}_f\{\tau_3 \to \tau_4\}$, can be meaningfully integrated against *locally constant* $\mathbb{C}$-valued functions on $\mathbb{P}_1(\mathbb{Q}_p)$. Taking one's cue from equation (2–6) of Section 2 defining Coleman's p-adic line integral, it would be tempting to define

$$\int_{\tau_1}^{\tau_2}\int_{\tau_3}^{\tau_4} \omega_f \overset{?}{:=} \int_{\mathbb{P}_1(\mathbb{Q}_p)} \log\left(\frac{t-\tau_2}{t-\tau_1}\right) d\tilde{\mu}_f\{\tau_3 \to \tau_4\}(t), \tag{5–3}$$

were it not for the fact that the integrand is $\mathbb{C}_p$ and not $\mathbb{C}$-valued, and is merely locally analytic, not locally constant. To make sense of (5–3), it is necessary to modify the definition of the complex distribution $\tilde{\mu}_f\{\tau_3 \to \tau_4\}$ so that its values become p-adic, and bounded. This can be achieved, at the cost of restricting the complex endpoints τ_3, τ_4 to lie in the *boundary* $\mathbb{P}_1(\mathbb{Q})$ of the extended Poincaré upper half plane $\mathcal{H}^*$. Once this is done, the distribution $\tilde{\mu}_f\{\tau_3 \to \tau_4\}$ acquires the desired integrality properties, since it arises from a cocycle that takes values in the $\mathbb{Z}$-lattice Λ that occurs in the statement of Theorem 4.1 of Section 4. (Compare part 1 of Lemma 4.2.)

Motivated by this lemma, assume henceforth that τ_3, τ_4 belong to $\mathbb{P}_1(\mathbb{Q})$, so that $\tilde{\mu}_f\{\tau_3 \to \tau_4\}$ can be viewed as a Λ-valued p-adic measure.

For the sake of notational simplicity, let $\beta : \Lambda \longrightarrow \mathbb{Z}$ be the surjective group homomorphism which is zero on $(i\mathbb{R} \cap \Lambda)$ and maps positive real numbers in $(\Lambda \cap \mathbb{R})$ to positive integers. (It may be noted that any other choice of β will do equally well for our purposes.) Using β, define a $\mathbb{Z}_p$-valued measure on $\mathbb{P}_1(\mathbb{Q}_p)$ by the rule

$$\mu_f\{\tau_3 \to \tau_4\}(U) := \beta(\tilde{\mu}_f\{\tau_3 \to \tau_4\}(U)). \tag{5–4}$$

Equation (5–3), with $\tilde{\mu}_f$ replaced by μ_f, now makes perfect sense. Because the measure $\mu_f\{\tau_3 \to \tau_4\}$ is $\mathbb{Z}$-valued and not just $\mathbb{Z}_p$-valued, it is possible to apply the same multiplicative refinement as in the definition (2–6) of the multiplicative Coleman integral of Section 2 to define

$$\times\!\!\!\!\!\int_{\tau_1}^{\tau_2} \int_{\tau_3}^{\tau_4} \omega_f := \times\!\!\!\!\!\int_{\mathbb{P}_1(\mathbb{Q}_p)} \left(\frac{t-\tau_2}{t-\tau_1}\right) d\mu_f\{\tau_3 \to \tau_4\}(t). \tag{5–5}$$

Mimicking the definitions in Section 4, let $\tilde{d}_{\tau,x} \in Z^2(\Gamma, \mathbb{C}_p^\times)$ be the two-cocycle obtained as in equation (4–3) by choosing base points $\tau \in \mathcal{H}_p$ and $x \in \mathbb{P}_1(\mathbb{Q})$ and setting

$$\tilde{d}_{\tau,x}(\gamma_1, \gamma_2) := \times\!\!\!\!\!\int_{\tau}^{\gamma_1\tau} \int_{\gamma_1 x}^{\gamma_1\gamma_2 x} \omega_f. \tag{5–6}$$

Let $d \in H^2(\Gamma, \mathbb{C}_p^\times)$ be the cohomology class of $\tilde{d}_{\tau,x}$. A direct calculation shows that this class does not depend on the choices of base points τ and x that were made to define $\tilde{d}_{\tau,x}$.

It will be convenient in some of the calculations to relate the class d to an element in the first cohomology of Γ with values in a module of so-called M-symbols. If A is any abelian group, a function $m\{\ ,\ \} : \mathbb{P}_1(\mathbb{Q}) \times \mathbb{P}_1(\mathbb{Q}) \longrightarrow A$ denoted $(x, y) \mapsto m\{x \to y\}$ is called an (A-valued) M-symbol if it satisfies

$$m\{x \to y\} + m\{y \to z\} = m\{x \to z\},$$

for all $x, y, z \in \mathbb{P}_1(\mathbb{Q})$. Let $\mathcal{M}(A)$ (resp. $\mathcal{F}(A)$) denote the left Γ-module of A-valued M-symbols (resp. functions) on $\mathbb{P}_1(\mathbb{Q})$. In the special case where $A = \mathbb{C}_p$, it is suppressed from the notations, so that $\mathcal{M}$ and $\mathcal{F}$ denote the modules of $\mathbb{C}_p$ valued M-symbols and functions respectively.

The map $\Delta : \mathcal{F} \longrightarrow \mathcal{M}$ defined by

$$(\Delta f)\{x \to y\} := f(y) - f(x)$$

is surjective and has as kernel the space of constant functions. Taking the cohomology of the short exact sequence of $\mathbb{C}_p[\Gamma]$-modules

$$0 \longrightarrow \mathbb{C}_p \longrightarrow \mathcal{F} \xrightarrow{\Delta} \mathcal{M} \longrightarrow 0 \tag{5–7}$$

yields a long exact sequence in cohomology:

$$H^1(\Gamma, \mathcal{F}) \longrightarrow H^1(\Gamma, \mathcal{M}) \xrightarrow{\delta} H^2(\Gamma, \mathbb{C}_p) \longrightarrow H^2(\Gamma, \mathcal{F}) \tag{5–8}$$

and likewise with $\mathbb{C}_p$ replaced by any abelian group A.

If one fixes $\tau \in \mathcal{H}_p$ and defines

$$\tilde{c}_\tau(\gamma)\{x \to y\} := \times\!\!\!\!\!\!\int_\tau^{\gamma\tau} \int_x^y \omega_f,$$

then $\tilde{c}_\tau$ is an $\mathcal{M}(\mathbb{C}_p^\times)$-valued one-cocycle. A direct computation shows that its class $c \in H^1(\Gamma, \mathcal{M}(\mathbb{C}_p^\times))$ is independent of the choice of τ and satisfies

$$\delta(c) = d. \tag{5–9}$$

As is explained in [Da], sec. 3.1, the cohomology groups $H^1(\Gamma, \mathcal{M})$ and $H^2(\Gamma, \mathbb{C}_p)$ are finite-dimensional $\mathbb{C}_p$-vector spaces equipped with an action of the algebra $\mathbb{T}$ of Hecke operators as well as an action of an involution W_∞ induced by conjugation by the matrix $\begin{pmatrix} 1 & 0 \\ 0 & -1 \end{pmatrix}$. Let $H^1(\Gamma, \mathcal{M})^{f,+}$ denote the f isotypic part of $H^1(\Gamma, \mathcal{M})$ fixed by W_∞, and similarly for $H^2(\Gamma, \mathbb{C}_p)$.

Lemma 5.1. *The groups $H^1(\Gamma, \mathcal{M})^{f,+}$ and $H^2(\Gamma, \mathbb{C}_p)^{f,+}$ are one-dimensional $\mathbb{C}_p$-vector spaces.*

Proof. All the maps in the exact sequence (5–8) are equivariant under the natural actions of the Hecke operators and under the action of the involution W_∞. Furthemore, as explained in [Da], sec. 3.1, the modules $H^j(\Gamma, \mathcal{F})^{f,+}$ are trivial. Hence δ induces an isomorphism

$$\delta : H^1(\Gamma, \mathcal{M})^{f,+} \longrightarrow H^2(\Gamma, \mathbb{C}_p)^{f,+}.$$

The result now follows from [Da], corollary 3.3. □

Note that the assumption that p divides N exactly implies that E has (either split or nonsplit) multiplicative reduction at p, so that it is equipped with the Tate p-adic uniformisation

$$\eta_p : \mathbb{C}_p^\times / q^Z \longrightarrow E(\mathbb{C}_p)$$

where $q \in p\mathbb{Z}_p$ is Tate's p-adic period attached to $E_{/\mathbb{Q}_p}$. Let $w = \pm 1$ be the sign that was introduced earlier: thus $w = 1$ if E has split multiplicative reduction at p, and $w = -1$ if E has nonsplit multiplicative reduction.

The following proposition, referred to from now on as the p-adic period theorem, is a natural analoge of Proposition 4.3 (combined with Theorem 4.1) of Section 4.

Theorem 5.2. *There exists a lattice $\Lambda_p \subset \mathbb{C}_p^\times$ commensurable with $q^{\mathbb{Z}}$ and such that the natural images of d in $H^2(\Gamma, \mathbb{C}_p^\times/\Lambda_p)$ and of c in $H^1(\Gamma, \mathcal{M}(\mathbb{C}_p^\times/\Lambda_p))$ are trivial.*

Sketch of proof. Let $\mathrm{ord}_p : \mathbb{C}_p^\times \longrightarrow \mathbb{Q} \subset \mathbb{C}_p$ be the usual normalised p-adic valuation and let $\log_p : \mathbb{C}_p^\times \longrightarrow \mathbb{C}_p$ be a branch of the p-adic logarithm, chosen, to fix ideas, in such a way that $\log_p(p) = 0$. The one-cocycles $\mathrm{ord}_p(c)$ and $\log_p(c)$

each belong to $H^1(\Gamma,\mathcal{M})^{f,+}$, and $\mathrm{ord}_p(c)$ is nonzero (see [Da, Lemma 3.4]). By Lemma 5.1, there exists a constant $\mathfrak{L}$ in $\mathbb{C}_p$ such that

$$\log_p(\tilde{c}_\tau) = \mathfrak{L}\cdot \mathrm{ord}_p(\tilde{c}_\tau) \pmod{B^1(\Gamma,\mathcal{M})}. \tag{5–10}$$

To evaluate the constant $\mathfrak{L}$, let n be any positive integer which is relatively prime to N and let j be the least integer such that $p^{2j}\equiv 1 \pmod{n}$. If ν is in $(\mathbb{Z}/n\mathbb{Z})^\times$, let

$$\gamma_\nu := \begin{pmatrix} p^j & \nu\frac{p^j-p^{-j}}{n} \\ 0 & p^{-j}\end{pmatrix}$$

be the generator of the stabiliser in Γ of $(\infty,\nu/c)\in\mathbb{P}_1(\mathbb{Q})\times\mathbb{P}_1(\mathbb{Q})$. Note that for all 1-coboundaries $b\in B^1(\Gamma,\mathcal{M})$,

$$b(\gamma_\nu)\{\infty\to\nu/n\}=0.$$

Hence for all $\nu\in(\mathbb{Z}/n\mathbb{Z})^\times$, equation (5–10) implies

$$\log(\tilde{c}_\tau)(\gamma_\nu)\{\infty\to\nu/n\} = \mathfrak{L}\cdot\mathrm{ord}_p(\tilde{c}_\tau)(\gamma_\nu)\{\infty\to\nu/n\}. \tag{5–11}$$

Let $\chi:(\mathbb{Z}/n\mathbb{Z})^\times \longrightarrow \mathbb{C}_p^\times$ be a Dirichlet character of conductor n, viewed as taking values in $\mathbb{C}_p$, and suppose that $\chi(p)=w$, and that $\chi(-1)=1$. In [Da, Prop. 2.16], it is shown that

$$\sum_{\nu\in(\mathbb{Z}/n\mathbb{Z})^\times}\chi(\nu)\mathrm{ord}_p(\tilde{c}_\tau)(\gamma_\nu)\{\infty\to\nu/n\} = \frac{2jn}{\tau(\chi)}\frac{L(E/\mathbb{Q},\chi,1)}{\Omega^+}, \tag{5–12}$$

where $\tau(\chi)$ is the Gauss sum attached to χ and Ω^+ is an appropriate rational multiple of the real period of E. An independent calculation, which follows by combining Proposition 2.18 of [Da] with the definition of the Mazur–Swinnerton-Dyer p-adic L-function $L_p(E/\mathbb{Q},s)$ attached to $E/\mathbb{Q}$ shows that

$$\sum_{\nu\in(\mathbb{Z}/n\mathbb{Z})^\times}\chi(\nu)\log_p(\tilde{c}_\tau)(\gamma_\nu)\{\infty\to\nu/n\} = L'_p(E/\mathbb{Q},\chi,1). \tag{5–13}$$

The two equations (5–12) and (5–13) combined with (5–11) imply that the discrepancy between the (algebraic part of the) classical L-value $L(E/\mathbb{Q},\chi,1)$ and the first derivative $L'_p(E/\mathbb{Q},\chi,1)$, for even Dirichlet characters χ for which $\chi(p)=w$, is given by a constant $\mathfrak{L}$ which is *independent* of χ, a conjecture that was made by Mazur, Tate and Teitelbaum [MTT]. The article [MTT] went further than this, conjecturing on the basis of numerical evidence that

$$\mathfrak{L} = \frac{\log_p(q)}{\mathrm{ord}_p(q)}, \tag{5–14}$$

where q is the Tate p-adic period attached to $E/\mathbb{Q}_p$. This conjecture was later proved by Greenberg and Stevens [GS]. This crucial result — which is the principal ingredient in the proof of Theorem 5.2 — shows that

$$\log(c) = \frac{\log_p(q)}{\mathrm{ord}_p(q)}\mathrm{ord}_p(c), \quad \log(d) = \frac{\log_p(q)}{\mathrm{ord}_p(q)}\mathrm{ord}_p(d), \tag{5–15}$$

where the second equality follows from equation (5–9) relating the classes c and d. Now consider the two-cocycle

$$\tilde{e}_{\tau,x} := \tilde{d}_{\tau,x}^{\mathrm{ord}_p(q)} \div q^{\mathrm{ord}_p(\tilde{d}_{\tau,x})} \in Z^2(\Gamma, \mathbb{C}_p^\times). \tag{5–16}$$

and let e be its natural image in $H^2(\Gamma, \mathbb{C}_p^\times)$. The second part of equation (5–15) implies that

$$\mathrm{ord}_p(e) = \log_p(e) = 0,$$

so that e is torsion. Hence there exists an integer constant a such that

$$\tilde{d}_{\tau,x}^{a\cdot\mathrm{ord}_p(q)} = q^{a\cdot\mathrm{ord}_p(\tilde{d}_{\tau,x})} \qquad (\mathrm{mod}\ B^2(\Gamma, \mathbb{C}_p^\times)). \tag{5–17}$$

Theorem 5.2 follows. □

Assume, at the cost of eventually raising $\tilde{c}_\tau$ and $\tilde{d}_{\tau,x}$ to a common integer power, that $\Lambda_p = q^{\mathbb{Z}}$. Thanks to Theorem 5.2, one may as in Section 4 define an M-symbol $m_\tau \in C^0(\Gamma, \mathcal{M}(\mathbb{C}_p^\times/q^{\mathbb{Z}}))$ and a one-cochain $\xi_{\tau,x} \in C^1(\Gamma, \mathbb{C}_p^\times/q^{\mathbb{Z}})$ by the rules

$$\tilde{c}_\tau = dm_\tau, \quad \tilde{d}_{\tau,x} = d\xi_{\tau,x}.$$

It is useful to adopt the notation

$$⨍^\tau \int_x^y \omega_f := m_\tau\{x \to y\} \in \mathbb{C}_p^\times/q^{\mathbb{Z}},$$

which can be justified by noting that this "semi-indefinite" integral satisfies the identities

$$⨍^\tau \int_x^y \omega_f \times ⨍^\tau \int_y^z \omega_f = ⨍^\tau \int_x^z \omega_f \quad \text{for all } x, y, z \in \mathbb{P}_1(\mathbb{Q}),$$

$$⨍^{\gamma\tau} \int_{\gamma x}^{\gamma y} \omega_f = ⨍^\tau \int_x^y \omega_f \quad \text{for all } \gamma \in \Gamma,$$

$$⨍^{\tau_2} \int_x^y \omega_f \div ⨍^{\tau_1} \int_x^y \omega_f = ⨍_{\tau_1}^{\tau_2} \int_x^y \omega_f \quad \text{for all } x, y \in \mathbb{P}_1(\mathbb{Q}),\ \tau_1, \tau_2 \in \mathcal{H}_p.$$

Let $\mathcal{O}$ be a $\mathbb{Z}[1/p]$-order in the real quadratic field K. The notion of optimal embedding $\Psi : K \longrightarrow M_2(\mathbb{Q})$ is defined very much as in Sections 2 and 4. That is, we say that Ψ is optimal if $\Psi(K) \cap R = \Psi(\mathcal{O})$.

One likewise defines a notion of *oriented optimal embedding* by introducing orientations on $\mathcal{O}$ and on R (defined as ring homomorphisms to $\mathbb{Z}/M\mathbb{Z}$) and requiring that the embedding be compatible with these orientations. One requires in addition that the (unique) vertex on $\mathcal{T}_p$ fixed by $\Psi(K^\times)$ is $\mathrm{SL}_2(\mathbb{Q}_p)$-equivalent to v_0, i.e., its distance from this vertex is even. (This last requirement can be viewed as a condition of "orientation at p".)

We can then define, just as in Section 4:

$$J_\Psi := \xi_{\tau,x}(\gamma_\tau) = \times\!\!\!\!\!\int^{\tau}\int_x^{\gamma_\tau x} \omega_f \in \mathbb{C}_p^\times/q^{\mathbb{Z}}.$$

LEMMA 5.3. *The period J_Ψ depends only on the Γ-conjugacy class of τ (and therefore of Ψ), not on the choices of x and then $\xi_{\tau,x}$ made to define it.*

PROOF. This follows from manipulations identical to those that were used to prove the corresponding statement in Section 4. □

Let H denote the narrow ring class field of K attached to the order $\mathcal{O}$, (or, equivalently, to the $\mathbb{Z}$-order $\mathcal{O}_0 \subset \mathcal{O}$, since the prime p is inert in $K/\mathbb{Q}$ and hence splits completely in H/K). The reciprocity map of class field theory gives a canonical identification

$$\mathrm{rec} : \mathrm{Pic}^+(\mathcal{O}) \longrightarrow \mathrm{Gal}(H/K).$$

Here $\mathrm{Pic}^+(\mathcal{O})$ is defined just as the usual class group $\mathrm{Pic}(\mathcal{O})$, except that two fractional $\mathcal{O}$-ideals are said to be equivalent if they differ by multiplication by an element of positive norm.

Let η_p be the Tate uniformisation of E of equation (2–13). The counterpart of Theorem 4.8 in this setting is given by:

CONJECTURE 5.4. *For all Ψ in $\mathrm{Emb}(\mathcal{O}, R)$, the point $\eta_p(J_\Psi)$ belongs to $E(H)$. More precisely, for all $\mathfrak{b} \in \mathrm{Pic}^+(\mathcal{O})$,*

$$\eta(J_{\Psi^{\mathfrak{b}}}) = \mathrm{rec}(\mathfrak{b})^{-1} \cdot \eta(J_\Psi).$$

Note that, thanks to the change of notations and definitions entailed by our passage from a $\mathbb{C}$ to a $\mathbb{C}_p^\times$-valued integral and our change of quadratic field K, the statement of conjecture 5.4 is identical to that of Theorem 4.8. Yet the former appears to lie deeper than the latter, and no proof is known for it. What is lacking is an analogue of the theory outlined in Section 1, which was instrumental in proving Theorem 4.8.

6. Heegner Points and Integration on $\mathcal{H}_p \times \mathcal{H}_q$

This section considers an analogue of the conjecture formulated in Section 5, where one replaces the archimedean upper half plane $\mathcal{H}$ by a nonarchimedean upper half plane $\mathcal{H}_q$. Here one is led to revisit the setting of Section 2, since the group acting on the product $\mathcal{H}_p \times \mathcal{H}_q$ is defined in terms of a definite quaternion algebra B which is split at p and q. The period integral arises from a quadratic algebra K which embeds in B (so that K is an imaginary quadratic field), such that p is inert and q is split in K. In this case the analogue of the above mentioned conjecture can be reduced to Theorem 2.9 and therefore be proven.

Forms on $\mathcal{T}_p \times \mathcal{H}_q$. Assume that the conductor N of E can be written in the form

$$N = pqN^+N^-,$$

where p and q are primes, the factors are pairwise relatively prime, and N^- is squarefree and divisible by an odd number of prime factors. It follows that E has multiplicative reduction at p and q. Moreover, the above factorisation is of the type (2–1) considered in Section 2, in which the integer currently denoted N^+q was written N^+.

Let B be the (definite) quaternion algebra over $\mathbb{Q}$ of discriminant N^-, let R be an Eichler $\mathbb{Z}[1/p, 1/q]$-order in B of level N^+, and let $R_1^\times$ be the group of norm 1 elements in $R^\times$. Fix a embedding ι of B into $M_2(\mathbb{Q}_p) \times M_2(\mathbb{Q}_q)$, and define a group

$$\Gamma := \iota(R_1^\times) \subset \mathrm{SL}_2(\mathbb{Q}_p) \times \mathrm{SL}_2(\mathbb{Q}_q).$$

Write $\gamma = (\gamma_p, \gamma_q)$ for the elements of Γ. The group Γ acts discretely and properly discontinuously on the product $\mathcal{H}_p \times \mathcal{H}_q$ of the p-adic and q-adic upper half planes, by the rule $\gamma(\tau_p, \tau_q) = (\gamma_p\tau_p, \gamma_q\tau_q)$. It follows that Γ acts naturally also on the product $\mathcal{E}(\mathcal{T}_p) \times \mathcal{H}_q$.

DEFINITION 6.1. A cusp form of weight 2 on $(\mathcal{T}_p \times \mathcal{H}_q)/\Gamma$ is a function

$$f : \mathcal{E}(\mathcal{T}_p) \times \mathcal{H}_q \longrightarrow \mathbb{C}_q$$

satisfying

1. $f(\gamma_p e, \gamma_q z) = (cz+d)^2 f(e,z)$ for all $\gamma \in \Gamma$ with $\gamma_q = \begin{pmatrix} a & b \\ c & d \end{pmatrix}$.
2. For each $z \in \mathcal{H}_q$, the function $f_z(e) := f(e,z)$ is a $\mathbb{C}_q$-valued harmonic cocycle on $\mathcal{E}(\mathcal{T}_p)$.
3. For each $e \in \mathcal{E}(\mathcal{T}_p)$, the function $f_e(z) := f(e,z)$ is a q-adic rigid analytic modular form (see Section 2) on the stabilizer Γ_e of e in Γ.

Denote by $S_2((\mathcal{T}_p \times \mathcal{H}_q)/\Gamma)$ the $\mathbb{C}_q$-vector space of weight 2 cusp forms on $(\mathcal{T}_p \times \mathcal{H}_q)/\Gamma$. Define a base edge e^o in $\mathcal{E}(\mathcal{T}_p)$ by the property that Γ_{e^o} be equal via ι to the group Γ_q of norm 1 elements in a fixed Eichler $\mathbb{Z}[1/q]$-order in R of level pN^+. Denote by $S_2^{p\text{-new}}(\mathcal{H}_q/\Gamma_q)$ the subspace of weight 2 rigid analytic q-adic modular forms on Γ_q which are new at p.

LEMMA 6.2. *The assignment sending f to f_{e^o} induces an isomorphism from $S_2((\mathcal{T}_p \times \mathcal{H}_q)/\Gamma)$ to $S_2^{p\text{-new}}(\mathcal{H}_q/\Gamma_q)$.*

PROOF. The ideas in this proof are identical to those in the proof of Proposition 3.2. □

Note that the spaces appearing in Lemma 6.2 are equipped with the natural action of a Hecke algebra $\mathbb{T}$.

By combining Lemma 6.2 with Theorem 2.2, one obtains that the classical modular form f_E attached to E corresponds to a form f on $(\mathcal{T}_p \times \mathcal{H}_q)/\Gamma$. The form f is determined up to sign by the requirement that f_{e^o} be normalized as in Section 2.

Let $\mathcal{H}_q^{\mathrm{nr}}$ be the unramified q-adic upper half plane defined in Section 2. Given x_1 and x_2 in $\mathcal{H}_q^{\mathrm{nr}}$, reducing to v_1 and v_2 respectively, define a $\mathbb{Z}$-valued harmonic cocycle on $\mathcal{E}(\mathcal{T}_p)$ by the rule

$$\kappa_f\{x_1 \to x_2\}(e) = \sum_{\varepsilon : v_1 \to v_2} \kappa_{f_e}(\varepsilon), \tag{6–1}$$

where κ_{f_e} denotes the Γ_e-invariant harmonic cocycle attached to the q-adic modular form f_e as in Section 2, and the sum is taken over the set of edges $\varepsilon \in \mathcal{E}(\mathcal{T}_q)$ in the path joining v_1 to v_2. Write $\mu_f\{x_1 \to x_2\}$ for the measure on $\mathbb{P}_1(\mathbb{Q}_p)$ attached to $\kappa_f\{x_1 \to x_2\}$, defined on the compact open sets U_e by the equality

$$\int_{U_e} d\mu_f\{x_1 \to x_2\}(t) = \kappa_f\{x_1 \to x_2\}(e).$$

Let τ_1 and τ_2 be points in $\mathcal{H}_p$. The next definition is motivated by the use of Teitelbaum's theorem in Section 2: consider in particular formula (2–6). (See also the definitions of Sections 4 and 5.)

DEFINITION 6.3. Define the double integral

$$⨰_{\tau_1}^{\tau_2}\int_{x_1}^{x_2} \omega_f = ⨰_{\mathbb{P}_1(\mathbb{Q}_p)} \left(\frac{t-\tau_2}{t-\tau_1}\right) d\mu_f\{x_1 \to x_2\}(t) \in \mathbb{C}_p^\times. \tag{6–2}$$

Formula (6–2) should be regarded as the definition of the double definite integral of a "form ω_f" on $\mathcal{H}_p \times \mathcal{H}_q$. Such an integral can be defined solely in terms of the system of p-adic residues of ω_f, which is described in Definition 6.1.

Consider the assignment $(e, \varepsilon) \mapsto \kappa_{f_e}(\varepsilon)$: it defines a Γ-invariant harmonic cocycle

$$\lambda_f : \mathcal{E}(\mathcal{T}_p) \times \mathcal{E}(\mathcal{T}_q) \to \mathbb{Z}.$$

Note that λ_f completely determines the form f. More generally, the following statement holds. Let $S_2((\mathcal{T}_p \times \mathcal{T}_q)/\Gamma)_R$ denote the module of Γ-invariant harmonic cocycles on $\mathcal{E}(\mathcal{T}_p) \times \mathcal{E}(\mathcal{T}_q)$ with values in a ring R.

LEMMA 6.4. *The map from $S_2((\mathcal{T}_p \times \mathcal{H}_q)/\Gamma)$ to $S_2((\mathcal{T}_p \times \mathcal{T}_q)/\Gamma)_{\mathbb{C}_q}$, sending a form on $(\mathcal{T}_p \times \mathcal{H}_q)/\Gamma$ to a $\mathbb{C}_q$-valued harmonic cocycle on $(\mathcal{T}_p \times \mathcal{T}_q)/\Gamma$ via the natural residue map, is a Hecke-equivariant isomorphism of $\mathbb{C}_q$-vector spaces.*

PROOF. It follows from Lemma 6.2 combined with the theory of residues explained in Section 2. □

By interchanging the role of p and q, one defines the concept of a weight two cusp form

$$g : \mathcal{H}_p \times \mathcal{E}(\mathcal{T}_q) \longrightarrow \mathbb{C}_p$$

on $(\mathcal{H}_p \times \mathcal{T}_q)/\Gamma$. Let

$$\lambda_g : \mathcal{E}(\mathcal{T}_p) \times \mathcal{E}(\mathcal{T}_q) \to \mathbb{C}_p$$

denote the Γ-invariant harmonic cocycle attached to g by the rule $(e,\varepsilon) \mapsto \kappa_{g_\varepsilon}(e)$.

COROLLARY 6.5. *There exists a weight two cusp form $f^\sharp$ on $(\mathcal{H}_p \times \mathcal{T}_q)/\Gamma$ such that $\lambda_f = \lambda_{f^\sharp}$ (where f is the weight two cusp form on $(\mathcal{T}_p \times \mathcal{H}_q)/\Gamma$ attached to the elliptic curve E).*

PROOF. Use Lemma 6.4 together with the fact that λ_f is $\mathbb{Z}$-valued. □

In the sequel of this Section, denote by $f^\sharp$ the form satisfying the statement of corollary 6.5.

Let τ_1 and τ_2 be elements in $\mathcal{H}_p$, and let x_1 and x_2 be elements in $\mathcal{H}_q^{\mathrm{nr}}$, reducing respectively to vertices v_1 and v_2 in $\mathcal{V}(\mathcal{T}_q)$.

DEFINITION 6.6. Define the double integral

$$\int_{x_1}^{x_2} \times\!\!\!\!\!\!\int_{\tau_1}^{\tau_2} \omega_{f^\sharp} = \prod_{\varepsilon: v_1 \to v_2} \times\!\!\!\!\!\!\int_{\tau_1}^{\tau_2} f^\sharp_\varepsilon(z)\,dz \in \mathbb{C}_p^\times. \tag{6–3}$$

REMARK. Definition 6.6 is the (multiplicative) analogue in the current setting of the definition of double integral given in Section 4, whereas Definition 6.3 is the analogue of the definition of double integral given in Section 5. In the settings of Section 4 and 5, only one way of defining the double integral was available. The next result shows that equation (6–3) yields the same result as (6–2), so that one can use definition (6–3) in the arguments of the remaining part of this section.

PROPOSITION 6.7 (FUBINI'S THEOREM). *The equality*

$$\times\!\!\!\!\!\!\int_{\tau_1}^{\tau_2} \int_{x_1}^{x_2} \omega_f = \int_{x_1}^{x_2} \times\!\!\!\!\!\!\int_{\tau_1}^{\tau_2} \omega_{f^\sharp}$$

holds.

PROOF. By combining Definition 6.6 with Teitelbaum's Theorem 2.4, one obtains

$$\int_{x_1}^{x_2} \times\!\!\!\!\!\!\int_{\tau_1}^{\tau_2} \omega_{f^\sharp} = \prod_{\varepsilon: v_1 \to v_2} \times\!\!\!\!\!\!\int_{\mathbb{P}_1(\mathbb{Q}_p)} \frac{z-\tau_2}{z-\tau_1}\,d\mu_{f^\sharp_\varepsilon}(z). \tag{6–4}$$

In view of Definition 6.3 and equation (6–4), it suffices to show that

$$d\mu_f\{x_1 \to x_2\} = \sum_{\varepsilon: v_1 \to v_2} d\mu_{f^\sharp_\varepsilon}. \tag{6–5}$$

By equation (6–1), (6–5) reduces to the equality

$$\kappa_{f_e}(\varepsilon) = \kappa_{f^\sharp_\varepsilon}(e),$$

which follows from corollary 6.5. □

In order to define an analogue of the period integral J_Ψ considered in Sections 4 and 5, one may perform cohomological manipulations similar to those of these previous sections. The exposition here follows closely the arguments of Section 4, so that only the relevant changes are explicitly mentioned. Let q_T be the p-adic Tate period of E (denoted by q in the previous sections).

DEFINITION 6.8. For $\tau \in \mathcal{H}_p$ and $x, y \in \mathcal{H}_q^{\mathrm{nr}}$, define the semi-indefinite double integral

$$\int_x^y \!\!\!\!\times\!\!\!\! \int^\tau \omega_{f^\sharp} := \prod_{\varepsilon : r(x) \to r(y)} \times\!\!\!\!\!\!\int^\tau f_\varepsilon^\sharp(z)\, dz \in \mathbb{C}_p^\times / q_T^{\mathbb{Z}},$$

where the semi-indefinite integral in the right-hand side of definition 6.8 is defined in Section 2.

Recall from equation (2–14) that this definition involves a slight modification of the multiplicative definite integral.

Let $\tilde{d}_{\tau,x} \in Z^2(\Gamma, \mathbb{C}_p^\times)$ be the two-cocycle obtained by choosing base points $\tau \in \mathcal{H}_p$ and $x \in \mathcal{H}_q^{\mathrm{nr}}$, and setting

$$\tilde{d}_{\tau,x}(\gamma_1, \gamma_2) := \times\!\!\!\!\!\!\int_\tau^{\gamma_1 \tau} \int_{\gamma_1 x}^{\gamma_1 \gamma_2 x} \omega_f = \int_{\gamma_1 x}^{\gamma_1 \gamma_2 x} \times\!\!\!\!\!\!\int_\tau^{\gamma_1 \tau} \omega_{f^\sharp}. \tag{6–6}$$

The cohomology class $d \in H^2(\Gamma, \mathbb{C}_p^\times)$ of $\tilde{d}_{\tau,x}$ does not depend on the choices of base points τ and x.

PROPOSITION 6.9. *The natural image of* $\tilde{d}_{\tau,x}$ *in* $H^2(\Gamma, \mathbb{C}_p^\times / q_T^{\mathbb{Z}})$ *is zero.*

PROOF. Define a one-cochain $\xi_{\tau,x} : \Gamma \to \mathbb{C}_p^\times / q_T^{\mathbb{Z}}$ by the rule

$$\xi_{\tau,x}(\gamma) := \int_x^{\gamma x} \!\!\!\!\times\!\!\!\! \int^\tau \omega_{f^\sharp}. \tag{6–7}$$

By a direct computation similar to that of Proposition 4.3, one can show that $d\xi_{\tau,x}(\gamma_1, \gamma_2)$ is equal to the image modulo $q_T^{\mathbb{Z}}$ of $\tilde{d}_{\tau,x}(\gamma_1, \gamma_2)$. □

REMARKS. **1.** In the current setting, the analogue of the module of M-symbols studied in Section 5 is defined as follows. If A is any abelian group, a function $m\{\to\} : \mathcal{H}_q^{\mathrm{nr}} \times \mathcal{H}_q^{\mathrm{nr}} \longrightarrow A$ denoted $(x, y) \mapsto m\{x \to y\}$ is called an (A-valued) M-symbol if it factors through the reduction map $r : \mathcal{H}_q^{\mathrm{nr}} \to \mathcal{V}(\mathcal{T}_q)$ (so that it depends only on the images of x and y in $\mathcal{V}(\mathcal{T}_q)$) and satisfies

$$m\{x \to y\} + m\{y \to z\} = m\{x \to z\},$$

for all $x, y, z \in \mathcal{H}_q^{\mathrm{nr}}$. Let $\mathcal{M}(A)$ denote the left Γ-module of A-valued M-symbols on $\mathcal{H}_q^{\mathrm{nr}}$. If one fixes $\tau \in \mathcal{H}_p$ and defines

$$\tilde{c}_\tau(\gamma)\{x \to y\} := \times\!\!\!\!\!\!\int_\tau^{\gamma\tau} \int_x^y \omega_f,$$

then $\tilde{c}_\tau$ is an $\mathcal{M}(\mathbb{C}_p^\times)$-valued one-cocycle whose class $c \in H^1(\Gamma, \mathcal{M}(\mathbb{C}_p^\times))$ is independent of the choice of τ. A direct computation shows that c satisfies

$$\delta(c) = d,$$

where δ is the natural coboundary map $H^1(\Gamma, \mathcal{M}(\mathbb{C}_p^\times)) \longrightarrow H^2(\Gamma, \mathbb{C}_p^\times)$.

2. The proof of the triviality modulo $q_T^{\mathbb{Z}}$ of the cohomology class d (and also of c) given in Proposition 6.9 is based crucially on the Fubini theorem of Proposition 6.7, which allows to imitate the arguments of Section 4. An alternate strategy consists in using directly definition 6.3 of the double integral, and imitating the arguments of Section 5. This method can be carried out, although it turns out to be less elementary. It leads to the study of the relation between $\log_p(c)$ and $\mathrm{ord}_p(c)$, which requires in particular the results of Vatsal (see [Va1] and [Va2]) on the nonvanishing of certain combinations of Gross points.

The period integral. Let $K/\mathbb{Q}$ be a quadratic algebra, and assume that it satisfies the following modified Heegner hypothesis:

1. K is an imaginary quadratic field;
2. all the primes dividing N^- are inert in K;
3. all the primes dividing N^+ are split in K;
4. p is inert in K;
5. q is split in K.

Let $\mathcal{O}$ be a $\mathbb{Z}[1/pq]$-order order of K of conductor prime to N, and let R be the Eichler $\mathbb{Z}[1/pq]$-order R of level N^+ fixed at the beginning of this section. The above conditions ensure the existence of an oriented optimal embedding

$$\Psi : \mathcal{O} \to R.$$

(The notions of optimality and orientation are derived by adapting those that were introduced in Sections 2 and 4.)

Write K_p and K_q for the local algebras $K \otimes \mathbb{Q}_p$ and $K \otimes \mathbb{Q}_q$, respectively. The torus $K_p^\times$ acting on $\mathcal{H}_p$ via the embedding $\iota\Psi$ has a unique fixed point τ_Ψ normalized as in equation (2–19). The point τ_Ψ belongs to K_p. Furthermore, let $\gamma_\Psi \in \Gamma$ be the image by $\iota\Psi$ of a generator modulo torsion of the group $\mathcal{O}_1^\times$ of norm 1 elements in $\mathcal{O}^\times$.

DEFINITION 6.10. Fix an element x in $\mathcal{H}_q^{\mathrm{nr}}$. Let $\xi_{\tau,x}$ be the one-cochain defined in equation (6–7). The (multiplicative) period integral associated to the embedding Ψ is defined to be

$$J_\Psi := \xi_{\tau_\Psi,x}(\gamma_\Psi) = \int_x^{\gamma_\Psi x} \times\!\!\!\!\!\!\int^{\tau_\Psi} \omega_{f^\sharp} \in \mathbb{C}_p^\times / q_T^{\mathbb{Z}}.$$

LEMMA 6.11. *The period J_Ψ belongs to $K_p^\times / q_T^{\mathbb{Z}}$. Moreover, J_Ψ depends only on the Γ-conjugacy class of Ψ, and not on the choice of x that was made to define it.*

PROOF. The first statement follows from the calculation in the proof of Theorem 6.13 below. The second statement is proved by mimicking the arguments of the previous sections. □

The period J_Ψ is related to the period

$$I_\Psi := \times\!\!\!\!\!\!\int_{\tau_\Psi}^{\tau_\Psi} \int_x^{\gamma_\Psi x} \omega_f \in \mathbb{C}_p^\times$$

by the relation $J_\Psi/\bar{J}_\Psi = I_\Psi \pmod{q_\tau^{\mathbb{Z}}}$.

Let w_q, respectively, w_p, be equal to 1 if E has split multiplicative reduction at q, respectively, at p, and let w_q, respectively, w_p be equal to -1 otherwise. Write $\mathcal{O}_0$ for the maximal $\mathbb{Z}[1/p]$-order contained in $\mathcal{O}$, $\mathfrak{q}$ for the class in $\mathrm{Pic}(\mathcal{O}_0)$ of a prime above q, and h for the order of $\mathfrak{q}$ in $\mathrm{Pic}(\mathcal{O}_0)$.

LEMMA 6.12. *If w_q is equal to -1 and h is odd, then J_Ψ is equal to ± 1.*

PROOF. Let v be an element of $\mathcal{O}_0$ whose norm is equal to uq^h for a unit u in the ring of integers of K, and let δ_Ψ be the image of v by $\iota\Psi$. A direct calculation shows that

$$\int_{\delta_\Psi x}^{\delta_\Psi \gamma_\Psi x} \times\!\!\!\!\!\!\int^{\delta_\Psi \tau_\Psi} \omega_{f^\sharp} = \left(\int_x^{\gamma_\Psi x} \times\!\!\!\!\!\!\int^{\tau_\Psi} \omega_{f^\sharp} \right)^{w_q^h}. \tag{6–8}$$

On the other hand, one has

$$\int_{\delta_\Psi x}^{\delta_\Psi \gamma_\Psi x} \times\!\!\!\!\!\!\int^{\delta_\Psi \tau_\Psi} \omega_{f^\sharp} = \int_{\delta_\Psi x}^{\gamma_\Psi \delta_\Psi x} \times\!\!\!\!\!\!\int^{\tau_\Psi} \omega_{f^\sharp}. \tag{6–9}$$

The claim follows by combining (6–8) and (6–9) with the second statement in Lemma 6.11. □

The next theorem states that J_Ψ defines a global point on the elliptic curve E. In view of Lemma 6.12, assume from now that $w_q = 1$ when h is odd. Let H_0 denote the ring class field of the order $\mathcal{O}_0$, and let $\sigma_\mathfrak{q}$ be the element of $\mathrm{Gal}(H_0/K)$ corresponding to $\mathfrak{q}$ by the reciprocity map. Write H for the subfield of H_0 which is fixed by $\sigma_\mathfrak{q}^2$. (Thus, when h is odd, H is simply the ring class field of the order $\mathcal{O}$.) Recall the element $\sigma_\mathfrak{p} \in \mathrm{Gal}(H_0/\mathbb{Q})$ appearing in the statement of Theorem 2.9.

THEOREM 6.13. *The point $\eta_p(J_\Psi)$ is a global point P_Ψ in $E(H)$, on which the involution $\sigma_\mathfrak{q}$ acts via w_q. In particular $\eta_p(I_\Psi)$ is equal to $P_\Psi - w_p\sigma_\mathfrak{p} P_\Psi$. Furthermore, for all $\mathfrak{a} \in \mathrm{Pic}^+(\mathcal{O})$,*

$$\eta_p(J_{\Psi^\mathfrak{a}}) = \mathrm{rec}(\mathfrak{a})^{-1} \cdot \eta_p(J_\Psi).$$

REMARK. The stringent analogy between theorems 2.9 and 6.13. In fact, the proof of Theorem 6.13 reduces to Theorem 2.9.

PROOF. By Definition 6.8,

$$J_\Psi = \prod_{\varepsilon : v \to \gamma_\Psi v} \times\!\!\!\!\!\!\int^{\tau_\Psi} f_\varepsilon^\sharp(z)\, dz. \tag{6–10}$$

Let $R_0 \subset R$ be an Eichler $\mathbb{Z}[1/p]$-order of level N^+q in B. Denote by

$$\Psi_0^i, \quad i = 1, \ldots, h$$

a full set of representatives for the conjugacy classes of oriented optimal embeddings from $\mathcal{O}_0$ to R_0 which give rise, by extension of scalars, to the conjugacy class of Ψ. Note that the subgroup of $\mathrm{Pic}(\mathcal{O}_0)$ generated by $\mathfrak{q}$ permutes simply transitively the classes of the embeddings Ψ_0^i. Let $\tau_{\Psi_0^i} \in K_p$ be the normalized fixed point for the action of $K_p^\times$ on $\mathcal{H}_p$ via $\iota\Psi_0^i$. Let f_o be the rigid analytic modular form on $\mathcal{H}_p$ corresponding to $f^\sharp$. By a calculation similar to the proof of proposition 4.7, one finds that if h is even, the right hand side of (6–10) can be written as

$$J_\Psi = \prod_{i=1}^{h} \left(\times\!\!\!\!\!\!\int^{\tau_{\Psi_0^i}} f_o(z)\, dz \right)^{w_q^i}. \tag{6–11}$$

Likewise, if h is odd (so that $w_q = 1$ by our assumptions) one finds the formula

$$J_\Psi = \prod_{i=1}^{h} \left(\times\!\!\!\!\!\!\int^{\tau_{\Psi_0^i}} f_o(z)\, dz \right)^{2}. \tag{6–12}$$

But the factors

$$\times\!\!\!\!\!\!\int^{\tau_{\Psi_0^i}} f_o(z)\, dz$$

appearing in (6–11) and (6–12) are the p-adic period integrals $J_{\Psi_0^i}$ defined in Section 2. Hence the statement about J_Ψ (including the description of the Galois action) is a consequence of the analogous statement about the $J_{\Psi_0^i}$ contained in Theorem 2.9. □

7. Periods of Hilbert Modular Forms

Let E be an elliptic curve defined this time over a real quadratic field F. It is assumed for notational simplicity (to allow ourselves to continue to use classical instead of adelic language, and thereby stress more strongly the parallel with the constructions of the previous sections) that F has narrow class number one. In particular, the conductor of E over F is generated by a totally positive element N of $\mathcal{O}_F$.

Choose an ordering v_1, v_2 of the real embeddings of F, and given $a \in F$, write a_j for $v_j(a)$. Given an integral element n of $\mathcal{O}_F$, denote by $|n| := n_1 n_2$ its norm, and denote likewise by $|n|$ the norm of an integral or fractional ideal of F.

For each prime ideal p of F, the integer a_p is associated to E just as in the case where E is defined over $\mathbb{Q}$. Let

$$L(E,s)=\prod_{p\nmid N}(1-a_p|p|^{-s}+|p|^{1-2s})^{-1}\prod_{p|N}(1-a_p|p|^{-s})^{-1}=\sum_n a_n|n|^{-s}$$

be the Hasse–Weil L-function attached to E/F, where the product (resp. the sum) is taken over the prime (resp. all) ideals of $\mathcal{O}_F$.

Let $\Gamma\subset \mathrm{SL}_2(\mathcal{O}_F)$ be the group of matrices which are upper-triangular modulo N. It acts on the Poincaré upper half plane in two different ways via the embeddings v_1 and v_2, and the induced diagonal action of Γ on $\mathcal{H}\times\mathcal{H}$ is discrete. A *Hilbert modular form* of weight $(2,2)$ and level N is a holomorphic function $f(\tau_1,\tau_2)$ on $\mathcal{H}\times\mathcal{H}$ satisfying

1. The differential two-form $\omega := f(\tau_1,\tau_2)d\tau_1 d\tau_2$ is invariant under Γ;
2. The form ω is holomorphic at the cusps.

The analogue of the Shimura–Taniyama conjecture in this setting predicts that the holomorphic function on $\mathcal{H}\times\mathcal{H}$ given by the absolutely convergent Fourier series

$$f(\tau_1,\tau_2)=\sum_{n\gg 0} a_{(n)}e^{2\pi i(\frac{n_1}{d_1}\tau_1+\frac{n_2}{d_2}\tau_2)}, \tag{7–1}$$

is a *Hilbert modular form* of weight $(2,2)$ and level N. Here the sum is taken over all totally positive elements of $\mathcal{O}_F$, and d is a totally positive generator of the different ideal of $\mathcal{O}_F$. Let ε be a unit of $\mathcal{O}_F$, such that $\varepsilon_1<0$ and $\varepsilon_2>0$, and define

$$\begin{aligned} \omega_{++}&=-4\pi^2 f(\tau_1,\tau_2)d\tau_1 d\tau_2, & \omega_{-+}&=4\pi^2 f(\varepsilon_1\bar\tau_1,\varepsilon_2\tau_2)d\bar\tau_1 d\tau_2,\\ \omega_{+-}&=4\pi^2 f(-\varepsilon_1\tau_1,-\varepsilon_2\bar\tau_2)d\tau_1 d\bar\tau_2, & \omega_{--}&=-4\pi^2 f(-\bar\tau_1,-\bar\tau_2)d\bar\tau_1 d\bar\tau_2. \end{aligned} \tag{7–2}$$

Assume that E satisfies the conclusion of the Shimura–Taniyama conjecture. The periods attached to ω have been much studied classically and can be computed numerically in practice from the Fourier expansion of equation (7–1). Define a Γ-invariant $\mathbb{R}$-valued double integral on $\mathcal{H}\times\mathcal{H}$ by the rule:

$$\int_{\tau_1}^{\tau_2}\int_{\tau_3}^{\tau_4}\omega_f := \int_{\tau_1}^{\tau_2}\int_{\tau_3}^{\tau_4}(\omega_{++}+\omega_{+-}+\omega_{-+}+\omega_{--}),$$

where the integral on the right is the usual complex integral which can be computed using the Fourier series expansion (7–1) for f.

With this integral in hand, we can mimick the definitions in Sections 4, 5, and 6, letting $\tilde d_{\tau,x}\in Z^2(\Gamma,\mathbb{R})$ be the two-cocycle obtained as in equation (4–3) by choosing base points $\tau,x\in\mathcal{H}$ and setting

$$\tilde d_{\tau,x}(\gamma_1,\gamma_2):=\int_{\tau}^{\gamma_1\tau}\int_{\gamma_1 x}^{\gamma_1\gamma_2 x}\omega_f.$$

Let $d\in H^2(\Gamma,\mathbb{R})$ be the cohomology class of $\tilde d_{\tau,x}$. It does not depend on the choices of base points τ and x that were made to define $\tilde d_{\tau,x}$.

If one fixes $\tau \in \mathcal{H}$ and defines, for $x, y \in \mathbb{P}_1(F)$ any cusps on the Hilbert modular surface:

$$\tilde{c}_\tau(\gamma)\{x \to y\} := \int_\tau^{\gamma\tau} \int_x^y \omega_f,$$

then $\tilde{c}_\tau$ is an $\mathcal{M}(\mathbb{R})$-valued one-cocycle, whose image c in $H^1(\Gamma, \mathcal{M}(\mathbb{R}))$ is related to the 2-cocycle $\tilde{d}_{\tau,x}$ as in Section 5:

$$\delta(c) = d. \tag{7–3}$$

Choose a Néron differential ω_E for E/F, and let Ω_1 and Ω_2 be the real periods attached to this choice. Note that ω_E is only well-defined up to multiplication by a power of ε, so that Ω_1 and Ω_2 are only well-defined up to multiplication by a power of ε_1 and ε_2 respectively. But the product $\Omega_1\Omega_2$ is well-defined up to sign.

The following conjecture, referred to from now on as the period conjecture, is a natural analogue of Proposition 4.3 Theorem 5.2, and Proposition 6.9.

CONJECTURE 7.1. *There exists a lattice $\Lambda \subset \mathbb{R}$ which is commensurable to the lattice generated by $\Omega_1\Omega_2$, and such that the natural images of d in $H^2(\Gamma, \mathbb{R}/\Lambda)$ and of c in $H^1(\Gamma, \mathcal{M}(\mathbb{R}/\Lambda))$ are trivial.*

REMARK. Conjecture 7.1 predicts in particular that if $\alpha/\nu \in \mathbb{P}_1(F)$ is a cusp (with $\alpha, \nu \in \mathcal{O}_F$ and $\gcd(\alpha, \nu) = 1$) and $\gamma = \gamma_{\alpha/\nu} \in \Gamma_0(N)$ is a generator for the stabiliser of α/ν and ∞, then the period

$$J_{\alpha/\nu} := \tilde{c}_\tau(\gamma)\{\infty \to \alpha/\nu\} = \int_\tau^{\gamma\tau} \int_{\alpha/\nu}^{\infty} \omega_f \text{ belongs to } \Lambda. \tag{7–4}$$

This is consistent with (a special case of) the Birch and Swinnerton-Dyer conjecture for the L-function of E/F and its twists by abelian characters of F. Recall that if $\chi : \mathrm{Gal}(\bar{F}/F) \longrightarrow \mathbb{C}^\times$ is a finite order character, the twisted L-series is defined in the usual way by the rule

$$L(E/F, \chi, s) = \sum_n \chi(n) a_n |n|^{-s}.$$

A direct calculation shows that the integral on the right of (7–4) is equal to the integral of ω_f on the cycle denoted $\gamma_{\alpha/\nu,E}$ in § 15 of [Oda]. (In Oda's notation E denotes the subgroup of $\mathcal{O}_F^\times$ generated by the eigenvalues of γ.)

Let c_1 and c_2 denote the conjugacy classes of complex conjugation attached to the infinite places v_1 and v_2 respectively. (These are the "Frobenius elements at ∞" attached to the real places of F.) A character χ of $G_F = \mathrm{Gal}\bar{F}/F)$ is called *even* if $\chi(c_1) = \chi(c_2) = 1$, so that χ factors through the Galois group of a totally real abelian extension of F.

Theorem 16.3 (combined with Lemma 15.4) of [Oda] asserts (in the special case where $N = 1$, but the calculations explained there are readily adapted to more general N) that for all characters $\chi : (\mathcal{O}_F/\nu)^\times/(\mathcal{O}_F^{\times 2}) \longrightarrow \mathbb{C}^\times$ (which can be viewed as characters of G_F in the usual way)

$$\sum_{\alpha\in\mathcal{O}_F/\nu} \chi(\alpha)J_{\alpha/\nu} \doteq \begin{cases} G(\bar{\chi})L(E/F,\chi,1) & \text{if } \chi \text{ is even,} \\ 0 & \text{otherwise,} \end{cases} \tag{7–5}$$

where $G(\bar{\chi})$ is a Gauss sum and where the symbol $\doteq$ denotes equality up to rational multiples with denominator bounded independently of χ and ν. On the other hand, a natural equivariant extension of the Birch and Swinnerton-Dyer conjecture predicts that for all even characters χ the ratios

$$G(\bar{\chi})L(E/F,\chi,1)/(\Omega_1\Omega_2)$$

are algebraic (in the field generated by the values of χ, and with denominators bounded independently of χ). Thus the Birch and Swinnerton-Dyer conjecture leads one to expect the vanishing of all the periods $J_{\alpha/\nu}$ modulo a lattice $\Lambda \subset \mathbb{R}$ which is commensurable with $\mathbb{Z}\cdot\Omega_1\Omega_2$.

Assume, at the cost of eventually multiplying $\tilde{c}_\tau$ and $\tilde{d}_{\tau,x}$ by a common integer, that the lattice Λ of conjecture 7.1 can be chosen to be equal to $\mathbb{Z}\cdot\Omega_1\Omega_2$. Thanks to conjecture 7.1, we may then define an M-symbol $m_\tau \in C^0(\Gamma, \mathcal{M}(\mathbb{R}/\mathbb{Z}\cdot\Omega_1\Omega_2))$ and a one-cochain $\xi_{\tau,x} \in C^1(\Gamma, \mathbb{R}/\mathbb{Z}\cdot\Omega_1\Omega_2)$ by the rules

$$\tilde{c}_\tau = dm_\tau, \quad \tilde{d}_{\tau,x} = d\xi_{\tau,x}.$$

We adopt as before the useful notation

$$\int^\tau\!\!\int_x^y \omega_f := m_\tau\{x \to y\} \in \mathbb{R}/\mathbb{Z}\cdot\Omega_1\Omega_2.$$

Let K be a quadratic extension of F, which is imaginary at v_1 and real at v_2. (Thus, in particular, K is not itself Galois over $\mathbb{Q}$.) Let $\mathcal{O}$ be the ring of integers of K (or a more general $\mathcal{O}_F$-order in K). The notion of an oriented optimal embedding $\Psi : K \longrightarrow M_2(F)$ is defined exactly as in the previous sections. One can then associate to any oriented optimal embedding Ψ a period by letting $\tau \in \mathcal{H}_1$ be the unique fixed point of $\Psi(K^\times)$ acting on $\mathcal{H}_1$, letting $\gamma_\tau = \Psi(\varepsilon) \in \Gamma$, and setting

$$J_\Psi := \xi_{\tau,x}(\gamma_\tau) = \int^\tau\!\!\int_x^{\gamma_\tau x} \omega_f \in \mathbb{R}/\mathbb{Z}\cdot\Omega_1\Omega_2,$$

for any $x \in \mathbb{P}_1(F)$. As in the previous discussions, the period J_Ψ depends only on the Γ-conjugacy class of τ (and therefore, of Ψ), not on the choices of x and then $\xi_{\tau,x}$ that were made to define it. This follows from manipulations identical to those that were used to prove the corresponding statement in Section 4.

Let H denote the ring class field of K attached to the order $\mathcal{O}$. The reciprocity map of class field theory gives a canonical identification

$$\mathrm{rec} : \mathrm{Pic}(\mathcal{O}) \longrightarrow \mathrm{Gal}(H/K).$$

Let η be the Weierstrass uniformisation of E_1 over $\mathbb{R}$, associated to the differential ω_E that was chosen earlier in defining the periods Ω_1 and Ω_2. For all Ψ in $\mathrm{Emb}(\mathcal{O}, R)$, the period J_Ψ/Ω_2 is a well defined element of $\mathbb{R}/\mathbb{Z}\cdot\Omega_1$ on which it is natural to compute the uniformisation η. The following conjecture is the counterpart to Theorem 4.8, conjecture 5.4, and Theorem 6.13 in a purely archimedean setting:

CONJECTURE 7.2. *For all Ψ in $\mathrm{Emb}(\mathcal{O}, R)$, the point $\eta(J_\Psi/\Omega_2)$ belongs to $E(H)$. More precisely, for all $\mathfrak{b}\in \mathrm{Pic}(\mathcal{O})$,*

$$\eta(J_{\Psi^{\mathfrak{b}}}/\Omega_2) = \mathrm{rec}(\mathfrak{b})^{-1}\cdot \eta(J_\Psi/\Omega_2).$$

References

[BCDT] C. Breuil, B. Conrad, F. Diamond, and R. Taylor, "On the modularity of elliptic curves over **Q**: wild 3-adic exercises", *J. Amer. Math. Soc.* **14**:4 (2001), 843–939.

[BD1] M. Bertolini and H. Darmon, "Heegner points on Mumford–Tate curves", *Invent. Math.* **126**:3 (1996), 413–456.

[BD2] M. Bertolini and H. Darmon, "Heegner points, p-adic L-functions, and the Cerednik–Drinfeld uniformisation", *Invent. Math.* **131**:3 (1998), 453–491.

[BD3] M. Bertolini and H. Darmon, "The p-adic L-functions of modular elliptic curves", pp. 109–170 in *Mathematics unlimited: 2001 and beyond*, Springer, Berlin, 2001.

[BDIS] M. Bertolini, H. Darmon, A. Iovita, and M. Spiess, "Teitelbaum's exceptional zero conjecture in the anticyclotomic setting", *Amer. J. Math* **124** (2002), 411–449.

[BoCa] J.-F. Boutot and H. Carayol, "Uniformisation p-adique des courbes de Shimura: les théorèmes de Cerednik et de Drinfeld", pp. 45–158 in *Courbes modulaires et courbes de Shimura* (Orsay, 1987/1988), Astérisque **196–197**, Soc. math. de France, Paris, 1992.

[Bu] D. Bump, *Automorphic forms and representations*, Cambridge Studies in Advanced Mathematics **55**, Cambridge University Press, Cambridge, 1997.

[CF] J. W. S. Cassels and A. Fröhlich (ed.), *Algebraic number theory*, Academic Press, London, and Thompson Book, Washington, DC, 1967.

[Da] H. Darmon, "Integration on $\mathcal{H}_p\times\mathcal{H}$ and arithmetic applications", *Ann. of Math.* (2) **154** (2001), 589–639.

[DG] H. Darmon and P. Green, "Elliptic curves and class fields of real quadratic fields: algorithms and evidence", *Experimental Mathematics* **11**:1 (2002), 37–55.

[GS] R. Greenberg and G. Stevens, "p-adic L-functions and p-adic periods of modular forms", *Invent. Math.* **111**:2 (1993), 407–447.

[GvdP] L. Gerritzen and M. van der Put, *Schottky groups and Mumford curves*, Lecture Notes in Mathematics **817**, Springer, Berlin, 1980.

[I] Y. Ihara, *On congruence monodromy problems*, vol. 1, Department of Mathematics, University of Tokyo, 1968.

[JoLi1] B. W. Jordan and R. Livné, "Local diophantine properties of Shimura curves", *Math. Ann.* **270** (1985), 235–248.

[JoLi2] B. W. Jordan and R. Livné, "Integral Hodge theory and congruences between modular forms", *Duke Math. J.* **80** (1995), 419–484.

[La] S. Lang, *Introduction to modular forms*, Grundlehren der mathematischen Wissenschaften **222**, Springer, Berlin, 1976.

[Man1] Y. Manin, "Parabolic points and zeta functions of modular curves" (Russian), *Izv. Akad. Nauk SSSR Ser. Mat.* **36** (1972), 19–66; translation in Math. USSR Izv. **6** (1972), 19–64.

[Man2] Y. Manin, "p-adic automorphic functions, *J. Sov. Math.* **5** (1976), 279–334.

[Maz] B. Mazur, "Courbes elliptiques et symboles modulaires", pp. 277–294 (exposé 414) *Séminaire Bourbaki* (1971/1972), Lecture Notes in Mathematics **317**, Springer, Berlin, 1973.

[MS] Y. Matsushima and G. Shimura, "On the cohomology groups attached to certain vector valued differential forms on the product of the upper half planes", *Ann. of Math.* (2) **78** (1963), 417–449.

[MTT] B. Mazur, J. Tate, and J. Teitelbaum. "On p-adic analogues of the conjectures of Birch and Swinnerton-Dyer", *Invent. Math.* **84**:1 (1986), 1–48.

[Oda] T. Oda, *Periods of Hilbert modular surfaces*, Progress in Mathematics **19**, Birkhäuser, Boston, 1982.

[Sch] P. Schneider, "Rigid-analytic L-transforms", pp. 216–230 in *Number theory* (Noordwijkerhout, 1983), Lecture Notes in Math. **1068**, Springer, Berlin, 1984.

[Se1] J.-P. Serre, "Complex multiplication", pp. 292–296 in *Algebraic number theory* (Brighton, 1965), Thompson Book, Washington, DC, 1967.

[Se2] J.-P. Serre, "Le problème des groupes de congruence pour SL_2", *Ann. of Math.* (2) **92** (1970), 489–527.

[Sh1] G. Shimura, "Class-fields and automorphic functions", *Ann. of Math.* (2) **80** (1964), 444–463.

[Sh2] G. Shimura, "Construction of class fields and zeta functions of algebraic curves", *Ann. of Math.* (2) **85** (1967), 58–159.

[Sh3] G. Shimura, "Algebraic number fields and symplectic discontinuous groups", *Ann. of Math.* (2) **86** (1967), 503–592.

[Si] J. Silverman, *The arithmetic of elliptic curves*, Graduate Texts in Mathematics **106**, Springer, New York, 1986.

[St] H. M. Stark, "Modular forms and related objects", pp. 421–455 in *Number theory* (Montreal, 1985), CMS Conf. Proc. **7**, Amer. Math. Soc., Providence, 1987.

[Te] J.T. Teitelbaum, "Values of p-adic L-functions and a p-adic Poisson kernel", *Invent. Math.* **101**:2 (1990), 395–410.

[Va1] V. Vatsal, "Uniform distribution of Heegner points", *Invent. Math.* **148** (2002), 1–46.

[Va2] V. Vatsal, "Special values of anticyclotomic L-functions", preprint.

[Vi] M-F. Vignéras, *Arithmétique des algèbres de quaternions*, Lecture Notes in Mathematics **800**, Springer, Berlin, 1980.

[Wi] A. Wiles, "Modular elliptic curves and Fermat's last theorem", *Ann. of Math.* (2) **141**:3 (1995), 443–551.

[Zh] S. Zhang, "Heights of Heegner points on Shimura curves", *Ann. of Math.* (2) **153**:1 (2001), 27–147.

MASSIMO BERTOLINI
DIPARTIMENTO DI MATEMATICA
UNIVERSITÀ DEGLI STUDI DI MILANO
VIA SALDINI 50
20133 MILANO
ITALY
Massimo.Bertolini@mat.unimi.it

HENRI DARMON
MCGILL UNIVERSITY
MATHEMATICS DEPARTMENT
805 SHERBROOKE STREET WEST
MONTREAL, QC, H3A 2K6
CANADA
darmon@math.mcgill.ca

PETER GREEN
HARVARD UNIVERSITY
MATHEMATICS DEPARTMENT
1 OXFORD STREET
CAMBRIDGE, MA 02138
UNITED STATES
green@math.harvard.edu